The boreal forests of the world, geographically situated to the south of the Arctic and generally north of latitude 50 degrees, are considered to be one of the earth's most significant terrestrial ecosystems in terms of their potential for interaction with other global scale systems, such as climate and anthropogenic activity. This book, developed by an international panel of ecologists, provides a synthesis of the important patterns and processes which occur in boreal forests and reviews the principal mechanisms which control the forests' pattern in space and time. The effects of cold temperatures, soil ice, insects, plant competition, wildfires and climatic change on the boreal forests are discussed as a basis for the development of the first global scale computer model of the dynamical change of a biome, able to project the change of the boreal forest over timescales of decades to millennia, and over the global extent of this forest.

A systems analysis of the global boreal forest

A Systems
Analysis
of the Global
Boreal Forest

Edited by

Herman H. Shugart
University of Virginia, Charlottesville, Virginia, USA

Rik Leemans
National Institute for Public Health and Environmental Protection, Bilthoven, The Netherlands

Gordon B. Bonan
National Center for Atmospheric Research, Boulder, Colorado, USA

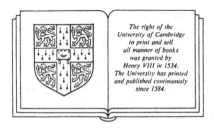

The right of the
University of Cambridge
to print and sell
all manner of books
was granted by
Henry VIII in 1534.
The University has printed
and published continuously
since 1584.

CAMBRIDGE UNIVERSITY PRESS
Cambridge

New York   Port Chester   Melbourne   Sydney

PUBLISHED BY THE PRESS SYNDICATE OF THE UNIVERSITY OF CAMBRIDGE
The Pitt Building, Trumpington Street, Cambridge, United Kingdom

CAMBRIDGE UNIVERSITY PRESS
The Edinburgh Building, Cambridge CB2 2RU, UK
40 West 20th Street, New York NY 10011–4211, USA
477 Williamstown Road, Port Melbourne, VIC 3207, Australia
Ruiz de Alarcón 13, 28014 Madrid, Spain
Dock House, The Waterfront, Cape Town 8001, South Africa

http://www.cambridge.org

First published 1992
First paperback edition 2005

A catalogue record for this book is available from the British Library

Library of Congress cataloguing in publication data

A systems analysis of the global boreal forest / edited by Herman H. Shugart,
Rik Leemans, and Gordon B. Bonan.
    p.   cm.
Includes bibliographical references and index.
ISBN 0 521 40546 7 hardback
1. Taiga ecology.   2. Taiga ecology – Computer simulation.
3. Forest dynamics.   4. System analysis.   I. Shugart, H. H.
II. Leemans, Rik.   III. Bonan, Gordon B.
QK938.T34S87   1991
574.5′2642–dc20   91-7760   CIP

ISBN 0 521 40546 7 hardback
ISBN 0 521 61973 4 paperback

Contents

Contributors

M. Ya. Antonovski
International Institute for Applied Systems Analysis, A-2361
Laxenburg, Austria

G. B. Bonan
National Center for Atmospheric Research, PO Box 3000, Boulder,
CO 80307, USA

P. Duinker
School of Forestry, Lakehead University, Thunder Bay, Ontario, P7B
5E1, Canada

V. V. Furyaev
Institute of Forestry and Wood, Siberian Branch, USSR Academy of
Sciences, Krasnoyarsk, 660036 USSR

F. Z. Glebov
Institute of Forestry and Wood, Siberian Branch, USSR Academy of
Sciences, Krasnoyarsk, 660036 USSR

H. Helmisaari
Institute of Ecological Botany, Uppsala University, Box 559,
S-751 Uppsala, Sweden

C. S. Holling
Department of Zoology, University of Florida, 223 Bartram Hall,
Gainesville, FL 32611, USA

M. D. Korzukhin
Goskohydromet Natural Environment and Climate Monitoring
Laboratory, Glebovskaya St 20B, Moscow, 107258 USSR

R. Leemans
Global Change Department, National Institute for Public Health and
Environmental Protection, PO Box 1, 3720 BA Bilthoven,
The Netherlands

D. Mladenoff
Natural Resources Research Institute, University of Minnesota,
Duluth, MN 55811, USA

N. Nikolov
Institute of Ecology, Gagarin str.2, 1113 Sofia, Bulgaria

S. Nilsson
International Institute for Applied Systems Analysis, A-2361
Laxenburg, Austria

M. Nygren
University of Helsinki, Forestry Field Station, Korkeakoski,
Finland

J. Pastor
Natural Resources Research Institute, University of Minnesota,
Duluth, MN 55811, USA

S. Payette
Université Laval, Centre d'études nordiques, Cité universitaire, Québec,
G1K 7P4, Canada

I. C. Prentice
Institute of Ecological Botany, Uppsala University, Box 559,
S-751 Uppsala, Sweden

K. J. Ranson
NASA Goddard Space Flight Center, Biospheric Sciences Branch,
Greenbelt, MD 20771, USA

O. Sallnäs
College of Forestry, Swedish University of Agricultural Sciences,
S-77073 Garpenberg, Sweden

T. L. Sharik
Michigan Technological University, Houghton, MI 49931, USA

H. H. Shugart
Department of Environmental Sciences, Clark Hall, University of
Virginia, Charlottesville, VA 22903, USA

L. Sirois
Département de Biologie, Université du Québec à Rimouski, 300
allée des Ursulines, Rimouski, Québec, G5L 3A1, Canada

A. M. Solomon
Lake Superior Ecosystems Research Center, Michigan Technological
University, Houghton, MI 49931-1262, USA

M. T. Ter-Mikaelian
Goskohydromet Natural Environment and Climate Monitoring
Laboratory, Glebovskaya St 20B, Moscow 107258, USSR

D. L. Williams
NASA Goddard Space Flight Center, Biospheric Sciences Branch,
Greenbelt, MD 20771, USA

J. C. Zasada
USDA Forest Service, PNW Research Station, Forest Science
Laboratory, 3200 SW Jefferson Way, Corvallis, OR 97331, USA

1 Introduction

Herman H. Shugart, Rik Leemans and Gordon B. Bonan

This is an era of increased interest in the function and interaction of the major geophysical, geochemical and ecological systems of the earth. The interest in these large spatial scale studies has had diverse origins: the success of the 'International Geophysical Year' of global observations (1957–8) and a shared comprehension of just how much time has passed since this effort; the characterization of the surface of the earth from an ever increasing availability of images from space; the realization that humans are altering the composition of the atmosphere; a relative warming in international political tensions and the increased likelihood of sustained international scientific exchanges; an improved understanding of the past dynamics of the earth's surface resulting from radioisotope dating and analysis of paleoecological data; and the ramifications of computers with the power to solve complex equations of the fluid motion of the atmosphere and oceans. The conjunction of these and many other developments have turned the interests of many scientists in different disciplines to the issue of increasing the level of understanding of the earth as an interacting, dynamical system.

However, for all of these exciting developments, scientists in one of the important disciplines contributing to the study of the working of the planet, ecological sciences, are focused largely on the understanding of biota – environment interactions at very short time and space scales. Kareiva & Andersen (1988) found that about 80% of the studies in a sample of the journal *Ecology* were developed on areas less than 100 m^2. Weatherhead (1986) sampled studies in the three areas of ecology, evolutionary biology and behavior and found the average duration of study to equal 2.5 years.

If, indeed, most ecological studies are conducted over a relatively short period of time and on very small areas, and given that it has proven difficult to 'scale up' observations made at relatively small space and time scales to larger-scale consequences, then one is prone to wonder how the body of present-day ecological observations and experiments can be made to fit with the increased interest in global studies or global ecology. We share this concern, and this book represents an attempt to assuage it.

Initial development of a systems analysis of the global boreal forest

In 1984 one of the editors of the present book (Shugart) and Professor Mikhail Ya. Antonovski (see Chapters 13 and 14) worked for a summer in the Department of Meteorology at the University of Stockholm developing a book chapter (Shugart *et al.* 1986) that assessed the potential effect of CO_2-related climatic change on the world's forests. At that time, as is the case today, uncertainty was the key word in any global-scale interpretation of CO_2-related changes. A great list of important considerations contributed to this uncertainty: the actual projections of the anthropogenic production of CO_2, the estimations of the principal fluxes of the global carbon cycle, the reliability of the general circulation models (GCMs) that were used to compute and predict climatic change, the direct effects of CO_2 on the biota, the climate effects on the biota, and a myriad of other sources. For all this uncertainty, it seemed clear at the time that the boreal forests of the earth were among the most important potential target zones of what was (and is) a recurrent prediction from most of the GCMs: a global warming that is most pronounced in the higher northern latitudes.

In the course of the review of the chapter that was developed in Stockholm (Shugart *et al.* 1986), a panel of experts met in Stockholm with two purposes. The first purpose was to provide review and commentary on a book chapter but the second purpose was to outline the physical and ecological functions that would need to be developed in a computer simulator that would be capable in some sense of simulating the dynamics of the world's boreal forests. This review group was the seed of an informal, international group that was to meet again two more times to develop the present book.

The meeting in Stockholm in 1985 was attended by a number of boreal forest ecologists, many of whom are contributors of chapters to this book. All of these scientists contributed to the knowledge base that was used in the model development and their help is greatly appreciated.

The format of the Stockholm meeting was to be repeated at the two subsequent meetings and in the outline of this book. The agenda was simple. It was to review the state of knowledge of major processes in the boreal forests, to discuss what important patterns a boreal forest model should be capable of simulating, and to construct an agenda for developing what was talked about as a 'Unified Boreal Forest Simulator.' At the end of the Stockholm meeting, one of the editors (Bonan) took the charge of developing a boreal zone simulation model centered on intensively studied sites near Fairbanks, Alaska (Van Cleve *et al.* 1983*a*). This work would emphasize physical dynamics such as permafrost formation, evapotranspiration, and soil temperature. Linked to this project was a parallel study coordinated by I. C. Prentice at the University of Uppsala that would emphasize the biological aspects of a boreal forest simulation model. Work on this model was soon taken up by another of the editors (Leemans).

By 1987 sufficient progress had been made in developing the two pieces of a unified boreal forest simulator that a second meeting was held to communicate and to assess the progress before a larger forum. This meeting was held in Laxenburg, Austria, at the International Institute for Applied Systems Analysis (IIASA) and followed the aforementioned pattern of discussions of processes, patterns and modeling agenda. The results of this second conference were severalfold. It was decided that an active working session to compile the silvicultural information on Eurasian species in a format that matched that of the silvicultural descriptions of the North American species found in Fowells (1965) and Harlow, Harrar & White (1979) was necessary to estimate all the species parameters needed for a panboreal forest model. This effort was coordinated by IIASA and took the form of a coordinated effort by a number of young scientists from several different countries (USA, Canada, Finland, Sweden, Bulgaria, USSR, People's Republic of China) during the summers of 1988 and 1989 and resulted in Chapter 2 by Nikolov and Helmisaari.

It was further decided that the domain of time scales and space scales that attended some of the major controlling processes in the boreal forest implied a range of different models. Hence, the present book includes several model approaches in the final section to indicate some of this diversity. A final conclusion was that the basic boreal simulator was sufficiently developed to warrant publication.

The third meeting of the informal boreal zone group was held in Sopron, Hungary, in August of 1989 under the sponsorship of IIASA and supported locally by the Hungarian Academy of Sciences. This book is a product of that meeting.

Uniting features

We are documenting the central considerations of ecologists attempting to develop what we think is a first global-scale model of the dynamical change of a biome. There are 'community models' developed by scientists in the atmospheric sciences and there is a long tradition of coordinated collection and sharing of data by several of the subdisciplines in the geophysical sciences. In general ecological scientists have been less able (or willing?) to develop such closely coordinated modeling and experimental projects. There are several reasons why the boreal forest zone of the northern hemisphere is a logical ecosystem upon which to focus an initial attempt. These uniting features associated with the boreal forests arise from the nature of the ecological systems themselves, the interests of ecologists working in this region and the proximity of the forests to industrial nations.

The boreal forests in some senses are floristically rather simple (compared with many tropical rain forests, for example). In most boreal forest stands, one or two species dominate. Most of the dominant trees over a region in the boreal zone can be described in a list of fewer than ten tree species. In comparisons across disjunct continents, one finds the same limited number of genera (*Picea*, *Abies*, *Betula*, *Populus*, *Larix*, *Pinus*) to be repeated elements. The temporal pattern of replacement following a disturbance among the species that represent these genera is similar. Even though the species names (but not the genus names) may change among boreal scenes in Europe, Asia, and North America, the ecological 'look' of the landscape reads similarly.

Further, the boreal forests of the world are well-studied forests. They are the holdings of developed nations and are a repository of valuable timber reserves already well incorporated into national and international economies. We are likely to know more about these forests than any others at a global level.

In our analysis of the boreal forests, we are using models with fundamental bases that arise from a rather rich fount of detailed information. We have attempted to include that synthesis of this detailed information in the presentations that follow. We feel that it is important for the reader to be able to see how, for example, provided that a given species is capable of producing root sprouts following a disturbance (Chapters 2 and 3), this information is scaled up in the applications of models (Chapter 15) through to the landscape (Chapters 7, 8, and 9) and larger-scale (Chapter 11) consequences of this biological feature. We would like to indicate in the chapters on ecological processes the rich complexity of possible interactions in the boreal forests of the world and the attendant difficulty of the task of incorporating these into a single model.

The relationship between form and function, or pattern and process, is a classic ecological theme (Lindeman 1942; Watt 1947; Whittaker & Levin 1977). Often the pattern–process interaction is discussed in terms of processes causing pattern in such familiar examples as understanding how ecological energetics and the thermodynamic constraints shape food-webs (Elton 1927; Lindeman 1942; Odum 1969), or interpreting the processes that cause a landscape vegetation pattern to have a given appearance (Watt 1947; Whittaker & Levin 1977). It is also clear that patterns can influence ecological processes to a great degree. For example, Bormann & Likens (1979a) pointed out the effects of changes in forest pattern on processes affecting productivity and nutrient cycling. Many ecologists recognize that pattern and process are mutually causal, with changes in ecosystem processes causing change in pattern, and modifications in ecosystem pattern changing processes. Nonetheless, it is difficult to investigate directly the feedback between pattern and process. Some of the functional complexity that can result from pattern–process mutual interactions are discussed in detail in the contribution of Holling (Chapter 6).

In considering the patterns in the boreal forests, one finds a richness of response that belies the seeming constancy of processes, and taxonomic similarities across the boreal forests. While the landscape patterns in the boreal forests of the world may be similar in their gross appearance, the generation of these similar patterns may be the consequences of complex, nonlinear reactions among processes and patterns. Several of the chapters directly discuss the expression of the parametric value of a process (for example, fire frequency (Chapters 5, 7, and 14), probability of insect outbreak (Chapter 6), rate of heat flux through the soil (Chapter 4) and nitrogen mineralization rate (Chapter 8), altering the pattern of the forest and the effect of the change in pattern altering the process in return.

On developing a 'Unified Boreal Forest Simulator'

Modeling is playing an ever-increasing role in the development of ecological theory at several scales, from understanding the mechanisms of carbon fixation (Farquhar & Sharkey 1982; Farquhar & von Caemmerer 1982b) and plant water balance (Cowan 1982, 1986); to scaling of physiological processes to whole-plant function (Reynolds et al. 1986); to exploring how ecosystem processes of carbon and nitrogen cycling operate at continental to global scales (Emanuel et al. 1984, 1985). Of particular importance is the role of modeling in exploring phenomena which occur at spatial and temporal scales at which extensive direct observation and experimentation are prohibitive, if not impossible. Recent examples include the role of spatial and temporal variation in competition on ecosystem functioning (Wu et al. 1985; Sharpe et al. 1985, 1986; Walker & Sharpe 1989), extrapolation of the processes of carbon fixation and water balance to the landscape scale to enable linking ecosystem models with remotely sensed data (Running & Coughlan 1988) (Chapter 10), and exploring the implications of the evolution of plant adaptations to varying environmental conditions on current patterns of ecosystem structure across environmental gradients (Tilman 1988).

The diversity of extant modeling approaches proscribes the designation of *the* model of virtually any ecological system. Appropriateness of model structures depend strongly on the objectives of the model users. The scientific challenge in modeling tends to lie on the proper selection of the phenomenena that attend a question's solution and that are appropriate to the time and space domain of the problem. There are six different models of boreal-forest-related phenomena in the chapters that follow and almost all of the other chapters mention some aspects of ecological models (model parametrization, model testing, theoretical results that relate to models) as a focus. The models are interwoven in their relationships. For example, the population models for forests presented in Chapter 13 provide insights into the dynamic responses of the more biological but less analyzable simulation models (Chapters 15, 16 and 17). The landscape level models for forest fires, paludification and forest harvest all require an underlying knowledge of expected forest dynamics that could be obtained from other models. The biological and physical models of forest dynamics (Chapters 15 and 16) are sufficiently closely allied that these two models have been merged to create a panboreal forest simulator that has been named the BOFORS (for BOreal FOReSt model). Availability of this model is discussed in the final chapter.

Overall considerations

This book is divided into three parts. The first is composed of chapters reviewing important processes in boreal forests. The second treats major patterns in boreal forests. The third presents computer models developed to simulate boreal forest dynamics. Each section is preceded by a brief summary of the reviews and discussions that follow. The chapters have all been subjected to anonymous peer review by at least two reviewers.

The rich diversity of interests and opinions of the authors and the breadth of the scientific literature that has been considered in developing this project have been one of the exciting aspects of this project for the editors. In most scientific endeavor, research leads to more and often richer questions, rather than to tidy answers. It is likely the case with this project. A theme that recurs in this book is the potentially rich dynamics that stem from the feedbacks between patterns and processes in these forest ecosystems. We are pleased to present a globally extensive review of these pattern–process interactions and first documentation of dynamic models for the panboreal forests. It is our hope that the next generation of forest studies and forest models will profit from this synthesis.

Acknowledgements

Support for the individual research projects that are associated with particular chapters are acknowledged in the relevant chapters. Needless to say, this support is greatly appreciated by the editors as well as the individuals responsible for the chapters.

The 1985 Stockholm meeting was supported by the World Meteorological Organization and the United Nation's Environmental Program. The meeting was hosted by the Department of Meteorology at the University of Stockholm and the Swedish Academy of Sciences. Meeting attendees were: B. Bolin (University of Stockholm, Sweden), G. W. Bonan (University of Virginia, USA), B. R. Döös (University of Stockholm, Sweden), S. Kellomaki (University of Joensuu, Finland), I. C. Prentice (University of Uppsala, Sweden), J. C. Ritchie (University of Toronto, Canada), H. H. Shugart (University of Virginia, USA), C. J. Tucker (Goddard Space Flight Center, USA), K. Van Cleve (University of Alaska, USA), R. W. Wein (University of New Brunswick, Canada), T. Wigley (University of East Anglia, UK), and O. Zachrisson (University of Umea, Sweden).

The 1987 Laxenburg meeting was supported and hosted by the International Institute for Applied Systems Analysis. Meeting attendees were: M. Ya. Antonovski (IIASA), A. N. Auclair (Atmospheric Sciences Service, Canada), C. Binkley (Yale University, USA), G. B. Bonan (University of Virginia, USA), A. E. Carey (USDA-Forest Service), W. Cramer (Trondheim University, Norway), P. Duinker (IIASA), L. Kairiukstis (IIASA), P. Kauppi (IIASA), S. Kojima (Toyama University, Japan), P. Kolosov (USSR Academy of Sciences), M. Korzukhin (USSR Academy of Sciences), V. Koski (Forest Research Institute, Finland), Y. Kuznetsov (USSR Academy of Sciences), R. E. Munn (IIASA), W. Oechel (San Diego State University, USA), S. Payette (Laval University, Canada), H. H. Shugart (University of Virginia, USA), A. M. Solomon (IIASA), R. B. Street (Canadian Climate Centre), Mikhail Ter-Mikaelian (USSR Academy of Sciences), B. Vlasiuk (State Forest Committee of the USSR), and R. W. Wein (University of Alberta, Canada).

The 1989 Sopron meeting was also supported by IIASA and, in part, by the Global Systems Analysis Program of the University of Virginia. We thank the Hungarian Committee for Applied Systems Analysis for hosting this meeting. Zsofia Zamora's (Hungarian Committee for

Applied Systems Analysis) and Lyndele McCain's (University of Virginia) work in the organization of the logistics of the meeting was invaluable. Meeting attendees were: G. B. Bonan (National Center for Atmospheric Research, USA), N. Chebakova (Siberian Division of the USSR Academy of Sciences), P. Duiker (Lakehead University, Canada), A. D. Friend (IIASA), Ruiping Gao (Academia Sinica, PRC), H. Helmisaari (Uppsala University, Sweden), C. S. Holling (University of Florida, USA), M. Korzukhin (USSR Academy of Sciences), B. McLaren (IIASA), L. Nedorezov (Siberian Division of the USSR Academy of Sciences), N. Nikolov (Institute of Forestry, Bulgaria), M. Nygren (Society of Forestry, Finland), J. Pastor (University of Minnesota, USA), E. J. Schwarz (IIASA), J. Sendzimir (University of Florida, USA), H. H. Shugart (University of Virginia, USA), L. Sirois (University of Virginia, USA), A. M. Solomon (IIASA), and J. Zasada (USDA-Forest Service).

Several people provided what has proven to be invaluable help in the development of this project. We thank our authors for their patience and their willingness to work and rework their manuscripts; the contagious enthusiasm that came from these colleagues cannot be understated. A. M. Solomon and Bo Döös worked behind the scenes to support the meetings that brought us together to develop this project. We are indebted to a multitude of anonymous reviewers who contributed valuable time and energy to review manuscripts. Erica J. Schwarz was technical editor, schedule developer, and task master for the production of this book. Lyndele von Schill (*née* McCain) organized travel, manuscripts and the editors. Their hard work and cheerful attitudes made the task of producing this book both pleasant and possible.

We appreciate the support of the International Institute for Applied Systems Analysis (Laxenburg, Austria), the US National Aeronautics and Space Administration (Grant NAG-5-1018), the US Environmental Protection Agency (Grant CR-816267-01-0), the US National Science Foundation (Grants BSR-8702333 & BSR-8807882), and the National Center for Atmospheric Research (Boulder, Colorado) during the development of this book.

Processes in boreal forests

Introduction

Gordon B. Bonan

The boreal forest environment is characterized by a wide range of site conditions. Climatic conditions range from extremely cold, dry continental regimes in interior Alaska and Siberia to warmer, moist, oceanic regimes in eastern Canada and Fennoscandinavia (Hare & Hay 1974). Extreme temperatures as low as $-70\,°C$ are not uncommon in interior Alaska and Siberia (Rumney 1968). In more moderate regions such as Saint John, Canada, extreme lows average $-23\,°C$ (Hare & Hay 1974). Annual precipitation can be as little as 10–20 cm in the dry continental regions of interior Alaska and Siberia, but as much as 50–90 cm in eastern Canada (Rumney 1968). Annual global solar radiation varies from less than $3352\,MJ\,m^{-2}$ at the tree-line in western Canada to over $5028\,MJ\,m^{-2}$ in the south (Hare & Hay 1974). Maximum day length varies from 16 hours at the southern edge of the boreal forest to 24 hours at the northern tree-line (Hare & Hay 1974).

In North America, soil moisture ranges from xeric jack pine (*Pinus banksiana* Lamb.) and black spruce (*Picea mariana* (Mill.) B.S.P.) forests to lowland black spruce and tamarack (*Larix laricina* (Du Roi) K. Koch) bogs (Larsen 1980). Soil temperature ranges from widespread warm, permafrost-free soils to scattered permafrost soils in interior Alaska and western Canada to extensive cold, permafrost soils in central and eastern Siberia (Larsen 1980). Local soil temperature gradients can be large. In the discontinuous permafrost zone of interior Alaska, growing season soil temperature sums above 0 °C range from as low as 483 in wet, permafrost soils to as high as 2217 in dry, permafrost-free soils (Viereck *et al.* 1983). Nutrient availability is generally low. For example, in interior Alaska, nitrogen mineralization ranges from $8\text{–}9\,kg\,ha^{-1}\,yr^{-1}$ in cold, wet black spruce forests (Van Cleve, Barney & Schlentner 1981) to $24\text{–}58\,kg\,ha^{-1}\,yr^{-1}$ in birch (*Betula papyrifera* Marsh.) forests on warmer, permafrost-free soils (Flanagan & Van Cleve 1983).

9

Recurring disturbances are common in the boreal environment. Fires are ubiquitous, and the natural fire cycle in the North American boreal forest averages 50–200 years (Heinselman 1981*a*,*b*; Viereck 1983). However, the fire cycle can be as short as 26 years in extremely dry deciduous forests in interior Alaska (Yarie 1981) or as long as 500 years in the moist regions of eastern Canada (Foster 1983). The vast majority of fires in northern North America burn less than 5 ha (Rowe & Scotter 1973; Barney & Stocks 1983), but in extreme fire years individual fires can cover 50 000 to 200 000 ha (Dyrness, Viereck & Van Cleve 1986). Fire frequencies for the boreal forests of northern Sweden are similar to those of North America, ranging from 50 to 270 years and averaging 110–155 years (Zackrisson 1977; Engelmark 1984).

Severe defoliating insect outbreaks that kill or damage forests over wide areas are also prevalent. Periodic spruce beetle (*Dendroctonus rufipennis*) outbreaks have caused widespread white spruce (*Picea glauca* (Moench) Voss) tree mortality in south-central Alaska (Werner & Holsten 1983). Repeated spruce budworm (*Choristoneura fumiferana*) outbreaks in eastern Canada have been documented as far back as the early 1700s (Blais 1968), and it is estimated that as much as 720 000 000 m^3 of wood was killed during severe outbreaks between 1910 and 1920. In West Siberia, 3 000 000 ha of forest was destroyed as a result of a Siberian silkworm (*Dendrolimus sibiricus*) outbreak between 1954 and 1957 (Isaev & Krivosheina 1976).

These environmental factors are thought to contribute to the mosaic of forest types and the wide range in stand productivity characteristic of the boreal forest biome (Bonan & Shugart 1989). For example, near Fairbanks, Alaska, above-ground biomass in mature forests ranges from 26 t ha^{-1} on cold, nutrient-poor soils to 250 t ha^{-1} on warmer, more fertile soils (Van Cleve *et al.* 1983*b*). Further north near Fort Yukon, above-ground biomass of mature stands ranges from 1 to 158 t ha^{-1} depending on soil conditions (Yarie 1983). In the nutrient-poor black spruce-lichen woodlands of subarctic Quebec, woody biomass in mature stands ranges from 10 to 29 t ha^{-1} (Moore & Verspoor 1973). Yet on more fertile soils in the same region, stand biomass ranges from 78 to 163 t ha^{-1} (Moore & Verspoor 1973). In warmer climates such as the Boundary Waters Canoe Area of Minnesota, the landscape is a mosaic of vegetation types, where the biomass of mature stands ranges from 121 to 268 t ha^{-1} (Ohmann & Grigal 1985). In the jack pine forests of New Brunswick, above-ground tree biomass ranges from 0.8 t ha^{-1} for recent burns to 91.1 t ha^{-1} for more mature stands (MacLean & Wein 1976). Limited biomass data are available for the boreal forests of the Soviet Union. Maximum reported biomass for Norway spruce (*Picea abies* (L.) Karst.) stands ranges from 209 t ha^{-1} (DeAngelis, Gardner & Shugart

1981) to 280 t ha^{-1} (Cannell 1982). In the open larch (*Larix gmelinii* (Rupr.) Litv.) woodlands on the Taimyr Peninsula, above-ground tree biomass is as little as 3.3 t ha^{-1} (Ignatenko *et al.* 1973).

In this section, various authors examine the role of key environmental factors and ecological processes in creating this wide range in stand productivity and the mosaic pattern of vegetation in the landscape. First, Nikolov & Helmisaari and Zasada, Sharik & Nygren examine the ecological characteristics of the major boreal forest tree species. Though the silvics of the North American boreal forest tree species have been well documented (see, for example, Fowells 1965), similar data have not been collated for the boreal tree species of the Soviet Union. In Chapter 2, Nikolov & Helmisaari review the silvics of these species with particular reference to their geographic distribution, climatic and soil conditions for optimal growth, life-history characteristics that define reproductive behaviour and growth patterns, and their ecological response to environmental factors such as light, moisture, nutrients, and fire. Many factors affect the distribution of species in the landscape. However, such patterns are ultimately determined by the seed and vegetative reproduction characteristics of trees and the interactions of these characteristics with site conditions. Zasada, Sharik & Nygren (Chapter 3) examine the reproductive process in boreal tree species with emphasis on the population biology of sexual and vegetative regeneration.

The remaining chapters show how specific environmental factors such as soil temperature, fire, and insect outbreaks interact with the silvics of tree species to control forest productivity and to organize the pattern of vegetation in the landscape. Soil temperature is an important factor organizing boreal forest structure and function (Van Cleve & Viereck 1981; Van Cleve, Barney & Schlentner 1981; Van Cleve *et al.* 1983*a,b*; Van Cleve & Yarie 1986); Bonan (Chapter 4) examines soil temperature as an ecological factor in boreal forests. Emphasis is placed on a systems-level perspective, i.e. not only the effects of soil temperature on ecological processes, but also the factors regulating soil temperature. Much of the floristic diversity and mosaic vegetation patterns in the boreal forest are attributable to recurring forest fires (Rowe & Scotter 1973; Viereck 1973, 1975, 1983; Viereck & Schandelmeier 1980; Heinselman 1981*a,b*; Rowe 1983). In Chapter 5, Payette examines the role of fire as an organizer of vegetation patterns. Detailed reviews of the ecological consequences of fire have been published elsewhere (Rowe & Scotter 1973; Viereck 1973, 1983; Viereck & Schandelmeier 1980; Heinselman 1981*a,b*; Wein & MacLean 1983; Dyrness, Viereck & Van Cleve 1986). Instead, Payette uses a latitudinal gradient in northern Quebec to document the effects of climate on the fire regime and post-fire regeneration and, through these, stand development and landscape vegetation patterns. Defoliating insect population dynamics and boreal forest development are intricately linked (Baskerville 1975), and in the final chapter of Part 1 Holling (Chapter 6) examines the role of these insects in structuring the boreal landscape.

2 Silvics of the circumpolar boreal forest tree species

Nedialko Nikolov and Harry Helmisaari

1. Introduction

The development of mechanistic approaches in the study of forest ecosystem dynamics over the past 16 years has increasingly drawn scientific attention to autecological characteristics of forest tree species. Model exercises have shown that emergent properties of forest ecosystems are predictable from interactions of species-specific life history attributes (e.g. Shugart 1984). This chapter presents silvical data for the dominant boreal tree species in the northern hemisphere to be used with ecosystem analysis and modeling studies of the circumpolar boreal forests. The selection of the species is based on distribution maps provided by Sokolov, Svyaseva & Kubly (1977) for the Eurasian species and by Fowells (1965) for the North American ones, as well as on boreal zone maps by Hämet-Ahti (1981). Fourteen tree species have been found to dominate the boreal zone in Fennoscandia and the USSR (Fig. 2.1); fifteen dominate North America. Section 2 of this chapter presents autecological reviews for the boreal tree species in Eurasia. Detailed reviews for the North American boreal tree species have been omitted since they are already available (e.g. Fowells 1965; Harlow, Harrar & White 1979) and because of publication limitations. However, North American boreal tree species have been included in Section 3 of this chapter, which provides species life-history data for parametrization of boreal forest simulation models. Data for silvics of the Eurasian boreal tree species have been extracted from both published and unpublished literature sources in six languages (Bulgarian, English, Finnish, German, Russian and Swedish) and have been collected according to the following scheme.

1. *Systematic classification.* This category provides the genus, species and authorities for the names as well as the most commonly used

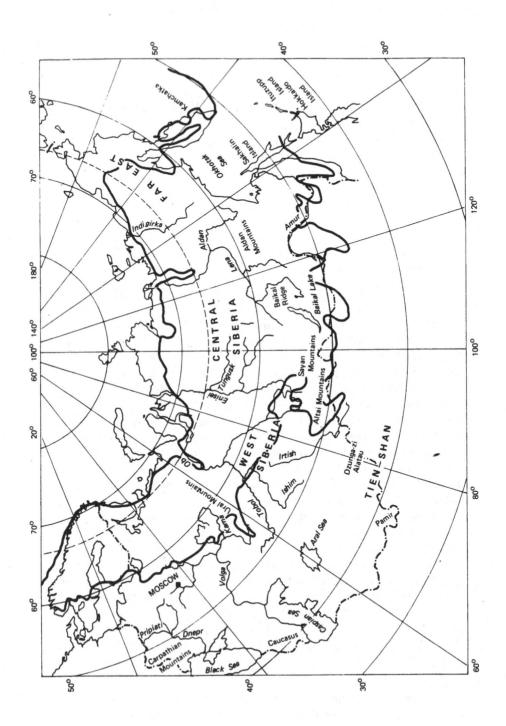

synonyms for the species. The nomenclature follows mainly Czerepanov (1981).

2. *Spatial distribution*. This category describes the geographical range
   of natural growth of the species (planted trees and urban ornamen-
   tals are excluded). Some data concerning altitudinal limits are also
   presented.
3. *Habitat*. This category describes climate and soil conditions where
   the species is found, and its associated species.
4. *Life history*. This category describes species reproductive behavior
   and growth pattern.
5. *Responses to environmental factors*. This category characterizes
   species' responses to light, soil moisture, nutrients, fire, frost,
   permafrost, wind and flooding.
6. *Races and hybrids*. This category gives examples of closely related
   species, races and hybrids.
7. *Enemies and diseases*. This category describes potential threats to
   the species.

It should be pointed out that this chapter is a result of an international
scientific collaboration study carried out during the summers of 1988 and
1989 at the International Institute for Applied Systems Analysis in
Laxenburg, Austria. The study was a part of the Biosphere Dynamics
Project.

2. Ecological characteristics of boreal tree species in Eurasia

2.1. *Abies sibirica* Ledeb. Syn. *Abies pichta* Forb., *Abies semenovii B.* Fedtsch.

Distribution

A. sibirica is a continental species. It is found in the north-eastern
European USSR as well as in the mountains of Ural and Altay, in central
and southern parts of western Siberia, and in Kazakhstan (Tarbagatay
and Dzungazi Alatau Mountains). Outside the USSR, it grows in northern Mongolia and in the north-western part of China. A map of its
distribution is shown in Fig. 2.2. In the polar region of the Ural
Mountains it grows up to between 400 and 600 m above sea level (a.s.l.);
in the northern Urals and on the plateau of Aldansk it is found up to
between 700 and 900 m, in the central Urals, up to 1000 m, and in the

Fig. 2.1. Boreal zone of Fennoscandia and the USSR.

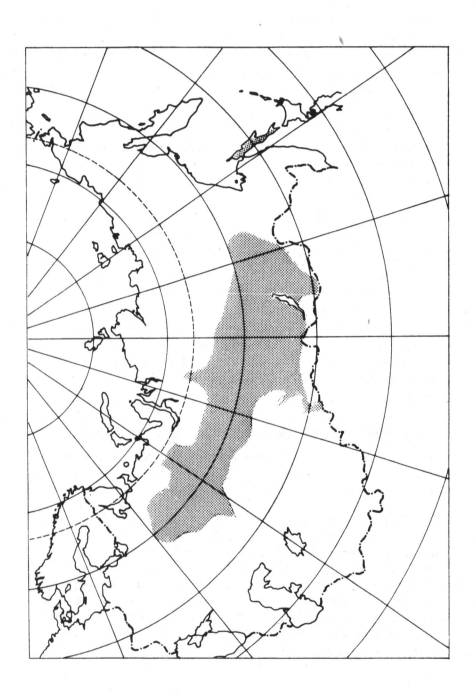

southern Urals, up to 1300 m. It forms high-mountain woodlands together with *Picea obovata* and *Betula pubescens* and on some sites with *Betula pendula*. In Kuznetsky Alatau *A. sibirica* dominates stands from 300 m up to 1400 m and is also found at 2200–2400 m. In the southern Ural Mountains it forms mountain woodlands with *Pinus sibirica* and *Larix sibirica* (Sokolov, Svyaseva & Kubly 1977). In the USSR, *A. sibirica* dominates a forest area of more than 10 million hectares (Sokolov, Svyaseva & Kubly 1977).

Habitat

A. sibirica has a much smaller ecological range than other Eurasian coniferous tree species (Sokolov, Svyaseva & Kubly 1977). The northern and southern limits of its range coincide with the 16.5 °C and 20 °C July isotherms respectively. It is found on river banks (Blomqvist 1887) and in mountainous areas in Siberia (Bärner 1961).

A. sibirica dominates forest only when the air humidity is high. Dry winds and relative humidity below 50% may even cause tree death (Polikarpov, Tchebakova & Nazimova 1986).

A sibirica requires the following for optimal growth:

> A climatic growing season of 120 days (Falaleev 1982; Polikarpov, Tchebakova & Nazimova 1986). Climatic growing season is defined as the period of a year with air temperature above 10 °C.
> A mean temperature for the climatic growing season of 12 °C, and for July 16 °C (Falaleev 1982).
> Not less than 700 mm precipitation over the growing season and not less than 900–1500 mm over the whole year (Falaleev 1982).
> Maximum snow cover not exceeding 120 cm (Falaleev 1982).
> A hydrothermal index (HTI) greater than 1.8 (Polikarpov, Tchebakova & Nazimova 1986). HTI is evaluated as the ratio of annual 10 °C base temperature sum to precipitation.

A. sibirica thrives on well-drained, moist, fertile, sand–loamy and slightly podzolized soils (*Trees and Shrubs of the USSR* 1956; Polikarpov, Tchebakova & Nazimova 1986). On nutrient-rich soils it grows well even far from its climatic optimum. Under favorable climatic conditions (e.g. HTI greater than 2, annual precipitation between 600 and 1000 mm, and a snow cover of 50–60 cm) it also grows successfully on shallow rocky soils (Polikarpov, Tchebakova & Nazimova 1986). *A. sibirica* does not survive

Fig. 2.2. Distribution map of *Abies sibirica* Ledeb.

on bogs or marshy sites (Sokolov, Svyaseva & Kubly 1977; Falaleev 1982).

Associated species

A. sibirica is associated over large areas with *Picea obovata* and *Pinus sibirica*. These species are the major components of dark-coniferous taiga forests. *A. sibìrica* is also found in the understorey of *Pinus sylvestris*, *Larix sukaczewii*, *Larix sibirica* and *Larix gmelinii* forests. It also grows in secondary forests of *Betula pubescens*, *Betula pendula* and *Populus tremula*. It competes successfully with *Pinus sylvestris* on fertile soils but the latter is a stronger competitor on poor soils and on paludified sites (Lindholm & Tiainen 1982). In the European part of the USSR, *A. sibirica* grows in mixed forests of *Quercus robur*, *Tilia cordata*, *Acer platanoides*, *Ulmus laevis* and *Ulmus glabra*. In the Kuznetsky Alatau and Altay Mountains it is found together with *Tilia sibirica*, while in the flood-plains of some Siberian rivers it occurs under the canopy of *Populus laurifolia* (Sokolov, Svyaseva & Kubly 1977).

Life history

A. sibirica is a monoecious and anemophilous species (Sokolov, Svyaseva & Kubly 1977). Its seeds are dispersed by wind in autumn (Blomqvist 1887; Falaleev 1982). Seed production seems to occur more often during dry and hot years (Lagerberg & Sjörs 1972). In closed stands, seed production begins at 60–70 years; solitary trees already produce seeds at 15–18 years (Sokolov, Svyaseva & Kubly 1977). Good seed crops occur at intervals of 2–3 years in the southern USSR (Sarvas 1964; Falaleev 1982) and at 4–6 year intervals in the northern parts of the country (Sarvas 1964); on the western Siberian plains seed crop intervals are 6–7 years, and 10 years in the region of the Enisei River (Falaleev 1982). In Kazakhstan, good seed years occur every 2–3 years, and in the Urals every 3–4 years (Falaleev 1982). *A. sibirica* reproduces better than *Pinus sylvestris*, *Picea abies* and *Larix* species on shady and grassy sites (Sarvas 1964; Lindholm & Tiainen 1982). It is also able to reproduce by layering (Lagerberg & Sjörs 1972; Sokolov, Svyaseva & Kubly 1977). However, vegetative regeneration occurs only at the limits of its distribution and near the treeline (*Trees and Shrubs of the USSR* 1956; Sarvas 1964). *A. sibirica* has a foliage retention time of about 7–10 years (Sokolov, Svyaseva & Kubly 1977).

Saplings of *A. sibirica* grow very slowly in the first 5–8 years (*Trees and Shrubs of the USSR* 1956). Young trees continue to grow slowly, reaching one meter in height in 15–25 years. Thereafter, the growth rate increases (Falaleev 1982). Near the mountain timberline and close to the polar forest limit, its growth form is shrubby (Sarvas 1964).

Maximum values for height, diameter at breast height (DBH) and age are

Height: 32 m (Polikarpov, Tchebakova & Nazimova 1986), 30–35 m (Sokolov, Svyaseva & Kubly 1977), 30–40 m (Beisser 1891; *Trees and Shrubs of the USSR* 1956; Bärner 1961; Krussman 1971; Drakenberg 1981); for a good site in the USSR, 37–38 m (Falaleev 1982).

DBH: 30 cm at a height of 38 m (Sarvas 1964), 50 cm (*Trees and Shrubs of the USSR* 1956), 55 cm (Beisser 1891). The largest DBHs quoted are 60–80 cm (Sokolov, Svyaseva & Kubly 1977; Falaleev 1982) and 68 cm (Polikarpov, Tchebakova & Nazimova 1986).

Age: Range is between 150 and 200 years (*Trees and Shrubs of the USSR* 1956; Sarvas 1964; Polikarpov, Tchebakova & Nazimova 1986) and 300 years (Falaleev 1982). Maximum age is dependent on climatic conditions: in wet climates, individuals seldom reach 220–260 years because they are attacked by fungi (*Trees and Shrubs of the USSR* 1956). About 60% of trees older than 70–80 years are damaged by a stem rot, which makes them sensitive to windthrow (Sokolov, Svyaseva & Kubly 1977). In the western Siberian plains and on periodically flooded sites near rivers, maximum age is 160 years, while on drier sites in the same area it is 220–240 years (Falaleev 1982).

Response to environmental factors

Light
A. sibirica is one of the most shade-tolerant of the boreal tree species. Among the taiga tree species, it is the most shade-tolerant. It can withstand full canopy shading for more than 60 years (Falaleev 1982), which enables it to develop extremely mixed-aged forest stands (Sokolov, Svyaseva & Kubly 1977). Its light requirements are similar to those of *Picea abies* (Sarvas 1964; Falaleev 1982).

Soil moisture
A. sibirica is drought-intolerant (Sokolov, Svyaseva & Kubly 1977). It favors moist and well-aerated soils (Lindholm & Tiainen 1982) and does not grow on peat sites (Sarvas 1964) or bogs (*Trees and Shrubs of the USSR* 1956). In mountainous areas, it prefers well-drained and slightly elevated sites with moist soils (*Trees and Shrubs of the USSR* 1956).

Nutrients
A. sibirica is characterized as nutrient-stress-intolerant (Sokolov, Svyaseva & Kubly 1977), being less tolerant than *Picea abies*, *Picea obovata* and *Pinus sibirica* (*Trees and Shrubs of the USSR* 1956) and than *Larix*

sibirica, *Larix gmelinii* and *Pinus sylvestris* (Drakenberg 1981; Polikarpov, Tchebakova & Nazimova 1986).

Fire
Its fire-tolerance is poor because it has thin bark (Falaleev 1982; Polikarpov, Tchebakova & Nazimova 1986), but fires are rare in the moist forests where it grows (Falaleev 1982). After forest fire, *A. sibirica* dominates stands in the middle phases of succession, after the forest has been dominated by deciduous trees and herbs (Lindholm & Tiainen 1982). It also suffers from sunburns when forests are thinned (Polikarpov, Tchebakova & Nazimova 1986)

Frost
Although *A. sibirica* can withstand winter temperatures as low as −50 °C without damage (Falaleev 1982), it is sensitive to spring and fall frosts (Beisser 1891; Bauch 1975; Polikarpov, Tchebakova & Nazimova 1986), especially in the western part of its range (Sokolov, Svyaseva & Kubly 1977).

Flooding
A. sibirica has been characterized both as tolerant of occasional flooding (Blomqvist 1887) and as having good flooding tolerance (Falaleev 1982).

Wind
It is tolerant of disturbance by wind until old age, when its tolerance is reduced by insect damage causing stem rotting (*Trees and Shrubs of the USSR* 1956; Sokolov, Svyaseva & Kubly 1977). According to Polikarpov, Tchebakova & Nazimova (1986), it is sensitive to windfall because of its superficial root system and its tendency to rot during old age.

Races and hybrids
A. sibirica is closely related to *Abies nephrolepis* and *Abies sachalinensis* (Sarvas 1964).

2.2. *Betula pendula* Roth. Syn. *Betula verrucosa* Ehrh., *Betula alba* L., *Betula alba* var. *vulgaris* Rgl., *Betula alba* subsp. *pendula* var. *vulgaris* Rgl.

Distribution
B. pendula is found in western Europe, and in the European part of the USSR, except in the northern- and southernmost areas. It also occurs in the Caucasus and west and eastern Siberia, except in the northernmost areas, and in the far east of the USSR, except in Tchukotka, and in the

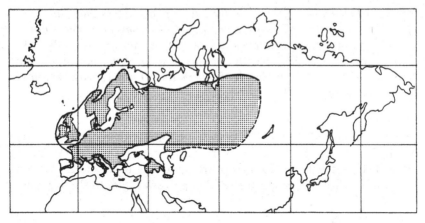

Fig. 2.3. Distribution map of *Betula pendula* Roth.

lower part of the Anadir River and the Koryanski Mountains. It is also found in Kazakhstan, in the northern part of the Dzungazi Alatau, in minor and central Asia, in western Tien-Shan, in Mongolia, in China, and on the Korean Peninsula (Sokolov, Svyaseva & Kubly 1977). Its distribution is shown in Fig. 2.3.

Habitat

B. pendula is a pioneer species on burned, cleared or old field sites (*Trees and Shrubs of the USSR* 1956). It does not occur as far north as *Betula pubescens* (Bonnemann & Röhrig 1971). It grows best on drained loam-sand and podzol soils in the forest and in the forest-steppe zones (Haritonovitsh 1968). It is tolerant of acidic and saline soils but not of paludified soils (Haritonovitsh 1968). In the steppe region, *B. pendula* becomes larger and older than in other parts of the USSR (Haritonovitsh 1968).

Associated species

B. pendula is found as an accompanied species in many coniferous, coniferous–broad-leaved, and broad-leaved forest types (Sokolov, Svyaseva & Kubly 1977). *B. pendula* occurs in association with *Betula pubescens* on drained sites (Sokolov, Svyaseva & Kubly 1977). In West and central Siberia *B. pendula* forms highly productive forests with *Populus tremula*, *Picea obovata*, *Abies sibirica* and *Pinus sibirica*. The sparse understorey is composed of *Sorbus sibirica*, *Padus racemosa*, *Rosa acicularis*, *Spirea media*, *Lonicera altaica* and other shrubs. After disturbance of mixed broadleaved–spruce and broadleaved–pine forests, *B. pendula* forms the main canopy with *Populus tremula*. In the understorey are found *Picea abies*, *Quercus robur*, *Acer platanoides*, *Carpinus betulus*, *Fraxinus excelsior*, *Sorbus aucuparia*, *Corylus avellana* and *Euonymus verrucosa*. On northern slopes of the main Caucasus ridge, *B. pendula* grows in association with *Populus tremula*, *Betula latwinowii* and *Tilia caucasica*. In the Far East it often occurs in *Larix* forests (Sokolov, Svyaseva & Kubly 1977).

Life history

The wind-dispersed seeds of *B. pendula* are formed annually (Kellomäki 1987). Good seed crops occur every third year (Hempel & Wilhelm 1897). The species forms seed crops from the ages of 15 and 25 years in closed and open stands respectively (Hempel & Wilhelm 1897). Seeds are spread during autumn, and usually germinate within 4–5 weeks after dispersal (Hempel & Wilhelm 1897). This species regenerates both from seed and from stump sprouts (*Trees and Shrubs of the USSR* 1956; Kellomäki 1987). Regeneration from stump sprouts can continue until 30 years of age (Haritonovitsh 1968). *B. pendula* regenerates well on grassy sites (Blomqvist 1887; *Trees and Shrubs of the USSR* 1956). Fire is important for its regeneration, as it is unable to regenerate under a closed canopy (Haritonovitsh 1968).

It is one of the fastest-growing boreal tree species (Hempel & Wilhelm 1897). Young *B. pendula* grow faster than *Pinus sylvestris* (Haritonovitsh 1968). The increase in height and diameter of *B. pendula* in steppe regions is highest between 5 and 15 years (Haritonovitsh 1968). The following height growth pattern can be seen: slow growth during the first five years, increasing to a maximum at the age of 10–15 years, when the height increment can reach 0.75–1 m per year; thereafter, the growth rate is reduced and at 50–60 years growth has ended (Hempel & Wilhelm 1897). Table 2.1 shows the growth of pure *B. pendula* stands on the most favorable sites in the Ukrainian SSR.

Maximum values for height, diameter and age are

Height: 30 m in central Europe (Leibundgut 1984), 27–30 m in Finland (Kellomäki 1987) and 20–25 m in the USSR (*Trees and Shrubs of the USSR* 1956; Sokolov, Svyaseva & Kubly 1977).

DBH: More than 60 cm in central Europe (Leibundgut 1984), 40–45 cm in Finland (Kellomäki 1987).

Age: 125–135 years in moist moss forest in southern Finland (Kellomäki 1987).

Response to environmental factors

Light
B. pendula is shade-intolerant (Bonnemann & Röhrig 1972; Sokolov, Svyaseva & Kubly 1977; Leibundgut 1984). Only *Larix* species are more

Table 2.1. *Growth of fully stocked* B. pendula *stands on the most favorable sites in the Ukrainian SSR*

Age (years)	Mean stand height (m)	Mean stand DBH (cm)	Trees per ha	Basal area ($m^2\,ha^{-1}$)	Stem volume (m^3)
10	6.3	3.5	13 832	13.5	56
15	9.8	6.4	5 960	18.9	103
20	13.2	9.4	3 293	22.8	150
25	16.2	12.5	2 101	25.7	196
30	19.0	15.6	1 471	28.1	239
35	21.6	18.6	1 100	30.0	280
40	23.9	21.6	863	31.6	318
45	25.9	24.4	703	32.9	352
50	27.8	27.1	590	34.0	383
55	29.4	29.6	507	34.9	400
60	30.8	32.0	444	35.7	453
65	32.0	34.2	396	36.3	456
70	33.1	36.2	358	36.9	475
75	34.0	38.1	327	37.3	490
77	34.7	39.8	303	37.7	503

Source: Shvydenko *et al.* (1987).

Table 2.2. *Light level (% of full sunlight) and CO_2 uptake (mg g^{-1} dry matter) for* Betula pendula

Light level (% of full sunlight)	CO_2 uptake (mg g^{-1} dry matter)
1	0.18
30	6.00
100	9.40

Source: Haritonovitsh (1968).

light-demanding (Hempel & Wilhelm 1897; Haritonovitsh 1968). Table 2.2 shows the relationship between rate of photosynthesis of *B. pendula* and illumination (Haritonovitsh 1968).

Soil moisture
B. pendula is drought-tolerant (Hempel & Wilhelm 1897) but does not tolerate wet soils (Kellomäki 1987) and sites with standing water (Sokolov, Svyaseva & Kubly 1977). It requires higher air humidity and soil moisture than *Pinus sylvestris* (Haritonovitsh 1968).

Nutrients

B. pendula can grow on very poor soils (Hempel & Wilhelm 1897; Bonnemann & Röhrig 1972) but is has a higher nutrient demand than *Pinus sylvestris* (Haritonovitsh 1968).

Fire and frost

B. pendula can survive forest fires (Kellomäki 1987) and is very frost-tolerant (Hempel & Wilhelm 1897); Haritonovitsh 1968; Bonnemann & Röhrig 1972). It is also resistant to frosts in late spring and early autumn (Haritonovitsh 1968).

Flooding and windstorm

B. pendula does not tolerate prolonged flooding (Sokolov, Svyaseva & Kubly 1977). It is often damaged by strong winds (Hempel & Wilhelm 1897).

Races and hybrids

B. pendula forms hybrids with *Betula pubescens* (Hempel & Wilhelm 1897; Bonnemann & Röhrig 1971).

2.3. Betula pubescens Ehrh. Syn. Betula alba L. sensu Roth., Betula alba subsp. pubescens Rgl.

Distribution

B. pubescens is found in western Europe, in the European USSR (except in the northern- and southernmost parts), in west and central Siberia except in the northernmost areas, and in northern Kazakhstan (Sokolov, Svyaseva & Kubly 1977). It is also found in the Caucasus (Bonnemann & Röhrig 1971). Its distribution is shown in Fig. 2.4.

Habitat

B. pubescens occurs further north than *B. pendula* (Bonnemann & Röhrig 1971). In the north, on poor and acidic soils, it forms the forest–tundra border (Sokolov, Svyaseva & Kubly 1977). It also grows on loamy sand and podzolic, marshy soils (Haritonovitsh 1968) as well as on lime soils and peat bogs (Sokolov, Svyaseva & Kubly 1977). It is a pioneer species in the arboreal colonization of non-forested land (Tseplyaev 1961).

Associated species

In the forest–tundra transition zone *B. pubescens* is found with *Pinus sylvestris*, *Picea obovata* and *Larix* species. In this area it also forms sparse stands with a mean height of 3–6 m, with an understorey of *Betula nana* and *Juniperus sibirica*. In the forest zone *B. pubescens* is an associate species in many coniferous and mixed forest types. It forms pure stands only in western and central Siberia; in these areas it is one of the main components of the birch–aspen subzone on wet sites. After fire or in forest clearings *B. pubescens* often dominates stands in association with *Betula pendula*. The proportion of *Betula pendula* decreases with increased soil paludification. In larger river valleys and on wet sites *B. pubescens* forms almost pure stands with a small proportion of *Picea abies* and *Pinus sylvestris*. In the Belorussian SRR *B. pubescens* forms pure stands on bogs. On wet soils with good aeration it forms stands with *Alnus glutinosa*. On peat bogs it forms pure stands or grows in association with *Pinus sylvestris* and *Salix* species. In forest–steppe and in northern steppe zones *B. pubescens* forms low-productivity stands on wet saline soils together with *Salix cinerea*, *Salix rosmarinifolia* and *Salix pentandra* (Sokolov, Svyaseva & Kubly 1977).

Life history

B. pubescens is a monoecious species and its seeds are dispersed by wind. It is able to regenerate every year from seed and also from stumps. Regeneration from stumps is especially common among young individuals (Kellomäki 1987). Trees also regenerate well on grassy sites (Blomqvist 1887). Seedlings of *B. pubescens* survive better under the forest canopy than seedlings of *Betula pendula* (Haritonovitsh 1968).

Maximum values for height, diameter and age are

Height: Ranges from 20 m (*Trees and Shrubs of the USSR* 1956; Haritonovitsh 1968; Eiselt & Schröder 1977; Sokolov, Svyaseva & Kubly 1977; Leibundgut 1984) to 30 m (Drakenberg 1981; Kellomäki 1987).

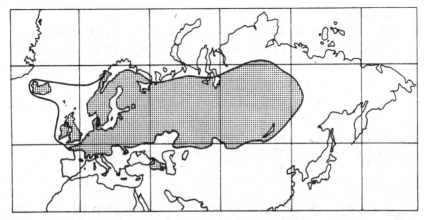

Fig. 2.4. Distribution map of *Betula pubescens* Ehrh.

DBH: Seldom more than 30–40 cm (Leibundgut 1984; Kellomäki 1987). In the Federal Republic of Germany (FRG), diameters of 40–60 cm have been found (Gode 1986).

Age: Seldom more than 100 years (Kellomäki 1987).

Response to environmental factors

Light
B. pubescens is shade-intolerant (Tseplyaev 1961; Bonnemann & Röhrig 1972; Sokolov, Svyaseva & Kubly 1977) but less so than *Betula pendula* (Haritonovitsh 1968; Bonnemann & Röhrig 1971).

Soil moisture
B. pubescens grows on a range of dry to wet soils (Kellomäki 1987). It tolerates wetter growing conditions and prefers moister sites than *Betula pendula* (*Trees and Shrubs of the USSR* 1956; Bonnemann & Röhrig 1971; Sokolov, Svyaseva & Kubly 1977), and is also found on sphagnum bogs (Haritonovitsh 1968; Bonnemann & Röhrig 1971; Sokolov, Svyaseva & Kubly 1977). According to N. P. Polikarpov (unpublished), it is drought-intolerant but not as much as *Picea abies*, *Picea obovata* and *Alnus glutinosa*.

Nutrients
B. pubescens is able to grow on very poor soils (Tseplyaev 1961; Bonnemann & Röhrig 1972). It is nutrient-stress-tolerant but less so than *Pinus sylvestris* and *Pinus pumila* (N. P. Polikarpov, unpublished).

Fire and frost
B. pubescens is able to survive forest fires (Kellomäki 1987) and is very frost-tolerant (Bonnemann & Röhrig 1972). Extreme frost tolerance allows it to reach the forest–tundra zone in the north, and to reach the mountain timberline (Haritonovitsh 1968).

Flooding and windstorm
B. pubescens is tolerant of flooding since it is also found in swamps (Hempel & Wilhelm 1897; Bonnemann & Röhrig 1971).

Races and hybrids
B. pubescens forms hybrids with *Betula pendula* (Hempel & Wilhelm 1897; Bonnemann & Röhrig 1971).

2.4. *Chosenia arbutifolia* (Pall.) A. Skvorts. Syn. *Chosenia macrolepis* (Turcz.) Kom. (Source: Sokolov, Svyaseva & Kubly 1977)

Distribution

Ch. arbutifolia is found in eastern Siberia, in the Far East, in north-eastern China, on the peninsula of Korea and in Japan. Its distribution is shown in Fig. 2.5.

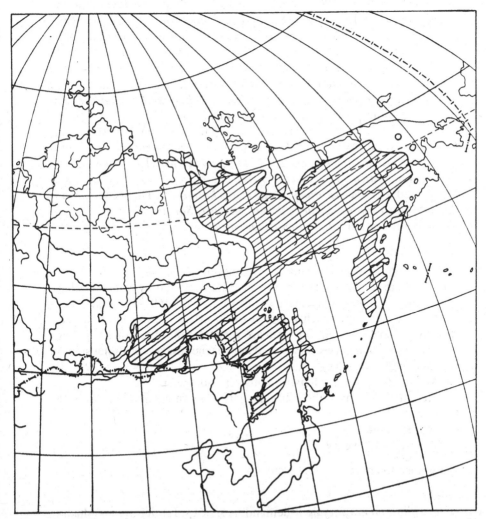

Fig. 2.5. Distribution map of *Chosenia arbutifolia* (Pall.) A. Skvorts.

Habitat

Ch. arbutifolia occurs mainly on alluvial soils of mountain river flood-lands. It is a pioneer species of new alluvials.

Associated species

Ch. arbutifolia forms both pure and mixed stands. In northern Yakutia it grows in mixed stands (mean height of 8–10 m) with *Populus suaveolens*, in which the understorey is composed of *Salix schwerinii* and *Salix udensis*. Along mountain rivers of Tchukotka it is found on alluvial soils in association with *Populus suaveolens* and *Betula pendula*; in the understorey *Salix schwerinii*, *Salix udensis*, *Alnus hirsuta*, *Sorbus kamtchatcensis* and *Sorbus anadyrensis* grow. In the northern part of its range it also occurs in association with *Larix gmelinii*. On Sakhalin Island *Ch. arbutifolia* forms productive stands with *Populus maximowiczii* and it is also found in association with *Abies sachalinensis*; the understorey consists of *Salix schwerinii*, *Salix rorida*, *Alnus hirsuta*, *Crataegus chlorosarca* and some other species. In the central part of its range it is found with *Picea obovata* and *Picea ajanensis*. On the Sikhote–Alin Mountains *Ch. arbutifolia* grows in mixed stands with *Populus maximowiczii*, *Ulmus laciniata*, *Fraxinus mandshurica*, *Tilia mandshurica*, *Tilia taquetii*, *Juglans mandshurica* and *Phellodendron amurense*. On wetter sites it is associated with *Picea koraiensis* and *Pinus koraiensis*. The second layer is composed of *Padus asiatica*, *Sorbus sibirica*, *Sorbus amurensis*, *Ligustrina amurensis* and *Malus mandshurica*.

Life history

Ch. arbutifolia is a dioecious and anemophilous species. It regenerates mainly from seeds, which are dispersed by wind and by water. On bare new alluvial soils it may produce up to 130 000–160 000 seedlings per hectare, which may reach a height of 30–40 cm during the first year. Seedlings develop deep roots and are very tolerant of seasonal flooding.

   Ch. arbutifolia grows very fast. It approaches maximum dimensions at the age of 30–40 years.

   Maximum values for height, diameter and age are

   Height: 35–37 m in the southern part of its range and 8–10 m in the northern.

   DBH: 80 cm.

   Age: 120–130 years.

Response to environmental factors

Light

Ch. arbutifolia is a very light-demanding species. It tolerates less shading than *Larix* and *Populus*.

Soil moisture
It is very drought-intolerant. It occurs only on periodically flooded and wet sites.

Frost
Ch. arbutifolia tolerates low winter temperatures but does not survive on permafrost. The tree also tolerates large diurnal temperature variations.

Wind
Ch. arbutifolia is very tolerant of windthrow because of its deep root system.

2.5. *Larix gmelinii* (Rupr.) Litv. Syn. *Larix dahurica* Turcz., *Larix cajanderi* Mayr.

Distribution

L. gmelinii is widely distributed in north-eastern Asia. Its range covers wide areas in eastern Siberia and the Far East (Dylis 1981). It is also found in northern Mongolia and north-eastern China. It forms continuous forests over almost its whole range (Dylis 1981). The western border of its geographic range coincides with the eastern extent of *Larix sibirica*. In the north, it reaches 72°40′ latitude (River Hatangi) (Sokolov, Svyaseva & Kubly 1977). Its distribution is shown in Fig. 2.6.

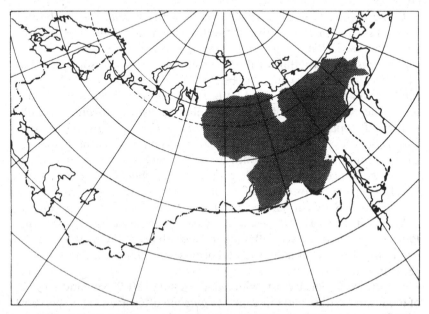

Fig. 2.6. Distribution map of *Larix gmelinii* (Rupr.) Litv.

Habitat

L. gmelinii grows under many different conditions: in mountainous areas, on plains, along river valleys, along sea-shores, in the taiga, on tundra and in forest–steppe regions (Dylis 1981). Over most of its distribution, the climate is continental and the soil has permafrost (Sokolov, Svyaseva & Kubly 1977; Dylis 1981). Its occurrence is characterized by areas having peat bogs and permafrost (*Trees and Shrubs of the USSR* 1956). *L. gmelinii* often forms a thick litter layer (Dylis 1981).

L. gmelinii thrives on scree or talus slopes, on alluvial soils, on humus-carbonic soils and on acidic podzol soils. *L. gmelinii* is the only forest-forming species on these soils under severe climatic conditions. *L. gmelinii* is less tolerant of dry soils, but more tolerant of cooler and infertile soils, than *Larix sibirica* (Polikarpov, Tchebakova & Nazimova 1986).

Associated species

L. gmelinii usually forms pure stands (Sokolov, Svyaseva & Kubly 1977), especially north of 66°N, where the tree layer is more discontinuous (Schotte 1917). Most stands are even-aged, with an age difference of approximately 20–40 years. A reason for this may be the absence of competitors during colonization in severe climates (Dementev 1959; Dylis 1961).

On new alluvial soils, it is found together with *Chosenia arbutifolia* and *Populus suaveolens*, while on old as well as on lime soils and wet sites, it is associated with *Picea obovata*, *Picea ajanensis*, *Abies sibirica*, *Abies nephrolepis* and sometimes with *Pinus sibirica*. On moist soils, it grows with *Pinus sylvestris*. In many forest types, *L. gmelinii* occurs with *Betula pendula* and *Pinus pumila*. The latter forms dense understorey on some sites. The understorey of forests dominated by *L. gmelinii* typically consists of *Alnus fruticosa*, *Rhododendron dauricum* and *Betula midden-dorffii* (Sokolov, Svyaseva & Kubly 1977). On mountains and plains, *L. gmelinii* forms large forests both as pure stands and together with *P. sylvestris*, *Picea obovata* and other species (*Trees and Shrubs of the USSR* 1956). *L. gmelinii* often forms monospecific stands with only a few individuals of *Betula* in warmer regions, whereas in southern Yakutia and Zabaikal, it grows with *P. sylvestris*. In mountainous areas, it grows with *Pinus sibirica* and in river valleys with *Picea obovata* (Dylis 1981). In the Amur area and on Sakhalin Island it occurs with *Picea ajanensis* (Beisser 1891; Bärner 1961).

On sites with permafrost, with a thaw depth of 50–90 cm, and on soils of moderate fertility and moisture, *L. gmelinii* often grows in association with *Picea obovata*, *Pinus sibirica*, *Pinus sylvestris* and *Betula pendula*.

The understorey of these forests consists of *Pinus pumila*, *Alnus fruti-cosa*, *Betula middendorffii* and *Rosa acicularis*. On moist leached carbonate soils, it is associated with *Picea obovata*, *Pinus sibirica* and *Betula pendula*. The understorey is composed of *Pinus pumila*, *Alnus fruticosa*, *Sorbus sibirica*, *Juniperus sibirica*, *Rosa acicularis*, *Spirea media*, *Rhododendron dauricum* and *Lonicera altaica*. On soils of moderate fertility and near standing permafrost at 1200–1300 m a.s.l., *L. gmelinii* forms forests with a dense understorey of *Rhododendron dauricum* and *Alnus fruti-cosa*. On dry rocky soils, *L. gmelinii* is found in sparse stands sometimes with an understorey of *Pinus pumila*, *Betula middendorffii*, *B. exilis*, *Alnus fruticosa*, *Juniperus sibirica*, *Salix glauca*, *S. lanata*, *Dasiphora fruticosa*, *Spirea media* and *Spirea dahurica*. On marshy sites, *L. gmelinii* often forms the understorey with *Pinus sibirica* and *Betula middendorffii*. On peaty sites, *L. gmelinii* often forms pure stands with a dense shrub layer of *Ledum palustre*, *Betula middendorffii*, *Betula exilis*, *Vaccinium uliginosum*, *Betula fruticosa*, *Salix myrtilloides* and *Alnus fruticosa*. *L. gmelinii* grows well in river valleys and flooded sites. In these stands it occurs in association with *Picea obovata* or *Picea ajanensis*, *Pinus sibirica* and sometimes with *Populus suaveolens* and *Chosenia arbutifolia*. The understorey consists of *Cornus alba*, *Crataegus sanguinea*, *Alnus fruti-cosa*, *Sorbaria sorbifolia*, *Sorbus sibirica*, *Lonicera edulis*, *Ribes pallidi-florum*, *Spirea salicifolia* and *Spirea media*.

Life history

L. gmelinii is a coniferous, deciduous species. It is monoecious and anemophilous. Seed production starts at 15–30 years. Seeds are wind-dispersed and can be found hundreds of kilometers from the seed source (Sokolov, Svyaseva & Kubly 1977). Good seed years occur at least every third year (Sarvas 1964). Regeneration is best on bare, burned sites, but the species is also able to regenerate under closed canopies. Regeneration is poor under its own canopy because of the thick litter layer on the forest floor (Dylis 1981). If burned plots are seeded successfully, hundreds of young trees can establish and a larch forest develops. If not, burns are occupied by grass, shrubs and sometimes birch forests, which delays the restoration of *L. gmelinii* forests for some decades (Pozd-nyakov 1983).

    L. gmelinii grows slower than *Larix sibirica* but faster than *Pinus sylvestris* (*Trees and Shrubs of the USSR* 1956). Most of its height growth takes place before it is 60 years old; afterwards the growth decreases to nearly zero by 100–120 years (Dylis 1981).

    The growth pattern of *L. gmelinii* on the most favorable sites in Yakutia (*Trees and Shrubs of the USSR* 1956) is shown in Table 2.3.

    On peat bogs, it becomes stunted: 4–6 m high (Sokolov, Svyaseva &

Table 2.3. *The growth pattern of* Larix gmelinii
on the most favorable sites in Yakutia

Age (years)	Height (m)
10	1–1.5
20	2–4.0
50	6–11.0
100	15–23.0
200	20–29.0

Source: *Trees and Shrubs of the USSR* (1956).

Kubly 1977). In the north of its distribution and at the treeline in mountains, *L. gmelinii* becomes shrubby, with a maximum height of 0.2 m (Sarvas 1964). On marshy sites with nearby permafrost, its root system becomes flat; on deep light soils it develops a pivotal root system. When growing on bogs or permafrost, *L. gmelinii* produces adventitious roots coming up from the stem (Sokolov, Svyaseva & Kubly 1977; Pozdnyakov 1983).

Maximum values for height, diameter and age are

Height: between 20 m (Beisser 1891; Lagerberg & Sjörs 1972; Dylis 1981) and 30–40 m (*Trees and Shrubs of the USSR* 1956; Bärner 1961; Sarvas 1964; Sokolov, Svyaseva & Kubly 1977); 40 m (Pozdnyakov 1983). Maximum reported height is 45 m (Sokolov, Svyaseva & Kubly 1977).

DBH: 0.5–1 m (Beisser 1891; *Trees and Shrubs of the USSR* 1956; Bärner 1961; Lagerberg & Sjörs 1972; Dylis 1981), 0.8–1 m (Sokolov, Svyaseva & Kubly 1977). DBH of 1.40 m is rare (Sarvas 1964).

Age: between 250 and 400 years in the USSR (*Trees and Shrubs of the USSR* 1956; Dylis 1981; Pozdnyakov 1983).

Response to environmental factors

Light
L. gmelinii is one of the most shade-intolerant of the boreal tree species (Sarvas 1964; Sokolov, Svyaseva & Kubly 1977; Dylis 1981). According to N. P. Polikarpov (unpublished), it is the most light-demanding boreal tree species in Eurasia.

Soil moisture
L. gmelinii grows best on moist to well-drained soils (*Trees and Shrubs of the USSR* 1956; Dylis 1981), but it also grows on bogs (Sarvas 1964). On swampy sites, it becomes shrubby (Beisser 1891). It grows better on bogs than do *Pinus sylvestris* and other *Larix* species because of its ability to

produce adventitious roots (*Trees and Shrubs of the USSR* 1956). According to N. P. Polikarpov (unpublished), *L. gmelinii* tolerates more drought than *Carpinus betulus*, *Fraxinus excelsior*, *Populus tremula* and *Larix sukaczewii* but less than *Larix sibirica*, *Betula pendula*, *Quercus mongolica*, *Quercus robur* and *Pinus sylvestris*.

Nutrients

L. gmelinii is very nutrient-stress-tolerant (*Trees and Shrubs of the USSR* 1956; Dylis 1981). It grows on many different substrates (Sarvas 1964), including saline soils (*Trees and Shrubs of the USSR* 1956). However, it tolerates less nutrient stress than *Pinus sylvestris*, *Pinus pumila*, *Betula pendula* and *Betula pubescens* (N. P. Polikarpov, unpublished).

Fire

L. gmelinii is classified as fire-tolerant because of its thick bark (Kabanov 1977). *L. gmelinii* is found in very dry climates and forms large amounts of litter, which favors the occurrence of forest fires. Low-intensity fires usually dominate. *Larix* forests are killed by intense fires. *L. gmelinii* has a superficial root system, which is often damaged by fires. Damaged root systems rot and are infected by fungal diseases. After fire the thickness of summer thawing increases dramatically. If fires occur on steep slopes the thin soil layer is eroded, becoming a rocky stream, and is never recolonized by forest (Pozdnyakov 1983).

Frost

L. gmelinii is very frost-tolerant (Sarvas 1964; Sokolov, Svyaseva & Kubly 1977). It grows at higher latitudes and altitudes than all other conifers except *Pinus pumila* (Sokolov, Svyaseva & Kubly 1977). It can withstand extended periods of cold (Dylis 1981), and is able to grow on permafrost (Sarvas 1964; Sokolov, Svyaseva & Kubly 1977). It is the most permafrost tolerant of the boreal tree species (Polikarpov, Tchebakova & Nazimova 1986). The temperature gradient between air and soil which is associated with permafrost may even be necessary for the normal growth of *L. gmelinii* (Dementev 1959; Dylis 1961).

Flooding

L. gmelinii tolerates seasonal flooding (Sokolov, Svyaseva & Kubly 1977).

Races and hybrids

L. gmelinii has many local races (Lagerberg & Sjörs 1972). Two local races have been distinguished on the basis of climatic tolerance, one eastern and one western type (Dylis 1981).

Enemies

The Siberian silkworm reduces the growth in both height and diameter of *L. gmelinii* (*Trees and Shrubs of the USSR* 1956).

2.6. *Larix sibirica* Ledeb. Syn. *Larix europaea* var. *sibirica* Lonnd, *Larix decidua* var. *rossica* Henk et Horst, *Larix decidua* var. *sibirica* Rgl., *L. sibirica* var. *russica* Endl.

Distribution

L. sibirica is distributed over a range of 3000 km from west to east (Dylis 1981). In the east, its range borders that of *Larix gmelinii*, and in the west, that of *Larix sukaczewii* (Dylis 1981). It is found from 58° W longitude to 119° E (Lake Baikal), from tundra in the north (70° latitude) to the mountains of Altay and Sayan in the south (46° latitude). *L. sibirica* is the dominant species of the north, where it forms the polar forest timberline (Dylis 1981) in the Ural Mountains and in western Siberia (north of 63° latitude) as well as in southern Siberia, in Altay, Tanu-Ola, and on the Baikal Ridge (*Trees and Shrubs of the USSR* 1956; Dylis 1981). In the Altay Mountains, it reaches the upper timberline at 2000–2400 m (*Trees and Shrubs of the USSR* 1956). In the south, it reaches the steppe and arid regions of Kazakhstan and central Asia. It is most abundant in the Altay and Sayan Mountains, in the Baikal region and on the central Siberian Plain. In western Siberia, the abundance of *L. sibirica* increases only in the polar zone, where it forms the forest-tundra vegetation (Dylis 1981). Its distribution is shown in Fig. 2.7.

Paleodata show that the distribution range of *L. sibirica* was broader in the past. During the postglacial hypsithermal, it grew 2° latitude further north than today. At the end of the Pliocene and during the Pleistocene, its eastern limit was several thousand kilometers further east than it is today (Dylis 1981).

Habitat

L. sibirica has a very broad ecological amplitude. Its potential distribution area is the largest of all major boreal tree species in Asia (Polikarpov, Tchebakova & Nazimova 1986). It is found from steppe lowlands to tundra uplands. It is a dominant tree in dry and continental climates with permafrost (Drakenberg 1981; Gode 1986; Polikarpov, Tchebakova & Nazimova 1986). It is adapted to dry climates with an HTI ranging from 1.1 to 1.6, annual precipitation of 300–450 mm, and air humidity of 40%. Increasing relative humidity up to 65% causes a reduction in transpiration rate. Because of this, wet climates slow down its growth rate and reduce seed production (Polikarpov, Tchebakova & Nazimova 1986).

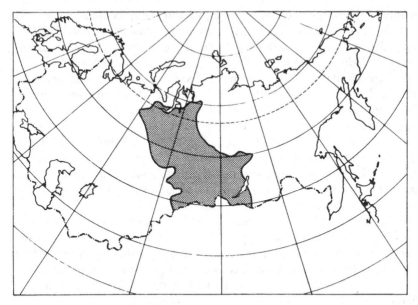

Fig. 2.7. Distribution map of *Larix sibirica* Ledeb.

L. sibirica occurs on many different soil types. In the upper part of the Lena River valley, it grows on loamy soils, and near the Ob and Poluy Rivers it grows on well-drained podzol soils. Near the polar timberline, it grows on dry sandy as well as on sand–loam soils, and in the lower parts of the Sayan Mountains from the River Enisei to the Lake of Baikal, it is found only on nutrient-rich rendzina soils (Dylis 1981). Generally it prefers well-drained loamy and sandy soils on limestone. Sometimes it can be found in dry lichen and sphagnum forest types. *L. sibirica* requires more fertile and aerated soils than *Pinus sylvestris*. Its litter is rich in nitrogen, potassium and sulphur, which are released when the litter decays. The decaying litter increases the element content of the humus and decreases soil acidity (Kalinin 1965). *L. sibirica* grows poorly on dry, sandy and marshy soils (Kalinin 1965; Haritonovitsh 1968): on such sites it is found on the taiga wetlands (Dylis 1981). In artificial plantations, it grows successfully on any kind of soil (Dylis 1981).

Associated species
L. sibirica often grows in pure stands or in mixed stands with *Pinus sylvestris*, on well-drained and potassium-rich soils (Dylis 1981). It is seldom found in forests dominated by *Picea obovata*, *Pinus sibirica*, and *Abies sibirica* (*Trees and Shrubs of the USSR* 1956; Sokolov, Svyaseva & Kubly 1977; Dylis 1981). It forms mixed associations with *Pinus sibirica*

mainly in the border area between light- and dark-needled forests, and with *Picea obovata* in river valleys that are seasonally flooded (Polikarpov, Tchebakova & Nazimova 1986). It also associates with *Betula pendula,* *Betula microphylla, Populus tremula, Populus laurifolia, Sorbus sibirica* and some other shrub species. On dark-gray soils, where *L. sibirica* shows the best growth, it forms mixed stands with *Pinus sibirica, Picea obovata, Abies sibirica, Betula pendula* and *Populus tremula. L. sibirica* occurs on dry soils in western Siberia and in the northernmost areas of the Ural Mountains. In these areas it forms sparse stands together with *Picea obovata, Pinus sibirica, Pinus sylvestris* and *Betula pendula*. The understorey is composed of *Betula nana, Juniperus sibirica* and other shrub species. On wet podzol–gley soils, *L. sibirica* forms nearly pure stands with a small proportion of *Pinus sylvestris, Pinus sibirica, Abies sibirica* and *Picea obovata;* its growth there is poor. The understorey is sparse and composed of *Sorbus sibirica* and *Lonicera altaica*. In the forest–tundra region and in the northern part of western Siberia, *L. sibirica* is found on sites with permafrost and on peatbogs. In this region it forms sparse stands with *Pinus sylvestris* in the canopy and *Betula nana* in the understorey. On alluvial, wet soils in larger river valleys of central and southern taiga, *L. sibirica* forms productive mixed forests with *Pinus sibirica, Picea obovata, Abies sibirica, Populus tremula, Populus laurifolia* and *Betula* species. The understorey is composed of *Padus racemosa, Sorbus sibirica, Sambucus sibirica, Ribes nigrum, Ribes hispidulum* and others (Sokolov, Svyaseva & Kubly 1977).

Life history

L. sibirica is a coniferous deciduous species. It is a monoecious and anemophilous species. Its seeds are wind-dispersed but they are not carried far from the mother tree (Sarvas 1964). Good seed years occur every 6–7 years in the north and every 3–4 years in the south (Blomqvist 1887; Dylis 1981). Seed production starts at the age of 12–15 years. The most intensive seed production occurs at 30–40 years and between 70 and 100 years. Seeding continues to a very old age (*Trees and Shrubs of the USSR* 1956). *L. sibirica* regenerates best after forest fire (Blomqvist 1887) or on bare soils (*Trees and Shrubs of the USSR* 1956). Regeneration and the broadening of the area occupied by *L. sibirica* is strongly dependent on fire regimes (Polikarpov, Tchebakova & Nazimova 1986). It also regenerates well on moist sites with moss cover. Thick litter or grass cover prevents regeneration (Dylis_1981). *L. sibirica* does not regenerate at all under *Pinus sibirica* and *Pinus sylvestris* canopies, and its regeneration is weak under its own canopy (*Trees and Shrubs of the USSR* 1956).

Table 2.4. *Growth of fully stocked, even-aged* Larix sibirica *plantations in the Ukrainian SSR on favorable sites*

Age (years)	Mean stand height (m)	Mean stand DBH (cm)	Trees per ha	Basal area (m^2 ha^{-1})	Stem volume (m^3)
10	6.1	7.9	2898	14.2	52
20	12.3	14.1	1545	24.1	152
30	17.4	19.3	1072	31.3	268
40	21.6	23.8	822	36.6	382
50	25.0	27.8	667	40.5	485
60	27.8	31.3	563	43.3	574
70	30.2	34.4	489	45.4	651
80	32.3	37.1	434	46.9	781
90	34.2	39.5	392	48.0	777
100	35.9	41.6	360	48.9	829
110	37.4	43.5	334	49.6	874
120	38.7	45.2	312	50.1	913

Source: Shvydenko *et al.* (1987).

Table 2.5. *Age and height increase for* Larix sibirica *on sites with fertile soils*

Age (years)	Height (m)
10	3.6
20	9.9
50	22.7
100	31.2
150	36.2

Source: Trees and Shrubs of the USSR (1956).

The litter of *L. sibirica* decays much faster than that of other coniferous species because it contains fewer toxic substances which inhibit decay (Dylis 1981).

L. sibirica is a fast-growing species (Sokolov, Svyaseva & Kubly 1977). It grows faster than the other coniferous species occurring in the USSR, but slower than *Larix decidua*. Its growth rate is highest between 80 and 100 years (Dylis 1981), but also peaks at the age of 20–40 years (*Trees and Shrubs of the USSR* 1956). Increase in height and diameter continues until 300 years (Dylis 1981). The height growth of *L. sibirica* in fully stocked, even-aged plantations on favorable sites in the Ukrainian SSR is shown in Table 2.4. Table 2.5 shows another growth pattern of the species on fertile soils.

Maximum values for height, diameter and age are

Height: Between 30 and 45 m (Blomqvist 1887; *Trees and Shrubs of the USSR* 1956; Bärner 1961; Sarvas 1964; Dylis 1981) and 40 and 45 m (Sokolov, Svyaseva & Kubly 1977).

DBH: 80–100 cm (Sarvas 1964; Dylis 1981); 1.8 m has been recorded as an extreme value (*Trees and Shrubs of the USSR* 1956; Sokolov, Svyaseva & Kubly 1977).

Age: 120–150 years in natural forests (Blomqvist 1887) and 182 and 187 years in plantations (*Trees and Shrubs of the USSR* 1956), 400–500 years (Sokolov, Svyaseva & Kubly 1977).

Response to environmental factors

Biologically, *L. sibirica* is related to *Larix sukaczewii* but its ecological features are slightly different (Sokolov, Svyaseva & Kubly 1977):

> *L. sibirica* tolerates cooler climates.
> It has higher drought tolerance.
> Its ecological amplitude is broader: its habitats range from permafrost to the steppe border.
> It is better adapted for growth on peatbogs. Depending on the thickness of the peat layer, it can produce adventitious roots.

Light

L. sibirica is a shade-intolerant species (*Trees and Shrubs of the USSR* 1956; Sarvas 1964; Sokolov, Svyaseva & Kubly 1977). Only *Larix gmelinii* has higher light requirements than *L. sibirica* (Polikarpov, Tchebakova & Nazimova 1986). *L. sibirica* occupies mainly sites that are not suitable for the shade-tolerant Siberian species *Picea obovata*, *Pinus sibirica* and *Abies sibirica* (Dylis 1981).

Soil moisture

L. sibirica is drought-tolerant, less so than *Pinus sylvestris*, and more so than *Larix decidua*, *Picea obovata* and *Pinus sibirica* (*Trees and Shrubs of the USSR* 1956). However, it prefers moist soils. On boggy soils or soils with standing ground water, its growth is poor (Sarvas 1964). For good growth, it needs well-aerated soils; for this reason it does not grow well in the Siberian taiga or on peatlands around rivers (Dylis 1981).

Nutrients

The nutrient requirements of *L. sibirica* are somewhere between those of *Pinus sylvestris* and *Picea abies* (Blomqvist 1887). It can grow under poorer conditions than *Larix decidua* (Dylis 1981). According to N. P. Polikarpov (unpublished), *L. sibirica* tolerates more nutrient stress than

Larix sukaczewii, Pinus sibirica, Picea obovata and *Populus tremula* but less than *Larix gmelinii, Betula pendula, Betula pubescens, Pinus pumila* and *Pinus sylvestris*. *L. sibirica* is able to grow on different kinds of soils, but at the margins of its distribution requires fertile soils (Sarvas 1964). On podzolized soils its growth response to fertilizer application is not as strong as for *Betula pendula* but stronger than for *Picea abies* (*Trees and Shrubs of the USSR* 1956).

Fire
L. sibirica is a fire-tolerant species (Sarvas 1964) because of its thick bark (*Trees and Shrubs of the USSR* 1956). It is more fire-tolerant than *Pinus sylvestris* (Sokolov, Svyaseva & Kubly 1977).

Frost
L. sibirica is adapted better than other species to spring frost, winter temperature inversions, thin snow cover, and cold soils (Polikarpov, Tchebakova & Nazimova 1986). The species is able to grow on permafrost (Sarvas 1964; Lagerberg & Sjörs 1972; Sokolov, Svyaseva & Kubly 1977), but it occupies regions characterized by insular permafrost rather than areas with continuous permafrost (Pozdnyakov 1983).

Flooding
L. sibirica occurs on flood-lands along rivers (Sokolov, Svyaseva & Kubly 1977; Polikarpov, Tchebakova & Nazimova 1986) as well as on wet and boggy sites (*Trees and Shrubs of the USSR* 1956; Sokolov, Svyaseva & Kubly 1977); this indicates that the species tolerates seasonal flooding.

Wind
L. sibirica develops a strong tap root and long branch roots, up to 10–12 m, which make it very wind-tolerant. However, on permafrost, boggy soils, or limestone beds, where the root system is superficial, it becomes wind-intolerant (Polikarpov, Tchebakova & Nazimova 1986).

Races and hybrids
L. sibirica forms hybrids with *Larix sukaczewii* (Sokolov, Svyaseva & Kubly 1977).

Enemies and diseases
L. sibirica is damaged by a rot caused by *Polyporus schweinitzii* Fr., or *Trametes pini* Fi. Lateral and upper shoots are occasionally damaged by *Steganoptycha diniana* Gn. As a result, trunks become crooked, and annual height increments decline (Kalinin 1965).

2.7. *Larix sukaczewii* Dylis

Distribution

L. sukaczewii occurs in the north-eastern part of the European USSR, in the Ural Mountains, and in the south-western part of western Siberia. To the east, it reaches the Ob and Irtish valleys (Dylis 1981). Its distribution is shown in Fig. 2.8. Borders between the ranges of *L. sukaczewii* and *Larix sibirica* in the Urals and western Siberia are rather vague because of the occurrence of hybrids (Sokolov, Svyaseva & Kubly 1977).

Habitat

L. sukaczewii is found on soils of various textures, but its growth is best on loamy-sandy and sandy soils (Sokolov, Svyaseva & Kubly 1977) as well as on slightly podzolized and humus-rich soils. It seldom occurs on bogs. Optimal climatic conditions for its growth are found in the southern taiga and in the mixed forest zone (Dylis 1981).

Associated species
L. sukaczewii rarely occurs in pure stands (Sokolov, Svyaseva & Kubly 1977; Dylis 1981). Typically, it grows in forests of *Pinus sylvestris*, *Picea abies* and *Picea obovata* with some occurrence of *Betula pubescens*,

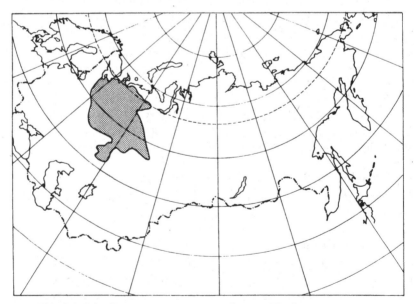

Fig. 2.8. Distribution map of *Larix sukaczewii* Dylis.

Betula pendula and *Populus tremula*. In the forest–tundra transition zone it forms sparse forest stands with *Picea obovata* and *Betula pubescens* on sandy soils. The understorey is composed of *Betula nana* and *Juniperus sibirica*. In the north, on dry loamy soils, it grows poorly and sometimes has an understorey of *Betula nana*. In northern and central taiga, *L. sukaczewii* forms the main canopy. The understorey in these forests consists of *Pinus sylvestris* and *Picea abies*, and *Sorbus aucuparia*, *Juniperus communis*, *Lonicera pallasii* and *Rosa acicularis*. On carbonate (rendzina) soils, *L. sukaczewii* grows in mixed-age forests with *Pinus sylvestris* and *Picea* species. The understorey is often dense and consists of spruce saplings and individuals of *Sorbus aucuparia*, *Juniperus communis*, *Lonicera pallasii*, *Cotoneaster melanocarpus*, *Daphne mezereum* and *Spirea media*. On alluvial and lime soils along rivers. *L. sukaczewii* has optimal growth and associates with spruce, birch and aspen, with a sapling layer of *Picea obovata*. The understorey consists of the same species as in the northern and central taiga (see above), but *Padus racemosa*, *Salix phylicifolia*, *Salix caprea*, *Ribes pubescens*, and in the north *Betula nana*, are also found. *L. sukaczewii* is also found on marshy sites (peatbogs) growing with *Betula nana*, *Andromeda polifolia* and *Empetrum nigrum*. In the southern part of its range and in the southern Urals, *L. sukaczewii* grows in mixed stands with *Pinus sylvestris*, *Picea obovata* and *Abies sibirica*. The understorey consists of saplings of *Picea obovata*, *Abies sibirica*, *Tilia cordata*, *Quercus robur*, *Acer platanoides*, *Ulmus glabra* and *Corylus avellana* (Sokolov, Svyaseva & Kubly 1977).

In the Archangelsk region, *L. sukaczewii* often grows on steep river slopes, where it can form up to 70% of the two-layered canopy. The upper stratum consists of *L. sukaczewii* and some *Pinus sylvestris*, and the lower stratum consists of *Betula* (Dylis 1981).

Life history

L. sukaczewii is a coniferous deciduous species. It is basically mono-ecious, but about 10% of all individuals produce predominantly male or female flowers. The species is anemophilous. Seed production starts at 10–15 years for solitary trees and at 25–30 years for trees growing in closed stands (Sokolov, Svyaseva & Kubly 1977). The species requires bare mineral soils for successful regeneration. These conditions are found on mountain slopes, along river valleys, and in the plains after forest fires (Dylis 1981). Although rare, *L. sukaczewii* can also regenerate by layering. In some cases young trees produce stem sprouts after being cut (Sokolov, Svyaseva & Kubly 1977).

L. sukaczewii is a fast-growing species. Height growth is fastest during the first 50–70 years, reaching up to 100 cm per year (Sokolov, Svyaseva & Kubly 1977). On good sites, it grows faster than any *Picea* or *Pinus* species. When growing on bogs with a thick sphagnum layer, it can produce adventitious roots (Sokolov, Svyaseva & Kubly 1977).

Maximum values for height, diameter and age are

Height: 40 m (Sokolov, Svyaseva & Kubly 1977; Dylis 1981).

DBH: 100–120 cm (Sokolov, Svyaseva & Kubly 1977; Dylis 1981).

Age: 350 years (Sokolov, Svyaseva & Kubly 1977).

Response to environmental factors

Light
L. sukaczewii is very shade-intolerant (Dylis 1981). However, it is less light-demanding than *Larix gmelinii* and *Larix sibirica* (N. P. Polikarpov, unpublished).

Soil moisture
L. sukaczewii grows best on well-drained, aerated soils. According to N. P. Polikarpov (unpublished), it is intermediately water-stress-tolerant, being more tolerant than *Abies* and *Picea* but less so than *Pinus sylvestris*, *Quercus robur*, *Quercus mongolica*, *Larix sibirica* and *Larix gmelinii*.

Nutrients
According to N. P. Polikarpov (unpublished), *L. sukaczewii* is relatively nutrient-stress-tolerant. It is more tolerant than *Pinus sibirica*, *Picea obovata*, *Picea ajanensis*, *Picea abies* and *Populus tremula* but less so than *Larix sibirica*, *Larix gmelinii*, *Betula pendula*, *Betula pubescens* and *Pinus sylvestris*.

Fire
L. sukaczewii is fire-tolerant, because old trees have thick bark (Dylis 1981).

Frost
Its occurrence at the northern forest limit is a sign of its ability to tolerate frost (Dylis 1981). However, it can sometimes be damaged by spring frosts (Sokolov, Svyaseva & Kubly 1977).

Flooding
The occurrence of the species in river valleys (see, for example, Sokolov, Svyaseva & Kubly 1977) indicates that it can tolerate at least occasional flooding.

Wind
L. sukaczewii is windstorm-tolerant (Dylis 1981).

Races and hybrids
L. sukaczewii is a subspecies of *Larix sibirica* (Lagerberg & Sjörs 1972).

2.8. *Picea abies* (L.) Karst. Syn. *Picea excelsa* Link.

Distribution
P. abies is found in the mountains of central and southern Europe (Sarvas 1964), in Fennoscandia, and in the north-western, western and central parts of the European USSR (Sokolov, Svyaseva & Kubly 1977). In the east, it reaches the Kama River, in the south the Pripiati River and Ukrainian Carpathians (Kazimirov 1983). Its distribution is shown in Fig. 2.9.

Habitat
P. abies grows naturally in areas with a continental climate and high amounts of precipitation (Bonnemann & Röhrig 1972). Its northern limit is determined by the duration of the growing season: it needs at least 2–2.5 months with mean temperature above 10 °C. The southern limit is determined by drought (Schmidt-Vogt 1977) and coincides well with the

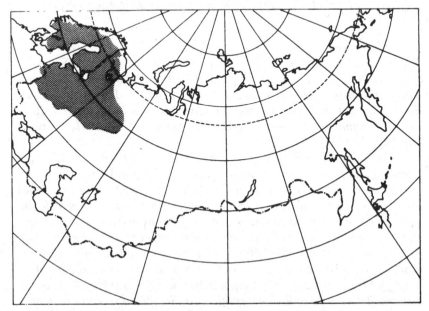

Fig. 2.9. Distribution map of *Picea abies* (L.) Karst.

northern border of the chernozem region (Sokolov, Svyaseva & Kubly 1977). The western limit is determined by oceanic climate and the eastern one by drought and competition from other trees and grasses (Schmidt-Vogt 1977). *P. abies* is, however, characterized as a very flexible species (Bonnemann & Röhrig 1972) which is able to grow under many different climatic conditions (Schmidt-Vogt 1977).

It is found where:

the length of the frost period is at least 3 months;
rainfall is 600–800 mm from May to August (Mayer 1909; data from the Alps). Another author sets the lower limit to 230 mm (Schmidt-Vogt 1977); the limit is dependent on the temperature conditions at the site;
the effective temperature sum is on average 600 to 800 °C, calculated on a 5 °C base (Siren 1955; data from northern Finland).

The soil moisture availability determines the temperature requirements of *P. abies* (Schmidt-Vogt 1977). In northern parts of Sweden it prefers areas with a higher air humidity (Aminoff 1909).

P. abies grows best on loamy and loam–sandy soils, where it develops a deep root system. It is also found on marshy sites and sphagnum bogs but it cannot survive on oligotrophic peatbogs where anaerobic conditions dominate. *P. abies* is considered to be a major contributor to soil podzolization (Aminoff 1909; Sokolov, Svyaseva & Kubly 1977).

Associated species

P. abies is a dominant forest tree species of the European taiga (Sokolov, Svyaseva & Kubly 1977). Throughout its entire range it associates with *Pinus sylvestris*, *Populus tremula*, *Betula pendula*, *Betula pubescens*, *Sorbus aucuparia*, *Padus racemosa*, *Salix caprea*, *Salix pentandra*, *Juniperus communis* and others (Sarvas 1964; Schmidt-Vogt 1977; Sokolov, Svyaseva & Kubly 1977).

In the northern part of its range it is found with *Betula nana*, *Betula fruticosa* and *Juniperus sibirica* and in the eastern part it grows in mixed forests with *Picea obovata*, *Abies sibirica*, *Larix sukaczewii* and *Pinus sibirica*. On marshy sites it grows with *Salix phylicifolia*, *Salix myrtiloides* and *Salix lapponum*. In southern taiga *P. abies* forms mixed coniferous–broadleaved forests with *Tilia cordata*, *Acer platanoides*, *Ulmus glabra*, *Ulmus laevis*, *Ulmus carpinifolia*, *Quercus robur*, *Fraxinus excelsior*, *Corylus avellana* and *Euonymus verrucosa*. In the western part of its range it also associates with *Carpinus betulus*, *Fagus sylvatica*, *Quercus petraea*, *Tilia platyphyllos*, *Malus sylvestris*, *Sambucus racemosa*, *Sambucus nigra* and *Euonymus europea*. In the Carpathian Mountains it is

Table 2.6. *Growth of even-aged* Picea abies *stands on favorable sites in the Ukrainian Carpathian Mountains*

Age (years)	Mean stand height (m)	Mean stand DBH (cm)	Trees per ha	Basal area $(m^2 ha^{-1})$	Stem volume (m^3)
20	7.2	6.1	5784	16.9	71
30	13.9	12.3	2736	32.5	229
40	19.5	17.7	1755	43.2	411
50	24.2	22.4	1279	50.4	587
60	28.1	26.5	1003	55.3	741
70	31.3	30.1	822	58.5	868
80	33.8	33.3	696	60.6	967
90	35.8	36.2	601	61.8	1042
100	37.2	38.8	530	62.7	1098
110	38.4	41.2	474	63.2	1138
120	39.1	43.4	429	63.4	1160

Source: Shvydenko *et al.* (1987).

found with *Acer pseudoplatanus*, *Tilia tomentosa*, *Tilia petiolaris*, *Abies alba*, *Prunus spinosa*, *Larix polonica* and *Alnus viridis* (Sokolov, Svyaseva & Kubly 1977).

Life history

P. abies is a monoecious and anemophilous species (Sokolov, Svyaseva & Kubly 1977). Seed production starts at an age of 40–60 years in a closed canopy, and earlier in open conditions (Lagerberg & Sjörs 1972). Seeds are light: 4.8 g per 1000 seeds in northern Europe and 8.0 g per 1000 seeds in central Europe (Sarvas 1964). Seeds are dispersed mainly by wind and partly by birds and animals (Sokolov, Svyaseva & Kubly 1977) during March and April in the year following flowering (Schmidt-Vogt 1977). Seed crops occur every 4–8 years in the USSR (Kazimirov 1983) and every 12–13 years in Finland (Sarvas 1964). The species can also regenerate by layering, although this is rare (Siren 1955; Sokolov, Svyaseva & Kubly 1977; Kazimirov 1983). Seedlings are very shade-tolerant and can survive under a closed canopy for a long period of time (Sylven 1916; Siren 1955). *P. abies* grows slowly at early ages (Sokolov, Svyaseva & Kubly 1977). The height growth rate increases after 5–10 years and can be intensive until 150–170 years (Kazimirov 1983). Table 2.6 shows the growth of even-aged *P. abies* stands on favourable sites in the Ukrainian SSR.

Maximum values for height, diameter and age are
Height: From 30 to 60 m (Beisser 1891; Schmidt-Vogt 1977). In the Russian plains it is 35–40 m, while it reaches 50 m in the Carpathian

Mountains (Sokolov, Svyaseva & Kubly 1977; Kazimirov 1983); in Sweden, maximum height is 45 m (Lagerberg and Sjörs 1972).

DBH: In Sweden, 1.7 m (Lagerberg & Sjörs 1972); in the FRG, 2 m (Bärner 1961; Drakenberg 1981).

Age: Seldom more than 400–500 years in the USSR (Kazimirov 1983); 570 years (Sokolov, Svyaseva & Kubly 1977). However, 1000-year-old trees have been found in primeval forests in the Alps (Bärner 1961). In managed forests, the oldest trees are 150 years old (Bärner 1961). The oldest trees found in Sweden are seldom more than 400 years old (Lagerberg & Sjörs 1972).

Response to environmental factors

Light
P. abies is very shade-tolerant but less so than *Abies sibirica* and *Taxus baccata* (Sokolov, Syvaseva & Kubly 1977). The ability to tolerate shading depends both on its age and on site conditions: young individuals are more shade-tolerant than older trees (Ellenberg 1952, 1978; Siren 1955; Walter 1960).

Soil moisture
P. abies grows preferably on moist or wet soils (Bonnemann & Röhrig 1972) and does not tolerate drought (Sokolov, Svyaseva & Kubly 1977). In the south-eastern part of its range, even large trees die after periods of drought (Sokolov, Svyaseva & Kubly 1977). However, it does not thrive on wet anaerobic soils (Sylven 1916). *P. abies* outcompetes *Pinus sylvestris* on wet humid soils where there are no fires (Aminoff 1909).

Nutrients
P. abies is characterized as intermediately nutrient-stress-tolerant (Sokolov, Svyaseva & Kubly 1977) or as a tolerant species (Bonnemann & Röhrig 1972). According to N. P. Polikarpov (unpublished), *P. abies* tolerates nutrient stress better than *Populus tremula*, *Quercus mongolica*, *Quercus robur*, *Tilia cordata* and *Abies sibirica* but worse than *Picea ajanensis*, *Picea obovata*, *Pinus sibirica*, *Larix* and *Betula* species and *Pinus sylvestris*. *P. abies* prefers acidic soil, pH 4–5 (Schmidt-Vogt 1977), but needs good nutrition to grow well (Sarvas 1964).

Fire
Fire tolerance of *P. abies* is very poor (Drakenberg 1981).

Frost
The species is quite frost-tolerant (Bonnemann & Röhrig 1972). Ability to survive unseasonal frosts depends on the place of origin. Siberian

subspecies are very resistant to spring frosts (Schmidt-Vogt 1977). At early ages, however, damage can occur from spring frosts (Sylven 1916; Kazimirov 1983). Young trees growing under open conditions or in larger canopy gaps are especially prone to spring frosts (Sokolov, Svyaseva & Kubly 1977). The species tolerates winter frosts well (Sylven 1916).

Flooding
The occurrence of the species on wet sites and near flowing waters (Sokolov, Svyaseva & Kubly 1977) indicates that it can tolerate at least occasional flooding.

Wind
P. abies is basically windthrow-intolerant because of its shallow root system (Sokolov, Svyaseva & Kubly 1977).

Races and hybrids

P. abies has formed many races with different ecological requirements (Bärner 1961; Schmidt-Vogt 1977). Based on differences in cone scales the following tree varieties are recognized (Schmidt-Vogt 1977):

> *Picea abies* var. *acuminata*, found in the Alps.
> *Picea abies* var. *europaea*, found in northern Europe as far east as the Ural Mountains.
> *Picea abies* var. *obovata*, found east of the Ural Mountains.

2.9. *Picea ajanensis* (Lindl. et Gord.) Fisch. ex Carr. Syn. *Picea jezoensis* (Sieb. et Zucc.) Carr.

Distribution

P. ajanensis is found in the south-eastern parts of the USSR, on the Shantar Islands, on the peninsula of Kamchatka, and on Sakhalin Island (Bärner 1961; Sokolov, Svyaseva & Kubly 1977; Drakenberg 1981). Outside the USSR it is found in the mountains of north Japan, North Korea and north-eastern China (Bärner 1961; Sokolov, Svyaseva & Kubly 1977; Drakenberg 1981; Kazimirov 1983). Its distribution is shown in Fig. 2.10. In the Far East, *P. ajanensis* is a forest-dominant tree species (Sokolov, Svyaseva & Kubly 1977). On Hokkaido Island, where it forms pure stands, it is the most important *Picea* species (Sarvas 1964).

In the northern part of its range, along the coast of the Okhotsk Sea, it occurs up to 100–120 m a.s.1.; further inland it reaches 1000–1200 m. In the central and southern Sikhote–Alin Mountains it is a dominant tree species from 600 to 1400 m. On northern Sakhalin it is found from sea level up to 700–800 m. In the Zeisky–Priural Mountains it dominates a forest belt between 500 and 800 m. At lower elevations forests are dominated by *Larix gmelinii* while at higher altitudes *Pinus pumila* prevails (Sokolov, Svyaseva & Kubly 1977).

Habitat

P. ajanensis is found in similar habitats to *Picea abies* (Krussman 1971). It occurs mainly in mountains (Haritonovitsh 1968), but it needs more humid, maritime climates than *Picea abies* (Schmidt-Vogt 1977). *P. ajanensis* is favored by a monsoon climate (Sokolov, Svyaseva & Kubly

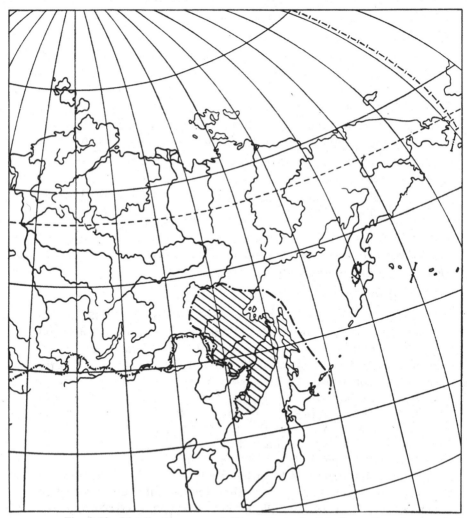

Fig. 2.10. Distribution map of *Picea ajanensis* (Lindl. et Gord.) Fisch. ex Carr.

1977). A large part of its range is characterized by continental climate and presence of permafrost. Incoming radiation is 80–100 kcal yr^{-1}(335–420 KJ yr^{-1}); precipitation is 460–1850 mm yr^{-1}. Relative humidity is more than 60–75%. Thickness of snow cover is 40–100 cm. Growing season (i.e. the period of the year with air temperature above 10 °C) is 112 days in the northern part of the distribution area, and 190 days in the southern part (Manko 1987).

P. *ajanensis* prefers well-drained, moist soils, but it also grows on boggy soils. The best growth occurs on alluvial soils with good drainage. The species can grow on shallow and deep soils and on a variety of parent materials (Manko 1987). P. *ajanensis* is mainly found on brown or podzolized, moist, relatively thin, rocky soils (Sokolov, Svyaseva & Kubly 1977; Kazimirov 1983). It grows best on moist, sand–loam soils and in a moist climate near the sea (Haritonovitsh 1968). It also grows well on deep delluvial soils and on occasionally flooded wet alluvial soils (Sokolov, Svyaseva & Kubly 1977). Its growth is poor on sandy or marshy soils (Haritonovitsh 1968; Sokolov, Svyaseva & Kubly 1977). P. *ajanensis* cannot tolerate paludification and standing ground water. On boggy sites it is often replaced by *Larix gmelinii* (Sokolov, Svyaseva & Kubly 1977). P. *ajanensis* is well adapted for growth on permafrost (Pozdnyakov 1985).

Associated species
P. *ajanensis* is associated with many tree and shrub species (Sokolov, Svyaseva & Kubly 1977). It occurs in both pure and mixed stands (Haritonovitsh 1968). In northern parts of its range and at higher altitudes, P. *ajanensis* grows mainly with *Larix gmelinii*, *Picea obovata*, *Betula ermanii* and *Betula pendula*. The forest understorey is sparse and consists of *Alnus fruticosa*, *Pinus pumila*, *Sorbaria sorbifolia*, *Padus asiatica*, *Rosa acicularis*, *Alnus hirsuta* and *Spirea media*. In the north-western part of its range P. *ajanensis* dominates forests on sphagnum sites. In this area the understorey is composed of *Pinus pumila*, *Ledum hypoleucum*, *Ledum macrophyllum*, *Ledum palustre* and others. In the valley of the Kamchatka River P. *ajanensis* forms low-productivity stands with *Larix gmelinii*, *Larix kurilensis*, *Betula ermanii*, *Betula pendula* and *Populus tremula*.

In northern and central parts of Sakhalin Island it grows with *Abies sachalinensis*; in the south it is associated with *Abies mayriana*. Other associated species on Sakhalin Island are *Abies nephrolepis*, *Larix kurilensis* and *Taxus cuspidata*. In southern Prymorie P. *ajanensis* grows in mixed forests with *Abies nephrolepis*, *Abies holophylla*, *Carpinus cordata*, *Tilia taquetii*, *Tilia mandshurica*, *Tilia amurensis*, *Betula costata*, *Betula ermanii*, *Betula pendula*, *Ulmus lacinata*, *Quercus mongolica*, *Fraxinus mandshurica* and *Populus tremula*. The secondary layer of these forests is composed of the canopy species in admixture with *Acer mono*, *Alnus hirsuta*, *Padus asiatica* and *Maackia amurensis*. In river valleys of the Sikhote–Alin Mountains *P. ajanensis* occurs in mixed coniferous–broadleaved forests with *Pinus koraiensis*, *Abies holophyla*, *Fraxinus mandshurica*, *Ulmus japonica*, *Juglans mandshurica*, *Tilia taquetii* and *Picea koraiensis*. The understorey has the same composition as the main layer and *Phellodendron amurense*, *Acer mandshuricum*, *Acer tegmentosum* and *Padus asiatica* are also found. In the central and southern Sikhote–Alin Mountains, *P. ajanensis* is found on marshy sites together with *Betula pendula*, *Betula costata*, *Fraxinus mandshurica*, *Abies nephrolepis*, *Pinus koraiensis* and *Populus tremula*. The understorey consists of *Sorbus amurensis*, *Sorbaria sorbifolia*, *Spirea salicifolia* and *Eleutherococcus senticosus*. On flood-plains *P. ajanensis* is often associated with *Picea obovata*; both species regenerate successfully under canopies of *Populus maximowiczii* and *Chosenia arbutifolia* in the east, and of *Populus suaveolens* in the west (Sokolov, Svyaseva & Kubly 1977).

Life history

P. ajanensis is a monoecious and anemophilous species (Sokolov, Svyaseva & Kubly 1977). It usually reproduces by seeds but at higher altitudes and in the north it also regenerates by layering. Solitary trees start seed production at the age of 20–25 years (Manko & Voroshilov 1978). Seeds weigh 2–2.9 g per 1000 (Sarvas 1964). Good seed crops occur once in 2–4 years in the Far East and on Sakhalin Island. The number of seeding trees and the amount of seed crop decline in highlands where typically 40–80% of the seeds are sterile (Manko & Voroshilov 1978). Seeds are dispersed mainly by wind and partly by birds and animals (Sarvas 1964; Sokolov, Svyaseva & Kubly 1977). Seedlings of *P. ajanensis* are found under different canopy densities including under an open canopy (Haritonovitsh 1968); they are, however, sensitive to high light levels. Their growth is best under a light intensity which is one-third of the normal maximum one (Sokolov, Svyaseva & Kubly 1977). The growth pattern of *P. ajanensis* on a favorable site (Haritonovitsh 1968) is shown in Table 2.7.

Maximum values for height, diameter and age are

Height: From 30 m (Bärner 1961) to 50–60 m (Krussman 1971; Schmidt-Vogt 1977; Kellomäki 1987); some varieties 50 m (Drakenberg 1981), 40–45 m in USSR (Kazimirov 1983), 33.5 m on Kamchatka (Manko & Voroshilov 1978).

DBH: 100–150 cm (Sokolov, Svyaseva & Kubly 1977); 65–87 cm on Kamchatka (Manko & Voroshilov 1978).

Age: 300 years (Kazimirov 1983); 350 years (Sokolov, Svyaseva & Kubly 1977); 250–300 years on Kamchatka (Manko & Voroshilov 1978); 400–450 years (Manko 1987).

Table 2.7. *The growth pattern of* Picea ajanensis *on a favorable site*

Age (years)	Height (m)
10	2
50	12
100	25

Source: Haritonovitsch (1968).

Table 2.8. *Size characteristics of the oldest* P. ajanensis *individuals found*

Location	Age (years)	DBH (cm)	Height (m)
Kamchatka	430	44.5	21.8
Konder Ridge	400	25.6	17.3
Konder Ridge	355	40.9	23.0
Konder Ridge	378	28.0	19.9
Kyzy Lake	439	120.0	35.0
Northern Sakhalin	520	40.0	20.0
Shmith Peninsula	420	57.2	23.4
Feklistof Island	318	31.6	16.2
Golets M. Nanak	452	39.0	23.8
Golets Lukinda	434	33.0	18.7

Source: Manko (1987).

Size characteristics of the oldest *P. ajanensis* individuals found (Manko 1987) are presented in Table 2.8.

Response to environmental factors

Light
P. ajanensis is very shade-tolerant, especially at early ages (Sokolov, Svyaseva & Kubly 1977; Manko 1987). It tolerates more shade than other *Picea* species (Schmidt-Vogt 1977; Sokolov, Svyaseva & Kubly 1977) but less than *Abies nephrolepis* and *Abies sachalinensis* (Manko & Voroshilov 1978; Manko 1987). Suppressed individuals can survive under a closed canopy for 100 years (Sokolov, Svyaseva & Kubly 1977) and even 210–250 years according to Manko & Voroshilov (1978) and Manko (1987). It is possible that trees of the same diameter and size class have an age difference of 230 years (Sokolov, Svyaseva & Kubly 1977).

Soil moisture

P. ajanensis does not tolerate water stress. Mature stands may be killed by periods of drought (Sokolov, Svyaseva & Kubly 1977). The species usually occupies sites with good water supply provided by slowly melting permafrost. It can tolerate a drier climate only on well-drained sites (Manko & Voroshilov 1978; Manko 1987).

Nutrients

P. ajanensis is considered to be nutrient-stress-tolerant under conditions of favorable water supply (Manko & Voroshilov 1978; Manko 1987). According to N. P. Polikarpov (unpublished), *P. ajanensis* has intermediate nutrient requirements, being more stress-tolerant than *Picea abies*, *Populus tremula*, *Quercus robur*, *Tilia cordata* and *Abies sibirica* and less so than *Picea obovata*, *Pinus sibirica* and *Larix* and *Betula* species.

Fire

P. ajanenis does not tolerate fires (Sokolov, Svyaseva & Kubly 1977) because of its thin bark and shallow root system: even ground fires can kill it. It is less tolerant of fire than *Pinus sibirica* and *Larix* species, but more tolerant than *Abies nephrolepis* and *Abies sachalinensis* (Manko & Voroshilov 1978; Manko 1987). After fires it forms even-aged stands but often its regeneration starts only after establishment of and canopy closure by pioneer species such as *Betula pendula*, *Populus tremula* and *Larix gmelinii* (Sokolov, Svyaseva & Kubly 1977).

Frost

P. ajanensis tolerates winter frosts. The tree does not occur in the lowlands because buds and young trees are sensitive to spring frosts (Haritonovitsh 1968; Sokolov, Svyaseva & Kubly 1977; Manko & Voroshilov 1978; Manko 1987). Frosts sometimes damage trunks (Manko & Voroshilov 1978; Manko 1987).

Flooding

P. ajanensis occurs on sites that may be seasonally flooded. It can withstand flooding for a short time, but continuous accumulation of alluvial deposits brought by floodwater reduces soil aeration, causing some trees to die (Manko & Voroshilov 1978; Manko 1987).

Wind

P. ajanensis is susceptible to windthrow because of its shallow root system (Haritonovitsh 1968; Kazimirov 1983; Manko 1987). It suffers more from wind on wet soils than on drained soils. On rocky soils, it becomes tolerant of windstorms, because it braids its roots between large pieces of rock (Manko 1987).

Races and hybrids

Picea ajanensis var. *hondoensis* Rehn. is found on Honshu Island (Sarvas 1964).

Enemies and diseases

On Kamchatka, the main insects damaging *P. ajanensis* are *Urocerus gigas taiganus* Bens., *Melanophila acuminata* De Geer., *Acmaeops septentrionis* Thoms., *A. smaragdula* F. and *Xylechinus pilosus* Ratz. Fungal diseases are caused most often by *Chrysomixa woroninii* Tranz. Nine species of fungus are known to destroy living trees. The most common fungus is *Phellinus pini* (Fr.) Pil. var. *abietis* (Karst.) Pil. The degree of damage depends on the site. Most trees occurring on poorly drained soils and on soils frozen for a long time are susceptible to damage (Manko & Voroshilov 1978).

2.10. *Picea obovata* Ledeb. Syn. *Picea abies* var. *obovata*

Distribution

P. obovata is commonly found in continental areas (Schmidt-Vogt 1977). It occurs from 70–71° N to 54–56° N latitude (Kazimirov 1983), from north-eastern parts of the European USSR to the Okhotsk Sea in the east (Haritonovitsh 1968). To the north, it reaches the tundra–forest region, and to the south, the taiga zone (Haritonovitsh 1968). It grows on the Siberian Plains and in the upper Ural, Altay and Sayan Mountains. In the northern Urals it grows up to 800 m; in the Altay and Sayan Mountains it occurs up to 1800–2000 m a.s.l. Outside the USSR it occurs in Fennoscandia and in northern Mongolia (Sokolov, Svyaseva & Kubly 1977). Its distribution is shown in Fig. 2.11.

On the Kolsk Peninsula and in the northern part of Europe it forms the northern treeline; in Siberia, it grows further south and the timberline is formed by *Larix sibirica* and *Larix gmelinii*. The geographic range of the species is found to have been broader in the recent geological past both to the west and to the south but it has been reduced by fires and human activities (Sokolov, Svyaseva & Kubly 1977).

Habitat

P. obovata is able to grow in severe continental climates. It tolerates lower humidity and more continentality than any other shade-tolerant boreal tree species. It grows in steppe regions, with an annual temperature amplitude of more than 40 °C, minimum relative humidity of

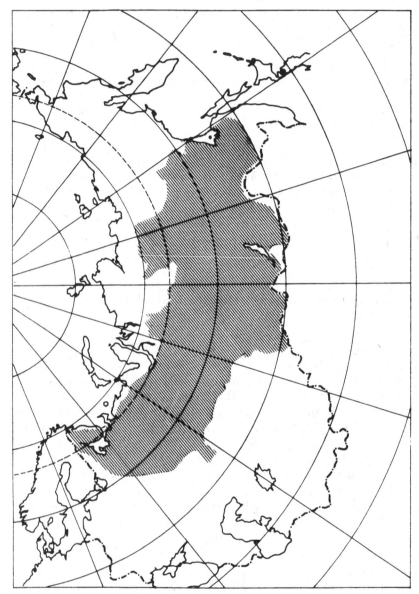

Fig. 2.11. Distribution map of *Picea obovata* Ledeb.

30–35% and annual precipitation of 300–350 mm. It thrives in both wet and dry climates. Its distribution is more dependent on soil conditions than on climate. *P. obovata* requires well-drained to moist, sandy–loam soils and has poor growth on marshy soils in central and southern taiga (Haritonovitsh 1968; Polikarpov, Tchebakova & Nazimova 1968). It tolerates cold soils and grows on permafrost. Only *Larix sibirica* and *Larix gmelinii* are reported as being more tolerant to frozen soils (Polikarpov, Tchebakova & Nazimova 1986).

Associated species
P. obovata forms pure and mixed stands (Kazimirov 1983). It is a typical forest-dominant species of the boreal zone (Polikarpov, Tchebakova & Nazimova 1986). In Siberia it occurs mainly in mixed forests. It forms pure stands mainly in river valleys. *P. obovata*, *Abies sibirica* and *Pinus sibirica* are the major tree species of the Siberian dark taiga forest. In the western part of its geographic range it associates with *Picea abies*, *Betula pubescens*, *Larix sukaczewii*, *Padus racemosa*, *Sorbus aucuparia*, *Alnus glutinosa*, *Frangula alnus*, *Salix pentandra*, *Salix triandra*, *Salix lapponum* and others. In the southwestern part of its range, it grows with *Tilia cordata*, *Acer platanoides*, *Ulmus laevis*, *Ulmus glabra*, *Quercus robur* and *Corylus avellana*. In the central part of its range it is found with *Larix sibirica*, *Betula pendula*, *Betula fruticosa*, *Betula nana*, *Caragana arborescens* and others. In the east it grows in mixed forests with *Larix gmelinii*, *Pinus pumila*, *Betula pendula*, *Betula ermanii*, *Betula fruticosa*, *Betula exilis*, *Populus suaveolens*, *Populus maximowiczii*, *Abies nephrolepis*, *Picea ajanensis*, *Padus asiatica* and others (Sokolov, Svyaseva & Kubly 1977).

Life history
P. obovata is a monoecious and anemophilous species. It has good seed years once every 12–13 years (Kellomäki 1987). The seeds are dispersed in September of the year of flowering (Schmidt-Vogt 1977), mainly by wind and partly by birds and animals (Sokolov, Svyaseva & Kubly 1977). It regenerates on bare, burned soil and under canopies of all forest types; however, regeneration is poor on sphagnum bogs (Kazimirov 1983).

Maximum values for height, diameter and age are

Height: 30 m (Beisser 1891; Haritonovitsh 1968; Kazimirov 1983), 30–35 m (Sokolov, Svyaseva & Kubly 1977), 40 m (Bärner 1961; Kellomäki 1987).

DBH: 40–50 cm (Kellomäki 1987).

Age: In southern Finland, 250–350 years; in northern Finland, 400–500 years (Kellomäki 1987). Similar values have been reported for the USSR: 200–300 and 500 years (Sokolov, Svyaseva & Kubly 1977).

Response to environmental factors

Light
P. obovata is very shade-tolerant (Sokolov, Svyaseva & Kubly 1977; Kellomäki 1987), but tolerates less shade than *Abies sibirica* (Haritono-vitsh 1968), *Abies nephrolepis* and *Picea ajanensis* (Sokolov, Svyaseva & Kubly 1977).

Soil moisture
P. obovata is characterized as one of the most drought-intolerant tree species in the boreal zone of Eurasia. It tolerates less water stress than other *Picea* species as well as *Abies, Betula, Larix* and *Pinus* species (N. P. Polikarpov, unpublished). According to Sokolov, Svyaseva & Kubly (1977), it is more drought-tolerant than *Picea abies*. P. obovata does not grow on soils with standing water and poor aeration (Kellomäki 1987). On marshy sites its growth is poor (Polikarpov, Tchebakova & Nazimova 1986).

Nutrients
P. obovata is classified as intermediately tolerant to nutrient stress. It is less tolerant than *Pinus sylvestris, Pinus sibirica* and *Larix* and *Betula* species, but more so than *Picea ajanensis, Picea abies, Populus tremula, Quercus robur, Tilia cordata* and *Abies sachalinensis* (N. P. Polikarpov, unpublished). Of the south Siberian species it has the second highest nutrient requirements: only *Abies sibirica* has higher requirements (Polikarpov, Tchebakova & Nazimova 1986).

Fire
Picea species are usually reported as fire-intolerant, owing to their thin stem bark (Kabanov 1977). However, fires seldom occur in forests dominated by P. obovata since it usually occupies wet and shaded sites with a hydrophilic moss layer (Voropanov 1950). Seedlings of P. obovata are particularly sensitive to fires (Haritonovitsh 1968).

Frost
P. obovata is tolerant to low winter temperatures (Sokolov, Svyaseva & Kubly 1977), but it is sensitive to spring frosts (Polikarpov, Tchebakova & Nazimova 1986). It also grows on permafrost, but its distribution on permafrost is azonal and it does not dominate forests (Pozdnyakov 1985).

Flooding
The occurrence of P. obovata on wet sites and in river valleys (Sokolov, Svyaseva & Kubly 1977) indicates that it tolerates at least occasional flooding.

Wind

Picea species are usually reported as windthrow-intolerant because of their shallow root system.

Races and hybrids

P. obovata has many different varieties (Bärner 1961). *P. obovata* itself is a 'cold climate variety' of *Picea abies* (Pravdin 1975).

Enemies and diseases

In the boreal zone of the European USSR, 60–90 species of insect harmful to *P. obovata* are found. In forests of the boreal zone, significant loss of seeds occurs owing to cone damage by 15 species of insect and fungus. Sometimes seed crops can be completely destroyed. The most weakened trees are plagued by *Poligraphus poligraphus* L., *Ips typographus* L., *Pitygenes chalcographus* L. and others. In young stands, 64 damaging insect species are known. Leaves are often damaged by insects from the orders Diptera and Homoptera and by ticks of the genus *Sryophyes*: sometimes ticks damage 70–80% of leaves. Fungal diseases are widespread in spruce forests: twenty species of fungus are found. *Fomitopsis pinicola*, *F. annosa* and *Phellinus pini* var. *abietis* are the most common fungal diseases. The degree of fungal damage depends on environmental conditions: it decreases from south to north. Diseases of leaves are widespread. In the north European USSR, mass infection of leaves of young trees is caused by the fungus *Chrysomixa* Ledi D.B.: 70–100% of trees can be damaged. The fungus *Chrysomixa abietis* Wint. may damage as much as 40% of young trees in the middle of the boreal zone in Europe. The fungus *Lophodermium macrosporum* Hart. may damage 50–60% of young trees. In spruce stands, 10–20% of cones are frequently damaged by *Thecopsora padi* Kleb. and *Chrysomixa pirolae* Rostr. (Chertovsky 1979).

2.11. *Pinus pumila* (Pall.) Regel. Syn. *Pinus cembra* var. *pumila* Pall. Fl. Ross.

Distribution

P. pumila is found in eastern Siberia and in the far eastern USSR: Sikhote–Alin, Kamchatka, Sakhalin and the Kuril Islands (Bärner 1961; Sokolov, Svyaseva & Kubly 1977). Outside the USSR, it occurs in Japan, northern Mongolia, and on the northern peninsula of Korea. In northeastern China, it occurs in the Daxingan, Xiaoxingan and Changbai Mountains (Uanjun 1958; Sokolov, Svyaseva & Kubly 1977). Its distribution is shown in Fig. 2.12. On most mountains, it grows up to 1600–

2000 m a.s.1., but in Japan it occurs up to 3000 m (Sokolov, Svyaseva & Kubly 1977).

Habitat

In mountainous areas *P. pumila* is found on rocky soils and on igneous rock outcrops. On plains, it grows on peat and peat–gley soils and in river valleys on alluvial podzol soils as well as on river sand deposits (Sokolov, Svyaseva & Kubly 1977). In China, *P. pumila* always grows on ridges or mountain tops. There it forms dense, pure stands of short stature, or it grows under the canopy of *Larix* (Uanjun 1958).

Associated species

P. pumila often forms shrubwoods in association with *Alnus fruticosa, Betula middendorfii, Betula exilis, Rhododendron dauricum, Rhododendron auretum, Rhododendron kamtschaticum, Arctous alpina, Vaccinium vitis-idaea, Empetrum nigrum, Rosa acicularis* and others. On marshy sites, it associates with the species mentioned above, as well as with *Ribes triste, Lonicera coerulea, Spirea salicifolia, Linnaea borealis* and *Ramischia secunda*. Throughout its entire range *P. pumila* forms the understorey of *Larix gmelinii* forests. Over wide areas it grows under

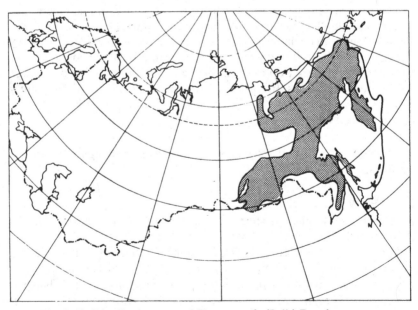

Fig. 2.12. Distribution map of *Pinus pumila* (Pall.) Regel.

sparse canopies of *Pinus sylvestris*, *Pinus sibirica*, *Picea obovata*, *Betula pendula* and *Betula pubescens*. In the eastern part of its range, *P. pumila* grows as an understorey tree in forests of *Picea ajanensis*, *Betula ermanii* and *Betula pendula*. On Kamchatka, it also associates with *Alnus kamtschatica* and *Populus suaveolens*. It is occasionally found with *Chosenia arbutifolia*. In the river basin of Amgun, it associates with *Quercus mongolica* and *Juniperus davurica* (Sokolov, Svyaseva & Kubly 1977).

Life history

P. pumila is a monoecious and anemophilous species (Sokolov, Svyaseva & Kubly 1977). It has good seed years nearly every year. One thousand seeds weigh 97 g. Seeds are dispersed by animals and birds, as well as by water. Close to the tree limit, it is able to regenerate by layering (Sarvas 1964; Sokolov, Svyaseva & Kubly 1977).

   P. pumila usually grows in groups of up to eight individuals (Sokolov, Svyaseva & Kubly 1977). Its growth form is shrubby (Sarvas 1964). Its growth is slow: in 10 years the species can reach a height of 0.3 m, in 50 years 1.5 m and 1 cm in diameter, and in 100 years 3.6 m in height and 5 cm in diameter. At lower altitudes and under relatively sparse canopies *P. pumila* reaches a height of 7 m. However, at higher elevations and in the north it typically forms dense bushwoods of creeping individuals. The length of a stem can reach 10–12 m but they rise only 1–2 m above the ground (Sokolov, Svyaseva & Kubly 1977).

   Maximum values for height, diameter and age are

   Height: 7 m (Sokolov, Svyaseva & Kubly 1977), 6–10 m (Uanjun 1958).

   DBH: 20 cm (Uanjun 1958).

   Age: 150–200 years (Sokolov, Svyaseva & Kubly 1977).

Response to environmental factors

Light
The species is intermediately shade-tolerant (Sokolov, Svyaseva & Kubly 1977). According to N. P. Polikarpov (unpublished), *P. pumila* tolerates more shade than *Quercus robur*, *Quercus mongolica*, *Pinus sylvestris* and *Betula* and *Larix* species but less than *Pinus sibirica*, *Tilia cordata* and *Picea* and *Abies* species.

Soil moisture
P. pumila tolerates poorly drained and moisture-limited soils (Uanjun 1958). It is characterized as relatively intolerant to water stress. It is more tolerant than *Picea ajanensis*, *Pinus sibirica*, *Betula pubescens*, *Picea*

abies and Picea obovata but less so than Abies sibirica, Populus tremula, Betula pendula and Larix species (N. P. Polikarpov, unpublished).

Nutrients
P. pumila is nutrient-stress-tolerant (Trees and Shrubs of the USSR 1956). It is classified as more tolerant than Betula pubescens, Betula pendula and Larix, Picea and Abies species (N. P. Polikarpov, unpublished).

Fire
P. pumila is fire-intolerant (Sarvas 1964).

Frost
The species is very frost-tolerant (Uanjun 1958).

Flooding
The occurrence of the species on boggy sites and in river valleys (Sokolov, Svyaseva & Kubly 1977) indicates that it can tolerate at least occasional flooding.

Races and hybrids
P. pumila is closely related to Pinus cembra var. sibirica (Drakenberg 1981).

2.12. Pinus sibirica Du Tour. Syn. Pinus cembra subsp. sibirica (Rupr.)

Distribution
P. sibirica is found in the European Alps (Sarvas 1964), but its main distribution area is in the USSR. Its northern border is between 66°30′ N and 57° N latitude, crossing the Ural Mountains. The southern border passes through northern Mongolia, the county of Tobolsk and the Altay Mountains, and its western limit is at the River Vitschegda (Trees and Shrubs of the USSR 1956). To the east it is found in central and southern eastern Siberia (Sokolov, Svyaseva & Kubly 1977). Its distribution is shown in Fig. 2.13.

In the northern Urals P. sibirica occurs as a tree up to 500 m a.s.l., but as a creeping plant it can be found up to 800 m. In the Altay Mountains it is found from 400 to 2400 m, and in the Sayan Mountains from 500 up to 1800 m and only as a dwarf tree at 2500 m (Sokolov, Svyaseva & Kubly 1977). In the USSR, P. sibirica dominates a forest area of about 23 million hectares (Sokolov, Svyaseva & Kubly 1977). Its distribution range has become broader in the past 80–100 years (Krylov 1957; Nepomilueva 1974).

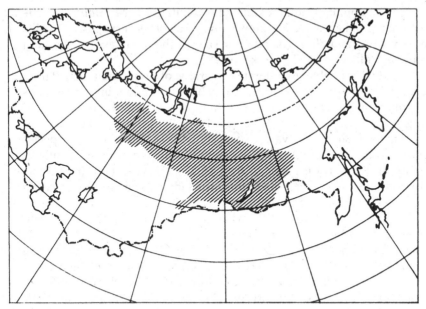

Fig. 2.13. Distribution map of *Pinus sibirica* Du Tour.

Habitat

P. sibirica is able to grow in cold regions with a growing season (i.e. a period of the year with air temperature above 10 °C) of only 1.5 months. However, for optimal growth it requires a more humid and less continental climate. Its highest transpiration rate is observed under a relative humidity of 45–50%. The optimum for growth in terms of HTI is between 2.0 and 3.5: this explains why *P. sibirica* is found mainly on mountains and uplands in central and eastern Siberia. In southern Siberia it forms forest stands on wet windward slopes at 300–400 m a.s.l.; on drier leeward slopes it is found only at 1500–1700 m. *P. sibirica* has its ecological optimum in the moutain regions of Altay and Sayan where it has the highest growth rates and longevity (Polikarpov, Tchebakova & Nazimova 1986). The climatic requirements of *P. sibirica* have been described as follows.

(1) A mean annual relative humidity of 62% seems to be optimal for the species (Hohrin 1970). According to Polikarpov & Tchebakova (1982), humidity above 70% during the growing season limits its normal growth, but other authors (e.g. Sukachev 1934; Gortchakovsky 1956) state that it prefers sites with higher humidity. The lowest relative humidity tolerated by *P. sibirica* without affecting its growth is 45% (Polikarpov & Nazimova 1963).

(2) The optimal amount of annual precipitation for *P. sibirica* is 500–600 mm (Polikarpov & Nazimova 1963; Talantsev, Pryajnikov & Mishukov 1978) and 800–1400 mm per year in the mountains of southern Siberia (Polikarpov & Tchebakova 1982).

P. sibirica is able to tolerate a variety of soil conditions. Parent material, soil structure and soil fertility have only minor influences on its distribution. It occurs on sites with a thin soil layer as well as on sphagnum bogs (*Trees and Shrubs of the USSR* 1956; Polikarpov, Tchebakova & Nazimova 1986; Katayeva & Korzukhin 1987). However, the species seems to have different soil requirements at its northern and southern limits. In the north it grows in river valleys (Sarvas 1964), but prefers drier, drained slopes with sandy-loam to sandy soils and seldom grows on bogs (Povarnitsin 1955; Gortchakovsky 1956; Beh 1974; Talantsev, Pryajnikov & Mishukov 1978), whereas in the south it prefers drained moist soils and grows on sphagnum bogs (Povarnitsin 1955; Beh 1974; Talantsev, Pryajnikov & Mishukov 1978). *P. sibirica* and *Picea obovata* tolerate swamp soils only if they are mesotrophic bogs. On oligotrophic bogs, *P. sibirica* is replaced by *Pinus sylvestris* (P'yavchenko 1979). *P. sibirica* also tolerates cold and frozen soils (Polikarpov, Tchebakova & Nazimova 1986). In eastern Siberia it often grows on permafrost (Sokolov, Svyaseva & Kubly 1977).

P. sibirica dominates forests on the following soils (Sokolov, Svyaseva & Kubly 1977):

Turf–podzol, relatively dry soils;
Gray and brown or dark-brown soils
Gray podzol–gley soils typical of plains of the northeastern European USSR;
Sites with alluvial soils and seasonal flooding;
Boggy sites with peaty and peat-gley soils. On these sites the species often develops stilt roots.

Associated species
P. sibirica is mainly associated with *Picea obovata* and *Abies sibirica*. These three species are major components of the dark-coniferous taiga forests, which are typical of western Siberia (Sokolov, Svyaseva & Kubly 1977). *P. sibirica* favors drier sites with a more continental climate. Under wetter and warmer conditions *Abies sibirica* is usually a stronger competitor than *P. sibirica* except on eroded dry mountain slopes, in frost pockets or on lowland swampy soils (Polikarpov, Tchebakova & Nazimova 1986).

Over large areas *P. sibirica* is also found with *Larix sibirica, Larix gmelinii, Larix sukaczewii, Pinus sylvestris, Populus tremula, Betula pendula* and *Betula pubescens*. Forests dominated by *P. sibirica* have a second layer of *Sorbus sibirica, Padus asiatica* and *Salix caprea*. The understorey is composed of *Juniperus communis, Juniperus sibirica, Lonicera altaica, Caragana arborescens, Sambucus sibirica, Sorbaria sorbifolia, Rosa acicularis, Ribes nigrum, Ribes rubrum, Salix phylicifolia, Spirea chamaedrifolia, Spirea media, Cotoneaster multiflora, Alnus fruticosa, Betula nana* and others (Sokolov, Svyaseva & Kubly 1977).

Life history

P. sibirica is a monoecious and anemophilous species. In the central and southern taiga, seed production starts at 30–40 years for solitary trees and at 70–80 years in closed stands (Beh & Tarpan 1979). In the European USSR solitary trees form seed crops when they are 25–30 years old (*Trees and Shrubs of the USSR* 1956; Sarvas 1964). In closed forests seed crops are formed at 60–80 years (*Trees and Shrubs of the USSR* 1956). When seed production begins depends on the amount of light received by young trees: the better the light conditions, the earlier they begin to produce seeds. Under suppression, seed production may start at 140–160 years of age. Both cold and rainy and hot and dry weather conditions are unfavorable for seed development. Good seed crops occur every 8–9 years in the European USSR, every 4–6 years in western Siberia (Danilov 1952) and every 3–4 years in the Baikal area (*Trees and Shrubs of the USSR* 1956). In southern taiga, at the climatic optimum of *P. sibirica*, good seed crops occur every 5–6 years. High seed production can last for 4–5 years (Beh & Tarpan 1979). Seed-producing trees are found in stands every year but in some years the percentage of seeding trees is 80–90% compared with 50% in non-seeding years (Nekrasova 1962). On bogs the seeding frequency is reduced (Beh 1971). The average annual seed production on good sites is about 220–259 kg ha^{-1}, on marshy sites 50–70 kg ha^{-1}, and on sphagnum bogs 10–20 kg ha^{-1}. The number of cones per tree ranges from 71 to 120 (Nekrasova 1960). Mean number of seeds per cone is 80 (Saeta 1971).

Seed production reaches its peak at the age of 80–160 years (Sarvas 1964) but seeding continues until the age of 400–450 years (Kirsanov 1974). The seed-bearing cones fall to the ground and the seeds are released when cones decay (Sarvas 1964). *P. sibirica* seeds are eaten and dispersed mainly by the cedar bird, *Nucifraga caryocatactes* L. (Sokolov, Svyaseva & Kubly 1977) and partly by animals (*Trees and Shrubs of the USSR* 1956; Sokolov, Svyaseva & Kubly 1977) and humans (Sarvas 1964). The seeds may be dispersed 5–7 km from the parent tree (Beh & Tarpan 1979). *P. sibirica* seeds germinate at a soil temperature of 5 °C (Beh 1974). Light requirements of *P. sibirica* increase with age (Kataeva & Korzukhin 1987). The relation between seedling age and minimum

Table 2.9. *The relation between age and minimum light requirements for growth of* Pinus sibirica *seedlings*

Age (years)	Light level (% of full sunlight)
1–2	1–3
3–5	3–6
6–10	6–9
11–15	9–13

Source: Polikarpov & Babintseva (1963).

Table 2.10. *The growth of fully stocked* Pinus sibirica *stands on the most favorable sites in southern Siberia*

Age (years)	Mean stand height (m)	Mean stand DBH (cm)	Trees per ha	Basal area (m^2 ha^{-1})	Volume (m^3 ha^{-1})
40	10.3	9.1	1800	11.7	77
60	14.5	15.4	1048	19.5	160
80	18.2	24.7	646	30.9	299
100	21.4	32.0	478	38.4	413
120	24.1	38.9	378	44.8	520
140	26.5	45.0	315	50.1	622
160	28.4	50.0	275	53.9	703
180	30.1	54.5	244	56.9	774
200	31.5	58.2	222	59.0	826
220	32.5	61.1	204	59.9	859
240	33.4	63.8	189	60.5	877

Source: Isaev (1985).

light requirements for growth (Polikarpov & Babintseva 1963) is shown in Table 2.9.

Under dense canopies of *P. sibirica*, seedlings can survive for 10–15 years; some individuals survive up to 30–35 years (Katayeva & Korzukhin 1987). *P. sibirica* is a rather slow-growing tree species. The height growth rate is very low especially during the first 5–7 years. The growth rates reach maximum after 100–120 years of age. Table 2.10 shows the growth pattern of fully stocked *P. sibirica* stands on the most favorable sites in southern Siberia (Isaev 1985).

P. sibirica is a shallow-rooted tree (Zubov 1960) and it is able to produce adventitious roots, enabling growth on wet sites (Zubov 1960).

Table 2.11. *Relation between altitude and maximum values for height and age of* Pinus sibirica *in the Sayan and Altay Mountains*

Altitude (m a.s.l.)	Maximum height (m)	Maximum age (years)
300–600	35	800–850
600–800	30–32	600–650
800–1000	24–25	500
1000–1500	16–18	300–400
1500–1800	10–12	250–300
1800–2000	6–8	100–120
2000–2400	0.5–2	40–80

Source: Polikarpov, Tchebakova & Nazimova (1986).

Maximum values for height, diameter and age are

Height: 30–40 m (Blomqvist 1887; *Trees and Shrubs of the USSR* 1956; Bärner 1961; Sarvas 1964; Krussman 1971; Beh 1974), 40–45 m (Sokolov, Svyaseva & Kubly 1977), 33 m (Talantsev 1981).

DBH: 1.80–1.90 m (Sarvas 1964; Beh 1974), 1–1.5 m (Sokolov, Svyaseva & Kubly 1977; 120–140 cm (Talantsev 1981).

Age: Usually 400–600 years (Sukachev 1934; *Trees and Shrubs of the USSR* 1956; Sergievskaya 1971; Beh 1974; Gode 1986), but 800-year-old trees have been found (Sukatchev 1934; Sergievskaya 1971; Beh 1974); 390 years (Talantsev 1981).

In Table 2.11 the relationship between altitude and maximum values for height and age of *P. sibirica* in the Sayan and Altay Mountains (Polikarpov, Tchebakova & Nazimova 1986) are shown.

Response to environmental factors

Light
P. sibirica is classified as shade-tolerant (*Trees and Shrubs of the USSR* 1956; Sokolov, Svyaseva & Kubly 1977; Katayeva & Korzukhin 1987). It tolerates less shade than *Abies* and *Picea* species but more than *Pinus pumila, Quercus robur, Alnus glutinosa, Populus tremula, Pinus sylvestris* and *Betula* and *Larix* species (Tkatchenko 1955; Krylov, Talantsev & Kozakova 1983; N. P. Polikarpov, unpublished). Under its own canopy it stands shading until 30–50 years of age; under canopies of *Pinus sylvestris, Betula, Populus* and *Larix* it may remain vital up to 70–100 years of age. The light requirements of the species increase with aging and under poor conditions (Polikarpov, Tchebakova & Nazimova 1986).

Soil moisture

P. sibirica is characterized as drought-intolerant. It tolerates more water stress than *Betula pubescens*, *Picea abies* and *Picea obovata*, but less than other *Pinus*, *Abies*, *Larix* and *Quercus* species (N. P. Polikarpov, unpublished). *P. sibirica* and *Picea obovata* tolerate swamp soils well only if they are nutrient-rich or intermediately nutrient-rich bogs. On nutrient-poor bogs, *P. sibirica* is replaced by *Pinus sylvestris* (P'yav-chenko 1979).

Nutrients

P. sibirica is classified as tolerant (Gorodkov 1916b; Talantsev, Pryaj-nikov & Mishukov 1978) or intermediately tolerant (N. P. Polikarpov, unpublished) of nutrient stress. It is more tolerant than *Picea*, *Quercus* and *Abies* species and less so than *Pinus sylvestris*, *Larix* and *Betula* (N. P. Polikarpov, unpublished).

Fire

P. sibirica is very sensitive to high temperatures. It is damaged by fires and sunburns (Polikarpov, Tchebakova & Nazimova 1986), but it is more fire-tolerant than *Picea abies* and *Betula* species (Blomqvist 1887).

Frost

P. sibirica is frost-tolerant (Blomqvist 1887; Sukachev 1934; Morozov 1947; *Trees and Shrubs of the USSR* 1956; Beh 1974). It tolerates minimum daily temperatures down to $-60\,°C$ (Beh 1974) as well as large temperature variations and spring frosts. However, when the mean annual temperature amplitude is higher than $40\,°C$, it is replaced by *Larix sibirica*, *Pinus sylvestris* and *Picea obovata* (Polikarpov, Tchebakova & Nazimova 1986). It also tolerates cold soils and grows well on permafrost (Sokolov, Svyaseva & Kubly 1977; Polikarpov & Tchebakova 1982).

Flooding

P. sibirica grows well on wet sites but poorly on undrained soil (Sarvas 1964; Beh 1974; Polikarpov, Tchebakova & Nazimova 1986). It tolerates periodic flooding: this feature, along with its high tolerance of low temperatures, permits growth in river valleys (Sarvas 1964; Beh 1974).

Wind

The species is intolerant of windstorms, possibly because of its shallow root system.

Races and hybrids

P. sibirica has formed several varieties within its range (Katayeva & Korzukhin 1987). *Pinus cembra* L. is usually divided into two geographically separated races: *Pinus cembra* L. var. *typica* in the European Alps, and *Pinus cembra* L. var. *sibirica* Loud. in the USSR (Sarvas 1964). *P. sibirica* f. *furfosa* (maximum height 6–7 m) is found on bogs (*Trees and Shrubs of the USSR* 1956). In the Altay, Sayan and Transbaikal Mountains a shorter variety, *Pinus cembra* var. *coronans*, is found (Sarvas 1964).

Enemies and diseases

About 125 insect species are known to thrive on *P. sibirica*. Twelve species damage seeds (e.g. *Dendrolimus superans sibiricus* Tsch. and *Ocneria dispar* L.) and cones (*Monochamussutor* spp. L.). Seedlings of *Pinus sibirica* are damaged by *Coleosporium pinicola* (Arth.) Jacks. and *Cronatrium ribicola* Dietr.; fungi of the genus *Fusarium* cause seedlings to become flattened, and the fungus *Thelephora terrestris* Fr. causes seedling suffocation. Twenty-eight species damage young trees (e.g. *Megasemum quadricostulatum* Krts.). Eight species affect shoots, e.g. *Monochamus urussovi* Fisch.), 14 species damage needles (e.g. *Criocephalus rusticus* L.), 76 species damage wood and bark, and 16 species damage roots (Krylov, Tarantsev & Kozakova 1983). About 100 species of wood-destroying fungus, including 58 pathogenic species, are found in association with *P. sibirica*. Needles of adult trees are damaged by the fungus *Lophodermium pinastri*, branches are damaged by *Melampsora pinitorqua* A. Braun, and the trunk is damaged mainly by *Phellinus pini* (Thore ex Fr.) Pil., *Fomitopsis officinalis* (Will) Bond et Sing. and *Laetiporus sulphureus* (Bull. et Fr.). *Fomitopsis officinalis* causes trunk rot and *Fomitopsis pinicola* Fr. causes serious wood rotting. Buttress decay is often caused by *Fomitopsis annosa* (Fr.) Karst, *Phaeolus schweinitzii* (Fr.) Pat. and *Armillaria mellea* Quell. (Krylov, Tarantsev & Kozakova 1983).

2.13. *Pinus sylvestris* L.

Distribution

P. sylvestris is the most widely distributed pine tree in the world (Drakenberg 1981). It occurs in almost all of Europe, from latitude 70° north to the mountains of southern Europe (Delkov 1984). In Asia it is found in western Siberia, in the Altay Mountains, in eastern Siberia, in the far eastern USSR and in northern Kazakhstan as well as in northern Mongolia and northern China (Sokolov, Svyaseva & Kubly 1977). Its distribution is shown in Fig. 2.14. *Pinus sylvestris* var. *mongolica* Litv. is

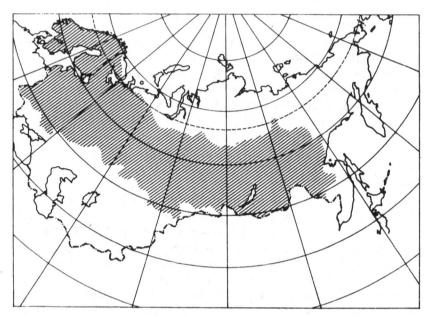

Fig. 2.14. Distribution map of *Pinus sylvestris* L.

the easternmost ecotype of *P. sylvestris*. It is found in the Daxingan Mountains, at elevations of 400–900 m. It also occurs south-east of the Hulunbeir grassland. Its southern boundary is 40°46′ N latitude in China. This ecotype has also been found in the USSR and Mongolia (Uanjun 1958).

The altitudial limit of *P. sylvestris* is 800 m a.s.l. in central Sweden, 600 m in southern Lapland and 450 m in northern Lapland (Aminoff 1912). The northern limit of *P. sylvestris* is further south than that of *Picea obovata* and *Larix sukaczewii* in Europe and of *Larix gmelinii* in Siberia. In Siberia the northern limit coincides with the southern border of the continuous permafrost region (Sokolov, Svyaseva & Kubly 1977). Permafrost is considered as the main obstacle to species migration further north in central Siberia and in the Far East (Zvetkov & Semenov 1985; Pozdnyakov 1986).

P. sylvestris dominates a forest area of about 108 million hectares within its range (Sokolov, Svyaseva & Kubly 1977).

Habitat

P. sylvestris occurs in areas with continental or submaritime climate (Sarvas 1964). In the boreal zone it grows best in lowland areas; on mountains it is found mainly on warm eastern and western slopes (Sarvas 1964). In northern China it is found where:

the period of the year with air temperature above 10 °C is 90–120 days (Uanjun 1958);

the mean annual temperature is −2 to −4 °C (*The Mountain Forests of China* 1988);

the January mean minimum temperature is −20 to −30 °C (*The Mountain Forests of China* 1988);

the July mean temperature is 17 to 20 °C (*The Mountain Forests of China* 1988);

the annual rainfall is between 350 and 500 mm (*The Mountain Forests of China* 1988);

the snow cover lasts for seven months (*The Mountain Forests of China* 1988).

P. sylvestris occurs from dry, nutrient-poor to wet, nutrient-rich sites (Sylven 1916; Uanjun 1958; Walter 1974; Kellomäki 1987). It thrives on drained alluvial humus as well as on loamy and sandy soils. In northern parts of the European USSR it is common on ferric podzol soils (Zvetkov & Semenov 1985). It also occurs on podzol, podzol–gley and chernozem soils and often grows on peat and rocky soils (Walter 1974; Sokolov, Svyaseva & Kubly 1977). The growth is poor on peat and rocky soils (Sokolov, Svyaseva & Kubly 1977). Near the northern treeline of the European USSR, *P. sylvestris* occurs on both rocks and bogs with a thick peat layer (Zvetkov & Semenov 1985). The species grows on wet and boggy sites only if soils are not too cold (Pozdnyakov 1986).

In Siberia it forms pure stands on nutrient-poor, sandy, rocky and boggy soils. In the mountains of southern Siberia the distribution of *P. sylvestris* is determined by soil temperature: in these mountains it occurs on sites with permafrost only if the depth of summer soil thaw exceeds 2 m (Polikarpov, Tchebakova & Nazimova 1986). In northern China it is found on a variety of soil types (Uanjun 1958).

Associated species
P. sylvestris is found together with most species of the boreal and boreonemoral zone of Europe and Asia. In the European USSR and in Siberia it is associated with *Picea abies*, *Picea obovata*, *Larix sukaczewii*, *Larix sibirica*, *Larix gmelinii*, *Abies sibirica*, *Pinus sibirica*, *Pinus pumila*, *Betula pubescens*, *Betula pendula*, *Betula ermanii*, *Populus tremula*, *Quercus robur*, *Quercus petraea*, *Acer platanoides*, *Corylus avellana*, *Fagus sylvatica*, *Padus racemosa*, *Sorbus aucuparia*, *Sorbus sibirica*, *Juniperus communis* and others. In the far east it grows with *Abies nephrolepis*, *Picea ajanensis*, *Quercus mongolica*, *Betula davurica* and *Alnus mandshurica*. On the Krim Peninsula it occurs in mixed forests

with *Carpinus betulus*, *Fagus orientalis* and *Pinus pallasiana*. In the Caucasus it is found mainly with *Betula pendula*, *Betula litwinowii*, *Betula pubescens*, *Betula raddeana*, *Picea orientalis*, *Abies nordmanniana*, *Quercus petraea*, *Quercus macranthera*, *Carpinus betulus*, *Populus tremula*, *Tilia caucasica*, *Tilia platyphyllos*, *Acer trautvetteri*, *Alnus incana*, *Carpinus orientalis* and others (Sokolov, Svyaseva & Kubly 1977).

Life history

P. sylvestris is a monoecious and anemophilous species (Sokolov, Svyaseva & Kubly 1977). It regenerates only from seed (Kellomäki 1987). Solitary trees start seeding at 8–20 years (Aminoff 1912; Uanjun 1958; Sarvas 1964). In closed stands, seeding starts at 30–50 years (Aminoff 1912; Lagerberg & Sjörs 1972). In the Far East, trees already produce seeds at 8–12 years of age (Zvetkov & Semenov 1985). In northern China, full seed production occurs after 25 years of age (Uanjun 1958). Seed production is most intensive between 60 and 100 years but seeding continues until the age of 250 years (Aminoff 1912). Good seed crops occur once every 5–7 years in the boreal zone (Sarvas 1964; Ageenko 1969; Kellomäki 1987), and every 2–4 years in northern China (Uanjun 1958). Seed production is higher in central Siberia and Transbaikalia than in the European USSR. Poor seed crops occur in Transbaikalia twice in 10 years, whereas in the European USSR they occur during 3–4 years in every decade. More abundant seed production in Siberia is attributed to favorable climatic conditions and lower stand densities (Pobedinsky 1979). Poor seed crops may occur simultaneously with abundant cone production (Zvetkov & Semenov 1985). Cones mature 18 months after flowering and the seeds are released in February and March. Each cone contains 20–30 seeds (Aminoff 1912). Seed mass and viability decrease with increased age and latitude (Aminoff 1912). Seeds are dispersed by wind (Uanjun 1958; Sokolov, Svyaseva & Kubly 1977). For successful regeneration on pine heaths, a good seed year has to be followed by one or two wet summers (Sarvas 1964). Forest fire is important for successful regeneration (Drakenberg 1981), since seedlings of *P. sylvestris* are light-demanding (Lagerberg & Sjörs 1972).

 P. sylvestris is a fast-growing species. One-year-old seedlings may reach a height of 10 cm. Height growth rate increases during the second and third year, reaching 10–15 cm per year. Maximum annual height growth is 70–100 cm (Delkov 1984). In southern Sweden, height growth is most rapid from 5 to 30 years (Lagerberg & Sjörs 1972). Table 2.12 shows the growth of fully stocked, even-aged *P. sylvestris* stands on the most favorable sites in the Ukraine and Moldavia (Shvydenko *et al.* 1987). Table 2.13 gives an example of a growth pattern of an ecotype of *P. sylvestris* in northern China (Uanjun 1958).

Table 2.12. *Growth of fully stocked, even-aged stands of* Pinus sylvestris *on the most favorable sites in the Ukraine and Moldavia*

Age (years)	Mean stand height (m)	Mean stand DBH (cm)	Trees per ha	Basal area (m^2 ha^{-1})	Stem volume (m^3)
20	9.6	9.6	3350	23.4	112
30	14.3	14.5	2050	33.8	224
40	18.4	19.6	1430	40.6	339
50	22.2	23.3	1052	45.0	447
60	25.3	27.2	820	47.9	538
70	27.9	30.8	670	50.0	616
80	30.0	34.1	562	51.4	680
90	31.9	37.2	483	52.6	736
100	33.6	40.0	423	53.3	785
110	34.8	42.4	384	54.2	825
120	36.0	44.5	350	54.6	857
130	36.8	46.0	331	55.0	881
140	37.5	47.0	317	55.0	895

Source: Shvydenko *et al.* (1987).

Table 2.13. *Growth pattern of* Pinus sylvestris *var.* mongolica *in natural forests on northern slopes of the Daxingan Mountains in China*

Age (years)	Mean stand height (m)	Mean stand DBH (cm)
100	21	25
150	24	32
200	26	36

Source: Uanjun (1958).

Maximum values for height, diameter and age are
Height: 40–48 m under optimal growth conditions in the FRG (Bärner 1961), 37 m in Sweden (Drakenberg 1981), 35.5 m in Finland (Sarvas 1964) and 38–40 m for *Pinus sylvestris* var. *mongolica* (Uanjun 1958; Sokolov, Svyaseva & Kubly 1977).

DBH: 1.9 m in Sweden (Lagerberg & Sjörs 1972); 5.9 m is the maximum measured, but trees 2.5 m or more in diameter are uncommon (Drakenberg 1981); 0.7–1.0 m in the USSR and in northern China (Uanjun 1958; Sokolov, Svyaseva & Kubly 1977).

Age: 250–300 years on dry morainic soils in southern Finland, 600 years on fresh soils, 700–800 years in northern Finland (Kellomaki 1987); 350–400 years in the USSR (Sokolov, Svyaseva & Kubly 1977).

Response to environmental factors

Light

P. sylvestris belongs to the group of light-demanding species, but it is also able to grow in shady habitats on humid soil (Sarvas 1964; Bonnemann & Röhrig 1972). On fertile soils, it is more shade-tolerant (Sylven 1916). According to N. P. Polikarpov (unpublished), the species demands more light than *Quercus mongolica*, *Quercus robur*, *Pinus pumila*, *Pinus sibirica*, *Carpinus betulus* and *Picea* and *Abies* species but less than *Populus tremula* and *Betula* and *Larix* species. *P. sylvestris* L. var. *mongolica* has been classified as shade-intolerant (Uanjun 1958).

Soil moisture

P. sylvestris is classified as extremely drought-tolerant. It tolerates more water stress than any other boreal and boreonemoral tree species in Eurasia (N. P. Polikarpov, unpublished; Uanjun 1958). It grows on both wet and dry soils (Bonnemann & Röhrig 1972), but on wet soils it is outcompeted by *Picea abies* (Kellomäki 1987).

Nutrients

P. sylvestris is a very nutrient-stress-tolerant species (Bonnemann & Röhrig 1972; N. P. Polikarpov, unpublished). It is the most tolerant among the boreal and boreonemoral species in Eurasia (N. P. Polikarpov, unpublished).

Fire

P. sylvestris has a thick bark, especially in dry climatic conditions, which protects it against fire (Pobedinsky 1979). It survives forest fires and regenerates on burned areas (Kellomäki 1987).

Frost

P. sylvestris is very frost-tolerant (Aminoff 1912; Bonnemann & Röhrig 1972). It also tolerates unseasonal frosts (Zvetkov & Semenov 1985).

Flooding

The occurrence of the species on wet sites near rivers and on bogs (Sylven 1916; Sokolov, Svyaseva & Kubly 1977; Zvetkov & Semenov 1985; Pozdnyakov 1986; Kellomäki 1987) indicates that it can tolerate at least seasonal flooding.

Wind

P. *sylvestris* is intolerant of windthrow on boggy and cold soils, where it has a shallow root system. However, on dry sites with no permafrost, pine develops a tap root, which enables it to endure windstorms (Zvetkov & Semenov 1985).

Races and hybrids

P. *sylvestris* has many different subspecies (Bärner 1961). Sarvas (1964) recognizes two clines of this species:

> *Pinus sylvestris* var. *lapponica* Hartm., found in northern Fenno-scandia, north of 66°.
> *Pinus sylvestris* var. *septentrionalis* Schott., found in central and southern Fennoscandia, reaching to southern parts of Sweden.

Pravdin (1964) distinguishes five subspecies within the range of *P. sylvestris*:

> *Pinus sylvestris* ssp. *lapponica* L., occurring north of latitude 62° N.
> *Pinus sylvestris* ssp. *sylvestris* L., occurring in the European USSR.
> *Pinus sylvestris* ssp. *sibirica* Ledeb., occurring in the eastern part of its range.
> *Pinus sylvestris* ssp. *kulundensis* Sukacz., found in eastern Kazakh-stan and the Pribaikal region.
> *Pinus sylvestris* ssp. *hamata* (Stev.), found on the Krim Peninsula and in the Caucasus.

Uanjun (1958) recognizes an eastern ecotype of the species, called *Pinus sylvestris* L. var. *mongolica*, which occurs in montane areas of northern China.

2.14. *Populus tremula* **L. Syn.** *Populus pseudotremula* **N. Rubtz.;** *Populus davidiana* **Dode.**

Distribution

P. *tremula* is geographically one of the most widely distributed tree species of the world. It occurs as a tree up to 70° N (Leibundgut 1984), from the western borders of the USSR to Kamchatka (Tseplyaev 1961). It is found all over Europe except on the islands in the western parts of the Mediterranean Sea (Hempel & Wilhelm 1897), as well as in northern Mongolia, China and Korea (Sokolov, Svyaseva & Kubly 1977). Its distribution is shown in Fig. 2.15.

Habitat

P. *tremula* has few habitat preferences and can essentially grow every-where in the northern hemisphere. It occurs from the northern timberline

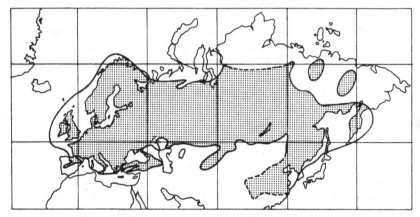

Fig. 2.15. Distribution map of *Populus tremula* L.

to the steppe (Sokolov, Svyaseva & Kubly 1977). In the forest–steppe and in the steppe zone it is used for landscape reforestation (Smilga 1986). In Sweden *P. tremula* is found from dry to wet and from fertile to nutrient-poor soils. On dry and poor soils it becomes dwarfed (Hesselman 1912). *P. tremula* grows best in Poland and in the Baltic states of the USSR, where optimal soil and climatic conditions are found (Hesselman 1931). In the harsh environments of eastern Siberia, it is found only in river valleys and on lower parts of mountain slopes (Tseplyaev 1961). Under the continental conditions of southern Siberia, it grows on nutrient-rich, moist soils (Polikarpov, Tchebakova & Nazimova 1986). In the mountains of southern Siberia, it dominates forests in warm and moist lowlands (Smilga 1986). In the European USSR, *P. tremula* requires soils of medium fertility. It thrives on fresh, loamy and nutrient-rich soils. It grows well on sandy and loamy soils and tolerates some soil salinity. It can also tolerate acidic and dense soils, which are not favorable for any other *Populus* species (Smilga 1986). *P. tremula* has a good growth on wet soils, but it does not tolerate marshy sites (Hesselman 1931; Sokolov, Svyaseva & Kubly 1977).

Associated species
Pure aspen stands are found only in Europe. *P. tremula* is associated with birch and spruce, and less frequently with pine (Smilga 1986). In European parts of the USSR, *P. tremula* is found in association with *Picea*, *Pinus*, *Betula*, *Alnus* and *Salix* species (Hesselman 1912). In Siberia it is associated with *Abies sibirica*. *P. tremula* is a co-dominant in dark-needled forests of middle elevations (Polikarpov, Tchebakova & Nazimova 1986). On permafrost *P. tremula* is associated with *Pinus sibirica* on fertile soils and with *Larix sibirica* on poorer sites (Pozdnyakov 1983). In

Table 2.14. *Growth of fully stocked* Populus tremula *stands on the most favorable sites in Latvia*

Age (years)	Mean stand height (m)	Mean stand DBH (cm)	Basal area (m^2 ha^{-1})	Volume (m^3 ha^{-1})
20	15	8	20.0	142
30	19	16	29.0	253
40	23	23	35.8	378
50	27	28	40.5	492
60	30	33	44.6	602
70	32	36	47.3	606

Source: Smilga (1986).

western Siberia *P. tremula* grows in mixed forest stands with *Betula pendula* and *Betula pubescens* (Sokolov, Svyaseva & Kubly 1977).

Life history

P. tremula is a dioecious tree. It regenerates from seed and from stump and root suckers (Hesselman 1931). Stump suckers do not form large trees (Kellomäki 1987). It flowers early in the spring, and seeds ripen in May (Hempel & Wilhelm 1897). They lose their viability a few weeks after maturation (Hesselman 1931). Seeds are dispersed long distances by wind: this enables the species to invade quickly sites where shade-tolerant species have been cut or killed by fires or insect invasions. Seed crops are formed from the age of 20 years (Hempel & Wilhelm 1897). *P. tremula* produces large seed crops, up to 0.5 billion per hectare, but only a small percentage of them develop into seedlings (Smilga 1986). Seed production shows a certain periodicity. Generally the size of the seed crop depends on climatic conditions of the previous year. In Norway it has been found that the seed crops are larger in hot and dry years (Smilga 1986). Seedlings are very sensitive to drought (Hesselman 1931).

P. tremula is a fast-growing species. The height growth rate during the first years is about 15–20 cm, but increases thereafter, rapidly reaching a maximum of 100–150 cm per year between 30 and 40 years (Delkov 1984). At 50–60 years the growth rate decreases significantly. *P. tremula* gets taller and has larger dimensions in northern and southeastern parts of its range (Hempel & Wilhelm 1897). Table 2.14 shows the growth of fully stocked *P. tremula* stands on the most favorable sites in Latvia (Smilga 1986).

Maximum values for height, diameter and age are

Height: 35 m in central Europe (Bärner 1961); 30 m in Finland (Bonnemann & Röhrig 1971; Leibundgut 1984; Kellomäki 1987), 42 m in the Latvian SSR (Smilga 1986).

DBH: From 35 cm to 100 cm in Europe and in the USSR (Bärner 1961; Bonnemann & Röhrig 1971); 35–40 cm in Finland (Kellomäki 1987), 100 cm in the Latvian SSR (Smilga 1986).

Age: 100 years (Bärner 1961); *P. tremula* rarely grows to an age of greater than 150 years (Kellomäki 1987).

Response to environmental factors

Light

P. tremula is light-demanding (Tseplyaev 1961; Leibundgut 1984). It is one of the most shade-intolerant tree species (Hempel & Wilhelm 1897). It does not survive under a closed canopy (Smilga 1986). According to N. P. Polikarpov (unpublished), only *Larix* and *Betula* species are more light-demanding.

Soil moisture

P. tremula is relatively drought-tolerant. It tolerates less water stress than *Pinus sylvestris*, *Quercus robur*, *Quercus mongolica*, *Larix sibirica*, *Larix gmelinii* and *Betula pendula* but more than *Larix sukaczewii*, *Fagus sylvatica*, *Pinus pumila*, *Pinus sibirica*, *Betula pubescens* and *Picea* and *Abies* species (N. P. Polikarpov, unpublished). It grows best on soils with good water supply and aeration (Leibundgut 1984). On soils with a high water table its growth form becomes shrublike (Hempel & Wilhelm 1897).

Nutrients

P. tremula tolerates moderate nutrient stress. It is less nutrient-stress-tolerant than *Picea abies*, *Picea ajanensis*, *Picea obovata*, *Pinus sibirica*, *Pinus sylvestris* and *Betula* and *Larix* species but more so than *Quercus*, *Fagus* and *Abies* species (N. P. Polikarpov, unpublished). It grows on all soil types (Hempel & Wilhelm 1897; Leibundgut 1984) except stony and marshy soils and loose sands (Tseplyaev 1961).

Fire

The species is fire-intolerant, possibly because of its relatively thin bark.

Frost

P. tremula tolerates summer frosts and cold winters (Leibundgut 1984). It tolerates both low and high temperatures (Tseplyaev 1961). According to Hempel & Wilhelm (1897), however, it is sensitive to late spring frosts and high temperatures.

Flooding
The occurrence of the species on wet sites and in river valleys (see, for example, Hesselman 1931; Tseplyaev 1961; Sokolov, Svyaseva & Kubly 1977) indicates that it can tolerate seasonal flooding.

Wind
P. tremula is sensitive to windthrow because its roots are shallow (Hempel & Wilhelm 1897; Smilga 1986). Its sensitivity increases after stand thinning (Smilga 1986).

Enemies and diseases

P. tremula is heavily grazed by animals (Hesselman 1912, 1931). Aspen forests are damaged by mice, deer, hares and other animals. The genus *Populus* is also damaged by 700 species of insect and about 100 species of fungus. Seeds are mostly damaged by larvae of the butterfly *Batracherda praengusta* Hwb. Catkins of aspen are damaged by the beetle *Ellschus scanicus* Pauk. The fungi *Pestalazzia hartigii* Tub. and *Phytophtora omnivora* De Bary. suffocate seedlings. Rust of leaves is caused by the fungi *Melampsora ulti-populina* Kleb., *M. lapici-populina* Kleb. and *M. laricis* Hart.; spotted leaves are caused by the fungus *Melampsora pinitorqua* A. Braun. Aspen leaves are often grazed and damaged by larvae of *Gladius viminalis* Fall., *Lygaeonematus compressicornis* F. and *Clanellaria amerinae* L. Rotting of trunks and roots is caused by *Armillaria mellea* (Quell.). Canker of roots is caused by the bacterium *Pseudomonas tumefaciens* Sm. et Tows. The fungi *Cytospora nivea* (Hoffm.) Sacc., *Napidadium tremulae* Aderh. and *Fusicladium radiosom* Lind. dry out the branches, especially under drought conditions. Trunk and branch canker is caused by various pathogens, e.g. *Hypoxylon holwayi* Ell. and *Dothichiza populea* Sacc. et Br. The most widespread and dangerous disease of the trunk is rot, caused by the fungi *Phellinus tremulae* (Bond.) et Boriss and *Phellinus igniarius* (L. ex Fr.) Quel. (Smilga 1986).

3. Silvical parameters of circumpolar boreal tree species

This section aims to assist parameterization of boreal forest simulation models (e.g. Bonan 1988*b*; Leemans & Prentice 1989). It summarizes autecological characteristics of the main circumpolar boreal forest tree species in 31 input model parameters (Tables 2.15 and 2.16).

Parameter explanation

AGEmax Maximum age recorded for the species (years)

Table 2.15. *Silvical parameters of boreal tree species in Eurasia*

Values for *ETSmax*, *ETSmin* and *Tcold* are estimated by overlapping of species distribution maps and climatic maps prepared by Dr. W. Cramer (personal communication). Values for *Hmax* in brackets are estimated in estimating *IS* from height and diameter data (see parameter explanation). Values for *B2* and *B3* are estimated so that the parabolic function for the diameter–height relation matches the asymptotic one (see explanation for parameter *IS*). Values for *Tl*, *Tdr* and *Tn* are based on unpublished data by N. P. Polikarpov. Unknown values are indicated by dashes.

Species	ETSmax	ETSmin	Tcold	SFrq	Disp	Sfire	Smsl	Sm	Sdr	Slmin	Rlr	SprTnd	SprMax	SprMin
Abies sibirica	1450	510	−35	30	T	F	F	T	F	4	T	0	0	0
Betula pendula	2300	410	−40	90	T	T	F	F	T	—	F	—	—	—
Betula pubescens	2050	340	−40	90	T	T	T	F	F	—	F	—	—	—
Chosenia arbutifolia	1750	240	−45	90	T	T	T	F	F	—	F	—	—	—
Larix gmelinii	1500	250	−45	33	T	T	T	F	T	—	F	0	0	0
Larix sibirica	1500	300	−33	20	T	T	T	F	T	—	F	0	0	0
Larix sukaczewii	1750	390	−22	20	T	T	T	F	T	5	F	0	0	0
Picea abies	2250	470	−17	24	T	F	T	F	T	—	T	0	0	0
Picea ajanensis	1800	750	−38	40	T	F	F	T	F	—	T	0	0	0
Picea obovata	1500	320	−40	8	T	F	F	F	F	—	T	0	0	0
Pinus pumila	1600	240	−45	90	F	F	F	F	T	6	T	0	0	0
Pinus sibirica	1450	490	−35	65	F	F	F	F	F	—	F	0	0	0
Pinus sylvestris	2350	450	−40	24	T	T	T	F	T	—	F	0	0	0
Populus tremula	3000	400	−40	35	T	T	T	F	F	—	F	—	—	—

Table 2.15 (cont.)

Species	Hmax	Dmax	Imax	AGEmax	IS	B2	B3	G	Tl	Tdr	Tn	Tfr	Tfl	Bog	Ptol
Abies sibirica	40 (42.4)	80	—	300	1.0329	91.18	0.5065	83.0	1	4	5	3	2	1	1
Betula pendula	30	60	100	135	1.0987	92.36	0.7448	235.0	4	1	1	1	1	1	2
Betula pubescens	30	60	—	100	1.0987	92.36	0.7448	235.0	4	5	1	2	2	2	2
Chosenia arbutifolia	37	80	—	130	—	89.06	0.5567	314.0	5	5	5	—	4	1	1
Larix gmelinii	40 (38.8)	140	120	400	1.0914	78.84	0.4022	164.0	5	2	2	1	3	2	3
Larix sibirica	45	180	—	450	1.4897	94.85	0.5155	165.0	5	1	2	1	3	2	2
Larix sukaczewii	40 (42.0)	120	—	350	1.4701	93.40	0.5368	161.0	5	3	2	1	2	2	1
Picea abies	63	170	70	570	1.0761	75.62	0.2320	152.0	2	5	3	3	2	2	1
Picea ajanensis	60	150	—	500	0.5649	57.20	0.1395	99.0	2	4	3	3	2	2	2
Picea obovata	40	50	—	500	1.0214	85.84	0.4769	97.0	2	5	3	3	3	2	3
Pinus pumila	7	—	4	200	0.5567	50.04	1.1121	13.5	3	4	1	2	3	2	2
Pinus sibirica	45	190	—	800	0.8955	69.81	0.2792	83.0	2	4	3	2	3	2	2
Pinus sylvestris	48	190	100	600	0.9772	81.81	0.3588	144.4	4	1	1	1	3	2	1
Populus tremula	42	100	150	150	1.3910	90.29	0.5016	225.0	4	2	3	3	3	1	1

Table 2.16. Silvical parameters of boreal tree species in North America

Values for ETSmax and ETSmin are based on data from Solomon et al. (1984), species distribution maps by Fowells (1965) and climatic maps prepared by Dr. W. Cramer (personal communication). Values for Tcold, Smsl, Sm, SprTnd, SprMin, SprMax, SprTnd, Tl are based on data by Fowells (1965). Values for SFrq, Disp, Sfire, Rlr and Tl are based on data by Fowells (1965). Values for B2, B3, G and FRT are taken from Pastor & Post (1985). Unknown values are indicated by dashes.

Species	ETSmax	ETSmin	Tcold	SFrq	Disp	Sfire	Smsl	Sm	Slmin	Rlr	SprTnd	SprMin	SprMax
Abies balsamea	2400	560	−25	40	T	F	F	F	15	T	0	0	0
Betula alleghanensis	3500	1320	−18	66	T	T	F	F	—	F	1	3	12
Betula papyrifera	2400	500	−28	50	T	T	F	T	—	F	2	1	10
Larix laricina	2660	280	−29	22	T	T	F	T	—	F	0	0	0
Picea glauca	2200	280	−30	25	T	F	F	F	—	T	0	0	0
Picea mariana	2200	247	−30	25	T	T	F	F	15	T	1	10	20
Picea rubens	3000	1400	−12	20	F	F	F	T	—	F	0	0	0
Pinus banksiana	2320	650	−30	28	T	T	F	F	—	F	0	0	0
Pinus contorta	3000	700	—	50	T	T	—	T	—	F	0	0	0
Pinus resinosa	2500	1100	−20	20	T	T	F	T	—	F	0	0	0
Pinus strobus	3400	1230	−20	25	T	T	F	T	20	F	0	0	0
Populus balsamifera	2700	440	−30	90	T	T	F	T	—	F	3	20	200
Populus tremuloides	2900	500	−30	22	T	T	T	F	—	F	3	20	150
Thuja occidentalis	2400	1000	−20	20	T	F	F	F	—	T	1	5	400
Tsuga canadensis	3800	1320	−12	40	T	F	T	F	—	F	0	0	0

Table 2.16 (cont.)

Species	Hmax	Dmax	AGEmax	B2	B3	G	Tl	Dry	FRT
Abies balsamea	15	50	200	54.52	0.5452	68.85	1	0.165	3
Betula alleghanensis	25	50	300	76.40	0.5013	106.40	4	0.343	1
Betula papyrifera	25	100	140	47.26	0.2363	159.70	4	0.347	1
Larix laricina	20	75	335	63.01	0.4201	66.11	5	0.267	1
Picea glauca	25	50	200	90.96	0.7105	132.30	2	0.309	3
Picea mariana	20	40	250	93.15	1.1600	70.49	2	0.170	3
Picea rubens	30	100	400	58.23	0.2912	89.34	2	0.237	3
Pinus banksiana	25	50	150	94.52	0.9452	5.50	5	0.511	2
Pinus contorta	32	184	600	33.57	0.0911	89.00	5	—	—
Pinus resinosa	25	75	310	63.01	0.4201	71.44	4	0.385	2
Pinus strobus	35	150	450	44.84	0.1495	68.37	3	0.267	2
Populus balsamifera	25	75	200	63.01	0.4204	147.60	4	0.267	1
Populus tremuloides	22	75	200	55.01	0.3668	157.80	5	0.267	1
Thuja occidentalis	35	100	400	46.04	0.2303	54.63	5	0.350	3
Tsuga canadensis	35	150	650	44.84	0.1495	47.00	1	0.288	3

B2   Tree form parameter. *B2* and *B3* (see below) are used in a parabolic function of height–diameter relation (see, for example, Shugart 1984):

$$H = 137 + B2\, D - B3\, D^2,$$

where H is tree height (cm) and D is tree diameter at breast height (cm)

B3   Tree form parameter (see above)

Bog·   Ability to grow on peatbogs (1 = no growth, 2 = poor growth)

Disp   Seed dispersal pattern (T = wind-dispersed, F = not wind-dispersed)

Dmax   Maximum diameter at breast height (1.3 m) recorded for the species (cm)

Dry   Maximum proportion of growing season days with soil moisture below the wilting point that the species can tolerate (%)

ETSmax   Maximum annual effective temperature sum (°C). This parameter characterizes temperature regime at the southern border of the species range. *ETSmax* and *ETSmin* (see below) are estimated by overlapping of species distribution maps and *ETS* maps prepared by W. Cramer. Effective temperature sums (*ETS*) for the maps have been estimated as follows:

$$ETS = \Sigma\,(D_i),$$

where $\Sigma\,(D_i)$ is the annual total of daily departures (D_i) of temperature above 5 °C, so that

$$D_i = (T_i - 5), \text{ if } T_i > 5 \text{ and } D_i = 0, \text{ otherwise.}$$

In the equation above, T_i are mean daily temperatures. Values for T_i are estimated from mean monthly temperature data by assuming that reported monthly values (MT_j) refer to the middle of each month, and applying a linear interpolation:

$$T_i = MT_j + i(MT_j - MT_{j+1})/days,$$

where *days* is number of days (i.e. 30 or 31)

ETSmin   Minimum annual effective temperature sum (°C). This parameter characterizes temperature regime at the northern border of the species range

FRT   Foliage retention time (years)

G   Scalar for species maximum diameter increment. This parameter is used in stand simulators to scale species-specific equations for optimal diameter growth (see, for example, Shugart 1984; Solomon *et al.* 1984; Pastor & Post 1985)

Hmax   Maximum height recorded for the species (m)

Imax   Maximum height growth rate (cm yr^{-1})

IS	Initial rate of increase of height with diameter (m cm^{-1}). This parameter is used in an asymptotic function of height–diameter relation (Leemans & Prentice 1989):

$$H = 1.3 + (Hmax - 1.3)\,(1 - \exp(-(D\ IS)/(Hmax-1.3))),$$

	where H is tree height (m), D is tree diameter at breast height (cm), and *Hmax* is maximum height (m) recorded for the species. Values for *IS* have been estimated from species-specific diameter and height data by nonlinear regression
Ptol	Ability of growing on permafrost (1 = no growth, 2 = poor growth, 3 = good growth)
Rlr	Ability of regenerating by layering (T = able, F = unable)
Sdr	Seedling germination with respect to drought (T = tolerant, F = intolerant)
Sfire	Seedling establishment with respect to fire (T = fire required, F = fire not required)
SFrq	Frequency of good seed crops (% of all years). The values refer to the boreal forest zone unless otherwise indicated
Slmin	Minimum light required for seedling growth (% of full sunlight)
Sm	Requires moss or litter layer for successful regeneration (T = true, F = false)
Smsl	Requires mineral soil for successful regeneration (T = true, F = false)
SprMax	Maximum diameter at breast height for sprouting (cm)
SprMin	Minimum diameter at breast height for sprouting (cm)
SprTnd	Tendency for stump or root sprouting (expected number of sprouts per tree after tree death)
Tcold	Mean temperature of the coldest month (°C) at the border of the species' geographical range
Tdr	Drought tolerance class. A nominal scale from 1 to 5 is used with 1 = drought-tolerant and 5 = drought-intolerant
Tfl	Flooding tolerance class (1 = intolerant, 2 = tolerant of occasional flooding, up to 1 month in a year, 3 = tolerant of seasonal flooding, up to 4 months in a year, 4 = tolerant of year-round flooding)
Tfr	Fire tolerance class. This parameter characterizes species' ability to recover after fire damage (1 = tolerant, 2 = intermediately tolerant, 3 = intolerant)
Tl	Shade tolerance class. A nominal scale from 1 to 5 is used with 1 = shade-tolerant and 5 = shade-intolerant. This and the following variable concern the reaction of mature trees
Tn	Nutrient-stress tolerance class. A nominal scale from 1 to 5 is

used with 1 = nutrient-stress-tolerant and 5 = nutrient-stress-intolerant.

We thank Dr Allen Solomon, who began the studies at the International Institute for Applied Systems Analysis (IIASA) concerning global vegetation change and gathered the Biosphere research team, for his comments on earlier versions of this chapter. We also thank the Library of the Higher Institute of Forestry and Forest Technology in Sofia, Bulgaria, for providing most of the Russian literature as well as the IIASA Library for providing the non-Russian sources. We thank Dr W. Cramer from the University of Trondheim, Norway, for providing climatic maps of Eurasia, and Dr Nedezda Tchebakova (USSR) and Ruiping Gao (China) for their contribution to data collection at IIASA during the summer of 1989.

3 The reproductive process in boreal forest trees

John C. Zasada, Terry L. Sharik and Markku Nygren

Introduction

The reproductive process plays an important role in determining species composition and distribution. At the landscape level, the variation in density and species composition of forests growing on similar, but geographically separated, sites is related to the quantity and type of reproductive material, the nature and severity of disturbance, and the growth requirements of the species comprising the forest. Within a given site, the spatial and age distribution of trees and associated plant species is, in part, a consequence of the nature of the substrate following disturbance and the ability of residual or new reproductive material to survive and develop under the prevailing biotic and abiotic conditions.

Regeneration of boreal forests varies in time and space. Temporal variation takes into account such variables as the periodic nature of seed production and changing seedbed conditions. Spatial variation includes factors such as seed dispersal distance and pattern of surface conditions as related to dynamics of sexual and asexual reproduction.

The state and dynamics of regeneration vary with the condition of the forest ecosystem. Natural disturbances, for example fire and major wind damage, remove or destroy variable amounts of the above-ground system, including the forest floor, and below-ground reproductive material, creating a continuum of conditions that determine the relative importance of sexual reproduction and vegetative regeneration to secondary succession. Colonization and forest development on newly formed or exposed sites (e.g. primary succession along rivers or following glacial retreat) is usually dominated by sexual reproduction, owing to the absence of pre-existing vegetation. Regeneration of forest gaps created by the death of individuals or small groups of trees and soil disturbance by animals or windthrow depends on yet another set of factors. Finally,

Seed reproduction cycle

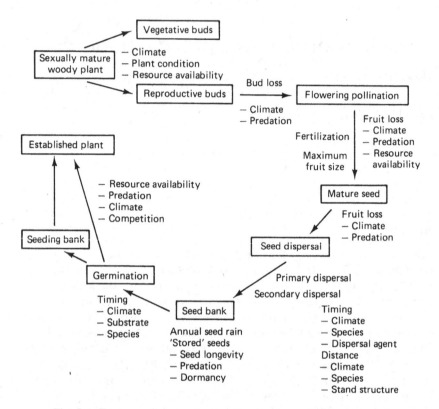

Fig. 3.1. Framework for evaluating the seed reproduction process in boreal forest trees.

human activities, for example forest harvesting, mining, and construction, create an array of site conditions and, thus, different patterns of forest recovery.

The process of forest recovery or site colonization as embodied in the concept of primary and secondary succession consists of a series of sequential steps which are discrete yet interrelated (Figs 3.1 and 3.2). The steps are independent in that they are regulated by different environmental factors. Yet, for successful regeneration to occur, each step must operate at a minimum level to have the process continue (for example, seeds must be produced before dispersal can occur and the latter two must occur before germination can occur). Models and concepts of a similar nature have been presented and discussed by Harper (1977), Grubb (1977), Grime (1979), Bormann & Likens (1979*a*), Zasada (1980), Willson (1983), Jeannson *et al.* (1989) and Maxwell (1990).

Sexual reproduction and vegetative regeneration are both important processes for the invasion, colonization, and maintenance of these species on sites in the boreal forest. Although similar in some respects (e.g. the presence of a potential pool of residual plant materials in seed and bud banks, which can respond quickly to disturbance), they differ in fundamental ways such as the genetic composition of the resultant population, growth rates, and dispersal potential. Thus, we will consider the two processes independently in our discussion realizing that there are similarities and that they frequently occur simultaneously on a site.

The general focus of this chapter is population biology. We will concentrate on describing changes in population numbers in terms of within- and among-genet or ramet dynamics and the natural history of the reproductive process. The following tree species will be considered because they provide good examples of the points we want to consider: *Picea glauca* (white spruce), *P. mariana* (black spruce), *P. abies* (Norway spruce), *Pinus sylvestris* (Scots pine), *Populus tremuloides* (trembling aspen), *P. balsamifera* (balsam poplar), *Betula papyrifera* (paper birch), *B. verrucosa* (silver birch) and *B. pubescens* (pubescent birch).

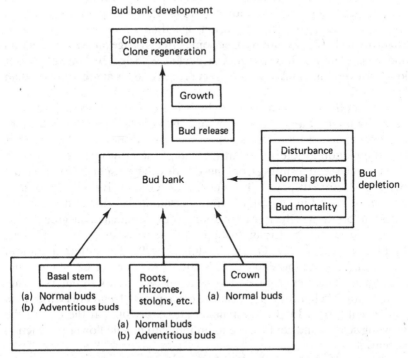

Fig. 3.2. Framework for evaluating the process of clone expansion and vegetative regeneration in boreal forest trees.

Although there are other important tree species and associated shrubs and herbs in the boreal forest that are of importance, space limitations preclude their treatment here. Ecological and silvicultural information on the boreal tree species not considered here can be found elsewhere in this book (Nikolov & Helmisaari, Chapter 2).

Sexual reproduction

Production and availability of reproductive buds

The production of reproductive buds depends on whether a tree is sexually mature and, if so, whether the environmental conditions trigger the differentiation of reproductive buds under a given set of physiological and morphological conditions for a tree or trees on a given site. Attainment of sexual maturity is determined by species and site conditions. All of the genera considered in this paper have been observed to flower sporadically at an age of 15 years or less (Fowells 1965; Schopmeyer 1974; Heinselman 1981b). The age at which consistent production is possible under good growing conditions is usually between 25 and 50 years (Koski & Tallquist 1978). In general, broadleaved species produce seeds at an earlier age than conifers. Among conifers, *P. mariana* and *P. sylvestris* tend to produce cones at an earlier age than *P. glauca* and *P. abies*. Silvertown (1982) has summarized some effects of age on seed production and concluded that there is a general decline with age. Because of greater longevity of individual trees, conifers produce seeds at greater ages than hardwoods.

Differentiation of reproductive buds is a complex process depending on many internal and external factors and is determined by conditions and events one or more years before seeds are dispersed. The potential bud population is determined at the end of the growing season prior to that in which seed resulting from these buds will be dispersed, or in *Pinus*, two years before dispersal. The number of reproductive buds in conifers can be estimated in late fall or winter based on external bud characteristics and position in the tree or by forcing development in the greenhouse or growth chamber (Eis & Inkster 1972; Sarvas 1972, 1974; Owens, Molder & Langer 1977; Owens & Molder 1979). The differentiation of large numbers of reproductive buds and subsequent excellent cone and seed years in conifers is related to relatively warm, dry summers (Krugman, Stein & Schmitt 1974); these are often years of high fire incidence in the boreal forest. In *Betula*, male catkins are visible at the end of the growing season and could serve as an index to female flower production potential.

Cone and seed crops produced in previous years affect future reproductive potential by reducing stored energy reserves and the number of sites available for differentiation of reproductive buds. Excellent seed crops are commonly followed by poor crops the following year. In *B. papyrifera*, for example, large seed crops resulted in crown deterioration (Gross 1972). Caesar & Macdonald (1984) concluded that development of flowers in *B. papyrifera* lowered growth potential and bud vigor and reduced crown expansion. Tirén (1935) has reported a decline in the bud population of *P. abies* as a result of heavy cone production. Silvertown (1982) concluded from a review of the literature that, in general, reproductive growth reduces vegetative growth in trees.

After the initiation of flower buds, reproductive potential declines due to a reduction in the bud population by biotic and abiotic factors. Abiotic factors include snow and ice breakage, and wind damage. They can affect whole trees through massive crown breakage (Van Cleve & Zasada 1976) or by affecting individual branches and buds. The magnitude of the effect of abiotic factors depends on the severity of the weather and tree and stand condition.

Moose, hares, ptarmigan, grouse, red squirrels, and other birds and mammals can have a significant impact on the reproductive bud population by browsing, girdling the main stem and branches, and selective removal of buds. Impact varies among stands in a given year, annually in a given stand and among species and individuals of the same species in a given stand (M. C. Smith 1967; Wolff & Zasada 1979; Bryant & Kuropat 1980; Neiland *et al*. 1981).

Flowering

Flowering (including pollination) is the first visual sign of the end of winter dormancy in boreal forest trees. Reproductive bud activity precedes or occurs at the same time as the onset of the development of vegetative buds. Although many factors can affect the onset of flowering, two effects of air temperature are of primary importance (Sarvas 1974). Sarvas (1962, 1968) has demonstrated that flowering in *P. sylvestris* and *P. abies* is closely associated with the accumulated annual heat sum (the sum of degree days above a 5 °C threshold temperature); flowering in *P. sylvestris* occurs at about 230 degree days and in *P. abies* at 140 degree days. In Alaska, *P. glauca* flowers fairly consistently at between 100 and 125 degree days and black spruce 10 days to two weeks later. For a detailed discussion of flowering and pollination in *Picea*, *Pinus* and *Betula*, refer to Sarvas (1952, 1962, 1968, 1972, 1974), Owens & Molder (1977, 1979) and Owens, Molder & Langer (1977). *Populus* and *Betula* species flower before conifers and have lower heat sum thresholds. We have observed reproductive bud activity in *P. tremuloides* during warm periods in February in interior Alaska.

Frost can be of overriding importance in the reproductive cycle when it occurs at the time of flowering. Flowers can be very susceptible to frost (Krugman, Stein & Schmitt 1974). Sarvas (1968), for example, demonstrated that the pollen drop system of *P. abies* is destroyed or hampered by frost. Severe frosts can kill entire flower crops. However, often only the most susceptible flowers are killed, and those flowers which are delayed or more advanced in development sustain less or no damage (Cram 1951; Cayford *et al.* 1959; Simak 1969; Zasada 1971).

Seed development

Seed development in relation to time of flowering varies considerably among these species. In *P. sylvestris*, fertilization and seed maturation occur the year after pollination. For the other species, maturation occurs in the year of pollination. *P. tremuloides* seeds mature in early summer, *P. balsamifera* in mid-summer, *Betula* species in late summer, and those of *Picea* and *Pinus* in late summer and early fall. The attainment of maximum cone or catkin size occurs from several weeks before seed maturation in the broadleaved species to a month or more in the conifers.

A critical consideration is whether or not seeds mature during the growing season. This is one aspect of the reproductive process that is more critical in the boreal forest than it is in temperate forests. The frequency of seed crops and the seed quality, as defined by the presence of mature seeds, become particularly critical at the treeline as they will be at least as important as factors related to vegetative growth in determining the location and dynamics of this boundary. This does not appear to be of importance for *Populus* and *Betula* seeds. Under normal circumstances, their seeds mature from early to late summer; a delay in maturation such as might occur in cold summers or at higher elevations would have the effect of delaying the time (calendar date) of maturation but would not jeopardize seed maturation except in the most extreme situations.

In conifers, however, the production of mature seeds can be a significant problem and becomes more acute with increasing latitude and elevation (Zasada *et al.* 1978; Tranquillini 1979; Remröd 1980; Bergman 1981; Henttonen *et al.* 1986; Zasada 1988). Henttonen *et al.* (1986) reported that 890 degree days (5 °C threshold temperature) were required to produce a *P. sylvestris* seed that was anatomically mature (that is, the embryo fills at least one third of the embryo cavity). Based on this information and long-term weather records, they concluded that there was a strong north–south gradient in seed maturity in Finland. For example, the probability of seeds reaching 50% anatomical maturity increased from 0.02 at 68–69°N latitude to 0.05 at 67–68°N latitude to 0.50 at 65–67°N latitude. Remröd (1980) and Bergman (1981), working in Sweden, reported a strong south–north gradient in seed maturity related

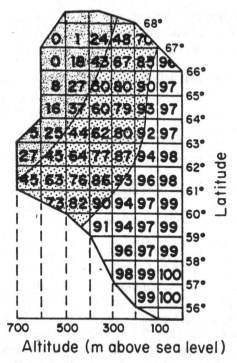

Fig. 3.3. Expected germination (per cent) for *P. sylvestris* seeds from different latitudes and elevations in Sweden in 1979–80 (from Remröd 1980).

to latitude (except at the lowest elevation) and an equally strong east–west gradient related to increasing elevation (0 to 700+ m) and distance from the Baltic Sea (Fig. 3.3). Zasada *et al.* (1978) observed that, in Alaska, immature *P. glauca* seeds were produced north of the Arctic Circle and at elevations of 600 m or more south of the Arctic Circle. Zasada (1988) found that 670–700 degree days (5 °C threshold temperature) were required for seeds to fill 75% of the embryo cavity in *P. glauca*. Because of the critical role of temperature in seed maturation in some parts of the boreal forest, any situation that may increase temperature conditions (e.g. warmer sites, location of seeds on the tree or in the cone) can have a favorable influence on seed maturation (Bergman 1981; Ryynänen 1982; Henttonen *et al.* 1986). Additionally, Ryynänen (1982) has observed significant variation among individual trees in *P. sylvestris* stands in their ability to produce anatomically mature seeds in cold summers.

Anatomically mature seeds may not be mature in terms of their biochemical composition (Edwards 1980). Seeds that are not biochemically mature may be lighter in weight, have lower vigor, and be less able to endure environmental stress as may occur during germination or handling and storage (Zasada 1973; Zasada *et al.* 1978; Edwards 1980; Bergman 1981; Winston & Haddon 1981).

Reductions in the seed population prior to seed dispersal

Reductions in the seed population prior to dispersal fall into two categories: those that are general in nature and those caused by predation of seeds. The general losses are caused mainly by climatic factors such as wind and rain storms. These have not been quantified to our knowledge, but general observations indicate they can be substantial in strong wind and rain storms.

Feeding by birds, animals, and insects and infection by disease results in variable losses depending on the predator population level, quantity of cones, and weather conditions (Willson 1983). One of the best-documented cases of animal predation on fruits or seeds in the boreal forest is that of the use of *P. glauca* cones and seeds by the red squirrel, *Tamiasciurus hudsonicus*. White spruce is a main source of winter food for *Tamiasciurus*. Cone consumption varies greatly and although the greatest number of cones may be harvested in excellent cone years, in these years the entire crop is usually not harvested as it can be in years of lower production (Brink 1964; Smith 1967; Searing 1975; Kelly 1978). *Tamiasciurus* also harvests *P. mariana* cones (Lutz 1956*b*; Kelly 1978). Birds can consume large quantities of seed as White & West (1977) have described in the case of redpoll consumption of *B. papyrifera* and *Alnus* species seeds.

Insects and diseases are commonly associated with reproductive structures and seeds. Insect damage will vary from year to year and may be highest in years of low production (Werner 1964; Hedlin 1974). In *P. glauca*, insect-damaged seeds per cone varied from 2 to 50% of the seed population over a 5-year period (Werner 1964). Insects are commonly associated with catkins of *Betula* and *Populus* species, but the amount of seed predation has not been quantified to our knowledge.

Diseases also affect seed availability. One example is the spruce cone rust (*Chysomyxa pirolata*). Sutherland (1981) observed that the reduction in the number of filled seeds per cone resulting from spruce cone rust varied from 7 to 100% over a 3-year observation period. Germinability and vigor of filled seeds may also be lower in diseased cones (McBeath 1981; Sutherland 1981). Entire seed crops during exceptionally good seed years can be affected (Ziller 1974); this is somewhat different from the case for insects and mammals where the percentage destruction appears to be less in good than in poor seed years.

Cone and seed abortion may also occur as a result of factors within the cone or seeds. Lack of pollination and lethal gene combinations result in the production of empty seeds (Koski 1973). The effect of poor polli-nation can result in conelet abortion in *P. sylvestris* (Sarvas 1962); in *Picea* it results in production of empty seeds and in extreme cases small and poorly developed cones.

Seed quantity and quality

The net effect of the various internal and external environmental factors on seed availability can be assessed on a per tree and per stand basis. Assessments on a per tree basis are made by estimating the number of fruits or cones present prior to dispersal and the number of seeds per fruit or cone. Stand-level estimates are most often made by measuring seed rain in stands. In both estimates, the seed population should be described in terms of the total number of seeds and the number of viable seeds.

The number of fruits or cones per tree varies annually and among trees and stands during a given year (Sarvas 1962; Waldron 1965; Zasada *et al.* 1978; Brown, Zobel & Zasada 1988). Table 3.1 provides examples of these levels of variation as well as an estimate of variation within and among regions of Alaska. These observations indicate that over relatively large areas, good to excellent cone crops of *P. glauca* can occur simul-taneously (e.g. 1970 and 1972). Some years are good in one area and poor in another (e.g. 1975). Years of no or low cone production can be widespread (e.g. 1971) or vary among areas (e.g. 1975). Within a fairly local area (see stands BC1, BC2, BC3) there can be relatively large variation in cone production between stands (e.g. 1975). The variability among trees within a given stand can be large. The years when general stand production is moderate seem to be those in which tree-to-tree variation is greatest (e.g. YR1 in 1978, BC2 in 1982).

The data in Table 3.1 provide information about the production by dominant and codominant trees, which produce the majority of seeds in a stand. Trees in the lesser crown classes also produce seed, but crops are usually less frequent and the quantity of seed lower (Waldron 1965; Kelly 1978; K. Brown 1983).

The hardwood species likewise exhibit annual variation in fruit pro-duction. The frequency of large seed crops tends to be greater in hardwoods than in conifers. There have been few attempts that we know of to monitor annual production of fruits by individual hardwoods in the boreal forest. Monitoring crops of these species is more difficult because fruits are not as conspicuous as cones are in conifers. Seed crop assess-ment in these species is more often at the stand level, and these estimates are discussed later. Furthermore, *Populus* species are dioecious, and, where male clones cover large areas, there will be no seed production.

The frequency of good cone and seed crops is related to climatic variables which are inversely related to latitude and elevation, as

Table 3.1. Cone production[a] is by individual trees in five interior Alaska white spruce stands (BC1–KR1) from 1970 to 1975

	1	2	3	4	5	6	7	8	9	10	11	12	13	14	15
1970															
BC1	350	800	330	350	750	650	600	600	500	100	480	430	300	435	420
BC2	500	1100	1300	700	550	585	415	775	420	360	270	400	850	900	470
BC3	420	405	340	230	120	180	415	310	530	805	950	930	310	420	520
YR1	435	135	620	360	600	375	405	330	475	410	900	585	800	390	500
KR1	445	415	280	505	320	645	500	205	470	260	170	210	375	260	985
1971	No cone production was observed in any of the five stands														
1972															
BC1	275	840	575	350	250	600	950	725	1500	500	990	510	200	190	400
BC2	750	1600	500	650	300	500	350	525	650	575	575	525	125	300	325
BC3	375	590	425	530	875	575	1075	625	675	1050	625	275	205	1325	775
YR1	550	300	225	425	150	350	625	150	1050	325	475	1750	375	950	350
KR1	750	800	400	450	275	1050	500	275	850	450	400	350	520	300	175

1973															
BC1	20	5	0	20	30	0	0	10	5	5	0	5	100	65	0
BC2	80	0	125	45	25	40	15	45	20	10	0	0	0	150	20
BC3	0	0	0	0	0	0	0	0	0	0	5	25	0	0	5
YR1	34	10	36	9	150	210	165	60	1	175	2	0	550	2	50
KR1	350	375	150	225	15	175	275	55	15	150	65	15	2	165	500
1974															
BC1	0	5	0	0	0	0	0	0	0	0	0	0	0	0	0
BC2	0	0	0	2	0	0	0	0	5	0	0	0	0	1	0
BC3	0	40	0	0	0	0	0	0	10	5	60	40	35	10	0
YR1	102	0	56	66	10	125	93	5	275	75	65	125	1.	102	45
KR1	0	0	50	5	10	0	0	0	0	5	50	10		15	0
1975															
BC1	105	125	0	145	75	40	25	80	10	10	45	15	10	15	30
BC2	500	600	375	275	225	225	200	175	425	85	50	175	0	275	65
BC3	35	235	25	60	40	80	80	125	230	450	250	200	40	525	375
YR1	210	0	75	305	275	310	230	30	390	140	5	950	350	240	550
KR1	0	0	8	2	2	0	5	0	0	0	0	0	3	50	0

Note: [a] Cones counted annually from a fixed point on the ground with a variable powered telescope. Cones which could not be seen from that point were not counted. Machaniček (1973) recommended that estimates of this type should be multiplied by 1.4 to obtain an indication of total cone production in years of low production and by a factor of 2.5 in years of good production.

discussed earlier. In Sweden (Fig. 3.3) and Finland there is a close correlation between production, and latitude and elevation. However, in North America *P. glauca* seed production appears to be quite similar on productive forest sites from 49 or 50 degrees latitude to 65 degrees in interior Alaska. In eastern Canada, there appears to be a stronger relationship with latitude than in western Canada and Alaska; this is probably due in part to the mountainous terrain in western North America. Within any geographical area, elevation can have a negative effect on cone and seed production. Because of these climate-related limitations, individual trees near the treeline may produce only 1–3 good (in terms of quantity and quality) seed crops during their lifetime. At lower latitudes and elevations, seed crops are more frequent, but even at these sites the number of good crops may not exceed 5–10 during the lifetime of a tree. Koski & Tallquist (1978) provide excellent information on long-term seed crop frequency (Fig. 3.4).

The maximum number of fruits or cones that a tree can produce is determined by the size of the tree and site conditions. Table 3.2 provides an indication of the maximum production which has been observed for these species.

Seed rain

Timing of seed rain

Timing of seed rain for different species can be placed into three categories: (1) species whose seeds mature in summer and disperse

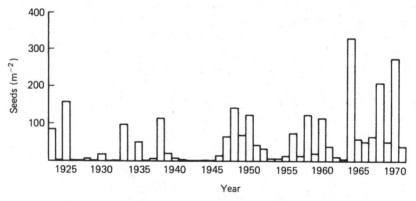

Fig. 3.4. Annual variation in *P. sylvestris* seed production in a stand at Vilppula, Finland over a 50-year period; the year refers to flowering year (reproduced with permission from Koski & Tallquist (1978)).

Table 3.2. *Cone/fruit and seed production by individual trees*

Values in this table provide the maximum reported values for these species

Species	Location	Cone/fruit production	Seeds per cone	Reference
P. glauca	Alaska	2000–3000	12–85	Zasada (unpublished); Zasada & Viereck (1970)
	Unknown	10 000	—	Fowells (1965)
P. mariana	Alaska	730	78	Kelly (1978); Zasada et al. (1978)
	Ontario	2865[a] (1146)	21	Haavisto (1975)
P. abies	Finland	1123	236	Mälkönen (1971); Sarvas (1968)
P. sylvestris	Finland	1473	54	Lehto (1956); Sarvas (1962)
B. papyrifera	Alaska	28 000	295–535	Zasada (unpublished)
P. tremula	Europe	40 000	1350	From Schreiner (1974)
P. tremuloides	Michigan	—	350	Henry & Barnes (1977)

Note: [a]Maximum production followed by average production by dominants in a good year. If cones of all ages are included, dominant trees had an average of 13 784 (Haavisto 1975).

during the current growing season; (2) species whose seeds mature in late summer and are dispersed during the dormant season; and (3) species whose seeds mature in late summer, but dispersal is delayed for an indefinite period of time or occurs over several to many years (i.e. those species with serotinous and semiserotinous cones) (Fig. 3.5). The general seasonal dispersal patterns illustrated in Fig. 3.5 depend on weather at the time of dispersal, damage to fruits and cones, predation by animals, and other factors. As a result, there will be annual variation and variation among sites in a given year in the inception and duration of the dispersal period (Fig. 3.6). For more detailed examples of variation in timing of seed rain among genera and among regions for the same genus, see Bjorkbom (1971), Fries (1985) and Zasada (1985) for *Betula*; Waldron (1965), Zasada & Viereck (1970), Dobbs (1976), Zasada et al. (1978) and Zasada (1985) for *P. glauca*; Haavisto (1975, 1978) and Zasada, Viereck & Foote (1979) for *P. mariana*; Heikinheimo (1932) for *P. sylvestris* and *P. abies*.

The quality of seeds dispersed over the course of the dispersal period may vary significantly. The causes of this variation will differ among species but will include, among others, insect and disease attack affecting cone opening and seed release, and activity of animals, particularly birds. Dobbs (1976) and Zasada (1985) observed essentially the same pattern for *P. glauca*, with the highest quality seed dispersed in September and a

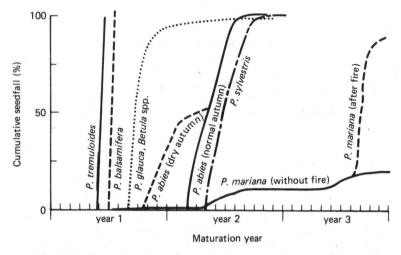

Fig. 3.5. Generalized annual seed rain patterns for selected boreal forest tree species. The amount of the annual seed crop dispersed in *P. mariana*, in the absence of fire, will vary depending on the level of cone serotiny in a given stand.

general decline in quality after that. Bjorkbom (1971) and Zasada (1985) observed variable patterns in the quality of *B. papyrifera* seed dispersed during the dispersal cycle. This variation included the full spectrum of variability from high-quality seed dispersed early in the cycle to no difference among seasons to the presence of higher quality seed later in the cycle.

The third dispersal type is characteristic of *P. mariana*. *P. mariana* cones are semiserotinous, and they open gradually in the absence of fire. When heated by fire, opening and dispersal can occur rapidly (Fowells 1965; Haavisto 1975). There does appear to be some variation among geographic areas in the degree of serotiny in black spruce. Morgenstern (personal communication, and in Haavisto 1975) has observed that cones are less serotinous in southern Ontario than in northern Ontario.

Fire may result in rapid opening of *P. mariana* cones and the release of large quantities of seed in a short time. Wilton (1963) reported that 370 seeds m^{-2} were released in a 60-day period following burning and concluded that this was 50% or less of the available seed bank occurring in the cones. In Alaska, the quantity of *P. mariana* seed released in a burned stand in a 70 day period following fire was 1.5 times greater than in an adjacent unburned stand (Viereck *et al.* 1979). The rate of seed release following fire will depend on the temperature to which cones are exposed during the fire and the weather conditions influencing cone opening after the fire. Seed rain during the 1–2-year period following a fire, not

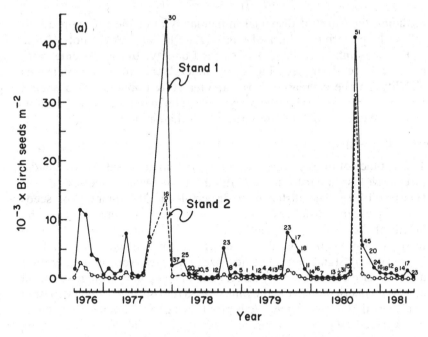

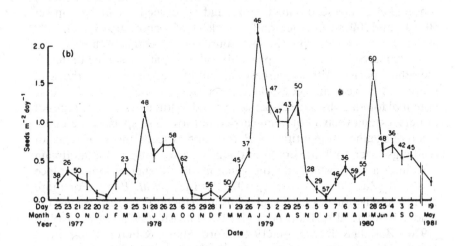

Fig. 3.6. (a) Periodicity of B. papyrifera seed rain in two stands near Fairbanks, Alaska. These stands were on upland sites in Bonanza Creek Experimental Forest. (b) Periodicity of P. mariana seed rain in an undisturbed upland stand near Fairbanks, Alaska. Numbers at the individual points indicate the percentage viability of the seed dispersed during that period.

including the burst of dispersal immediately after the fire, appears to follow the pattern in unburned stands (Zasada, Viereck & Foote 1979).

Fire may influence black spruce seed quality. In the extreme case, cones will be destroyed, but even under less severe fire intensities viability may be decreased. Zasada, Viereck & Foote (1979) observed that real germination (i.e. the percentage of filled seeds that germinated) was, on average, higher in unburned trees than in burned trees.

Seed dispersal distance

The distance of primary dispersal for species in dispersal classes 2 and 3 (see above) is similar to that described for most wind-dispersed species (Harper 1977). The distance of dispersal and the quantity of seeds reaching various distances from the seed source is dependent on the height of the release point (although in some cases, seeds are lifted to higher elevations by updrafts, as described for *P. glauca* by Zasada & Lovig (1983)), stand characteristics, size of the seed crop, air temperature and moisture content as they affect cone opening and closing, and wind and atmospheric stability at the time of seed release. The annual variation in dispersal of *B. papyrifera* seeds provides a good example of what may be expected (Fig. 3.7). There can be a substantial difference among geographic areas in the percentage of seeds reaching various distances from a seed source. For example, in British Columbia, Dobbs (1976) reported *P. glauca* seed rain of 19, 12, and 4% of seed rain in the stand at 50, 100, and 200 m from the stand; Zasada (1985) reported 9, 5, and <1% at the same distances from the seed source on a site in Alaska.

Dispersal of *Populus* seeds probably follows a pattern similar to that of the other species, with the largest amount of seed falling near the seed source. However, the seeds of these species are well known for their potential for long-distance dispersal by wind. Although standard types of dispersal studies have not been carried out for these species, the deposition of significant quantities of *P. tremuloides* seeds and occurrence of seedlings several kilometers from the nearest seed source on floodplain sites in Alaska substantiates the generally held belief of long distance dispersal (Zasada & Densmore 1979; Dyrness *et al.* 1988). For more detailed discussion of seed dispersal distance for boreal forest species, the reader is referred to Bjorkbom (1971), Haavisto (1975, 1978), Dobbs (1976), Zasada & Densmore (1979), Ford, Sharik & Feret (1983), Fries (1985), Zasada 1985), Walker, Zasada & Chapin (1986) and Brown, Zobel & Zasada (1988).

Secondary dispersal, that is dispersal occurring after primary dispersal, has the potential for very long-distance seed movement but more often results in the movement of seeds only a short distance. The potential for the longest-distance secondary dispersal is by water, particularly on

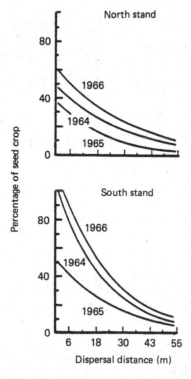

Fig. 3.7. *B. papyrifera* seed dispersal distance for three successive seed crops (Bjorkbom 1971).

rivers. This has not been documented, but we have observed seeds of all the Alaskan species considered here being dispersed by water. Seeds of *Populus* species readily germinate and root under water. The transport of seeds over snow surfaces also has the potential for fairly long-distance dispersal of seeds (Dobbs 1976; Matlack 1989). The importance of over-snow dispersal is dependent on the time of seed dispersal and quantity of seed dispersed in relation to the timing of annual snowfall (Dobbs 1976). Seeds dispersed prior to the first winter snow would be eliminated from this type of secondary dispersal. Although secondary seed dispersal may result in movement of seeds over a short distance, the movement may be very important in terms of moving seeds to a better microsite than would have resulted from primary dispersal.

Quantity and quality of seed rain

The amount of filled seed reaching the soil surface represents the portion of the seed crop that is available for seedling production. This varies substantially among species and years (Tables 3.3 and 3.4). *Betula* species

Table 3.3. *Seed rain in stands of selected boreal forest species*

The values shown as common indicate the range of more frequently observed seed rain. Maximum seed rain values are the highest observed for that species in a particular study.

Species	Location	Seed rain (seeds m^{-2})				Reference
		Common		Maximum		
		Total	Viable (%)	Total	Viable (%)	
P. glauca	Alaska	50–400	55	4000	70	Zasada (1980); Zasada & Viereck (1970)
	British Columbia	—		2546	—	Dobbs (1976)
	Manitoba	300–1500	57	1364	59	Waldron (1965)
	Finland	200–500	—	6346	—	Koski & Tallquist (1978)[a]
P. mariana[b]	Alaska	50–350	12	510	25	Zasada et al. (1979)
	Ontario	92–235		487		Haavisto (1975)
	Finland	300–800	—	3081	—	Koski & Tallquist (1978)
P. abies	Finland	200–1000	—	3487	—	Koski & Tallquist (1978)
P. sylvestris	Finland	75–250	—	530	—	Koski & Tallquist (1978)
B. papyrifera	Alaska	10 000–40 000	13	72 805	38	Zasada & Gregory (1972)
	Maine	190–315	18	6300	77	Bjorkbom, Marquis & Cunningham (1965)
B. verrucosa	Finland	5000–7000	—	177 413	—	Koski & Tallquist (1978)
B. pubescens	Finland	10 000–80 000	—	156 700	—	Koski & Tallquist (1978)
	Finland	15 000–75 000	—	225 158	—	Koski & Tallquist (1978)
P. tremuloides[c]	Alaska	571	—	—	—	Zasada (unpublished)
P. balsamifera[c]	Alaska	3107	—	—	—	Walker, Zasada & Chapin (1986)

Source: [a] These authors summarize years of seed rain data for native and introduced species in Finland. For species native to Finland, the maximum value is that reported for many stands and years of data. The seed rain for introduced species is for one stand in most cases.
[b] *P. mariana* seed rain is for unburned stands.
[c] Seed rain information for *Populus* spp. is very limited, and just how well these values represent seed rain is not known.

Table 3.4. *Total seed rain (seeds m⁻²) for selected species growing at the Punkaharju Experimental Forest in eastern Finland*

	B. verrucosa	B. pubescens	B. papyrifera	P. abies	P. glauca	P. mariana	P. sylvestris
1961	13 687	73 610	15 783	151	10	841	114
1962	51 770	81 188	169 577	55	227	953	24
1963	560	12 260	5137	8	8	848	1
1964	76 353	172 989	177 413	74	556	3081	308
1965	7150	11 483	7670	990	1259	866	71
1966	47 058	143 804	153 257	27	226	—	75
1967	26 234	45 433	69 817	1237	2079	—	68
1968	41 278	63 897	98 767	217	208	—	164
1969	71 172	194 612	168 353	188	140	—	31
1970	31 860	33 383	—	85	193	—	227

Source: Data from Koski & Tallquist (1978).

produce substantially more seeds and do so at shorter intervals than conifers (Table 3.4) (Sarvas 1952; Zasada & Viereck 1970; Zasada & Gregory 1972; Koski & Tallquist 1978; Zasada 1985). However, the quality of *Betula* seeds is generally lower than the quality of conifer seeds. Seed rain data for *Populus* species are too limited to draw any comparisons, but it should be equal to or greater than that of *Betula* species.

The seed bank

The seed bank represents the seed present on a site that is potentially viable and capable of producing seedlings. This seed population is dynamic in that it is increased by annual seed rain and reduced by seed death, predation, and germination (Harper 1977).

The seeds of species considered here vary significantly in their longevity under natural conditions and their potential for incorporation into the seed bank. *Populus* seeds are short-lived. They must germinate within a few days or weeks of dispersal or they die (Schreiner 1974; Zasada & Densmore 1977; Densmore & Zasada 1983). Cooler temperatures may extend seed life, but there is no evidence that seeds of these species can overwinter and germinate the spring following dispersal.

The other species all have longer periods of seed viability than the Salicaceae. There appears to be little doubt that seeds of some of these species do not germinate in the first growing season after dispersal but do so during the second growing season (see, for example, Hellum 1972; Black & Bliss 1980; Zasada *et al.* 1983; Granström & Fries 1985; Putman & Zasada 1986; Granström 1987; Perala & Alm 1989). Fraser (1976) reported that *P. mariana* seeds remained viable in the forest floor for at least 10 months but completely lost viability after 16 months. Granström (1987) reported similar results for *P. sylvestris*. In red squirrel middens, where large quantities of cones and seeds are buried, few viable seeds are found 1 year after burial (M. C. Smith 1967).

The situation is somewhat more interesting for *Betula* species. Granström & Fries (1985) reported that *B. papyrifera* seed viability dropped precipitously from 100 to 6% one year after sowing on natural seedbeds. However, viability dropped much less over the next two years, and about 3% of the seeds remained viable 36 months after sowing. In another study, Granström (1987) showed that *Betula* seed viability dropped from 100 to 60% over a 5-year period. In this study, seeds were placed in blocks of forest floor material before being artificially inserted into the undisturbed forest floor in stands. Based on the evidence available at this time, *Betula* species appear to be the most likely candidates to be present in the seed bank for an extended period of time. Granström (1987) concludes, however, that seeds of these species will probably not last much longer than 5–6 years in the seed bank.

Studies that have examined buried tree seed populations in the forest floor of boreal forest stands (Johnson 1975; Archibold 1979; Granström 1982; Payette, Deshaye & Gilbert 1982; Perala & Alm 1989) are not conclusive as to the importance of buried seed of these species in forest recovery following disturbance. The issue is complicated by factors such as predation, effects of insects and disease, inherent dormancy patterns of seeds, and the effects of environmental conditions on dormancy. Herb and shrub species (e.g. *Geranium* spp, and *Rubus* spp.) are good examples of seeds which are stored in the forest floor for many decades. For a more detailed consideration of this matter, refer to Granström (1987).

Black spruce presents a special case with regard to seed bank considerations because of the large quantities of seeds retained in semiserotinous cones for significant periods of time (up to 25 years in some cases) (Haavisto 1975). Although these seeds are gradually dispersed, fire triggers rapid dissemination of large quantities of seeds to potentially receptive seedbeds. Thus, one might consider the black spruce seed bank as 'suspended' rather than buried. This seed bank is also dynamic in that new seeds are added by annual seed crops and seeds are depleted by dispersal and predation.

Rather than a seed bank, some species may have a seedling bank. Payette, Deshaye & Gilbert (1982) proposed that seedling banks are more important than seed banks in the maintenance of tree populations in northern Quebec. They concluded that seedling banks are particularly important in areas where seed production is low. They suggest that rapid germination reduces the probability of seed mortality and increases the probability of recruitment into the tree population. *P. glauca* is often represented by a seedling bank on sites in interior Alaska. The presence and distribution of seedlings is related to seedbed conditions, microtopography of the forest floor, disturbance history in the stand, and seed availability (Zasada 1986).

Germination

Germination patterns are determined by the time of seed dispersal and environmental conditions (especially soil surface temperature and moisture). Spring- and summer-dispersed *Populus* seeds are not dormant and can germinate completely at any temperature between 5 and 30 °C, with only the rate of germination being affected by temperature. *Picea*, *Betula*, and *Pinus* seeds do not germinate at low temperatures occurring at the time of dispersal (Schopmeyer 1974; Densmore 1979). Germination of fall- and winter-dispersed seeds (Fig. 3.5) can occur earliest (Fig. 3.8). On warm sites in interior Alaska, germinants may appear in early to mid-May before *Populus* seed dispersal has started (Ganns 1977;

Zasada *et al.* 1978; Zasada & Wurtz 1990). On colder sites or in colder years, germination may be delayed for up to a month in these species (Clautice, Zasada & Neiland 1979; Black & Bliss 1980; Zasada *et al.* 1983; Walker, Zasada & Chapin 1986). The seasonal course of germination depends on the environmental conditions, particularly surface soil water content. If moisture is readily available, germination can be completed within several weeks to a month (Fig. 3.8, pattern 1). However, if water is not readily available, germination will stop or drop to low levels before increasing after the surface soil re-wets from summer rainfall (Fig. 3.8, pattern 2) (Zasada *et al.* 1978). The distribution of germinants on relatively uniform mineral soil seedbeds varies in both time and space as a result of the interaction of individual seed dormancy, seedbed microsite conditions, depth of seed burial, and random variation in seed rain (Fig. 3.9).

The summer-dispersed seeds of *P. tremuloides* and *P. balsamifera*, as well as those of other Salicaceae species, can begin to germinate within a matter of hours after landing on a suitable seedbed; germination can be

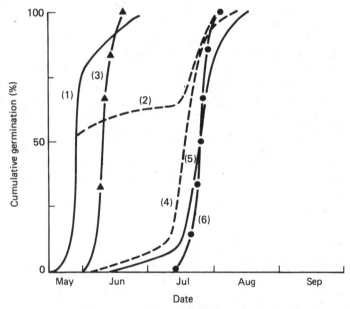

Fig. 3.8. Generalized field germination patterns for selected boreal forest trees. The two patterns for *P. glauca* and *B. papyrifera* indicate patterns observed under different summer precipitation regimes. Numbers indicate: (1) *P. glauca* and *B. papyrifera* with no extended dry period; (2) *P. glauca* and *B. papyrifera* with dry period; (3) *P. tremuloides*; (4) *P. abies*; (5) *P. sylvestris*; (6) *P. balsamifera*.

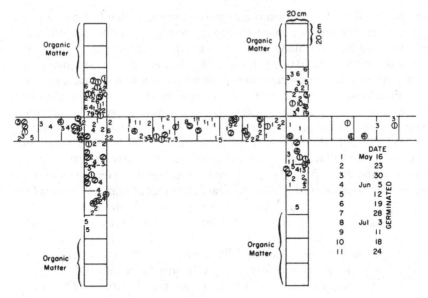

Fig. 3.9. Spatial and temporal distribution of *P. glauca* germinants on a mineral soil seedbed. Dates indicate when germinant was first observed. Circled numbers indicate germinants that died during the growing season. The greatest time separating the appearance of the first and last germinants in an individual 0.04 m² area was about 8 weeks (May 16–July 18).

completed within 24–48 hours (Densmore & Zasada 1983; Krasny, Vogt & Zasada 1988). Zasada *et al.* (1983) observed germinants of both *Populus* species over a period of about 1 month on plots sown artificially, indicating that seeds may remain viable for longer than generally believed; however, 90% or more of the germinants appeared within 2 weeks of sowing. The time at which individual seeds are dispersed within the annual dispersal cycle can be an important factor in the germination success of *Populus* seeds (Walker, Zasada & Chapin 1986).

Although seedlings are found growing on a variety of surfaces, there are large differences among surfaces in the efficiency of germination (total germination, rate of germination, and the number of seeds required to produce a germinant) on different substrates, and between species in their ability to utilize the range of substrates and microsites available (Yli-Vakkuri 1963; Waldron 1966; Black & Bliss 1980; Zasada *et al.* 1983; Walker, Zasada & Chapin 1986; Krasny, Vogt & Zasada 1988). On burned, upland, *P. mariana* sites in interior Alaska, heavily burned (exposed or ash-covered mineral soil) and moderately burned (residual organic matter burned to varying depths but mineral soil not exposed) substrates produced germinants, whereas lightly burned (organic layer scorched but little reduction in depth), and unburned surfaces did not produce germinants. Seed to germinant ratios on the heavily burned surface were 20 or less for *B. papyrifera* and *P. mariana* and 70 or more for the *Populus* species; on the moderately burned surfaces, seed to germinant ratios were as much as an order of magnitude greater (Zasada *et al.* 1983). Walker, Zasada & Chapin (1986) observed that percent germination differed significantly between *P. glauca* and *P. balsamifera* on natural and mineral soil flood plain surfaces in different stages of succession and that each species responded differently to surface condition in the five different successional stages studied. Germination was generally more successful in the two earliest stages of primary succession. Seedbeds in these stages were primarily exposed silt that had been recently deposited by flooding. In the later stages of succession, seedbeds consisted of organic matter of varying composition and depth. Krasny, Vogt & Zasada (1988) found no difference in germination between *P. balsamifera* and *P. tremuloides* on a given early successional flood plain surface, but large differences (range in germination of 0–72%) were observed in germination success among the types of surfaces studied. Germination on these essentially mineral soil surfaces was most affected by moisture content and electrical conductivity of the seedbed. Based on these and additional reports (Densmore 1979; McDonough 1979; Simak 1980) and our experience, we would rank the genera considered in this discussion as follows with regard to their ability to withstand at least moderate levels of environmental stress and still germinate successfully: *Picea* and *Pinus* > *Betula* > *Populus*.

Mineral soil seedbeds generally provide the most stable conditions for germination and early seedling establishment (Yli-Vakkuri 1963; Lees 1964; Jarvis *et al.* 1966; Dobbs 1972; Zasada & Grigal 1978; Walker, Zasada & Chapin 1986; Krasny, Vogt & Zasada 1988). However, organic surfaces are the most common before and after disturbance in the boreal forest and germination can be excellent on the surfaces provided that adequate moisture is available (Yli-Vakkuri 1963). Organic surfaces are highly variable and include, among others, rotted wood, various moss species (sphagnum mosses compared to feather mosses), and different types of litter and partly decomposed organic matter. Seedlings, particularly those of *Picea*, are found on various types of organic matter, particularly rotted wood, under natural stand conditions. Hellum (1972) found that *P. glauca* germination was 2 to 3.5 times greater on rotted *Picea*, *Populus* and *Betula* wood than it was on sphagnum peat or mineral soil under controlled field conditions. In field studies with *P. glauca* (Jablanczy & Baskerville 1969; Densmore 1979; Putman & Zasada 1986; Walker, Zasada & Chapin 1986), *P. mariana* (Johnston 1972; Black &

Bliss 1980; Zasada *et al.* 1983), *B. papyrifera* (Zasada & Grigal 1978; Zasada *et al.* 1983; Zasada & Wurtz 1990) and *P. balsamifera* (Zasada *et al.* 1983; Walker, Zasada & Chapin 1986), seedlings were found on various organic substrates. These studies and our field observations suggest that more detailed studies are needed to determine the types of organic seedbeds that are most likely to be receptive to germination and seedling establishment, some measure of seeding efficiency (e.g. seed to seedling ratios), and the relation of temperature and water availability to germination.

Mortality during germination can be significant. Because the germinating seeds and seedlings are small and succulent, they are susceptible to such abiotic factors as small-scale erosion and small changes in seedbed moisture and temperature conditions: factors that play little or no role in larger seedlings. Biotic causes of mortality at this stage include fungi (damping off), small mammals, birds, and insects.

Germination is not normally considered to be a limiting factor to tree regeneration. These conclusions are most often based on observations made on better seedbed types. However, on poorer seedbed types, both mineral soil and organic matter, germination may not occur as readily, and the pattern of germination may differ from that on good seedbeds. Before the overall influence of germination on reproductive success can be determined, information must be obtained on the reaction of different species to different seedbed types, the seed to germinant ratio on different seedbed materials, and the influence of seedbed type on initiation and pattern of germination.

Seedling establishment

Survival

Seedling survival can vary greatly among species (Lees 1964; Ganns 1977; Zasada *et al.* 1978, 1983; Clautice, Zasada & Neiland 1979; Black & Bliss 1980; Putman & Zasada 1986; Walker, Zasada & Chapin 1986; Zasada & Wurtz 1990; Youngblood & Zasada 1991) and is related to a host of independent and interacting biotic and abiotic factors whose importance may vary among sites and years and by species. First growing-season survival varies among cohorts in some cases but not in others (Fig. 3.10) (Waldron 1966; Zasada *et al.* 1978, 1983; Clautice, Zasada & Neiland 1979; Black & Bliss 1980). Success or failure of a particular cohort is determined by environmental conditions, commencing at the time of germination, and operating during the first growing season.

Survival during the period of establishment is generally characterized by a sharp decline in seedling density during the first two growing seasons and often even more specifically during the year of germination (Fig. 3.11).

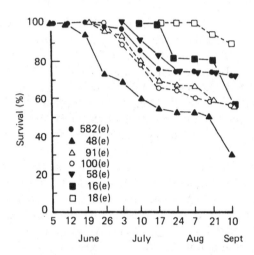

Fig. 3.10. Survival of *B. papyrifera* germinant cohorts on an upland *P. mariana* site in interior Alaska (adapted from Zasada *et al.* 1983). The date of cohort appearance is indicated by the first point for each hue. Numbers indicate size of cohort.

In the examples of *P. glauca* survival, winter mortality was significant in one case but not in the other two. In the comparison of *P. glauca* and *B. papyrifera*, survival for the two species was virtually identical after 5 years, but the timing of mortality differed in that *B. papyrifera* mortality tended to be fairly uniform over the period of observation while that of *P. glauca* occurred exclusively during the first summer and winter with little reduction in survival after the start of the second growing season (Fig. 3.11).

Regeneration of a disturbed site often occurs over a period of time rather than all at once (Gardner 1986). As a result seedlings on the same site but produced by different seed crops can have different probabilities of survival. In the example shown in Figure 3.11 (*b*) and (*c*), the seedlings produced in 1971 from the 1970 seed crop germinated on mineral soil that was exposed in the autumn of 1970; seedlings produced by subsequent seed crops germinated on seedbeds which were covered by varying amounts of litter and shade from plants developing after mineral soil exposure. This altered environment usually results in a different seed to seedling ratio and different probabilities of survival. Gregory (1966) and Waldron (1961) provide information on some limitations of *P. glauca* establishment under different types of vegetation.

The ratio of the number of seeds necessary to produce a seedling of a given age is a general measure of the difficulty of establishment on a given site. This ratio also provides a means of linking seed production with the

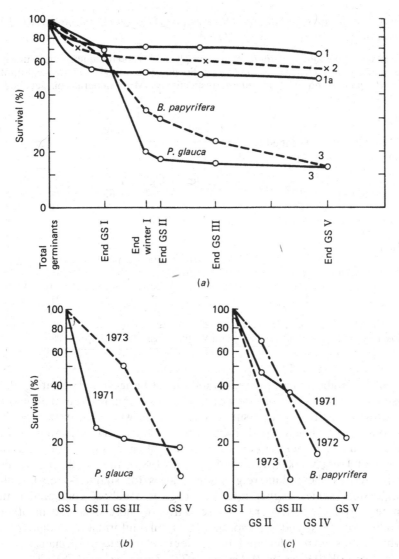

Fig. 3.11. *P. glauca* and *B. papyrifera* seedling survival on forest sites in interior Alaska. (*a*) *P. glauca* survival on newly created mineral soil seedbeds on upland and flood plain sites. Data for 1 and 1a from Ganns (1977) for mineral soil and mixed mineral soil-organic matter seedbeds on Tanana River flood plain sites. Data for 2 and 3 from Zasada & Wurtz (1990) and Zasada *et al.* (1978), respectively, for two upland sites. (*b*) *P. glauca* seedling survival on the same seedbeds for seedlings originating from the 1970 and 1972 seed crops (viable seeds in each year 400–600/m^2). Seed crops in 1971 (18 viable seeds/m^2) and 1973 (20 viable seeds/m^2) produced no seedlings. (*c*) *B. papyrifera* survival on the same seedbeds for seedlings originating from three consecutive seed crops.

Table 3.5. *Seed to seedling ratio for* P. glauca (*Pg*) *and* P. balsamifera (*Pb*) *on flood plain sites in interior Alaska*

The five vegetation types represent a primary successional sequence from newly formed siltbar to closed *P. glauca* forest. Sowing rate was 606 and 828 seeds/plot for *P. glauca* and *P. balsamifera*, respectively. MS = mineral soil seedbed; UT = untreated seedbed.

Vegetation type	Seedbed type	Seed:seedling ratio; growing season					
		0		1		2	
		Pg	Pb	Pg	Pb	Pg	Pb
Vegetated silt	MS	4	3	4	828	24	>828
	UT	8	4	7	69	21	>828
Salix	MS	6	4	15	882	43	>828
	UT	6	6	9	118	32	>828
Alnus incana	MS	36	6	200	828	151	>828
	UT	8	414	200	>828	>600	>828
P. balsamifera	MS	11	10	50	>828	151	>828
	UT	86	118	200	>828	>600	>828
P. glauca	MS	67	3	76	52	121	>828
	UT	303	44	200	>828	600	828

Source: Data from Walker (1985) and Walker, Zasada & Chapin (1986).

seedling establishment phase. Caution must be used in calculating and using these ratios because they do not reveal anything about the factors that are actually affecting survival, and these will often differ greatly among sites and years. In Table 3.5, these ratios were calculated for *P. glauca* and *P. balsamifera* growing on floodplain sites in interior Alaska (Walker 1985; Walker, Zasada & Chapin 1986). In general, the trends exhibited in these data (e.g. higher ratios for summer- versus fall-dispersing species, higher ratios for organic seedbeds compared with mineral soil seedbeds) are representative of those observed in other studies with these and other species. For similar information or data from which ratios can be calculated, the reader is referred to Ackerman (1957), Waldron (1966), Eis (1967), Ganns (1977), Zasada *et al*. (1978), Clautice, Zasada & Neiland (1979), Black & Bliss (1980), Zasada *et al*. (1983), Fries (1985), Putman & Zasada (1986) and Krasny, Vogt & Zasada (1988).

Seedling growth and development

Seedling development is affected by many biotic and abiotic factors and a full treatment of all parameters is beyond the scope of this paper. In order to fully describe growth during the seedling establishment period more than one measure of growth must be used. Brand & Janas (1988) concluded that a variable such as height or height increment could be used to assess late spring growing conditions while leaf length, basal area, and bud size might be used for early, mid-, and late summer growing conditions. We have chosen to compare height growth potential under good growing conditions, realizing that it may not be the best parameter to characterize growth. It is, however, one of the most commonly measured variables and one of the easiest to observe and measure. In addition, we will briefly summarize several studies that compared seedling height and height growth with other measures of growth to show how height compares with other variables in describing seedling growth.

The broadleaved species considered here grow much more quickly than the conifers (Fig. 3.12). Among the conifers, *P. sylvestris* grows more rapidly than the *Picea* species. On sites where it normally occurs, *P. mariana* grows most slowly, usually requiring 15–20 years to attain 1.5 m. When grown under less than desirable conditions, for example seedlings comprising an understorey seedling bank, height growth will be much slower. It is not uncommon to find *P. glauca* and *P. mariana* seedlings, in the understorey of conifer and hardwood stands, that are less than 1.5 m tall and more than 40 years old.

The relation of seedling height to other seedling parameters is compared below for several studies. Krasny, Vogt & Zasada (1984) reported that total height and annual growth of *P. glauca* seedlings did not differ among successional stages on a flood plain site in Alaska, but shoot

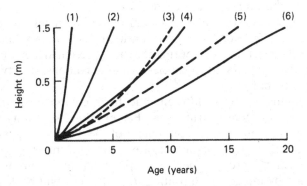

Fig. 3.12. Generalized height growth for open grown vegetative regeneration and seedlings of selected boreal forest tree species. Data for *Pinus sylvestris* and *Picea abies* from Ilvessalo (1917) and Lakari (1921) respectively. Numbers indicate: (1) *Betula* and *Populus* vegetative regeneration; (2) *Betula* and *Populus* seedlings; (3)–(6) *P. glauca*, *P. sylvestris*, *P. abies* and *P. mariana* seedlings, respectively.

biomass was 7–15 times greater and root biomass 8–70 times greater in open areas than on sites dominated by *Alnus*, *Salix* or *Picea*. Furthermore, the number of mycorrhizal root tips was as much as 10–20 times greater in the open areas when compared with those growing in an understorey dominated by trees or shrubs. Van den Driessche (1982) found that the height of year-old *P. glauca* seedlings decreased with increased growing space while diameter increased; root activity as measured by root growth capacity did not differ among seedlings with different amounts of growing space. Height, stem biomass, root biomass, mean leaf mass, and internode length increased in *P. tremuloides* and *P. balsamifera* seedlings as density declined in pure stands of these species (Morris & Farmer 1985; J. C. Zasada, unpublished data).

Vegetative regeneration

Vegetative regeneration in the form of clone expansion and vegetative regrowth from basal buds, roots, or stem segments following disturbance is important in primary and secondary succession in the boreal forest. Depending on species composition, site conditions, and severity of disturbance, reproduction from seed and vegetative regeneration may occur simultaneously on the same site. Stems originating from the two sources may have different temporal and spatial patterns. Competition for site resources will occur between the two, but vegetative regeneration may also serve as a nurse crop for seedlings and, thus, be a positive factor in seedling establishment on some sites.

The potential for vegetative regeneration is based on the species' ability to develop from dormant or adventitious buds present in the bud bank (Fig. 3.2). The relative importance of buds present in these different parts of the bank is related to species, stand conditions, and type and severity of disturbance.

Stands originating from different bud bank components will have different ramet or stem distributions. For example, sprouting from a basal bud bank results in the replacement of the dead tree in the same physical location; the form of the tree often changes from a single stem to a multiple-stemmed group. There is no potential for expansion of the area occupied by the genet. Sprouting from the root bud bank has the potential for producing new stems at any point within a site where roots with buds or the potential to produce buds occur. Stem production from the root bud bank results in a different spatial distribution of stems in the new stand. The area occupied by the genet or clone can increase immediately after disturbance and continue over time depending on site conditions and a species' growth characteristics.

Bud origin

Basal buds

The broadleaved species have the ability to regenerate from basal buds. The greatest capacity (in terms of individuals with the capacity to resprout and not the number of sprouts produced) appears to occur early in development and declines with age; the rate of decline is species dependent and may also be related to environmental conditions (Fig. 3.13).

P. tremuloides is the most limited in its ability to regenerate from basal buds. Perala (1979), working in Minnesota, reported that maximum resprouting occurred in 4-year-old stems, which themselves were of root bud origin. In Alaskan *P. tremuloides*, about 90% of 10- to 12-year-old stems produced one or more stump/root-collar sprouts after the parent stem was cut; in 25-year-old aspen, about 80% of the stems resprouted. In stems older than 50 years, less than 1% of the stems resprouted following cutting (J. C. Zasada, unpublished data). Weber (1990) reported that basal and root-collar sprouting predominated in 20-year-old *P. tremuloides* stands in Ontario.

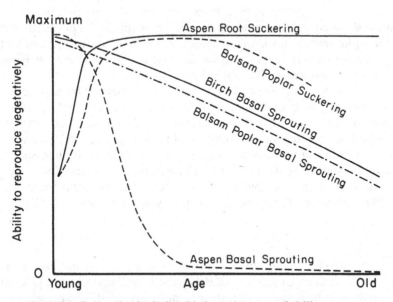

Fig. 3.13. Generalized relationship between age and ability to regenerate vegetatively. Based on unpublished data and experience from interior Alaska and information from other parts of the species ranges (see, for example, Perala 1979; Safford, Bjorkbom & Zasada 1990).

In *B. papyrifera* growing on upland sites in interior Alaska, 90–100% of trees less than 50 years old resprouted following death of the parent stem by disturbances such as cutting or fire. After age 50, the ability to produce basal sprouts declines; at age 125, 30–40% of the trees retain the ability to resprout (J. C. Zasada, unpublished). This decline in ability to sprout with increasing age appears to occur throughout the range of *B. papyrifera* (Safford, Bjorkbom & Zasada 1990) (Fig. 3.13).

The formation of the bud bank in *Betula* begins with development of the buds associated with the cotyledons and other buds formed on the lower stem of seedlings. Stone & Cornwell (1968) observed that basal buds in *B. populifolia* seedlings grew very slowly and, after several years, were flanked by daughter buds formed in the axils of the bud scales. Kauppi, Rinne & Ferm (1987) observed that basal buds of *B. pubescens* exhibited a short period of growth during which new buds formed at the base of the primary buds. These buds become embedded in the bark and may give rise to more buds. The primary buds may eventually die but will be represented by the secondary buds produced as a result of their growth. *B. pubescens* and *B. pendula* seedlings also have the ability to produce adventitious buds if the basal buds are destroyed or do not develop normally (Rinne, Kauppi & Ferm 1987). In mature trees of these species, stems formed from basal buds were formed from primary buds or those associated with them, suggesting that the capacity for production of adventitious buds is low in mature trees. For a detailed treatment of bud bank formation and dynamics in *Betula* refer to Kauppi, Rinne & Ferm (1987, 1988) and Rinne, Kauppi & Ferm (1987).

Mature *P. balsamifera* produced sprouts from dormant buds as well as from adventitious buds originating in the inner bark on the top of the stumps of harvested trees. Basal sprouting from dormant and adventitious buds occurred simultaneously on the same stumps. Production of shoots from adventitious buds occurred more consistently than shoots from basal buds (Zasada *et al.* 1981). The effects of age on the ability of *P. balsamifera* stumps to sprout is not well established, but trees older than 75–100 years have a lower sprouting capacity than trees less than 50 years old.

Root buds

P. tremuloides and *P. balsamifera* produce shoots from adventitious buds developed on the root system. These adventitious buds can form prior to disturbance or be initiated after the parent tree is killed. The presence of preformed buds is a function of genetic and stand history variables, and appears to result in more rapid sucker production in clones where they occur (Schier 1973a,b; Zasada & Schier 1973; Tappeiner 1982). Where the species are found on the same sites in interior Alaska, *P. tremuloides* seems to have a higher capacity for sucker production on intact root systems. Root pieces from *P. balsamifera* have a higher capacity for the initiation of new roots than *P. tremuloides*, indicating a greater potential for establishment following separation from the parent plant and transport, as occurs on flood plain sites. The origin of new shoots on root sections may differ between these species (Schier & Campbell 1976). Ability of seedlings to produce adventitious roots is greater in *P. balsamifera* than in *P. tremuloides* (Krasny, Zasada & Vogt 1988).

Production of root suckers can occur on the root systems of vigorous 1- to 2-year-old seedlings. However, the evidence presented by Perala (1979) suggested that *P. tremuloides* does not reach maximum sprout production capacity from the root bud bank until after its ability to basal sprout has declined. Rapid re-establishment of apical dominance by sprouts from the basal bud bank probably reduces the potential for sprout production from the root bud bank. In a 14-year-old *P. balsamifera* of seed origin growing on a flood plain site in interior Alaska, one sucker was produced from the time of germination to age 5 years, 14 from 6 to 9 years, and 8 from 10 to 14 years. The potential for root-bud bank expansion can be fairly great. In the above-mentioned seedling, root elongation was about 3 m during the first 5 years, 35 m from 6 to 9 years, and about 21 m from 9 to 14 years (Krasny, Vogt & Zasada 1988). Suckering capacity will remain high as long as the root system remains healthy and is capable of producing new roots (Schier & Campbell 1980). In Alaska, we have observed prolific sucker production after harvesting in 150-year-old stands. Repeated cutting at short intervals results in a decline in shoot production (Perala 1979). Cutting in summer results in lower sucker production than does winter cutting (Bella 1986).

Stem and branch buds

Vegetative regeneration can occur through layering of branches attached to living trees or from root and shoot production on branch segments broken from the tree in some way. *P. glauca* and *P. mariana* form clones by layering only while the parent stem is living. Layering is important in *P. glauca*, mainly at the treeline, but occurs throughout the range of *P. mariana* (Stanek 1968; Elliott 1979*b*; Densmore 1980; Legere & Payette 1981). Densmore (1980) observed that the density of *P. glauca* produced by layering was greater on a north aspect at the treeline (1830 layering ha^{-1}) than on south slope (200 ha^{-1}) and flood plain sites (110 ha^{-1}) in the upper Dietrich River Valley, Brooks Range, Alaska. Stanek (1968) has reported that stems in undisturbed *P. mariana* stands originated largely by layering. Site conditions determine the importance of layering in *P. mariana*. Lowland sites with sphagnum moss support more layering than drier, upland sites where feathermosses predominate (Sims *et al.* 1990).

Legere & Payette (1981) provide a good example of how *P. mariana* has maintained its presence by clonal growth on a forest–tundra transition site in northern Quebec. Layering also occurs in *P. balsamifera* but is not as important in this species because of its ability to regenerate vegetatively in other ways.

P. balsamifera is the only one of the species considered here which regenerates from detached pieces of stem or branch. This greatly increases the potential bud bank in this species and makes it possible for long-distance movement of clones through the dispersal of vegetative material. The greatest potential for long-distance dispersal occurs on flood plains, where broken branches are readily transported by water. Zasada *et al.* (1981) observed shoot and root production on stem pieces ranging from 6 to 300 cm long and from 11 to 50 cm in diameter. They also observed that branches broken and buried by winter logging activity produced more shoots than those buried as a result of summer logging activity. *P. balsamifera* does not appear to be able to regenerate from specialized shoots, having the capacity to produce new stems, which are produced and shed annually (termed cladoptsis) as has been observed for *P. trichocarpa* (Galloway & Worrall 1979).

Bud bank depletion

The bud bank is depleted through death of buds resulting from internal (e.g. bud longevity considerations) and external factors (e.g. browsing and disease) and activation of buds resulting from normal seasonal development or stimulation of dormant buds by disturbance. Kauppi, Rinne & Ferm (1987, 1988) have documented the occurrence of bud death in *Betula*. Schier (1975) and Schier & Campbell (1980) discussed *P. tremuloides* clone deterioration in the western United States and outlined the factors which might result in reduction of the root bud bank. In *P. glauca* and *P. mariana*, death of the lower branches eliminates the potential for layering; lower branches in *P. glauca* appear to have a shorter life than those of *P. mariana* as layering in the latter species is much more widespread than in the former.

Disturbance to individual trees or declining vigor of trees in stands results in the activation of buds. Girdling of stems by hares and removal of varying amounts of the stem through browsing by hares, moose, and beavers results in the outgrowth of basal buds. In mature and overmature *Betula* stands, basal sprouts are present on some trees that appear healthy, suggesting that the bud bank is under less internal control in some older trees. Tappeiner (1982) observed evidence of numerous dead shoots on the roots of *P. tremuloides*, indicating that on some sites suckers are frequently initiated but few if any survive. Insect defoliation of *P. tremuloides* stands may result in increased sprout production from the root bud bank in the understorey of stands of this species.

The best information on the depletion of the bud bank and estimates of its size is gained from the response to major disturbances such as fire and harvesting (Perala 1990; Safford, Bjorkbom & Zasada 1990; Zasada & Phipps 1990). Although response to disturbance provides an estimate of the potential, the number of buds or bud production potential may be greater. For example, Schier (1972) reported that developing shoots on *P. tremuloides* root systems re-established apical dominance. By removing the developing shoots, Schier (1972) was able to increase bud development by a factor of 2 over that on roots where developing suckers were not removed. Rinne, Kauppi & Ferm (1987) also suggest that basal sprout development in *Betula* suppresses development of other buds.

In *P. tremuloides* stands on upland sites in interior Alaska, sucker production following death of the parent stand by fire or removal by forest harvesting varies from about 3 to 35 suckers per square meter. The amount of residual organic matter affects root bud bank sprout (suckers) density (Fig. 3.14) (Bella 1986). Bella (1986) observed a reduction in sucker density of between 33 and 67% depending on the quantity and size of logging residue. About four times more suckers were produced on sites where all residual organic matter was removed than on sites where various amounts of logging residue remained (Fig. 3.14). The potential size of the bud bank in Alaskan *P. tremuloides* has been observed to be as high as 3 buds capable of producing shoots per centimeter of root in the diameter range of 1–2 cm (Zasada & Schier 1973). Combining this with Schier (1972), the bud bank for this species may be as high as 6 buds per centimeter of root.

Species that produce basal sprouts have the potential for producing dense multistem groups on individual stumps. On sites in interior Alaska, we have observed dense *B. papyrifera* and *P. balsamifera* basal sprout clumps with up to 200 sprouts, but most trees produce fewer sprouts. In *P. balsamifera*, 20–67% of cut trees produced more than 25 sprouts, 8–28% produced 10–25 sprouts, and 13–52% produced fewer than 10 sprouts (Zasada, Viereck & Foote 1979). In *B. papyrifera* stands on upland sites in interior Alaska, initial sprout production in a 50-year-old stand was as follows: 20% of stumps produced more than 25 sprouts after trees were cut in May, 33% produced 10–25 sprouts and 47% produced fewer than 10 sprouts. Ninety-eight per cent of the stumps in this stand sprouted. In two 125-year-old stands, 50% of the trees cut sprouted and sprout production was: 55% produced more than 25 sprouts; 9%, 10–25; and 36%, fewer than 10 sprouts (J. C. Zasada, unpublished). These observations indicated that the production of sprouts from the basal bud

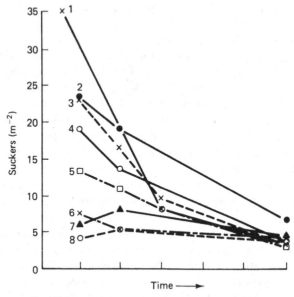

Fig. 3.14. Survival of shoots produced from *P. tremuloides* roots on upland sites in interior Alaska following disturbance. *Sources* are as follows. 1 From Viereck & Foote (1985 and unpublished); 2,6–8, all from same site; 2 had all logging debris and forest floor removed; 6–8 represent removal of different amounts of logging debris but no forest floor removal. 3 From Viereck & Dyrness (1979); 4 and 5, from harvested upland sites with no slash treatment.

bank in this species was highly variable within a stand for trees of the same diameter and age.

Survival and growth of vegetative regeneration

Survival

Survival of basal sprouts needs to be considered at two levels: survival of individual genets (parent tree) and shoots within a genet. Genet survival depends more on predisturbance vigor of the genet and the effect of disturbance on the portion of the stem remaining; sprout survival within genets, though dependent on genet vigor, will also be determined by stem density. In both *P. balsamifera* and *B. papyrifera*, genet survival differed among stands growing on similar sites (Zasada *et al.* 1981; J. C. Zasada, unpublished data). For *P. balsamifera*, survival was lower after 4 years on areas that had been harvested during the summer (9–14%) than on those harvested in fall and winter (49–64%). For birch, there was little difference in genet survival among trees harvested at different times during the growing season. However, genet survival in a 50- to 60-year-old stand declined from 89 to 70% and, in two 125-year-old stands, from 53 to 47% over a 7-year period (J. C. Zasada, unpublished data).

Survival of shoots originating from basal or root buds depends to a significant, but variable, degree on density of the shoots. In *P. tremuloides*, density of root origin sprouts after 6 years was almost identical even though initial density varied from about 3 to 35 suckers per square meter (Fig. 3.14). Bella (1986) reported similar levels of mortality over a range of initial densities for *P. tremuloides* in Saskatchewan.

Survival of sprouts from basal buds (within-genet) declined to varying degrees over time. In *P. balsamifera*, the extremes in survival over a 4-year period among 5 study areas were a decline in survival of more than 60% in the two highest sprout production classes (10–25 and more than 25 sprouts per stump) in the poorest case and in the best case only a 10% reduction in these classes (Zasada *et al.* 1981). In a 50-year-old *B. papyrifera* stand, the number of stumps in the 10–25 and greater than 25 sprouts per stump classes declined from 55 to 8% over a 6-year period. Decline in sprout survival in these classes in 125-year-old stands was from 64 to 24% (J. C. Zasada, unpublished).

Survival of stems originating from stem, root, and branch segments and buried by natural processes, such as silt deposition following flooding, has not been documented to our knowledge. The main information available is from studies with unrooted cuttings on disturbed sites, for example in intensively cultured plantations and burned or harvested sites. In the intensively cultivated *Populus* plantations, survival usually approaches 100%, indicating the high potential for survival that does exist. Under uncontrolled field conditions, survival is normally much lower (60% or less) and seems to be closely related to substrate quality, with mineral soil supporting higher survival than organic matter (Zasada, Holloway & Densmore 1977; Zasada *et al.* 1983; Zasada & Phipps 1990).

Growth and development of vegetative regeneration

Growth of basal sprouts and root suckers is rapid in areas where parent stems are killed by disturbances such as fire and forest harvesting and where resources necessary for growth are readily available (Fig. 3.12). Under these conditions vegetative reproduction will always grow more rapidly than seedlings of the same species.

Vegetative reproduction will attain heights of 1–2 m after one growing season, 3–5 m after 5 years and 7–10 m after 10–15 years. In *P. tremuloides*, the time of year during which disturbance occurs and the quantity of residual material on the forest floor can affect height growth as well as density of vegetative reproduction (Fig. 3.15). Generally, growth is better when death of the parent stem occurs during the dormant season; these effects may last for a decade or more (Bella 1986; Weber

1990). Height of vegetative regeneration can be significantly affected by browsing: low levels have little lasting effect whereas repeated browsing can reduce growth substantially.

In young *P. tremuloides* stands in north central Minnesota, which originated from the root bud bank, response to thinning and fertilization was followed for 11 years. At the end of this period, stem height was 20% greater in treated than in untreated stands. Basal area of stems responded more dramatically, increasing by about 170% in response to treatment; diameter increased by about 70% (Perala & Laidly 1989).

Reproductive processes and response of boreal trees to disturbance

Forest development in the boreal region is a complex process. Among all of the factors which interact to determine the spatial and temporal variation in forests at both large and small scale levels of resolution, we believe that reproductive considerations are of critical importance. No site can support species whose seeds or vegetative bud bank are not available for colonization or regrowth following disturbance. If the reproductive material of a desired species is not available soon after disturbance, the probability of that species being an important component on a particular site is reduced.

We will briefly consider forest reproduction following two types of disturbance: fire and forest harvesting. From a historical perspective both of these disturbances have been present in boreal forests since man inhabited them. Fire resulting from lightning and man's activities has decreased or remained about the same in importance depending on the intensity of forest management and the population density in and around forested areas. The tendency has been to reduce the frequency and extent of natural fires where possible. Forest harvesting, on the other hand, has increased greatly in frequency, extent, and intensity during the past 50–75 years when compared with historical use by native peoples. Boreal forests in the Scandinavian countries are some of the most intensively managed in the world; in other parts of the boreal zone there is little or no forest management other than that which occurs in the vicinity of small villages.

Fires in the boreal forest were characteristically stand-replacement fires. The mature trees of all species are relatively susceptible to fire; all of the above-ground stem, with the exception of the bud bank in some hardwoods, is killed by light to moderate underburning. In interior Alaska, for example, fire cycles are believed to be as low as 30 or as high as 400 years. The length of the fire cycle varies among geographic regions and sites within regions. *P. mariana* forests tend to have the shortest fire cycles and *P. glauca* longer cycles. Broadleaved deciduous species tend to sustain fires of lower intensity than conifers (Dyrness, Viereck & Van Cleve 1986).

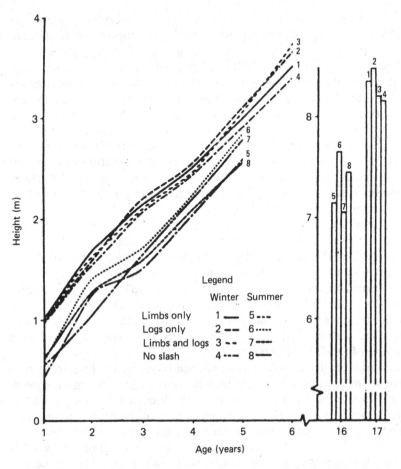

Fig. 3.15. Average height of tallest aspen by slash conditions (from Bella 1986 with permission).

The annual pattern of fire occurrence is determined by the weather conditions both prior to and during a given summer. As a result there can be great annual variation in the number of fires and the area burned. The characteristics of areas burned by fires can be highly variable and depend on the site, stand, and weather conditions at the time of burning. In spite of the potential for large variability among fires, two generalizations can be made. First, within the perimeter of most fires there are usually unburned or very lightly burned islands where all or a large percentage of the trees survive. Secondly, there is usually a large degree of variation in the amount of forest floor consumed by the fire. Both of these conditions significantly affect reproductive potential.

Predicting the potential for natural forest recovery on a site either before or after burning will depend on obvious factors such as distance to a suitable seed source, seedbed conditions, species composition before burning, and weather conditions after burning. Several examples of relatively subtle, yet important, considerations are time of the summer during which burning occurs in relation to phenological development (for example, the degree of seed maturity at the time of burning) and physiological activity (for example, the level of carbohydrate reserves available for sprout production and growth), intensity of the fire as it may affect rate of cone opening and seed viability in semiserotinous cones of *P. mariana*, the intensity and pattern of forest-floor burning as it relates to the small-scale pattern for seed reproduction and vegetative regrowth, and weather during years following the fire as it affects seed production and dispersal, and seedling establishment.

Post-fire recovery always consists of seed reproduction and vegetative regeneration. The impact of the fire on the various compartments shown in Figs 3.1 and 3.2 will determine the relative importance of each. The relative importance of seeds and the vegetative bud bank often varies on a small scale (meter by meter or less in some cases). Trees and associated species originating from these different sources will compete for space and resources and thus directly or indirectly affect species composition and rate of forest development.

Natural regeneration (that which depends on seeds or buds from trees on or adjacent to the disturbed site) following forest harvesting has many of the same considerations as were briefly discussed for post-fire recovery. However, there is more potential to control the degree, timing, and intensity of disturbance, seed availability, bud bank condition, and post-harvest environment by managing the harvest operation or by post-harvest treatments to improve conditions for one or more target species. Seed source considerations are critical because once a seed source is removed or a good seed year passes it may be impossible or much more difficult to have an adequate supply of seed for rapid regeneration. Regeneration methods can be designed to remove seed trees immediately after seed dispersal or (more commonly) to leave a seed source distributed over a site, as in a shelterwood, or adjacent to the harvested area as in clearcutting. Species with serotinous cones, such as *P. mariana*, have advantages over non-serotinous species in having some seeds always available. Branches containing cones can be distributed over the site to assure good seed distribution. Broadleaved species generally produce more seeds per seed crop, have more frequent seed crops, and disperse seeds farther than conifers, and tend to regenerate more readily.

Efficient use of the available seeds is a key consideration when using natural regeneration methods. Because seedbed conditions immediately after harvesting are often poor in terms of seed germination and seedling establishment, various methods have been used to improve them. Prescribed burning can be used to improve seedbeds, but it often does not create acceptable conditions. Mechanical methods, which include the use of specialized plows, discs, and patch scarifiers, have been developed to create mineral soil seedbeds to promote more efficient regeneration.

Harvesting practices, regeneration methods, and post-harvest practices are usually designed to promote one particular species, and in most parts of the boreal forest conifers are preferred. *Betula* and *Populus* species, although of commercial value, are frequently viewed as less desirable because they compete with conifers for site resources. It is possible to discriminate against these species at the time of harvesting by removal of seed sources or attempting to alter the size and viability of the bud bank.

Summary and conclusions

There are many factors that affect the distribution of forests at large and small scale levels of resolution. Species distribution within the landscape is to a large extent determined by local climate and soil conditions. However, differences in density, spatial distribution, and relative composition among stands and within-stand distribution are affected to a major extent by the seed reproduction and vegetative regeneration characteristics alone and their interaction with site conditions.

Within the context of current key issues such as biological diversity, climate change, and long-term site productivity, we need to understand the reproductive process of all boreal species. It will be vital to understand the reproductive process of each species and how the individual components of the process are affected by different climate change scenarios. For example, a climate change scenario in which winter temperatures are relatively more affected than summer temperatures will have a different effect on reproductive potential than if the opposite were true. Increased knowledge of the reproductive process will enhance the predictive capabilities and realism of current and future models used to assess spatial and temporal dynamics of the boreal forest.

4 Soil temperature as an ecological factor in boreal forests

Gordon B. Bonan

Introduction

The circumpolar boreal forest landscape reflects a combination of factors unique to high-latitude environments. A short growing season, strong seasonal fluctuations in air temperature and day length, low solar elevation angles, cold soil temperature, the presence of permafrost, poorly-drained soils, a thick forest floor, low nutrient availability, and recurring forest fires are thought to interact to produce the wide range in stand productivity characteristic of boreal forests (Bonan & Shugart 1989). Low soil temperatures and permafrost are perhaps the factors most unique to high-latitude environments. In this chapter, I examine the role of soil temperature and permafrost as ecological factors in the boreal forest.

Environmental controls of soil temperature and permafrost

Over 50% of Canada and the Soviet Union are underlain by permafrost, that is, the thermal condition of soil when its temperature remains below 0 °C continuously for two years or more (Brown 1969, 1970; Brown & Pewe 1973). The existence of permafrost is the result of the historical and current state of the surface energy balance and geothermal heat flow (Lunardini 1981). Even if present energy conditions are not conducive to the formation of permafrost, it may still exist as relic permafrost if past conditions have been favorable. However, the current dynamics of permafrost depends on the current surface energy balance, which primarily reflects air temperature as modified to a secondary degree by solar radiation, vegetation, snow cover, and soil characteristics (Brown 1970; Lunardini 1981; Rieger 1983).

Air temperature and solar radiation are correlated with the heat load received on a surface. In Canada, the southern limit of permafrost corresponds with the $-1.1\,°C$ mean annual air temperature isotherm (Brown 1969). Between $-3.9\,°C$ and $-6.7\,°C$, permafrost is discontinuous but widespread. North of $-6.7\,°C$, permafrost is virtually continuous. Solar radiation produces significant effects secondary to the broad pattern imposed by air temperature (Brown 1969, 1970; Lunardini 1981). In the continuous zone, permafrost is thicker and the soil active layer is thinner on north-facing slopes than on south-facing slopes (Brown & Pewe 1973; Ryden & Kostov 1980). In the discontinuous zone, permafrost is found on poorly drained, north-facing slopes or level sites, but not on well-drained, south-facing slopes (Brown & Pewe 1973; Dingman & Koutz 1974; Viereck et al. 1983; Slaughter & Viereck 1986).

Vegetation significantly affects the soil thermal regime, shielding permafrost from heat during the summer when the air temperature is above freezing. This insulating property is an important factor determining soil temperatures, depth of thaw, and thickness of the permafrost layer (Brown 1963; Brown & Pewe 1973; Linell 1973; Moskvin 1974; Van Cleve & Viereck 1981; Dyrness 1982; Viereck 1982). Removal of the vegetation cover reduces the insulating effect, increasing soil temperatures, deepening the soil active layer, and thawing the underlying permafrost (Brown 1963; Brown & Pewe 1973; Linell 1973; Dyrness 1982; Viereck 1982).

A forest cover influences soil temperature and the occurrence of permafrost because the presence of trees reduces wind speed and the turbulent transfer of heat between the ground and the air (Brown 1969; Tyrtikov 1973; Rosenberg, Blad & Verma 1983). Trees also shade the ground surface from solar radiation. The result of these factors is that in the summer, forested soils are cooler than open areas and thaw later and less deeply (Izotov 1968; Linell 1973; Moskvin 1974; Viereck 1982; Chindyaev 1987), and a dense tree cover can maintain permafrost in an otherwise unstable thermal regime (Linell 1973). However, in the winter, forested soils are warmer relative to open areas (Tamm 1950).

The presence of a thick forest floor also significantly lowers soil temperatures. In interior Alaska, soil temperature and depth to permafrost are directly related to the thickness of the forest floor (Van Cleve & Viereck 1981; Dyrness 1982; Van Cleve et al. 1983a,b). For example, Van Cleve & Viereck (1981) found an average $37\,°C$ decline in growing-season soil heat sum at the depth of 10 cm for each centimeter increase in the forest floor thickness. The low bulk density and low thermal conductivity of the organic mat effectively insulate the mineral soil, lowering soil temperatures and maintaining a high permafrost table.

Seasonal variations in the thermal properties of the organic layer create conditions conducive to the formation of permafrost (Tamm 1950;

Brown 1963, 1970; Brown & Pewe 1973; Tyrtikov 1973). In the summer, the surface of the organic layer dries out. Consequently, the thermal conductivity, which is a linear function of water content (de Vries 1975), is relatively low and warming of the underlying soil is impeded. In the autumn, decreased evapotranspiration demands combined with excess precipitation create wetter conditions. Thermal conductivity is increased relative to that during the summer, and the forest floor offers less resistance to cooling of the mineral soil. The net result is that a greater amount of heat is transferred in the winter from the ground to the air than is conducted in the summer from the air to the soil. This net negative imbalance of heat is conducive to the formation of permafrost.

With its high reflectivity and low thermal conductivity, a snow cover protects the soil from oscillations in air temperatures (Tamm 1950; Lunardini 1981; Rosenberg, Blad & Verma 1983). A snow cover in the fall significantly restricts winter frost penetration; conversely, in the spring, the snow cover acts as an insulating blanket to delay soil thawing (Izotov 1958; Brown 1969, 1970; Lunardini 1981). In addition, convective heat flow during snow melt may increase the input of heat to the soil and increase depths of soil thawing (Izotov 1968; Moskvin 1974; Kingsbury & Moore 1987). The overall effect of the snow cover on the soil thermal regime is quite noticeable in eastern Canada and western Siberia, where the presence of a thick, early snow cover is thought to preclude the occurrence of permafrost (Larsen 1980).

The warming and cooling of soil depends on its thermal conductivity and its heat capacity. These thermal properties depend on physical soil properties such as texture, porosity, water content, and the thermal state of the soil (Lunardini 1981; Rosenberg, Blad & Verma 1983). The effect of soil moisture on the thermal regime of soils is particularly important in the waterlogged soils found throughout the taiga. Though warm water puddled on the soil surface can increase depths of thawing through convective heat flow (Izotov 1968; Moskvin 1974; Kingsbury & Moore 1987), high soil moisture contents significantly increase the energy needed to freeze or thaw the soil. This is particularly important because the presence of a shallow permafrost table impedes soil drainage, causing the water table to be at or near the soil surface (Rieger, Dement & Sanders 1963; Dimo 1969; Tyrtikov 1973; Viereck *et al.* 1983).

Ecological effects of permafrost and soil temperature

Low soil temperatures and the presence of permafrost have several important effects on environmental site conditions. Soil movement caused by frost heaving can result in haphazardly leaning, 'drunken' trees

(Benninghoff 1952; Zoltai 1975*a,b*). Low soil temperatures reduce hydraulic conductivity and impede infiltration (Ford & Bedford 1987). Moreover, permafrost creates an impervious layer that impedes soil drainage (Rieger, Dement & Sanders 1963; Dimo 1969; Tyrtikov 1973; Viereck *et al.* 1983). Indeed, over much of the perennially frozen north, there is an abundant supply of soil water from seasonal snow melt that cannot drain due to underlying permafrost (Lutz 1956*a*; Wolff, West & Viereck 1977). Even in extremely arid regions, the slow thawing of permafrost soils is thought to provide sufficient moisture for tree growth (Walter 1979).

Plant metabolism is directly affected by low soil temperatures. The beginning and end of the growing season are closely related to soil thawing and freezing (Ryden & Kostov 1980). Root elongation in several boreal forest tree species is promoted by an increase in soil temperature (Lawrence & Oechel 1983*a*; Tryon & Chapin 1983). Low soil temperatures may inhibit biological activity in trees by increasing root resistance and the viscosity of water, thereby reducing water uptake (Wolff, West & Viereck 1977; Lawrence & Oechel 1983*b*; Goldstein, Brubaker & Hinckley 1985). Trees may also be subjected to severe water stress by high evaporation demands when their roots are still encased in frozen soil and cannot obtain water (Benninghoff 1952).

Low soil temperatures also influence forest productivity by restricting nutrient availability. The presence of permafrost restricts the rooting zone of trees and the amount of nutrients available for uptake. Slow rates of decomposition in cold soils result in low nutrient availability as organic matter and nutrients are tied up in the forest floor (Tamm 1953; Siren 1955; Weetman 1962; Tyrtikov 1973; Moore 1980, 1981, 1984; Van Cleve & Viereck 1981; Van Cleve *et al.* 1983*a,b*; Van Cleve & Yarie 1986). In interior Alaska, forest floor decomposition and soil respiration are directly correlated with soil temperature (Flanagan & Van Cleve 1977; Fox & Van Cleve 1983; Schlentner & Van Cleve 1985; Van Cleve & Yarie 1986). Van Cleve *et al.* (1983*b*) documented this effect by heating a *Picea mariana* forest floor in interior Alaska to approximately 9 °C above ambient temperatures for three summers. Heating resulted in a 20% reduction in forest floor biomass and increased available nutrient supplies. As a result of this more favorable soil temperature and nutrient regime, tree foliage showed increased rates of photosynthesis.

The importance of soil temperature and permafrost in influencing forest structure and function can be seen in geographic vegetation patterns. In the western Soviet Union, the boreal forest is bounded by isopleths of 5 °C and 10 °C mean thawing season soil temperatures (Shul'gin 1965). In this region, optimal root growth for *Larix sibirica* and *Picea abies* occurs on cooler and wetter soils than for *Pinus sylvestris*

(Korotaev 1987). Continuous permafrost is thought to be one of the major factors controlling the northern tree line in the eastern Soviet Union (Kryuchkov 1973). In Canada, the northern boundary of the forest-tundra ecotone corresponds with the approximate southern limit of continuous permafrost (Larsen 1980).

Comparisons of topographic vegetation patterns in the discontinuous permafrost zone of interior Alaska show the effect of soil temperatures on stand productivity and nutrient cycling (Fig. 4.1). With their high growth potentials and nutrient requirements (Chapin 1986), *Picea glauca* and the successional hardwoods form productive stands on warm, mesic, permafrost-free south slopes and floodplain sites, where soil temperature and soil moisture result in a high rate of nutrient availability (Van Cleve & Viereck 1981; Van Cleve *et al.* 1983*a,b*; Van Cleve & Yarie 1986). In contrast, *Picea mariana*, with its low growth potential and nutrient requirement (Chapin 1986), occupies the least productive, cold, wet, north-facing and bottomland sites. Low soil temperatures encountered in *Picea mariana* stands underlain by permafrost reduce organic matter decomposition and nutrient mineralization (Van Cleve & Viereck 1981; Van Cleve, Barney & Schlentner 1981; Van Cleve *et al.* 1983*a,b*; Van Cleve & Yarie 1986). *Picea mariana* may have adapted to these nutrient-poor conditions through high foliage longevity (25–30 years) in which even the oldest needles contribute positively to the carbon balance (Hom & Oechel 1983).

The importance of these factors can be seen in the dynamics of a *Picea mariana* forest growing on permafrost soil in interior Alaska. Over time, as the forest develops, the forest floor becomes the principal nutrient reservoir as nutrients are immobilized in undecomposed organic matter (Fig. 4.2). The accumulation of a thick forest floor reduces soil temperatures and increases soil moisture. This further reduces organic matter decomposition, nutrient availability, and stand productivity.

Fires are thought to interrupt this process, consuming the forest floor, mineralizing nutrients contained in raw organic matter, and improving the soil thermal regime (Viereck 1973, 1975, 1983; Viereck & Dyrness 1979; Viereck & Schandelmeier 1980; Dyrness, Viereck & Van Cleve 1986). Direct soil heating during burning is minimal and has little long-lasting effect on soil temperature (Uggla 1959; Viereck *et al.* 1979; Brown 1983). Mineral soil and organic matter are poor conductors of heat energy. Moreover, the lower portion of the thick forest floor often remains moist during burning (Uggla 1959; Viereck *et al.* 1979; Dyrness & Norum 1983). However, by reducing the thickness of the insulating forest floor, blackening the forest floor, and removing the forest canopy, fire drastically alters ground surface energy-exchange processes (Brown 1983; Dyrness, Viereck & Van Cleve 1986).

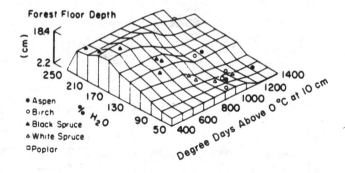

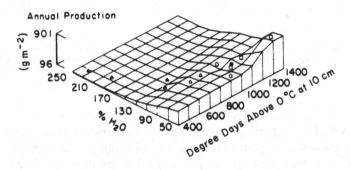

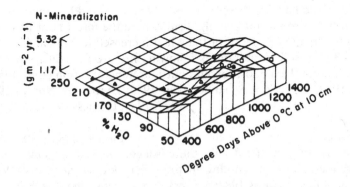

Fig. 4.1. Forest floor depth, annual above-ground tree production, and annual nitrogen mineralization in forests near Fairbanks, Alaska, in relation to forest floor moisture content and soil temperature (from Van Cleve & Yarie 1986).

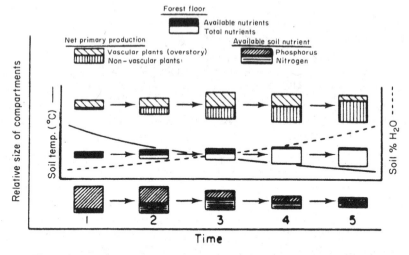

Fig. 4.2. Hypothesized changes in the structure and function of a *Picea mariana* forest in the uplands of interior Alaska for 150 years following fire (from Van Cleve *et al.* 1983*a*).

The main thermal effect of fire is an outcome of the complete or partial removal of the organic layer (Brown 1983). Reduction of the forest floor thickness reduces its insulative effect, allowing greater heat flow and resulting in increased soil temperatures that are directly related to the amount of organic matter removed (Viereck *et al.* 1979; Viereck & Dyrness 1979; Dyrness 1982; Viereck 1982). For example, four summers after experimental fires in an interior Alaska *Picea mariana* forest, depth of thaw where half the forest floor had been removed had increased to 85 cm from a pre-burn depth of 26 cm; where the forest floor had been completely removed, depth of thaw had increased to 138 cm (Dyrness 1982). Higher soil temperatures following fire are thought to increase organic matter decomposition and nutrient mineralization (Lutz 1956*a*; Uggla 1959; Viereck & Schandelmeier 1980; Dyrness & Norum 1983; MacLean *et al.* 1983; Viereck 1973, 1975, 1983; Dyrness, Viereck & Van Cleve 1986).

The long-term effects of fire on the soil thermal regime are poorly understood. In permafrost soils, there is a consistent annual increase in the active-layer thickness following fire (Viereck & Dyrness 1979; Viereck *et al.* 1979; Dyrness 1982; Viereck 1982). How long it will take the depth of thaw to stabilize or to return to pre-burn levels is unknown. Linell (1973) maintained sites free of ground cover for 26 years and found that depth of thaw increased yearly. Van Cleve & Viereck (1983) speculated that the permafrost table may return to its original depth

25–50 years after fire, when the forest canopy and forest floor are fully re-established. Heinselman (1981b) estimated this time as 10–15 years.

Simulation analyses of the soil thermal regime in boreal forests

Bonan (1989a) developed a model to examine the factors regulating the soil thermal regime in boreal forests. This model used easily obtainable climatic, topographic, soils, forest canopy, and forest floor data to calculate monthly solar radiation, soil moisture, and depth of soil freezing and thawing (Fig. 4.3). Solar radiation was calculated by attenuating incoming solar radiation for atmospheric effects, partitioning this into direct beam and diffuse components, and adjusting for slope and aspect (Fig. 4.4). Monthly potential evapotranspiration was calculated from solar radiation and air temperature using a modified form of the Priestley–Taylor equation. The soil was treated as a two-layered system consisting of a forest floor and mineral soil. The monthly water content of each layer was based on precipitation, actual evapotranspiration, water released in soil thawing, and drainage (Fig. 4.5). Depths of soil freezing and thawing in each soil layer were calculated from Stephan's formula. The number of degree-days required to freeze or thaw each soil layer was a function of thickness, thermal conductivity, water content, and bulk density (Fig. 4.6). The number of degree-days available to freeze or thaw the soil was a function of air temperature, solar radiation, and forest canopy cover.

The model simulated solar radiation, soil moisture, and soil thermal regimes that were consistent with observed data from boreal forest

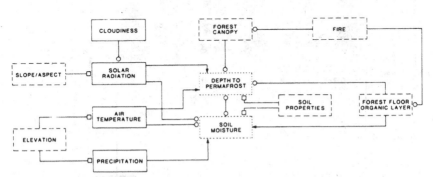

Fig. 4.3. Hypothesized interactions among environmental factors in boreal forests. Solid boxes represent climatic parameters. Dashed boxes indicate site parameters. Dotted boxes indicate important state variables thought to control the structure and function of boreal forests. Types of interaction: arrows are positive, circles are negative, squares are unspecified (from Bonan 1989a).

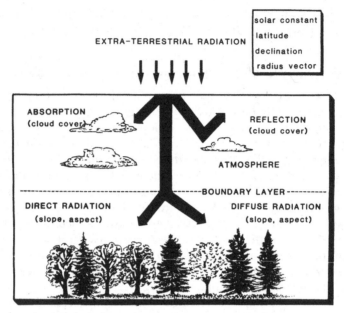

Fig. 4.4. Schematic diagram of solar radiation algorithm.

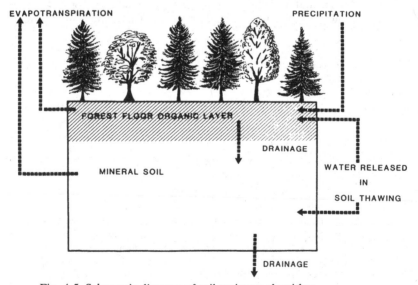

Fig. 4.5. Schematic diagram of soil moisture algorithm.

regions (Bonan 1989*a*). For example, simulated potential evapotranspiration was consistent with observed pan evaporation data for several sites in northern North America (Table 4.1). Simulated seasonal soil freezing and thawing (Fig. 4.7) corresponded with observed seasonal dynamics from various high latitude regions (Viereck & Dyrness 1979; Ryden & Kostov 1980; Viereck 1982; Chindyaev 1987; Kingsbury & Moor 1987). Simulated maximum depth of seasonal soil thaw provided good estimates of the soil thermal regime in relation to topography, elevation, and forest floor thickness (Table 4.2). Though depth of thaw is not necessarily the same as depth to permafrost (Ryden & Kostov 1980; Lunardini 1981), shallow depths of thaw were simulated on soils with shallow depths to permafrost and deep depths of thaw were simulated on permafrost-free soils. Permafrost is likely to exist where depth of soil freezing exceeds depth of soil thawing (Lunardini 1981). Simulated depths of soil freezing and thawing corresponded with geographic patterns of no permafrost, discontinuous permafrost, and continuous permafrost in northern North American and the Soviet Union (Fig. 4.8).

The soil moisture algorithm was developed so that soil moisture increased up to saturation as depth of thaw became shallower, thus mimicking the effects of permafrost on soil hydrology (Benninghoff 1952; Rieger, Dement & Sanders 1963; Dimo 1969; Kryuchkov 1973; Van Cleve & Viereck 1981; Viereck, Van Cleve & Dyrness 1986). As the soil

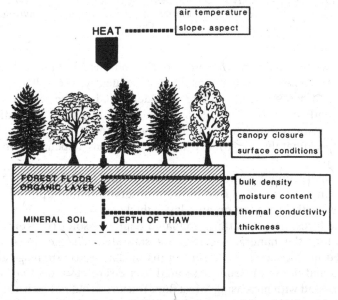

Fig. 4.6. Schematic diagram of soil freezing and thawing algorithm.

Table 4.1. *Observed pan evaporation and simulated potential evapotranspiration for selected sites in northern North America*

Month	Observed pan evaporation (cm)			Simulated potential evapotranspiration (cm)
	mean	minimum	maximum	
Fairbanks (64°49'N, 147°52'W)				
June	12.2	8.8	16.0	13.0
July	12.6	8.2	17.1	13.0
August	8.7	6.5	13.3	8.3
Moosonee (51°16'N, 80°39'W)				
June	14.2	11.6	18.5	12.7
July	15.4	11.8	18.6	15.7
August	10.9	9.2	12.6	12.4
September	8.0	6.2	9.9	6.9
Winnipeg (49°54'N, 97°15'W)				
May	19.1	13.1	26.1	16.0
June	21.2	18.7	23.5	19.8
July	22.6	20.0	25.3	24.6
August	17.5	14.4	19.4	19.8
September	13.5	11.2	19.2	11.0

Sources: Fairbanks: National Oceanic and Atmospheric Administration 1972–1986. *Climatic data*: *Annual Summary Alaska*, volumes 58–72. National Climatic Data Center, Asheville, North Carolina. Moosonee and Winnipeg: Environment Canada 1978–1982. *Monthly record*: *Meteorological Observations in Eastern/Western Canada*, volumes 63–67, numbers 1–12. Atmospheric Environment Service, Environment Canada, Downsview, Ontario.

became wetter, its thermal conductivity and the amount of energy required to thaw the soil increased. Consequently, the soil moisture content and depth of thaw in one year depended on the soil moisture content and depth of thaw in the previous year. In all simulations, however, stable estimates of soil thaw and soil moisture were obtained after approximately one to five annual iterations depending on site conditions.

This interaction between soil moisture and soil thermal regimes resulted in two effects. First, as with observed data, depth of thaw decreased with increased organic layer thickness (Fig. 4.9). However, when the forest floor was thin, simulated depths of soil thaw were deep enough that the mineral soil was not saturated. As the forest floor increased in thickness, shallower depths of thaw cause drainage to be impeded and the soil became saturated. Once saturated, depth of thaw still decreased with increasing forest floor thickness, but not nearly as fast as when the soil was unsaturated (Fig. 4.9).

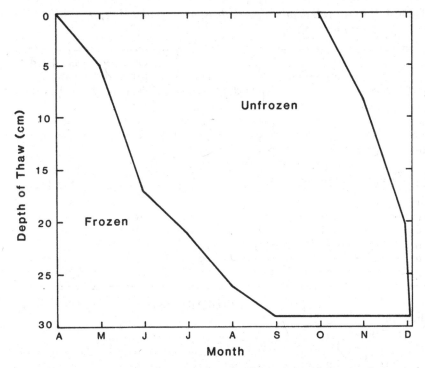

Fig. 4.7. Simulated seasonal mineral soil freezing and thawing for a closed-canopy forest, Fairbanks, Alaska (from Bonan 1989*a*). The simulated conditions are a 10% SE slope, altitude 350 m, poorly drained site and 25 cm forest floor.

A second consequence of the interaction between soil moisture and soil thawing was evident in simulated changes in the soil thermal regime following fire. When the forest canopy was removed and the forest floor partially reduced (i.e. as in a fire), the heat load increased, causing the mineral soil to thaw to a greater depth (Fig. 4.10). This, in turn, improved soil drainage, allowing for an even greater depth of thaw in the following year. Eventually a new equilibrium between depth of thaw and soil moisture was obtained and the depth of thaw stabilized. This suggests that for a given heat load onto a soil, there is a stable equilibrium between depth of thaw and soil moisture content. Moreover, these results are consistent with the reported annual increase in the active-layer thickness after removal of the forest canopy or forest floor (Linell 1973; Viereck & Dyrness 1979; Dyrness 1982; Viereck 1982), suggesting that this phenomenon is caused, in part, by improved drainage conditions as the permafrost table recedes.

Table 4.2. *Observed depth to permafrost and simulated depth of soil thaw for Fairbanks, Alaska*

Slope	Aspect	Elevation (m)	Drainage[a]	Forest floor (cm)	Observed permafrost depth (cm)	Simulated thaw depth (cm)
Uplands						
30%	N	427	PD	38	22	22
0	—	167	PD	14	55	59
10	SE	385	PD	23	35	31
25	S	396	WD	9	—[b]	129
18	SE	229	WD	6	—	154
Floodplain						
0	—	177	PD	25	16	30
0	—	122	PD	19	20	39
0	—	120	WD	18	—	116
0	—	177	WD	5	—	153
0	—	120	WD	15	—	125

Notes: [a] Drainage class: PD, poorly drained; MD, moderately drained; WD, well drained.
[b] No permafrost.

In these simulations, though topography, elevation, forest floor thickness, and soil drainage interacted to create the mosaic of permafrost patterns found in interior Alaska (Table 4.2), these patterns could not be reproduced without soil drainage differences. Using representative data for a north-facing and a south-facing slope in the uplands of Fairbanks, Alaska, Bonan (1989a) used the model to examine the influence of aspect, forest floor thickness, elevation, and soil drainage on depth of thaw.

On both poorly drained and well-drained soils, depth of maximum soil thaw was larger on south-facing slopes than on north-facing slopes (Table 4.3a). This primarily reflected the increased heat load on the south slope (Table 4.3b). In addition, increased potential evapotranspiration demands on the south slopes (Table 4.3b) resulted in relatively drier soils than the corresponding north slopes. Though thermal conductivity decreased with increased soil dryness, the amount of energy required to thaw the soil also decreased, making it easier to thaw the dry soils. However, on the whole, though changes in slope affected depths of thaw, the greatest proportional increases in depth of thaw were caused by soil drainage differences (Table 4.3c).

For all sites, as the forest floor increased in thickness, the depth of soil thaw decreased (Table 4.3a). This reflected the increased number of degree-days required to thaw the soil profile. However, for both the poorly drained and well-drained soil, the rate of decrease in depth of soil thaw with increasing forest floor thickness was greater on the south slope than on the north slope. This difference was so great on the poorly drained soils that when the forest floor was 20 cm thick, depth of thaw was less on the south slope than on the north slope. This reflected the increased dryness of the forest floor on the south slope and the corresponding decrease in the thermal conductivity of the forest floor.

These analyses have highlighted the importance of soil moisture as a regulator of the soil thermal regime simulated by Bonan's (1989a) model. Climatic parameters may also be important, and Bonan, Shugart & Urban (1990) used the model to examine the sensitivity of depth of thaw in an interior Alaska *Picea mariana* forest underlain with permafrost to climatic parameters. In a 3 × 3 factorial experiment, mean monthly air temperature and monthly precipitation were increased from current values in steps of 1 °C, 3 °C, and 5 °C and 120%, 140%, and 160%, respectively. All simulations were for 200 years in the absence of fire.

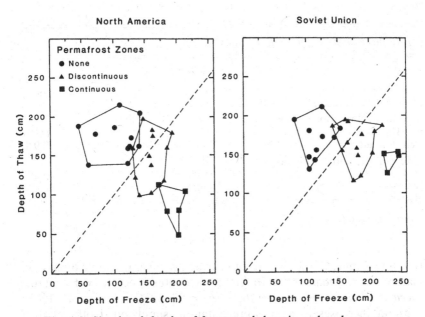

Fig. 4.8. Simulated depths of freeze and thaw in a closed-canopy, forested, well-drained, fine-textured soil for climatic sites located throughout northern North America and the Soviet Union (from Bonan 1989a).

Table 4.3. (a) *Simulated depths of soil thaw (cm), Fairbanks, Alaska*

	350 m elevation				200 m elevation		
	Forest floor (cm)				Forest floor (cm)		
Slope	0	10	20	Slope	0	10	20
Poorly drained soil							
30% S	126	82	38	30% S	139	95	40
30% N	100	72	43	30% N	109	79	46
Well-drained soil							
30% S	164	132	103	30% S	182	148	115
30% N	126	108	74	30% N	139	119	87

(*b*) *Site conditions*

	Heat load (°C day)		Potential evapotranspiration (cm)	
Slope	350 m	200 m	350 m	200 m
30% S	1002	1163	43.6	47.3
30% N	701	814	34.9	37.9

(*c*) *Relative changes in depth of thaw*

	Elevation	
	350 m	200 m
ratio south slope : north slope		
poorly drained	109%	112%
well-drained	130%	129%
ratio well-drained : poorly drained		
30% S	187%	192%
30% N	149%	156%

Source: Bonan (1989*a*).

Baseline climate simulations resulted in a 22 cm average depth of thaw for mature forests (years 100–200). Observed depths to permafrost in mature *Picea mariana* forests range from 16 to 55 cm (Viereck *et al.* 1983). Without increases in annual precipitation, simulated depths of thaw decreased with increases in air temperature (Table 4.4). Increased evapotranspiration with warmer climatic conditions resulted in drier forest floors. The decreased thermal conductivity of the thick forest

Table 4.4. *Simulated depth of soil thaw (cm) in mature* Picea mariana *forests in the uplands of Fairbanks, Alaska, as a function of increases in air temperature and precipitation*

Precipitation	Air temperature			
	base	1 °C	3 °C	5 °C
base	22	20	15	12
120%		24	18	14
140%		32	21	15
160%		40	28	25

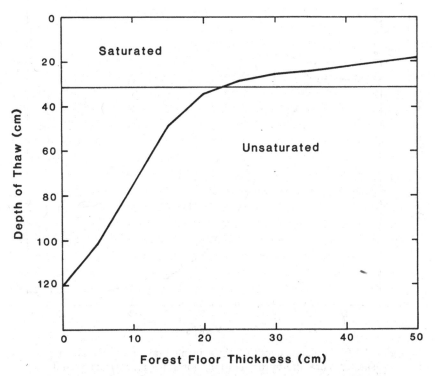

Fig. 4.9. Simulated depth of thaw in mineral soil at Fairbanks, Alaska, in relation to forest floor thickness (from Bonan 1989*a*). Simulated conditions: a 10% SE slope at 350 m, poorly drained.

floors offset the increased heat load onto the surface, with the result that the soils did not become warmer. With increased precipitation to offset the increased potential evapotranspiration, the soils became wetter. Thermal conductivity increased relative to the baseline precipitation simulations and depth of thaw increased (Table 4.4).

Conclusion

Field observations from various regions of the circumpolar boreal forests have indicated the importance of soil temperature as an ecological factor controlling forest dynamics and vegetation patterns. Interactions among solar radiation, soil moisture, the forest floor, and the forest canopy are important determinants of the soil thermal regime. A simulation model has quantified these interactions, and simulation analyses using this

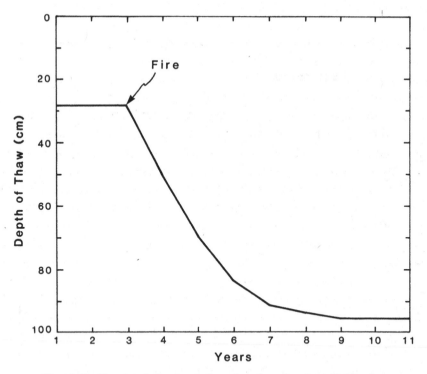

Fig. 4.10. Simulated depth of thaw in mineral soil at Fairbanks, Alaska, after fire. Pre-burn conditions: forest floor layer = 25 cm and closed forest canopy. Post-burn conditions: forest floor layer = 15 cm and no forest canopy (from Bonan 1989a). (Other conditions as for Fig. 4.9.).

model have highlighted the complexity of boreal environments. Inter-actions among soil moisture and soil thermal properties were required to adequately simulate the extant soil thermal regime in the taiga of interior Alaska. However, this interaction resulted in several unexpected feed-backs. In particular, the analyses suggested the possibility that soil temperatures in *Picea mariana* forests with a thick forest floor and underlain with permafrost may not increase with climatic warming unless accompanied by precipitation increases. Whether this result is an artifact of the model has not been evaluated. Climatic sensitivity experiments with a surface energy budget model indicate substantial soil warming in the Fairbanks area even when the thermal conductivity of the surface organic layer was decreased (Goodwin, Brown & Outcalt 1984). Appar-ently, ground surface temperatures may substantially increase under a warmer, drier climate. However, the model of Goodwin, Brown & Outcalt (1984) simulated bare surfaces without a forest canopy. More-over, their model did not simulate interactive soil temperature and soil moisture regimes, which was an essential feature of Bonan's (1989*a*) model. Clearly, a detailed surface energy budget model for forested sites must be developed to better understand the soil thermal regime in boreal forests.

This manuscript was prepared while the author was a post-doctoral fellow in the Advanced Study Program at the National Center for Atmospheric Research. The National Center for Atmospheric Research is sponsored by the National Science Foundation.

5 Fire as a controlling process in the North American boreal forest

Serge Payette

Introduction

Major patterns of plant communities and species distribution are induced by various disturbance regimes operating at different spatial and temporal scales (Loucks 1970; White 1979; Bormann & Likens 1979b; Delcourt, Delcourt & Webb 1983). The development of temperate forests is controlled by canopy disturbance associated with single and multiple treefalls creating small gaps (Henry & Swan 1974; Oliver & Stephens 1977) and, occasionally, larger gaps along windstorm tracks (Canham & Loucks 1984). In contrast, fire disturbance in the boreal forest creates all sizes of canopy gap, and constitutes one of the most important community processes in vegetation development along complex environmental gradients (Heinselman 1981b). At the landscape level, ecological disturbances result in the spatiotemporal development of the vegetation mosaic composed of an assemblage of stand patches of different age, areal extent, and floristic composition (Pickett & White 1985).

The boreal forest is one of the world's two largest forest belts; it covers more than 12^6 km² (Baumgartner 1979). The North American segment of the circumboreal forest, which constitutes a well-delineated biome both geographically and ecologically (Fig. 5.1), will be analyzed here with respect to fire disturbance. The boreal forest (excluding the Cordilleran region) spans more than 10° of latitude in eastern and western Canada; it is somewhat contracted south of Hudson Bay and James Bay, and in Alaska. At the continent scale, the boreal forest is a floristically poor biome (Takhtajan 1986) with only nine tree species dominating regionally or throughout the range, in coexistence with a subdued under-canopy flora in dense stands and an ubiquitous cryptogamic flora in open stands. Black spruce (*Picea mariana* (Mill.) BSP.) and white spruce (*Picea glauca*

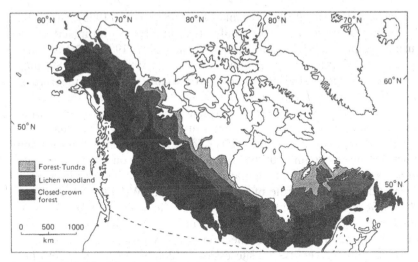

Fig. 5.1. The North American boreal forest.

(Moench) Voss.) are transcontinental in distribution; balsam fir (*Abies balsamea* L.) and tamarack (*Larix laricina* (DuRoil) K.Kock) have a predominantly central and eastern distribution. Lodgepole pine (*Pinus contorta* L.) is a Cordilleran species and jack pine (*Pinus banksiana* Lamb.) a continental species found more or less distant from maritime environments, except in New Brunswick and Nova Scotia. The other dominant species are deciduous trees, i.e. white birch (*Betula papyrifera* (Marsh.), aspen (*Populus tremuloides* Michx.) and balsam poplar (*Populus balsamifera* L.), which are distributed throughout the biome. The autecology of these tree species has been underlined recently by Ritchie (1987*a*). The depauperate boreal flora appears to be the result of sustained, severe climatic controls that occurred during the Quaternary. The bulk of the boreal flora is made up of robust, generalist species able to withstand recurrent, dramatic changes in the environment.

The boreal forest is a geologically young and ever-changing biome. In Alaska and western Canada, the boreal forest developed during the early Holocene (Ager 1983; Brubaker, Garfinkel & Edwards 1983; Ritchie 1984, 1987*a*; Ager & Brubaker 1985; MacDonald 1987); in the east it is much younger, about 4000–8000 years old, because of late deglaciation (Richard 1981*a,b*; Richard, Larouche & Bouchard 1982; Anderson 1985; Dyke & Prest 1987; Ritchie 1987*b*; Webb 1988). Except for rare examples of progressive range expansion throughout the Holocene, such as lodgepole pine in northwestern Canada (MacDonald & Cwynar 1985), most boreal tree species stopped their northward migration a few thousand years ago in reaction to climate change (Richard 1981*a*; Richard,

Larouche & Bouchard 1982; Ritchie 1984, 1987a; Lamb 1985). Similarly, the pollen increase of boreal conifer taxa in sites of the mixed forest during the past 2–3 thousand years (Richard 1977, 1981b; Green 1981, 1982, 1987; Comtois 1982; Holloway & Bryant 1985; Webb 1988) may represent a somewhat diffuse, southward expansion of the boreal forest.

In this chapter, I explore the multi-faceted influence of fire on the boreal forest in time and space, to explain the major patterns in terms of zonal types, plant species and community distribution. This analysis is based on the assumption that fire is one of the most important forcing functions in the dynamics of the boreal forest, acting on the biota and the environment as a climate-mediated transmitter. Several reviews have already dealt with the role of fire in this biome (Wright & Heinselman 1973; Viereck & Schandelmeier 1980; Heinselman 1981b; Wein & MacLean 1983); we should therefore confine ourselves to basic issues in addressing the fire–climate connection as a key process in boreal vegetation stability, diversity and succession. Much has been said about fire impact on the boreal forest, while empirical data are still lacking on many aspects of fire ecology and paleoecology. Two topics in the major developmental patterns and fate of the boreal forest associated with fire disturbance, all interrelated in time and space, will be considered and analyzed here.

(1) Along a latitudinal gradient, zonation of the boreal forest expresses a differential response of the major forest types to a large scale, post-fire regeneration process caused by climatically controlled fire regimes. Of the three zonal types, the closed-crown forest and the lichen woodland, located respectively in the southern and central boreal forest, are made of successfully regenerated post-fire tree communities, whereas the northern segment of the boreal forest – the forest–tundra – is predominantly composed of deforested post-fire communities. Dominant successional patterns throughout the boreal forest vary according to tree species distribution and regional and local fire regimes.

(2) Changes in climate and the associated fire regime may have modified the regional importance of fire-prone boreal communities during the Holocene.

Zonal vegetation and fire regime

Before attempting any discussion on these topics, it may be useful to describe briefly the elements of a fire regime (Heinselman 1981b), which are important to reconstruct the recent fire history in a given area. These elements have been defined more or less formally at a recent fire history workshop (Stokes & Dieterich 1980). The fire frequency refers to the number of fires per unit time in a given area. The fire interval (or fire-free interval or fire return interval) is the number of years between two successive fires in a given area; the mean fire interval (or mean fire-free interval or mean fire return interval) is the average of all fire intervals in a given area during a given period of time. Crown fires, surface fires, ground fires and spotting fires (Van Wagner 1983) are all different types of fire with variable intensity and severity. Fire intensity is used in a strict thermodynamic sense, i.e. units of energy released per unit area per unit time (Romme 1980; Van Wagner 1983); fire severity is used to indicate the ecological effects of a fire on the plant community and the habitat. The size of fires is equally important to consider: frequency distribution, mean size and modal size of fires are useful parameters to measure in a large area. The natural fire rotation period (Heinselman 1973) and the fire cycle (Van Wagner 1978) are synonyms expressing the number of years necessary to burn at least one time an area equal to the total area investigated. According to Heinselman (1981b), because fire is a semi-random process in many regions with sites having different fire intervals, some areas will burn more often than others during a rotation period.

Zonal types

Because the boreal forest covers such a huge continental area, a large array of climatic conditions is currently encountered across its range (Hare & Thomas 1979). A major moisture gradient is found from south to north and from west to east. Distinct fire climates are produced along these gradients, which affect the distribution of the dominant community types. Three plant zones are generally recognized in the boreal forest from south to north (Fig. 5.1): the closed-crown forest, the open-crown forest or lichen woodland, and the forest–tundra (Hare 1959; Hare & Ritchie 1972; Hustich 1949, 1966; Larsen 1980; Payette 1983; Rousseau 1968; Rowe 1972). Whether the forest–tundra is part of the boreal forest or not is a much debated matter not considered here. This plant zone is at the northern end of the boreal continuum, and it is analyzed as such.

Dominant community types

A schematic south–north transect crossing the boreal forest in eastern Canada illustrates the northward replacement of the dominant community types (Fig. 5.2). In this area, the boreal forest extends from about 47° N in the closed-crown forest (Grandtner 1966; La Roi 1967; Rowe 1972; La Roi & Stringer 1976) to 57° N at the treeline (Payette 1983). At the southern end of the transect, the boreal forest consists of closed-crown forests made of black spruce–balsam fir stands on mesic sites, with yellow birch (*Betula alleghaniensis*), white birch, and white spruce as co-dominants near the mixed forest transition to the south and the feather-moss black spruce forest to the north (Grandtner 1966). Major disturb-ances affecting the spruce–fire zone are massive spruce budworm out-breaks (Blais 1968; Hardy, Lafond & Hamel 1983) and occasional severe fires. The northern part of the closed-crown forest zone (Hare 1959) is composed of feathermoss black spruce stands located on mesic sites, with a tree cover varying between 40 and 80% in mature stands (Ducruc *et al.* 1976; Gerardin 1980). North of this zone on mesic sites, lichen–spruce woodlands with a tree cover <40% predominate. Trees cover about 25–40% of the ground surface in the southern part, and 5–25% in the

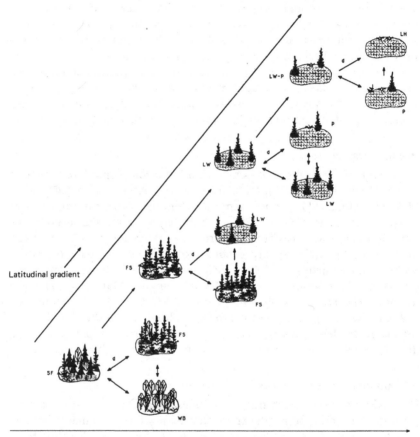

Fig. 5.2. Shifts of the dominant vegetation types of the boreal forest along a latitudinal and a post-fire successional gradients. SF: Spruce–fir forest. FS: Feathermoss spruce forest. WB: White birch forest. LW: Lichen woodland. P: Parkland. LH: Lichen-heath. D: diverging post-fire vegetation types.

northern part (Ducruc *et al.* 1976; Gerardin 1980; Morneau & Payette 1989). North of the lichen–spruce woodland zone, forest–tundra communities established on mesic sites are composed of patchy lichen–spruce woodlands and lichen–heath–dwarf birch (*Betula glandulosa* Michx.) stands. Lichen–spruce woodlands, which have a tree cover <25%, are extensive in the southern part of the forest–tundra but rare near the treeline; treeless communities are widespread in the northern part (Payette 1983). Lichen–spruce krummholz (stunted trees), with a spruce cover <25%, are mostly distributed in the northern part of the forest–tundra.

Fire regimes
Along the transect, fire regimes south of the forest–tundra are characterized by a natural mean rotation period <150 years (Table 5.1). In the southern boreal forest, Cogbill (1985) estimated the natural rotation period to be about 130 years, whereas most conifer stands analyzed by Gerardin (1980) in eastern James Bay area were generally <100 years old. In the northern part of the lichen–spruce woodland, the fire rotation period was about 100 years during the twentieth century (Payette *et al.* 1989*b*). In the forest–tundra, the fire rotation period was 180 years in the southern part, 1450 years in the northern part, and about 7800 years at the treeline during the same period. From the southern boreal forest to the northern limit of the lichen–spruce woodland zone, the oldest forests growing in mesic sites are of postfire origin and generally <250 years old (Gerardin 1980; Cogbill 1985; Morneau & Payette 1989), an indication that fire disturbance is the dominant process in community recycling at the landscape level. Most forest stands in these areas developed during a mean fire-free interval generally <150 years, suggesting that fire recurrence tends to perpetuate the floristic composition and structure of pre-fire communities (Carleton & Maycock 1978; Gerardin 1980; Cogbill 1985; Morneau & Payette 1989).

Zonal replacement in mesic sites
Zonal replacement of the dominant forest communities in the boreal forest is not a straightforward process. Lichen woodlands occur far south in the closed-crown forest zone, where they are generally considered as edaphic types developed on dry substrata such as sand and rock outcrops. More importantly, lichen woodlands also coexist with feathermoss forests in mesic sites in many parts of the closed-crown forest zone (Gerardin 1980). In contrast, closed-crown forests rarely develop in mesic sites of the lichen woodland zone, whereas spatially confined lichen woodlands are occasionally found throughout the northern part of the forest–tundra, even near the treeline. The expression 'closed-crown forest' may not be applicable to the entire zone where this forest type predominates. It is in

Table 5.1. *Fire rotation period and fire intervals from several boreal forest areas*

Zonal tree type	Fire rotation period or fire interval (yr)	Location	Type of record	Dominant species
closed-crown forest	100	Minnesota (Heinselman 1973)	fire scars and historic reports	northern pines
	130	southern Quebec (Cogbill 1985)	stand age profiles	boreal species (several)
	<100	central Quebec (Gerardin 1980)	post-fire stands	black spruce jack pine
	500	SE Labrador (Foster 1983)	fire scars, reports, air photos	black spruce
lichen woodland	<100	Mackenzie River (Black & Bliss 1978)	post-fire stands	black spruce
	100	Abitau Lake (Maikawa & Kershaw 1976)	post-fire stands	black spruce
	110	Great Slave Lake (Johnson & Rowe 1975)	fire reports	black spruce
	70–100	Great Slave Lake (Johnson 1981)	post-fire stands	black spruce
	<100	northern Quebec (Gerardin 1980)	post-fire stands jack pine	black spruce
	100	northern Quebec (Payette et al. 1989b)	fire scars jack pine	black spruce
southern forest–tundra	180	northern Quebec (Payette et al. 1989b)	fire scars	black spruce
northern forest–tundra	1450	northern Quebec (Payette et al. 1989b)	fire scars	black spruce

the southern part of the zone that the 'closed-crown' type is found, particularly in mesic and seepage sites occupied by dense spruce–fir, aspen and white birch stands. Thus, openness of the forest canopy is certainly a major characteristic of most boreal forests, particularly north of the spruce–fir belt where black spruce and pine take over in mesic and dry sites.

It is argued here that the patchy pattern observed in mesic sites within each zone, with the spatially dominant zonal type coexisting with other communities more commonly occurring northwards, is a long-term consequence of differential post-fire regeneration. Because zonation of the dominant plant communities is primarily controlled by the thermal environment, fire disturbance may have played a driving role in spatial partitioning of plant communities in each zone, and on the successional pathways leading to self-replacement or retrogression through the post-fire regeneration process. From south to north, the respective shifting dominance in mesic sites from fir to spruce stands, feathermoss forest to lichen woodland, and lichen woodland to lichen–heath represent similar, direct post-fire regeneration responses to reduced temperature on a yearly and seasonal basis, and also to particular fire regimes that occurred in the past.

Two major issues emerge from these observations. First, zonal re-placement in mesic sites from moss forest to lichen woodland, and to lichen–heath underlines the overwhelming control of climatic parameters on the dominant plant communities as one goes further north. This replacement suggests that differential competitive ability of the species forming the zonal types may play a role along the latitudinal gradient. Secondly, regional or local replacement from moss forest to lichen woodland in the closed-crown forest zone, and from relatively well-stocked lichen woodland to open or very open lichen woodland (park-land), or eventually to lichen–heath in the lichen woodland zone and the forest–tundra respectively, indicates retrogression or failure in post-fire regeneration during succession that leads normally to self-replacement of the zonal type. Consequently, factors associated with site parameters could also affect the course of succession and regional distribution of the main plant communities. The long-term fate of zonal types both latitudi-nally and regionally according to climatic and successional gradients is summarized in Fig. 5.2.

In this figure, it is suggested that lichen woodland replaces feathermoss forest on the northern segment of the climatic gradient, while the two forest types may develop and maintain themselves independently within the closed-crown forest zone. Current hypotheses explaining the forma-tion and maintenance of the lichen woodland are all based on the openness of the forest. The most cited explanations – radiation deficits

(Vowinckel, Oechel & Boll 1975), low soil temperature (Kershaw & Rouse 1976), nutrient-poor soils (Moore 1980), physiological drought of trees (Lucarotti 1976; Kershaw 1977), soil drought (Rowe 1984), allelopathic influence of lichens on conifer (Brown & Mikola 1974; Arsenault 1979; Fisher 1979; Cowles 1982) – represent some of the site and growth factors that characterize the plant formation and the habitat in which the lichen woodland develops. These site characteristics are not exclusive to the lichen woodland habitat, and allelopathic influence of lichens on trees is still a controversial hypothesis. As such, these factors are of little help to identify the ultimate causes responsible for inception of the lichen woodland within the feathermoss forest zone. Because both forest types are tightly tuned to cycles of fire disturbance, one must look at the conditions affecting success of the post-fire regeneration of forest stands currently growing at the time of the burn.

Successional pathways

Several reports have dealt with fire disturbance at the regional scale across the North American boreal forest. Post-fire recovery was described in several sites throughout the boreal forest, from Alaska (Lutz 1956a; Foote 1976; Van Cleve & Viereck 1981), Northwest Territories (Scotter 1964; Maikawa & Kershaw 1976; Black & Bliss 1978; Johnson 1981), Saskatchewan and Alberta (Dix & Swan 1971; Carroll & Bliss 1982), Ontario and Quebec (Carleton & Maycock 1978; Gerardin 1980; Clayden & Bouchard 1983; Morneau & Payette 1989), Labrador (Foster 1985) to Newfoundland (Ahti 1959; Damman 1964; Bergerud 1971). However, only a few detailed studies on post-fire chronosequences of the feathermoss forest and the lichen woodland types have been published (Table 5.2).

Succession in the spruce–fir forest type

In areas where the fire rotation period does not exceed 100–150 years, balsam fir rarely dominates post-fire stands, being able to regenerate more or less continuously in the forest floor of several stand types some time after fire (Carleton & Maycock 1978; Gerardin 1980; Cogbil 1985; Foster 1985). In Newfoundland, fir forests occupy sites not disturbed by fire during the past hundred years or more (Damman 1983). Failure to regenerate after fire explains the shift from fir forests to black spruce forests and to heathlands in many areas of the island (Damman 1964, 1971, 1983; Delaney & Cahill 1978; Meades 1983) and elsewhere (Gerardin 1980). Human influence favoring short fire-free intervals entails black spruce regeneration, because regenerating stands are too young to produce an adequate seed supply (Meades 1983). In Newfoundland, retrogression of black spruce forests to *Kalmia* heathlands may be

Table 5.2. *Selected post-fire chronosequences in lichen woodlands and feathermoss forests*

Post-fire age (yr)	Stage	Dominant species	Area	References
Lichen woodlands				
0-3	Bare ground	none	Newfoundland	Ahti 1959; Bergerud 1971
3-10	Crustose	*Trapeliopsis sp.*		
10-25	Horn lichens	*Cladonia crispata, C. deformis, C. cristatella*		
25-80	1st reindeer	*Cladina mitis, C. rangiferina, Cladonia uncialis*		
80+	2nd reindeer	*Cladina stellaris*		
4	Bare ground	*Polytrichum sp., Ceratodon*	Northern Quebec	Morneau & Payette 1989
15	Crustose	*Trapeliopsis sp.*		
25	Horn lichens	*Cladonia deformis, C. cornuta, C. cocciferra,* etc.		
35-100	Fruticose lichens	*Cladina mitis*		
100+	Fruticose lichens	*Cladina stellaris*		
1-20	Moss-crustose	*Polytrichum sp., Trapeliopsis sp.*	Northwest Territories	Maikawa & Kershaw 1976
20-60	Cladonia	*Cladonia uncialis, Cladina stellaris*		
60-130	Stereocaulon-spruce	*Stereocaulon paschale*		
130+	Spruce-moss	*Hylocomium, Pleurozium*		
Feathermoss forests				
1	Bare ground	*Polytrichum sp., Ceratodon*	South-eastern Labrador	Foster 1985
1-20	Crustose	*Trapeliopsis sp.*		
20-70	Fruticose lichens	*Cladina mitis, C. rangiferina*		
70-100	Fruticose lichens	*Cladina stellaris*		
100+	Moss	*Pleurozium, Ptilium*		

promoted by unsuitable seedbeds because of inadequate scarification of the raw humus associated with crown fires (Meades 1983). Although infrequent along maritime coasts, natural or man-caused fires are inducing dramatic changes in habitats dominated by extensive spruce and fir krummholz, or 'tuckamore' (Meades 1983). Krummholz clearance by fire and expansion of coastal tundra communities are common attributes of Atlantic coastal environments, either in Newfoundland (Meades 1983), along the Labrador coast (Hustich 1939), and along the Quebec North-Shore area (Gerardin 1981). Therefore, successional retrogression and deforestation associated with fire disturbance seem particularly important factors determining the major vegetational patterns of the Atlantic boreal forest, i.e. in a climatically marginal, cool and humid region characterized by a long fire rotation period.

Succession in the feathermoss forest type

Several post-fire chronosequences have been described for the closed-crown forest zone. Because of a shorter fire rotation period, fire recycling is more rapid in Alaska and western Canada, i.e. 50–100 years according to Viereck (1983), than in eastern Canada where it apparently increases to 500 years (Foster 1983). The first stage of the recovery sequence is characterized by the re-establishment of vascular species, with shrubs dominating in moderately burned sites owing to sustained sprouting from underground parts, and pioneer species (cryptogams and herbs) in severe burns where the mineral soil is exposed. Tree seedling establishment is also very active during this stage, and initiation of a tree canopy begins 25–30 years after fire (Viereck 1983). Shrub and tree density increases thereafter, promoting fast development of a feathermoss carpet, with species such as *Pleurozium schreberi* (Brid.) Mitt. and *Hylocomium splendens* (Hedw.) B.S.G. As the forest matures, the tree layer sometimes closes (Viereck 1983), but natural thinning and clonal spruce development tend to produce canopy openings in advanced successional stages (Gerardin 1980) where subdominant balsam fir individuals may develop (Carleton & Maycock 1978; Cogbill 1985; Foster 1985; Gerardin 1980). Feathermoss–spruce forests at least 80–100 years old in boreal Quebec generally have an extensive open canopy (Gerardin 1980), facilitating the growth of ericaceous shrubs, whereas in Labrador spruce–*Pleurozium* forests it is discontinuous but without a dense shrub understorey (Foster 1984). During fire-free intervals exceeding 200–250 years, old-growth spruce or fir–spruce moss forests may develop into moribund stands where tree regeneration is much reduced. These forests may change into a parkland type in the absence of fire disturbance (Dix & Swan 1971). Low productivity of old-growth stands is most likely caused by the accumulation of a thick organic mat, which enhances waterlogging, nutrient-poor soil conditions, and low soil temperatures (Viereck 1973; Van Cleve & Viereck 1981; Foster 1983). Paludification is thought to be a significant process in ageing conifer forests in many areas of the boreal forest (Heinselman 1981b; Van Cleve & Viereck 1981; Foster 1983, 1984; Engstrom & Hansen 1985).

Succession in the lichen woodland type
Relatively good correspondence exists between seral stages throughout the lichen–spruce woodland zone (Table 5.2), where two major post-fire chronosequences are generally recognized, i.e. the central-eastern sequence leading to mature *Cladina stellaris*–spruce woodland (Hustich 1951b; Ahti 1959; Bergerud 1971; Auclair 1983; Morneau & Payette 1989) and the western sequence formed by *Stereocaulon paschale*–spruce woodland (Maikawa & Kershaw 1976). However, whether the *Stereocaulon* woodland is representative of all western mature sites is not clear from the literature, maybe because of the lack of additional regional data. Ritchie (1959) suggested that *Stereocaulon* prominence in mature woodlands was caused by extensive caribou grazing, whereas Johnson (1981) stressed that the abundance of the species was best explained by habitat requirements.

During stand development, no real succession is occurring in vascular plants, because most species are present at the very beginning of the sequence and remain throughout with only subsequent changes in their abundance. Johnson (1981) found that most lichen species recolonized burned sites immediately after fire; most other studies indicated that post-fire lichen colonization necessitated several years to be completed. Although there are regional differences, succession occurs primarily in the lichen carpet, from inception a few years after fire to nearly complete establishment 10–15 years following a fire (Ahti 1959; Foster 1985; Morneau & Payette 1989). Lichen succession may be attributable largely to the dispersal ability and different growth rates of the species involved (Johnson 1981; Morneau & Payette 1989). During succession, a few bryophytes are early colonizers, such as *Polytrichum* and *Ceratodon*; late-arrival species (*Pleurozium*, *Ptilidium*, etc.) are influenced by the development of a tree layer. During stand initiation and development, vegetative regeneration seems to play an important role, whereas sexual reproduction is predominant only in tree species and herbaceous early colonizers such as *Epilobium* and *Corydalis*. Post-fire seedling recruitment is mostly restricted to the first decades of stand initiation (Cogbill 1985; Morneau & Payette 1989; Sirois & Payette 1989), except in Labrador (Foster 1985).

The feathermoss forest–lichen woodland connection
Fire disturbance is generally considered as the main factor preventing succession of lichen woodland to feathermoss forest (Kershaw 1977, 1978). Whether the lichen woodland is a successional stage towards feathermoss forest or a self-replacing post-fire community has been much debated (Maikawa & Kershaw 1976; Kershaw 1977, 1978; Johnson 1981; Foster 1985; Morneau & Payette 1989). On the basis of studies conducted in the Northwest Territories and Labrador, Maikawa & Kershaw (1976), Kershaw (1977, 1978) and Foster (1985) concluded that succession to closed-crown forest would occur in absence of fire. In Labrador, where long-term post-fire regeneration of black spruce and balsam fir is apparently occurring in lichen woodlands developed on dry sites, Foster (1985) hypothesized that successional establishment of the feathermoss forest is the rule. An estimated fire rotation period of about 500 years (Foster 1983) would provide the conditions for such a successional replacement, but higher fire frequency at fire-prone, dry sites could potentially maintain the lichen–spruce facies. Rowe *et al.* (1975) and Johnson (1981) argued that given present estimates of fire frequency in western Canada such a successional pathway has only little chance to occur. In fact, no field evidence has yet been presented to document a shift from lichen woodland to moss forest in subarctic areas. In cold boreal environments in northern Quebec, direct field evidence indicated that lichen–spruce communities may persist in the absence of fire for at least 250 years in the lichen woodland zone (Morneau & Payette 1989), and 800–900 years (Payette *et al.* 1985) or even more than 3000 years (S. Payette, unpublished data) in the forest–tundra. It is likely that maintenance of the lichen woodland in the central and southern part of the boreal forest is provided by climatic conditions promoting recurrent fires, while in the northern part, despite long fire-free intervals, seedling establishment and layering in old-growth lichen–spruce vegetation are slow regeneration processes only able to maintain marginal tree populations thriving in the lichen matrix.

Tree recruitment during post-fire succession
The age structure of most post-fire stands in mesic sites of the boreal forest is determined early after stand recovery. In species with serotinous or semi-serotinous cones, such as jack pine and black spruce, massive seedling establishment usually occurs the first years after a fire. Recruitment begins at year 1 and continues for several years when cones are not destroyed by fire (Wilton 1963; Wein 1975; Zasada, Viereck & Foote 1979). Using age-structure data, Cogbill (1985) found that 71% of all 1785 sampled trees (including several deciduous and conifer species) in the southern boreal forest of Quebec became established during the first

30 years after fire. In the lichen woodland zone (Morneau & Payette 1989) and the forest–tundra (Sirois & Payette 1989), most black spruce recruitment during lichen woodland recovery also occurred during the first 20–30 years of the post-fire period, as suggested by age-structure profiles, i.e. before full development of the lichen carpet. Early post-fire seedling establishment producing even-aged stands was also found by Ahlgren (1959), Wein (1975), and Black & Bliss (1980). The initial post-fire cohort in stands of the southern boreal forest dominates for up to 250 years; tree mortality becomes apparent only 130 years after fire (Cogbill 1985). During lichen woodland development, layering takes over seedling establishment as early as 65 years after stand initiation (Morneau & Payette 1989). It is also at that time that the tree cover reaches about 20%, a value that will remain fairly constant during the next 200 years of woodland development, natural tree mortality of the post-fire cohort being compensated for by layer expansion. From the southern to the northern part of the forest–tundra, a progressive post-fire exclusion of the lichen woodland and correlative expansion of the lichen–heath community are also currently observed, the so-called subarctic deforestation (Payette & Gagnon 1985; Sirois 1988), because of a failure in seedling establishment after fire disturbance. Thus, a gradient in tree density from highly stocked to treeless is likely to occur from south to north, depending on success of seedling establishment associated with minimal germination temperature (Black & Bliss 1980). Severity of the burn, seedbed conditions, importance of the seed source, quantity of viable seeds, air and soil temperatures are all important factors influencing success of tree regeneration in northernmost boreal environments.

Diverging post-fire types
In a particular region, the dominant zonal vegetation is usually composed of a mosaic of plant communities representing the whole spectrum of the post-fire recovery sequence leading to its maintenance within a definite fire rotation period. On the other hand, degraded or diverging post-fire types that preclude return to the so-called steady-state conditions are probably the result of catastrophic fire events occurring episodically. Lichen–spruce woodlands in sites where feathermoss–spruce forests usually predominate appear to be such diverging forest types initiated by one or several catastrophic fires, thereafter able to maintain their composition and structure through a regime of recurrent, light fires. Although most dominant plant communities in an area tend to maintain their facies during fire cycles, it is likely that the spatial importance of diverging post-fire types will increase with time under present climatic conditions, assuming the occurrence of occasional catastrophic fires on a longer timescale than current surface fires. Because the length of the fire

rotation period and the age of the oldest post-fire stands do not vary much south of the forest–tundra and north of the spruce–fir belt, only catastrophic fires appear to be responsible for stand initiation and expansion of diverging post-fire communities. Processes involved in stand shift are interfering at the post-fire regeneration level on all segments of the plant communities. Under the present conditions of most regional fire regimes, shifting dominance throughout the boreal forest, as described above, is a stepwise process leading to spruce expansion in the south and lichen expansion in the north. Corresponding expansion of the spruce forest in the spruce–fir belt, lichen woodland in the feathermoss-forest zone, and lichen–heath in the lichen-woodland zone appear to be typical examples of the influence of fire during the present cooling period (Neoglacial).

In the closed-crown forest zone, black spruce and birch expansion is promoted in mesic sites because of their ability to regenerate early after fire, while balsam fir becomes a secondary species assuming dominance only in late-successional sites (Fig. 5.2) with a longer fire rotation period (Carleton & Maycock 1978; Foster 1983, 1985; Gerardin 1980). Shifting dominance from black spruce to white birch forests may occur also as a result of severe fires that burn cones of standing conifers and soil humus, thus favoring seeding of birch, a stand conversion already suggested elsewhere (Lutz 1956a; Viereck 1973). Similarly, the presence of lichen woodlands in this zone, i.e. in mesic sites where feathermoss forests usually prevail, cannot be explained solely by openness of the tree cover, since both types have an open-crown canopy due to limited success in post-fire seedling recruitment. The origin of the lichen woodland must be explained, at least in part, by the elimination or substantial depletion of the shrub populations during catastrophic fires. Most shrub species of the closed-crown forest regenerate immediately after a fire, where the soil humus has not been combusted (Flinn & Wein 1977). Sprouting from rhizomes is an effective post-fire regeneration process in the boreal forest (Viereck 1973), whereas shrub establishment from seeds constitutes a rather minor component of the post-fire recovery process (Johnson 1975; Moore & Wein 1977). A severe fire that kills both aerial and subterranean parts of the dominant shrub species is likely to facilitate the expansion of lichens. Because most vascular species living in a fire environment regenerate early after fire, the size of the post-fire woody populations will be determined largely by the success of seedling establishment and vegetative propagation during the first years of the recovery sequence. Thus, the aggrading lichen mat will take over during succession at the expense of the depauperate shrub layer. Spatial restriction of the moss component may be a consequence of a failure in the post-fire regeneration of woody plants. The advent of recurrent, light fires will suffice eventually to maintain the lichen-woodland type within the feathermoss forest zone. The occurrence of subsequent, major fire events will likely affect the success of spruce regeneration and cause a drop in tree density of the lichen woodland and its conversion to parkland (Fig. 5.2).

In areas where jack pine occurs, this species will dominate over black spruce, particularly in dry sites where the soil humus is completely combusted (Carleton 1982a,b; Carroll & Bliss 1982; Desponts 1989). In this context, jack pine woodlands are positioned at the extreme end of the associated fire-severity and fire-frequency gradients (Cayford & McRae 1983) within the feathermoss-forest and lichen-spruce woodland zones. Multiple fire episodes are generally recorded in the stand structure of jack pine forests. Crown fires and severe surface fires produce even-aged forests, while light fires occurring in old-growth jack pine stands may induce complex age structures with concurrent postfire cohorts (Carleton 1982a,b, Desponts 1989). Black spruce will replace jack pine in sites where fire intervals exceed 150–200 years (Desponts 1989).

In the lichen woodland zone, the influence of catastrophic fires is also of great importance for the fate of the lichen–spruce woodland. Since the lichen carpet forms a permanent feature of the type while the shrub layer is somewhat reduced (Hustich 1950; Auclair 1985; Morneau & Payette 1989), severe fires will affect mostly the size of post-fire tree populations. Lichen woodlands developing after the passage of severe burns have a sparse tree cover (<10% coverage with trees several meters apart), corresponding to the parkland type. Such developmental pattern in the diverging recovery sequence of the lichen woodland culminates in the complete elimination of spruce from mesic sites in the forest-tundra (Payette & Gagnon 1985; Sirois 1988). From south to north, the increasing treelessness of mesic sites bearing spruce charcoal at the soil surface suggests that lichen–spruce woodlands, formerly widespread, have fragmented and retracted to their present distribution because of catastrophic fires during the past centuries and millennia (Payette & Gagnon 1985).

In conclusion, according to the large geographical distribution of zonal types, the closed-crown forest zone (excluding the spruce–fir type) and the lichen woodland zone are forming the active regenerative compartment of the boreal forest, induced by frequent fires typical of rotation periods <150 years favoring self-replacement, and occasional severe fires that produce diverging depauperate types usually found northwards. In contrast, the forest–tundra zone appears as a remnant of a once active, regenerative compartment, now undergoing a lengthy process of deforestation under a regime of infrequent, severe fires and frequent small surface fires during rotation periods largely in excess of 150 years (Payette et al. 1989b).

Fire disturbance during the Holocene

Attempts have been made to decipher the role of fire in boreal environments during the Holocene. Several approaches have been used to evaluate fire frequency from laminated lake sediments and peat deposits (Nichols 1975; Swain 1973, 1978; Cwynar 1978; Tolonen 1983; Gajewski, Winkler & Swain 1985; Patterson, Edwards & Maguire 1987), as well as from stratigraphic soil sections in other habitats (Filion 1984; Mathieu, Payette & Morin 1987; Desponts 1989). Concurrent changes in abundance of pollen species and fossil charcoal were also used to explain vegetation shifts in the Holocene record (Swain 1973; Cwynar 1978; Green 1981, 1982, 1987; Tolonen 1983).

Microfossil charcoal analysis

Microfossil charcoal is widespread and abundant in most lake sediments of northern environments, even in remote arctic areas such as Greenland where wildfire is a rare event (Fredskild 1967). Microfossil charcoal counted on pollen slides has a complex source area, with the local fire record strongly biased because of an unknown proportion of regional and extra-regional charcoal inputs (Patterson, Edwards & Maguire 1987). Improved sampling methods such as those reported by Clark (1988a) are needed to document the long-term fire history at the site scale. Charcoal values obtained with a petrographic thin-section method focusing on large charcoal fragments give much finer resolution than the pollen-slide method (Clark 1988a). Relatively good correspondence between charcoal content in dated varved sediments and fire-scar dates from a small lake catchment in Minnesota was used to assess the impact of climate change on fire regime (Clark 1988a,b). During the fifteenth and sixteenth centuries, fire frequency was apparently higher than during the Little Ice Age, with fire occurrence tuned to multiples of the 22-year drought cycle (Clark 1988b). Using a filtered series of the charcoal index, Clark (1988b) suggested a 44-year fire cycle during the fifteenth and sixteenth centuries and an 80–95-year cycle during the Little Ice Age, a shift from frequent, less intense fires to less frequent intense fires. Earlier studies on laminated lake sediments based on 10-year intervals of microfossil charcoal values from pollen slides interpreted charcoal peaks as evidence of major fire events, with an approximate mean fire frequency of 60–70 years in Minnesota (Swain 1973) and 80 years in Ontario (Cwynar 1978). Whatever the method used in this type of paleoecological study, it is virtually impossible to determine all the elements of the fire regime. Mean fire frequency and types of fire may be approximated to some extent (Clark 1988b), but fire frequency deduced from lake sediments is somewhat conservative compared with fire-scar data (Tolonen 1983). One needs comparisons among several core sites to evaluate changing fire regimes and their impact on regional vegetation through time.

Macrofossil charcoal analysis

Macrofossil charcoal analysis from stratigraphic sections in eolian sediments yields direct evidence of past fire activity at fire-prone sites. Regional correlation among several sites may give information on long-term fire frequency and periods of major fire activity with their associated vegetation change (Filion 1984). The most complete boreal forest fire chronology (5000 BP to present) using direct fire evidence from discrete charcoal layers in dune paleosols was provided by Filion (1984) along a transect from the lichen woodland zone (central boreal forest) to the Arctic tundra in northern Quebec. Three major eolian intervals closely associated with fire were identified during the past 3300 ^{14}C-years, i.e. 3250–2750 BP, 1650–1050 BP, and 750 BP to present. Dated fire events and eolian activity were used to deduce climatic episodes with particular fire impact on ecosystem development (Fig. 5.3). Two cold–dry periods with high eolian activity promoted by wildfires were identified at 3250–2750 BP and 1650–1050 BP. Warm–humid periods with only local eolian activity occurred at 4800–3250 BP, 2350–1650 BP, and 1050–750 BP. Low incidence or absence of significant fire activity between 2750 BP and 2350 BP was most likely caused by cool and humid climatic conditions.

In a detailed study of paleofire occurrences in jack pine sites of the lichen woodland zone where the species reached its northeasternmost distribution in northern Quebec, Desponts (1989) showed that fire has been a recurrent event since 6000 BP. A close correspondence was found between her fire data and the chronology of fire and eolian activities developed by Filion (1984), suggesting that well-bracketed fire–eolian periods from the mid-Holocene to the present were climatically controlled. An independent fire chronology built up from radiocarbon-dated charcoal sampled at the top of mineral soils or buried in soil by solifluction in sites of the forest–tundra also showed similar fire episodes during the past 3000 ^{14}C-years (Payette & Gagnon 1985; Millet & Payette 1987). The charcoal layers in treeless sites were interpreted as remnants of conifer stands unable to regenerate successfully after fire disturbance, and indicate cold periods (Fig. 5.3). Milder periods were deduced from dated charcoal in the topsoil of modern forest and krummholz stands, now forming old-growth communities derived from the post-fire cohorts established since the last fire, i.e. several hundred or thousand years ago. These fire chronologies appear to be representative of late Holocene (Neoglacial) fire–climate episodes throughout the lichen woodland and the forest–tundra zones. Failure of post-fire regeneration of the lichen

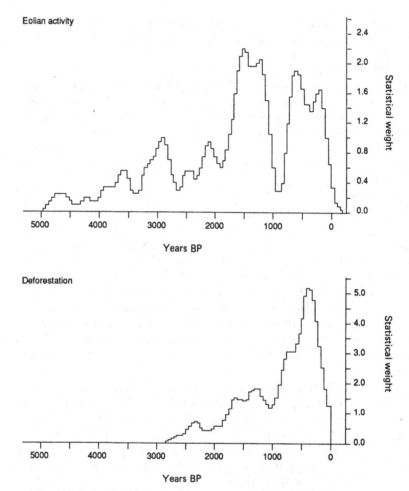

Fig. 5.3. Paleofire chronologies reconstructed from buried charcoal in eolian deposits of the lichen woodland and the forest–tundra zones (modified after Filion 1984) and from surficial and buried charcoal in lichen-heath sites of the forest–tundra (modified after Payette & Gagnon 1985) in northern Quebec. ($N = 86$ radiocarbon-dated samples in upper histogram and $N = 98$ in lower histogram). See original papers for calculation of statistical weight.

woodland, and retrogression towards treeless lichen–heath since 3000 BP, but most particularly during the past 1500 ^{14}C-years, seems to be the main factor responsible for inception and expansion of the neoglacial forest–tundra. Conversely, it may be hypothesized, as suggested in the first part of this chapter, that a simultaneous expansion of lichen–spruce woodland occurred in the central boreal forest at that time.

Charcoal layers in soil sections were also analyzed in the northern boreal forest west of Hudson Bay (Bryson, Irving & Larsen 1965; Sorenson *et al.* 1971). The geographical position of radiocarbon-dated charcoal relative to the modern forest–tundra boundary, defined as the limit where forests cover 50% of the soil surface (Larsen 1965), was used to assess past fluctuations of the boreal forest during the Holocene caused by fire–climate factors (Figure 2 in Sorenson *et al.* 1971). Charcoal samples from treeless sites, south of Dubawnt Lake where the modern treeline is located, were considered as remains of forests unable to regenerate after fire. Because the sampled sites were all located south of the present treeline, a situation that cannot provide evidence of past boreal forestline displacements, the alternative hypothesis is that charcoal samples from treeless sites are indicative of deforestation events, in a way similar to those reported for late Holocene development of the boreal forest east of Hudson Bay. The clustering of most charcoal dates between 1600 BP and 1000 BP (Sorenson *et al.* 1971) may be tentatively correlated with extensive deforestation that occurred during that period in northern Quebec, but lack of stratigraphic information on charcoal data and site conditions cannot provide a clear picture of the spatiotemporal influence of fire disturbance in this forest–tundra region. As a whole, the origin of the modern Canadian forest–tundra may be attributable to late Holocene deforestation induced by wildfires of a once larger lichen woodland zone during periods of climatic cooling.

Pollen records

Most patterns of tree pollen curves in published pollen diagrams dealing with the boreal forest environment (Ritchie 1987*b*) are interpreted in terms of climatic change, post-glacial migration, site conditions, and sedimentary processes, but rarely as a result of fire influence or gap-induced disturbance on the biota (Swain 1978; Green 1982, 1987). Unless clear-cut patterns are found between pollen and charcoal abundance, such as those showing, for example, striking reversible shifts between fire-prone *Eucalyptus* vegetation and humid rainforest in eastern Australia (Singh, Kershaw & Clark 1981), it is usually difficult to identify any fire influence on individual or multiple pollen curves. Some of the reasons may be that most boreal tree taxa are fire-adapted to some extent, and that pollen-curve fluctuations are polygenetic in origin. The sampling interval along the sediment cores is also an important factor influencing scale of data interpretation. Probably the best sampling sites to sort out the influence of fire disturbance on pollen assemblages are those located in ecotonal areas and in floristic regions dominated respectively by fire-evaders such as black spruce and fire-avoiders (Rowe 1983) such as balsam fir and white spruce. Two boreal regions of eastern Canada appear to provide such conditions. We here examine some post-glacial pollen sequences on a transect from north to south in Labrador–Newfoundland.

Forest–tundra

Pollen studies in sites of the forest–tundra in eastern Canada have emphasized substantial forest reduction after 3000 BP (McAndrews & Samson 1977; Short & Nichols 1977; Richard 1981*a,b*; Lamb 1985), one or two thousand years after initial woodland development. Although climatic cooling was considered as the ultimate cause of the decline, the ecological processes involved are somewhat obscured by the resolution of the pollen data and current knowledge of vegetation dynamics at different time scales. Because pollen analysis yields only indirect evidence for long-term vegetation change, factors responsible for the change are inferential and postulated rather than demonstrated.

A recent account of the postglacial history of the forest–tundra in Labrador (Lamb 1985) showed diachronous changes in spruce pollen since 3000 BP among sites positioned along a south–north transect. Spruce decline was interpreted 'as a potential record of lowering of the altitudinal tree limit' (Lamb 1985). Decline began earlier at the southernmost site (Gravel Ridge) located at a mean elevation of 565 m, and rather recently northwards (Caribou Hill site: mean elevation of 475 m). Lamb (1985) argued that diachronous spruce decline was caused by differences in catchment elevation and relief. However, present plant cover at both sites shows contrasting vegetation types (figure 4 in Lamb 1985), the southernmost site being dominated by lichen–heath and the northern site by more or less extensive lichen woodlands. As an alternative interpretation, one may suggest that woodland exclusion by fire disturbance started earlier at the south site, while deforestation is still progressing at the north site. Such diachronous decline could be an example of deforestation events associated with fire disturbance during critical climatic periods of the late Holocene, in a spatiotemporal patchy pattern similar to that recorded in northern Quebec (Payette & Gagnon 1985). Long-term deforestation and conifer exclusion in the area studied by Lamb (1985) do not seem to be related to distance from the present latitudinal treeline, an indication that expansion of forest–tundra through fire removal of lichen woodlands may proceed at some distance from the treeline without any significant treeline displacement (Payette & Gagnon 1985).

Closed-crown forest zone

In the warmer boreal sites of the closed-crown forest zone, near the Atlantic seaboard of southeastern Labrador, Engstrom & Hansen (1985) analyzed a detailed sequence of post-glacial vegetational change based on pollen and chemical stratigraphy of lake sediment cores. Successive shifts in tree vegetation, from white spruce to balsam fir to black spruce, were interpreted as a consequence of autogenic change associated with soil development, although the authors acknowledged the potential role of fire in the competitive replacement of white spruce and balsam fir by black spruce during the Holocene. The dominant vegetation type in the area is black spruce – feathermoss forest, with lichen–spruce woodland on upland sites considered as a transitional community (Foster 1984). Balsam fir and white spruce are fairly common in more productive mesic sites in protected locations (Foster 1984). Long fire-free intervals in excess of 250 years and poor post-fire seedbed conditions produce all-aged population structures associated with a slow post-fire recruitment.

Black spruce became dominant over balsam fir and white spruce between 6500 BP and 6000 BP in a period of maximum warmth, a situation apparently best explained by soil deterioration, so-called soil paludification. The accumulation of deep forest peat inducing cold, water-saturated and acidic soil conditions would be responsible for the change in about 2000 years (Engstrom & Hansen 1985). Although several workers have subscribed to the edaphic change hypothesis to explain the shift in regional tree dominance (Lamb 1980; Foster 1983, 1984; Engstrom & Hansen 1985; Ritchie 1987a), no direct evidence from past or present soil conditions has been provided on the paludification process, which is, in my view, a misnomer of the rather natural build-up of forest peat. There exist no general data on soil characteristics and raw-humus accumulation in each of the major forest types described in southeastern Labrador (Foster 1984). Under broadly similar climatic and edaphic conditions in central Newfoundland, Damman (1964) has shown a comparable distribution of forest types, with balsam fir forests occupying the better sites and black spruce forests restricted to nutrient-poor soils. Many balsam fir forests are replaced by black spruce forests after fire disturbance, but white birch and aspen forests are post-fire types succeeding to balsam fir stands in the better soils. Superimposed on the characteristics of parent material and topographic position (moisture regime), fire appears as a decisive factor influencing the distribution and composition of the major forest communities. Moreover, black spruce forests and balsam fir stands are often found next to each other in well-drained sites in Newfoundland, where both forests produce important amounts of litter (Damman 1971). Paludification leading to blanket-bog inception may occur locally in flat areas and in well-drained depressions, because of iron-pan formation in the B horizon associated with ericaceous raw humus. However, soil paludification does not occur on well-drained slopes (Damman 1971). Empirical data are not available to identify the ultimate causes of the shifting dominance from balsam fir to black spruce forests in southeastern Labrador, but fire and climate are the best candidates. The present distribution of white spruce and balsam fir coincides with mesic, nutrient-rich soils located in protected sites (Foster 1984), but also with environments where fire disturbance is far less frequent than on the exposed summits and slopes, dominated respectively by lichen–spruce woodlands and moss–spruce forests. According to the dominance of present forest types in the area, there is a possibility that the current fire rotation period of 500 years (Foster 1983) is somewhat different from past fire rotations. As an alternative to the edaphic change hypothesis, the dominance of black spruce over balsam fir since the mid-Holocene would be the result of sustained fire disturbances occurring in a progressively changing climate towards cooler conditions. At the Moraine Lake site, the maximum influx of spruce and ericaceous pollen between 2000 and 1000 BP may be a response to increased fire activity contemporaneous to that of northern Quebec (Filion 1984). In the Avalon Peninsula (eastern Newfoundland), a region presently dominated by an oceanic climate that restricts natural fire ignition, Brown-Macpherson (1982) has suggested that fire disturbance occurred more or less frequently since the mid-Holocene. Past, fire-prone climatic conditions similar to those deduced from macroscopic charcoal analysis in northern Quebec have possibly contributed to set the present vegetation mosaic in the Atlantic boreal forest.

Post-glacial plant dispersal and tree responses to environmental change
The changing spatial pattern of tree species since the late Pleistocene is the result of several biological and ecological factors, including the dispersal ability and ecological tolerance of tree species as well as environmental control. The modern forest assemblages are thus the historical outcome of the interplay between life-history traits of tree species and long-term environmental change, the individualistic behavior of tree species (Gleason 1926) being the basic tenet underlying interpretation of the post-glacial pollen sequences. Whether the dispersal ability of migrating species (Davis 1981) or overall climate change (Webb 1988) was the controlling process of Holocene tree distribution patterns remains a debated matter in what appears to be extreme views of the complex, interacting paleoecological phenomena that have produced changing vegetation patterns during post-glacial times. There is no reason to believe that post-glacial migration of tree species with their associated back-and-forth range limits was only a matter of dispersal ability or a direct response to climate forcing. Changing population growth rates and competition in relation to disturbance regimes must be also considered for a better understanding of post-glacial vegetation

development. Species range dynamics is basically a population phenomenon, controlled by active ecological factors, and includes migration, expansion, retraction, and sometime extinction of tree species depending on time and area. Studies in that direction are promising, by virtue of their capacity to integrate concepts of population ecology and paleoecology (Green 1981; Bennett 1983).

Large-scale tree species and vegetation patterns, migration routes and rates of spread of tree species during the late Pleistocene and Holocene have been deduced from pollen data by means of two different methods: the isopoll method and the 'pollen-increase' method (Bennett 1988). The first method is based on the changing pattern of pollen abundance through time, and the second method uses the first increase identified in each tree pollen curve of the diagram to determine the arrival time of a tree species at a site. Bennett (1988) has questioned the pollen-increase method because of subjectivity in the measure of initial rising of pollen curves. In addition, time-transgressive trends seen in some pollen maps may be more trends in population growth rates than arrival times of the species. The population-growth approach used by Bennett (1983) to interpret pollen data may be necessary to sort out changes in tree populations associated with natural disturbances during the Holocene. Provided there is sufficient dating of the post-glacial sequence, pollen abundances may be viewed as changing growth rates of tree populations caused by competition with or without climatic change. At long time-scales, natural disturbances creating all sizes of gap in the regional vegetation cover accelerate the rates of expansion of competitive tree species while restricting the range of others unable to adapt to the changing environment. Because of the prevalence of fire disturbance in the post-glacial record (Tolonen 1983; Filion 1984), range dynamics of most boreal tree species must have been controlled by ecosystem processes similar to those of present time where fire is a major controlling factor.

In this connection, Green (1981) has applied time-series analysis to pollen and charcoal data from the mixed forest in Nova Scotia to demonstrate responses of tree populations to fire recurrence. Using cross-correlograms between pollen and charcoal frequencies, he showed that pollen peaks of fire-adapted species occurred within less than 50 years of charcoal peaks and much longer time for late-successional species. In a further study in Nova Scotia, Green (1987) has suggested that early Holocene tree colonizers, especially conifers, appear also to have been fire-adapted species. As the climate became warmer, fire-adapted hardwood species expanded on sites cleared by burning, forming hardwood stands less susceptible to fire spread, thus promoting the expansion of late-successional temperate species. Green (1987) concluded that local environmental conditions, fire and competition were more important factors limiting the expansion of tree species than dispersal rates during the Holocene. Thus, it may be suggested, in a way similar to Mutch's (1970) hypothesis, that the dominance of conifer species throughout the Holocene in boreal areas has maintained a high probability of occurrence of large fires. The respective geographical dominance of fire-adapted conifer species in the boreal forest zone and late-sucessional hardwood species in the mixed and deciduous forest zones during pre-settlement times was probably the consequence of a dynamic equilibrium between life-history traits of tree species and climate-mediated disturbance regimes.

Of the three most important conifer species positioned along the fire-disturbance gradient in north-eastern boreal America, jack pine has a predominant continental distribution because of higher fire frequency, and balsam fir a maritime distribution associated with wetter conditions. The black spruce range overlaps the ranges of both jack pine and balsam fir as it can adapt to different fire–climate conditions. This gradient is rooted in the ecological history of this large area since at least the mid-Holocene. Dispersal mechanisms, climatic change and natural disturbance regimes have together been operative factors in setting the aggrading post-glacial limits of all three species. Past disturbance regimes have determined their respective regional dominance by population processes associated with success in post-fire regeneration. Jack pine and black spruce at their northern limit of distribution provide an interesting example in this respect. The post-glacial migration route of jack pine had a predominant northeasterly component (Webb 1988), most likely under slow dispersal conditions largely controlled by climate and fire disturbance. The species reached its northernmost limit in boreal Quebec around 3050 BP apparently at a low population density, which remained unaltered for several centuries because of low fire frequency (Desponts 1989). A significant population expansion occurred after 2400 BP exacerbated by increased fire activity (Desponts 1989). Of special interest is the fact that no range expansion accompanied the increase in regional population density because of sustained cooling conditions during much of the Neoglacial period, particularly after 2000 BP when dry-cold, fire conditions prevailed (Filion 1984). Black spruce showed contrasting population trends during the same period, this species experiencing large-scale post-fire exclusion especially at its outer limit in the forest–tundra (Payette & Gagnon 1985). The post-glacial history of both species was determined by the same controlling factors, i.e. dispersal, ecosystem disturbance and climate, but resulted in different population behavior at their outer distribution limit at the time when similar, climatically controlled fire events occurred. Jack pine was still a migrating species at the onset of the Neoglacial period, thus resulting in the demarcation and consolidation of its northern limit, in equilibrium with late-Holocene thermal and fire conditions. In contrast, black spruce reached its northernmost distribution before the Neoglacial period, under sustained warmer conditions, thereafter resulting in the spatial retraction of tree populations out of phase with late-Holocene climatic conditions.

I am grateful to K. Gajewski for commenting on the manuscript. Étienne Girard helped greatly to improve the illustrations. This paper is an outgrowth of my research program on the long-term dynamics of boreal ecosystems financially supported by the Natural Sciences and Engineering Research Council of Canada, the Fonds FCAR (Quebec), and the Geological Survey of Canada.

6 The role of forest insects in structuring the boreal landscape

C. S. Holling

Introduction

Human activities are increasingly affecting the relation between the biota and the physical environment. That has long been true of resource developments that have transformed vegetation on a regional scale. Now, however, the scale of human influence has increased to a planetary one because of the modification of the atmosphere by the accumulation of greenhouse gases and industrial pollutants (Clark & Munn 1986). The result could well be a significant increase in global temperature, as most general circulation models predict, exaggerated in the northern regions now occupied by the boreal forest. But our state of knowledge of such global-scale processes is sufficiently incomplete that the magnitude and location of those changes are highly uncertain. How, then, can we assess the impacts on vegetation, when we are so uncertain of the changes that might occur to the physical environment of northern regions? One way is to turn to the past to gain insight.

Certainly geophysical processes have led to planetary changes in the past that were extreme enough to trigger profound shifts in climate and in vegetation. When those produced pronounced shifts between glacial and interglacial conditions, the vegetation was transformed and individual species interactions became uncoupled to form a variety of transient assemblages very different from either those that preceded the shift (Wright 1987) or those that now characterize major biomes (Davis 1981).

Even modest changes in climate can be amplified if the frequency and extent of disturbance of vegetation is changed. That seems to have been the case when fire frequency declined along the prairie–forest border in Minnesota during the Little Ice Age, 400 years ago (Grimm 1983). Extensive areas changed from oak savanna to maple forest. Hence, however resilient ecosystems of the boreal zone are, there are limits to the changes they can absorb before being transformed into a different state.

With present knowledge, it is difficult to predict the future state into which the boreal forests might be transformed as a consequence of climate change. That depends on uncertain knowledge of species-specific dispersal rates and regeneration abilities. But transformation itself depends upon much better-known disturbance processes mediated by fire, wind, disease and insects. It is these processes that will critically determine how rapidly the boreal forest might be transformed from its present state into the radically altered assemblages of species that might characterize desert, shrub, savanna or different types of forest ecosystem (Davis 1986). Thus, although it is highly uncertain which of those states might lie in the future of the boreal landscape, it is possible to assess the sensitivity of existing boreal ecosystems to structural transformation into some qualitatively different state.

In this chapter, I concentrate on the role of forest insects – particularly defoliators and bark beetles – as one of the sets of agents causing extensive tree mortality in boreal forest ecosystems. My purpose is to organize what is known of forest–insect dynamics to assess how extreme climate change would have to be in order to initiate unexpectedly extensive and premature mortality of trees.

I shall do that by dealing with four topics. First, I will discuss the overall spatial structure and time dynamics of the boreal forest in order to demonstrate that forest insects have a specific role in affecting the structure of boreal ecosystems at landscape scales. Secondly, I will explore the degree to which organisms moderate variability of climate in order to emphasize that the response of forests to climate change is more likely to be sudden than gradual. Thirdly, I shall review the specific role of different outbreak classes of insect defoliators and bark beetles in precipitating such abrupt changes. Finally, I shall attempt to predict some of the impacts of climate change on forest break-up and transformation as mediated by insects.

Structure of the boreal forest

The way in which present boreal ecosystems structure both space and time will determine the rate and kind of response to climate change. The recent literature on ecosystems has led to major revisions in the original Clementsian view of succession as being a highly ordered sequence of species assemblages moving toward a sustained climax whose characteristics are determined by climate and edaphic conditions. This revision comes from extensive comparative field studies (West, Shugart & Botkin 1981), from critical experimental manipulations of watersheds (Bormann

& Likens 1981; Vitousek & Matson 1984) and from linked systems models and field studies (West, Shugart & Botkin 1981).

The revisions include four principal points. First, invasion of persistent species after disturbance and during succession can be highly probabilistic. Secondly, early and late successional species can be present continuously. Thirdly, large and small disturbances triggered by events like fire, wind and herbivores are an inherent part of the internal dynamics and in many cases set the timing of successional cycles. Fourthly, some disturbances can carry the ecosystem into quite different stability domains: mixed grass and tree savannas into shrub dominated semi-deserts, for example (Walker 1981). In summary, therefore, the notion of a sustained climax is a useful, but essentially static, equilibrium view. The combination of these advances in ecosystem understanding with studies of population systems has led to one version of a synthesis that emphasizes four primary stages in an ecosystem cycle (Holling 1986).

The traditional view of ecosystem succession has been usefully seen as being controlled by two functions: *exploitation*, where rapid colonization of recently disturbed areas is emphasized, and *conservation*, where slow accumulation and storage of energy and material is emphasized. But the revisions in understanding indicate that two additional functions are needed. One is that of *release*, or 'creative destruction', where the tightly bound accumulation of biomass and nutrients becomes increasingly fragile (overconnected) until it is suddenly released by agents such as forest fires, insect pests or intense pulses of grazing. The second is one of *reorganization*, where soil processes of mobilization and immobilization are organized so that nutrients become available for the next phase of exploitation, before they are leached away. That pattern is discontinuous and is dependent on the existence of multi-equilibria that are essential to the release and reorganization functions. Resilience and recovery are determined by the release and reorganization sequence and stability and productivity by the exploitation and conservation sequence. The cycle is summarized in Figure 6.1.

During this cycle, biological time flows unevenly. The progression in the ecosystem cycle proceeds from the exploitation phase (Box 1, Fig. 6.1) slowly to conservation (Box 2), very rapidly to release (Box 3), rapidly to reorganization (Box 4) and rapidly back to exploitation. Connectedness and stability increase, and a 'capital' of nutrients and biomass is slowly accumulated during the sequence from exploitation to conservation. The system eventually becomes overconnected so that rapid change is triggered. The agents of disturbance might be wind, fire, disease, insect attack or a combination of these. The stored capital is then released and the system becomes disconnected to permit renewal of the same stable state. If the disturbance destroys too much capital over too

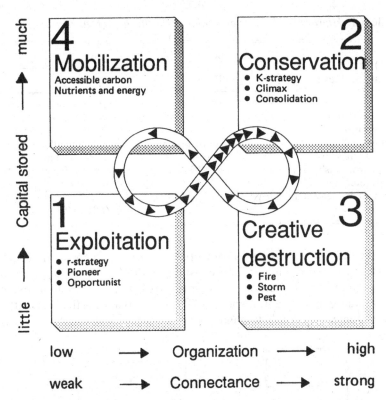

Fig. 6.1. The four ecosystem functions and their relationship to the amount of stored capital and the degree of connectedness. The arrowheads indicate a renewal cycle. The distance between arrowheads indicates speed, i.e. a short interval means slow change, a long interval, rapid change (reproduced with permission from figure 10.5 in Holling (1986)).

large an area, however, the system can flip into a qualitatively different stable state that persists until there is explicit rehabilitation by management. The critical question for the impact to climate change, therefore, is how much change does it take to release disturbances whose intensity and extent are so great as to destroy the renewal capital or prevent regeneration of the existing species?

Part of the answer lies in recognizing the hierarchical structure of forest ecosystems and the scale-dependent processes that shape that structure. Forested landscapes are organized in a hierarchy from needles, to crowns, to patches, to stands, to forest communities, to landscape elements (Table 6.1 and Fig. 6.2). Each of these levels is characterized by processes that operate on distinct time and space scales with each level

Table 6.1. *Hierarchical levels and their scales in the boreal region*

Hierarchical level	Scale	
	Time	Space
Landscape	1000+ yr	10 to 100+ km
Lakes, rivers, forest, clearings, communities	100+ to 1000+ yr	1000 to 10 000+ m (1 km to 10+ km)
Stands	10 to 100+ yr	100 m to 1000+ m
Patches and gaps	10 to 100 yr	10 m to 30 m
Treecrown, bushes, stumps, logs, etc.	1 to 10+ yr	1 m to 10+ m
Grass, forbs, leaves, needles, detritus, etc.	0.1 to 1 yr	0.01 to 0.1 m

nested within its next larger and slower one. The consequence is that each level has its own dominant time constant and its own scale of physical form or architecture (i.e. roughness or texture). These attributes provide the habitat template (Southwood 1977) for the organisms that live in the forest. Some, like defoliating insects, are directly coupled to a mosaic of

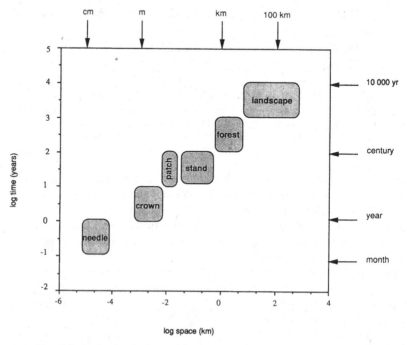

Fig. 6.2. Scales in the boreal forest.

needles in crowns, whereas others, like large ungulates, are directly coupled with the coarser mosaic of landscape elements that provide, in different places, browse, protection and seasonal ranges.

The physical structure of these levels moderates the variability of climate, transforming it into forms to which the animals adapt. Hence anything that modifies the physical or spatial characteristics of vegetation over scales from needles to forest communities is a critical agent in determining how climate change will affect forest ecosystems. That is why disturbance processes like fire and spreading insect outbreaks are so important in determining the vulnerability and resilience of forests to climate change.

Broadly speaking, three scale ranges can be identified whose attributes are determined by different classes of process. The micro-scale up to the patch is dominated by vegetative processes, from plant reproduction, growth and survival to succession. The other extreme, at a macro-scale, contains the landscape and its constituent elements of forest communities, clearings, lakes and rivers. These are shaped by much slower and larger geomorphological processes that determine topographic, hydrologic and edaphic properties. It is at the intervening meso-scale where a number of quite different biotic and abiotic processes operate.

These meso-scale processes transfer local events into large scale consequences because they have a spreading, or spatially contagious, character. Some of them are abiotic: water, fire and wind. But others are biotic, and are mediated by movements of animals as they disperse, migrate and move over hundreds to thousands of kilometers: insects that defoliate, ungulates that graze, mammals and birds that predate. Still others are the consequences of human transformations of regional landscapes. Meso-scale spatial dynamics and the distribution of vegetation are dominated by animals, people, water, wind and fire. It is at this scale that forest insects exert a role in facilitating a transformation of boreal ecosystems as a consequence of climate change. The degree of their influence depends upon the way the biota of the boreal zone has evolved to respond to variability.

Biotic responses to variability

The planet's biota has two strategies for dealing with variability. One is to adapt to the existing variability in time and space. The other is to control that variability, and by so doing, introduce stable states that are able to absorb specific ranges of variability without flipping into a different state. Forested ecosystems do demonstrate a measure of control of variability that gives them an inherent resilience. It is useful to identify the source and significance of that control by contrasting forested ecosystems with their opposite, i.e. pelagic ecosystems of the oceans.

A hierarchy of structures, similar to that in forests, occurs in the oceans. These structures are produced by physical forces and result in a nested set of levels that increase in scale from turbulence, to waves, to eddies, to gyres. As in the forest, animals respond by adapting to the structure of those levels: for example, phytoplankton to fine scales, predatory fish to coarse. The big difference between oceans and forests, however, is that the biota directly control the structure of forests over a wide range of scales, whereas the pelagic organisms of the ocean are stuck with the necessity of adapting to existing variability (Steele 1985, 1989). The physical properties of water leave no alternative except for the organisms to adapt to variability imposed by physical processes.

The terrestrial biota, on the other hand, has been able to exploit the second strategy of controlling variability because of the different spectral characteristics of physical variability in the atmosphere. At spatial scales of less than a few hundreds of kilometers, the atmosphere operates over considerably faster scales (of hours to months) than does the ocean (Fig. 6.3). As a consequence, forested ecosystems have been able to control the faster variability in the surrounding atmosphere by evolving the slower structures of vegetation biomass and texture that form the physical and biotic architecture of forest trees and soils. This means that the vegetation itself mediates the response of forested ecosystems to variability of climate.

That interaction is the focus of this book and because vegetation moderates both micro- and meso-climate, globally induced changes in climate will initially be strongly buffered by the physical structure of the boreal forest. Because there is a limit to that buffering, however, when the vegetation finally changes, the change will be abrupt. This is simply one facet of the inherent multi-stable nature of ecosystems that underlies the distinction that has been made between resilience and stability (Holling 1973). Resilience is the measure of how much change can be experienced before the behavior abruptly shifts into a different stability domain. In this case, those other domains define conditions that are identified as, and are maintained by, desert, savanna, grassland and forest ecosystems.

Those arguments see the global and regional climate as affecting the biota after being moderated by biotic processes. But they ignore the more speculative and intriguing idea that the biota themselves can affect atmospheric processes. If that is the case, then this analysis identifies where to look for such signs of GAIA (Lovelock 1988).

The hypothesis can be stated in two parts. First, do the albedos of major biomes that might be located in the north (e.g. coniferous, desert,

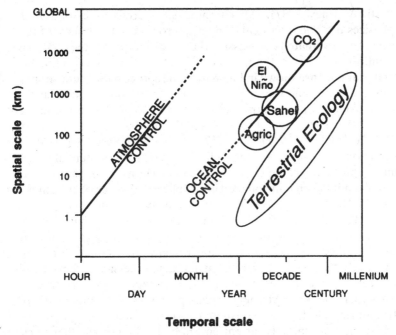

Fig. 6.3. Scale ranges for atmosphere, oceans and terrestrial ecosystems (Adapted from Steele (1989), courtesy of *Oceanus* magazine.
© 1989 Woods Hole Oceanographic Institution).

grassland) affect planetary heat balance differently, particularly in the
winter? Second, if so, does the climatic result of extensive cover by
coniferous forest in turn maintain those same forests through the function
of meso-scale processes of disturbance?

Both the effects of climate on vegetation and the possible effect of
vegetation on climate depend on what determines the shifts between
fundamentally different ecosystem types. This chapter examines how
forest insects can mediate such abrupt shifts in ecosystem type through
their impacts on vegetation over large areas. The question now is to assess
how their space and time dynamics intersect with other micro-, meso- and
macro-scale processes.

Forest and insect dynamics

Three separate groups of collaborators have developed similar and
reinforcing representations of the space and time dynamics of forest
insects (McNamee, McLeod & Holling 1981; Isaev *et al.* 1984; Berryman,

Stenseth & Isaev 1987). All combine an analysis of causation that emphasizes the role of critical ecological processes. The effects of those processes are evaluated by assessing their contributions to the occurrence of equilibria that shape the interactions between an insect and its host. Partial examples are shown in Fig. 6.4 for the spruce bark beetle in spruce forests of Norway and Sweden, and in Fig. 6.5 for the jack pine sawfly in jack pine stands in northern Quebec. Not all insects have recruitment curves of the types shown in Figs 6.4 and 6.5, however (summarized in McNamee, McLeod & Holling (1981)). The simplest have only one intersection with the zero growth line, implying one potentially stable equilibrium. More common are two intersections, such that a lower, unstable equilibrium is introduced implying the existence of an extinction threshold. But the more complex forms of the types shown in Figs 6.4 and 6.5 are typical of insects causing intense outbreaks, significant tree mortality and outbreaks that spread throughout large regions of forest. They are, therefore, the ones that are critically important as disturbance agents that can transform forests or initiate successional replacement. They are keystone species (Paine 1974) that shape ecosystem structure.

Such recruitment curves help explain part of the dynamics by temporarily ignoring or holding constant other time and space dependent processes so that, initially, attention can be focused on the dynamics of the insect alone. A more complete representation requires two additions. One is to include explicitly the interactions between two or, ideally, three sets of variables: the insect and the tree, or the insect, foliage and the tree. In this way, the interaction between the insects and vegetative structures can be described. This is the approach emphasized by one of the groups (Ludwig, Jones & Holling 1978; Clark, Jones & Holling 1979; McNamee, McLeod & Holling 1981). The other is to include the effects of spatial contagion, i.e. of interactions across space, as well as time. This is what the other two groups have emphasized (Berryman, Stenseth & Wollkind 1984; Isaev et al. 1984).

The results of including insect, foliage and tree can be summarized by zero-isocline surfaces of the kind shown in Figs 6.6 and 6.7 for the spruce budworm – fir forests of the south-eastern boreal forests of North America (from Figures 10.3 and 10.4 in Holling (1986)). These surfaces represent the conditions where variables are unchanging from one time period to another, assuming all other variables are fixed. The surfaces therefore separate regions of increase from regions of decrease. The way they intersect with each other makes it possible to predict distinct classes of time behavior, each driven by a distinct rate constant. McNamee, McLeod & Holling (1981), for example, predict four primary classes. In Class 1, none of the three primary variables oscillates. In Class 2, only the insect populations cycle, with the periodicity of months to two or three

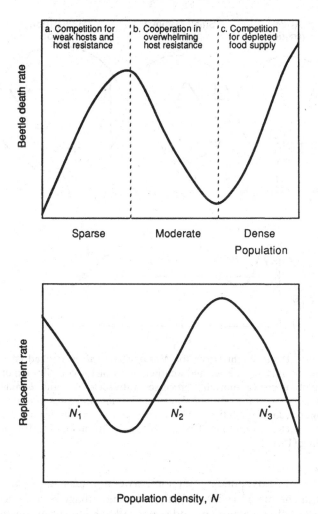

Fig. 6.4. *Upper*: Mortality processes and their effect on population densities of *Ips typographus*. *Lower*: Recruitment curve formed by those mortality processes under fixed conditions of fecundity, sex ratio, tree age and weather. The horizontal or zero growth line represents the conditions where the population density of the next generation is the same as the present generation. N_1^* represents a lower potentially stable equilibrium and N_3^* a higher one. The latter represents such a high density that the trees are killed and the outbreak collapses. N_2^* is an unstable equilibrium. If population densities exceed this value, the outbreak is released (modified with permission from Figure 7 in Berryman, Stenseth & Isaev (1987)).

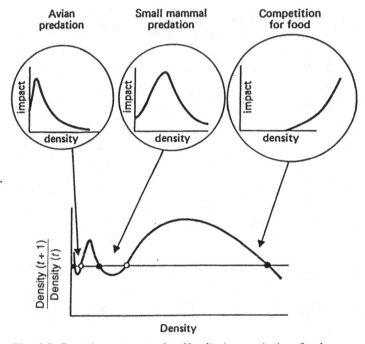

Fig. 6.5. Recruitment curve for *Neodiprion swainei* at fixed conditions of foliage, tree age and weather, showing the contribution of three of the critical mortality processes. Intersections of the recruitment curve with the horizontal zero growth line indicates potential equilibria, some potentially stable (closed circles) and some unstable (open circles) (reproduced with permission from Figure 10.1 in Holling (1986)).

years set by a fast dynamic interaction between insects and their associated natural enemies. In Class 3, both insect and foliage become coupled in a slower oscillation whose period is controlled by the time constants of the foliage, generating a basic periodicity of 10–15 years. In Class 4, all three variables become coupled and the periodicity is controlled by the slowest variable, the tree. Tree generation time hence sets the periodicity of multiple decades to one or two centuries.

The second addition requires a recognition of spatial processes. In contrast to the analysis of time dynamics, analyses of spatial dynamics have had a much shorter history. Levin & Paine (1974) and Paine & Levin (1981) established an important foundation when they represented spatial attributes in ecosystems as the birth and death of patches. That approach has proved to be of critical value in representing forest space–time dynamics in the FORET series of forest models (Shugart 1984). Spatial behavior above the scale of the patch, i.e. above 30 meters, is assumed to be a simple linear aggregation of patch dynamics. This is a useful step, but evaluation of the impacts of climate change has to include those spatial processes that initiate disturbances over larger meso-scales: fire, wind, water, ungulate grazers and insect defoliators. I am not aware, however, of many efforts to do this explicitly, for forested systems. One exception, however, explicitly includes dispersal of forest insects in a model as part of the spatial characterization of forest dynamics. That exception concerns the spruce budworm, an insect defoliator of balsam fir and spruce that has had a major role in shaping the structure of the south-eastern portion of the boreal forest from Manitoba to the Maritime Provinces of Canada and the New England states.

This system was extensively investigated in the classic study of Morris (1963), so that it became possible to develop a hierarchically structured model. Three levels were represented: (1) the interaction between insect, foliage and tree (as summarized above in Figs 6.6 and 6.7); (2) the aggregated effect in stands of even-aged trees; and (3) the regional consequences over some 50 000 square kilometers (Holling, Jones & Clark 1979; Jones 1979). For the purposes of this chapter, the key feature was the inclusion of dispersal of adult moths as an explicit meso-scale process that linked sites represented in 265 grid elements covering approximately 50 000 square kilometers (most of the province of New Brunswick).

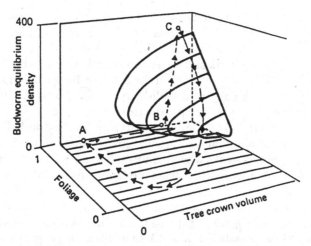

Fig. 6.6. Zero-isocline surfaces identifying equilibrium conditions for spruce budworm as a function of foliage and tree crown volume. The trajectory shows a typical unmanaged outbreak sequence (reproduced with permission from Figure 10.3 in Holling (1986)).

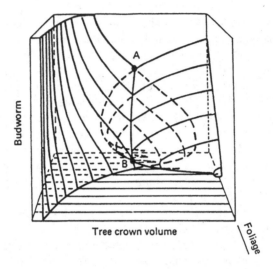

Fig. 6.7. Overlay of the zero-isocline surface of tree crown volume (a measure of forest age) against the budworm surfaces of Figure 6.6. The line AB shows where the two surfaces intersect. The third isocline surface for foliage is not shown, but does not intersect with line AB. Hence, there is no finite value where all three variables can be simultaneously at equilibrium. It is a condition of fundamental disequilibrium that generates a high-amplitude bounded cycle (reproduced with permission from Figure 10.4 in Holling (1986)).

The results of simulating the dynamics of the unmanaged system are shown in Figs 6.8 and 6.9. Two complete outbreak sequences are shown over the 84 years of the simulation. The figures demonstrate the contagious nature of an outbreak as it sweeps and oscillates across the region, causing extensive mortality of trees in the process. Between outbreaks, trees regrow and eventually re-establish conditions that make another outbreak increasingly likely. The proximate trigger can be the occurrence of warm dry springs that favor budworm survival (Wellington 1952), or a local invasion of moths that swamps the lower equilibrium. But such events only trigger outbreaks when the forest has matured sufficiently to dilute the effects of processes like avian predation that introduce a robust

Fig. 6.8. Spatial behavior of the budworm forest model resulting from an 84-year simulation. *X* and *Y* coordinates are geographical ones and the area embraces much of the Canadian province of New Brunswick. The *Z* axis represents the number of budworm eggs, with year zero representing the conditions in the field in 1953 (adapted with permission from Figure 8b in Holling, Jones & Clark (1979)).

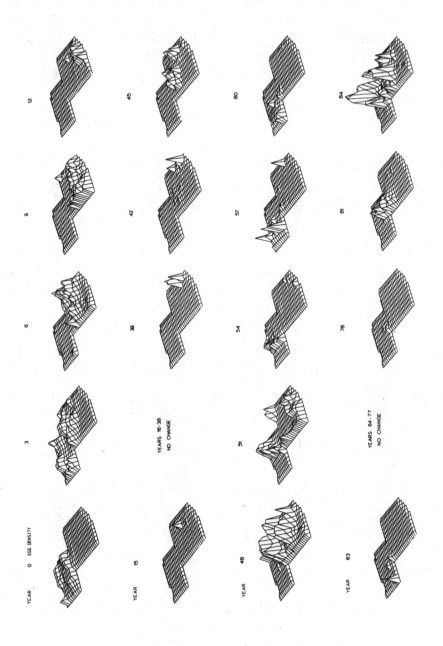

YEAR   0   EGG DENSITY

YEAR   15

YEAR   48

YEAR   63

YEARS 16-38
NO CHANGE

YEARS 64-77
NO CHANGE

3

51

12

45

60

84

9

42

57

81

6

39

54

78

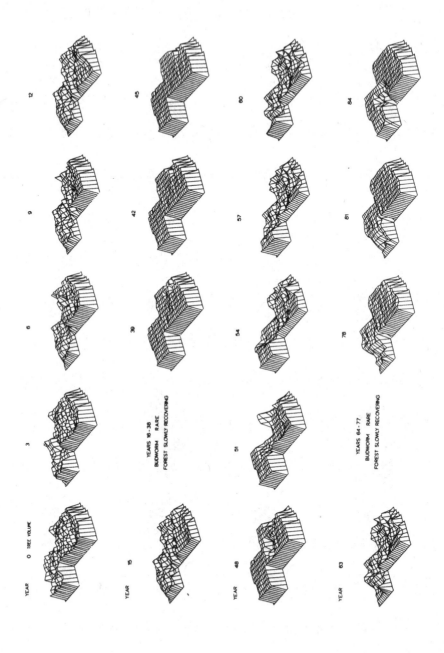

YEAR   0   TREE VOLUME

YEAR   15

YEAR   48

YEAR   63

YEARS 16 - 38
BUDWORM   RARE
FOREST SLOWLY RECOVERING

YEARS 64 - 77
BUDWORM   RARE
FOREST SLOWLY RECOVERING

low-density equilibrium (Holling 1988). The budworm, its associated natural enemies and the volume of foliage together interact to produce a 'boom and bust' pattern.

Once initiated in a sufficiently large patch, the outbreak becomes self-propagating and detached from the original weather conditions that might have been the initiating trigger. This interaction is a critical element in the successional cycle and in forming the spatial attributes of the forest. The result of dispersal of moths transfers a local impact within patches of the forest measured in tens of meters, to stands of trees measured in hundreds of meters, to whole forested landscapes of hundreds of kilometers. Together with the geomorphological processes that determine site conditions, the budworm-induced interactions produce a meso-scale pattern of stands of even-aged trees of tens to hundreds of hectares over regional landscapes of tens of thousands of square kilometers. Not only is the spatial distribution of age structure affected; the species composition is as well.

In any one outbreak cycle, a major portion of the balsam fir and some numbers of spruce died over extensive areas. The spruce–fir component of the region forms a major portion of the forested landscape, with birch as an early-successional species; the particular species composition depends upon competitive interactions that are very much influenced by budworm (Baskerville 1976). Although fir can outcompete spruce in many localities, it is particularly vulnerable and sensitive to attack by budworm. Spruce is less vulnerable and birch not at all. This results in a shifting balance in the competitive edge of the one coniferous species over the other. Between outbreaks, the rapidly growing balsam could begin to dominate spruce. Outbreaks reversed the competitive edge because balsam was so much more vulnerable, allowing spruce to sustain itself until the surviving understorey of balsam could again catch up.

Moreover, the proportion of fir to spruce is correlated with the intensity of outbreaks and this in turn reflects differences in regional climate (Clark, Jones & Holling 1979). The full geographical area of budworm-modified forest extends over a range of climatic regimes from continental in the west to maritime in the east. Spruce is more abundant in north-western Ontario, both coniferous species are present in approximately equal numbers in New Brunswick and, further to the east, balsam fir dominates in Newfoundland. The climate in each of those regions

Fig. 6.9. Same as in Figure 6.8 with the Z axis representing the tree foliage volume. Decrease in volume of foliage reflects tree mortality caused by budworm outbreaks (adapted with permission from Figure 8c in Holling, Jones & Clark (1979)).

differs: the warm dry springs that favor budworm survival are more common in the west, become less frequent in New Brunswick and still less so in the cool maritime climate of Newfoundland. When these different climate conditions were introduced into the model (Clark, Jones & Holling 1979), the simulated outbreaks followed the same progression, becoming progressively less severe from north-western Ontario, to New Brunswick, to Newfoundland. Thus the proportion of different tree species correlates well with intensity of outbreaks: the more intense the outbreak, the more spruce; the less intense, the more balsam fir. Differences in the frequency of warm dry springs thus affect tree species composition through the weather's influence on budworm as a disturbance process.

Other insects that cause spreading outbreaks associated with heavy tree mortality similarly form the structure and dynamics of forests. Outbreaks of the mountain pine beetle, interacting with fire, for example, maintain large stands of even-aged lodgepole pine much as does budworm in the spruce–fir forests (Peterman 1978; Peterman, Clark & Holling 1979). In the absence of fire, the outbreaks flip succession into more shade-tolerant species. Similarly, gypsy moth outbreaks have dramatically altered the composition of the hardwood forests in the eastern United States (Campbell 1979).

The critical feature of such spreading or contagious outbreak systems is that once a local outbreak exceeds a minimum patch size, it can become self-propagating, much like a forest fire (Holling 1980). So long as there is sufficient food in new territory, the outbreak can spread, largely independent of any weather condition that might have earlier triggered the outbreak. Some of the aggressive bark beetles, like the mountain pine beetle (Peterman 1978; Berryman 1982) and the spruce bark beetle (Worrell 1983), show the same behavior, killing extensive areas of coniferous trees in the process. Hence, even though quantitative representations and analyses of spatial dynamics have proved difficult, Berryman (1987) and Isaev et al. (1984) point out the necessity and value of even a qualitative classification of forest insect outbreaks into two classes: local and spreading. It is the latter, contagious properties of some forest insects that contribute to the meso-scale architecture of the boreal landscape.

The various time and space classifications of insect–forest interactions are synthesized and compared in Table 6.2. The ones of most significance for consideration of impacts of climate change are those classes that are spreading and either significantly impede growth (3B), or induce heavy tree mortality (4B). That focuses attention, therefore, on the aggressive bark beetles, like the mountain pine beetle, and the long-outbreak, spreading defoliators like spruce budworm.

Table 6.2. *Classification of insect, foliage and tree dynamics*

Class	Insect	Effect on foliage	Effect on tree	Hierarchical level and contagion	Berryman class	Isaev class
1A	persistent, low density	nil	nil	needles & local	non-outbreak	non-outbreak
1B	persistent, high density	affects buds, shoots, seeds	affects form and reproduction	needles, crown & local	sustained gradient outbreak	prodromal
2	short outbreaks at short (3–5 yr) intervals	minor	nil	needles & spreading	sustained eruptive outbreak	outbreak proper
3A	long outbreaks at moderate (8–15 yr) intervals	moderate on quality	growth inhibited	crown, patch & local	cyclic gradient outbreak	prodromal
3B	same as above	same as above	same as above	patch, stand & spreading	cyclic eruptive outbreak	permanent eruptive outbreak
4A	short outbreaks at irregular intervals	major	minor	crown & local	pulse gradient outbreak	prodromal
4B	long outbreaks at long (40+ yr) intervals	major	heavy tree mortality	stand, forest & spreading	pulse eruptive outbreak	outbreak proper

Let us pause now to review what I have concluded thus far. First, forest systems manage the variability of their physical environment because of the biotic architecture of trees and soils. This architecture develops as a consequence of slower biotic processes of tree growth, relative to insect dynamics, which are measured in terms of centuries and thousands of meters. Most organisms adapt to that structure, but a few become part of the dynamic that sustains and renews it. In particular, those insects that produce outbreaks that last several years, spread extensively and kill trees, are part of the renewal process of the forest ecosystem. They are a meso-scale process that affects the spatial patterns of the landscape by transferring fine-scale phenomena to landscape scales. The result is a landscape-scale pattern with resolution of ten to hundreds of hectares.

Now I can turn to the final objective for this chapter. How can this knowledge help us to predict and plan for possible climate changes?

Predicting the impacts of climate change

There is no doubt that the concentrations of greenhouse gases in the atmosphere have been increasing as a consequence of agricultural and industrial activities since early in this century (Clark 1986). There is little doubt that the concentrations will have doubled from their original post-glacial levels within the next 30–60 years. The exact timing is uncertain and will depend on the policy responses of nations and the importance of ecological processes that feed back to the atmosphere (Jaeger 1988). Some of those feedbacks could accelerate the rate of increase of carbon dioxide. For example, the large pool of carbon stored in the peat deposits of the boreal landscape and tundra could be suddenly released to the atmosphere because of drying. Other negative feedbacks could slow the buildup. The one being actively investigated now concerns the increase in fixation and storage of carbon because increased carbon dioxide concentrations can stimulate plant growth. On balance, however, it seems highly likely that the concentrations of these gases will continue to increase for the next few decades.

There is more uncertainty in predicting how climate will be changed by the doubling of greenhouse gas concentrations. The consensus of those developing General Circulation Models (GCMs) of the atmosphere is that global average temperature will increase by between 3 and 5 °C (Jaeger 1988). The regional distribution of that temperature change is much less certain, but there are good reasons to believe that the higher latitudes containing the boreal forest will be most affected. One model (Menabe & Wetherald 1986) generates predictions of an increase of 9 °C in the central part of the boreal landscape of North America. Even more extreme, the summer soil moisture is predicted to decrease by over 50% in the same region.

The various projections become progressively less certain as they proceed from questions of carbon dioxide buildup since the last glaciation, to buildup of greenhouse gases for the next few decades, to the effects of that increase on global average climate, to the changes in regional climate. But the uncertainty of even the most uncertain of those projections is minor compared to the uncertainty in predicting how such extremes of climate change could affect ecosystems of the boreal landscape. It would move the environmental conditions with such rapidity into such unfamiliar regimes, that the responses would be of an evolutionary character. And, just as we could not have predicted the disappearance of dinosaurs before the fact, only explained it after, so we can only assess changing vulnerabilities of the boreal biota and not the species-specific changes that could occur. There will be surprises hidden not only in the global 'greenhouse' because of unfamiliar interactions with the oceans, but in the regional 'gardens' (Clark 1986) as well, because of unfamiliar interactions with ecosystems.

Some sense of the way climate could change boreal forest–insect interactions can be deduced from the zero-isocline surfaces presented earlier in Figs 6.6 and 6.7 and in the more extensive sets described in McNamee, McLeod & Holling (1981) and in Berryman, Stenseth & Isaev (1987). The critical surfaces are those associated with outbreaks that spread and produce significant mortality of trees over extensive areas (Classes 3B and 4B, Table 6.2). Two groups of insects behave in this manner: defoliators, like the spruce budworm and jack pine sawfly, and bark beetles, like the mountain pine beetle and the spruce bark beetle.

Release of outbreaks is associated with escape from some lower equilibrium where populations are regulated. In the case of the bark beetles, host resistance is the process that controls low populations. Outbreaks occur when that resistance is overwhelmed by sufficiently intense attacks from other areas, or is reduced when trees are stressed by climatic extremes (e.g. *Dendroctonus ponderosae* (Peterman 1978; Raffa & Berryman 1983) and *Ips typographus* (Thalenhorst 1958)). In the case of the defoliators, predation by birds, small mammals, or ants maintains low populations. Outbreaks can occur when dispersal of adult moths overwhelms the regulation or when weather conditions boost insect survival (e.g. *Choristoneura fumiferana* (Clark, Jones & Holling 1979; Holling 1988), *Choristoneura occidentalis* (Campbell, Beckwith & Torgerson 1983), *Lymantria dispar* (Campbell 1975, 1979) and *Neodiprion swanei* (McLeod 1979)). The likelihood of that, however, depends on the volume of foliage and hence upon the age and condition of the trees. It is the slow expansion of the tree crowns that gradually dilutes the searching impact by predators to make the stand progressively more sensitive to an exogenous trigger of moth invasion, or to weather that enhances insect survival.

The effects of such low-density regulation are represented in the shape of the zero-isocline surfaces of the insect such as those in Fig. 6.6. It introduces a recurved undersurface, establishing a low-density stability domain that gradually contracts as the forest matures, until it disappears entirely. The upper surface of that isocline determines how high populations build up, once released, and that, together with the response of the tree, determines the intensity of tree mortality. Again, tree stress because of climate change or deteriorating site conditions can reinforce or intensify the amount and extent of tree mortality.

In summary, the responses of the insect and of the trees depend upon the shape of both the lower and upper surfaces of the zero-isoclines and their interaction with the tree zero-isocline surfaces (Fig. 6.7). The responses are determined by a set of interactions: predation, host resistance and foliage volume prior to release of an outbreak, and food volume and tree condition after release. The timing of the outbreak and the proximate trigger can be determined by climate. Increasing temperature and dryness in the spring, as forecast by some GCMs, could therefore increase insect survival and tree stress to precipitate premature stand break-up. If, at the same time, site conditions were permanently tipped into a drier and more eroded state, the conditions could be set for a radically altered ecosystem.

Explaining changes in ecosystem behavior from single causes like insect disturbances is, however, fundamentally wrong. Impacts of climate change would be likely to increase the incidence of fire, thereby reinforcing the likelihood of early stand break-up and of soil destruction and erosion resulting in a reduced ability to reorganize accumulated capital. But other anthropogenic changes, independent of climate change, are continually occurring.

In New Brunswick, for example, forest stand size has decreased from over 200 hectares in the 1950s to 10 hectares now (G. L. Baskerville, personal communication). That remarkable increase in heterogeneity is the consequence of human control of budworm and of harvesting activity. If the fragmentation continued to change not only the distribution of tree ages but also the land uses, the contagious character of outbreaks could be eliminated. But, again, by that time the forest would have been transformed by human activity in any case.

Even events distant from the boreal landscape could become part of the syndrome of causes precipitating change. One such event is the reduction in populations of migrating insectivorus birds, possibly as a consequence of deforestation in their Neotropical wintering grounds

(Terborgh 1989). They comprise the bulk of the species responsible for the low-density regulation described above. So far, however, it seems clear that tropical deforestation has yet to cause significant population decline in the summer breeding areas, although there is some evidence that three noted budworm predators – the wood thrush, chestnut-sided warbler and blackburnian warbler – might have started to be affected (Wilcove 1988).

A recent analysis of the possible consequences of declines in insectivorous bird populations shows that this regulation is surprisingly robust (Holling 1988) and that, if this were the only change, bird populations would have to be reduced by over two thirds to cause a qualitative change in budworm forest dynamics. That is simply one more demonstration of the considerable resilience of ecosystems and of their ability to persist in the same configuration in the face of fairly major disturbance.

This is not to say, however, that ecological systems are infinitely resilient nor that loss of robustness of regulation has no costs. If residual regulation becomes fragile, the ecosystem becomes more vulnerable to qualitative changes triggered by other causes such as climate change. Ironically, the great resilience and robustness of ecosystems and their regulation masks slow erosion of the capacity to renew, leaving society ill prepared for the inevitable surprises.

In conclusion, this syndrome of causes, of which insect-mediated changes are one, is largely reinforcing. It seems likely that even if climate change is less than that predicted by the more conservative GCMs, local and distant effects of land use change and reduced faunal diversity will precipitate a transformation of the boreal landscape. Both the southern and the northern limits of the forest would be affected, suggesting a fragmentation of the forest, and a decoupling of species-specific interactions. It is hardly possible to predict which species would inherit that landscape, but traditional boreal ecosystems would shrink to smaller, discontinuously distributed areas.

Patterns in space and time in boreal forests

Introduction

Herman H. Shugart

The perception that there is a relationship between the patterns observed on a landscape and the set of physical and biological processes that generate those patterns is central to modern ecology. The concept was perhaps best elucidated in the classic 1947 presentation of A. S. Watt but was also an important construct in earlier papers by Watt (e.g. 1925) and others. The basic premise is to view an ecosystem as a working mechanism (Tansley 1935; Watt 1947). Such a mechanism, as a consequence of its internal interactions and interactions with the environment, produces the patterns that we see in nature. When one inspects Tansley's (1935) original definition of the ecosystem, one finds the same concepts that one sees in hierarchy theory today (Allen & Starr 1982; Allen & Hoekstra 1984; O'Neill *et al.* 1986; Urban, O'Neill & Shugart 1987).

Of course, the Watt–Tansley ecosystem paradigm has been reintroduced as a major ecosystem construct in ecological studies. One conspicuous reintroduction of these concepts was Whittaker's (1953) review, which used the Watt pattern-and-process paradigm to redefine the 'climax concept'. These same ideas are also found in ecosystem concepts developed by Bormann & Likens (1979a,b) in their 'shifting-mosaic steady-state concept of the ecosystem', as well as in what Shugart (1984) called a 'quasi-equilibrium landscape'. Given the richness of concepts developed by ecologists in the first half of this century, it is foolish to propose that any idea is new, but we are now in a position to extend the pattern-and-process paradigm in what may be fundamentally important ways.

The categorization of controlling factors that are important at different space- and time-scales in particular ecosystems has been the topic of several reviews (Delcourt, Delcourt & Webb 1982; Pickett & White

1985). The processes that control pattern can change in importance as the relation between pattern and process is not a unidirectional one of cause (processes) and response (patterns). Patterns in space can influence the magnitude of important ecological processes; the relation can be a mutually causal or a feedback one.

If pattern can modify processes, which can generate new pattern, and so on, is this feedback process stable? Is there more than one state of the ecosystem in which the pattern–process feedbacks are mutually stabilizing? At what time- or space-scale is a given process stabilizing or destabilizing in its feedbacks to pattern?

The chapters in the following section provide some answers for these difficult questions. The first two chapters treat major transitions between the boreal forest and other biomes. Sirois discusses the factors controlling the boundary between boreal forest and tundra. Pastor & Mladenoff present a companion chapter on the southern boundary of the boreal forest. In both of these chapters the possibility of multiple stable states of boreal forests and other ecological systems is discussed and affirmed.

Sirois (Chapter 7) synopsizes the northern boundary of the boreal forest over its global extent. He furnishes a set of definitions of the different limits that could be considered at the northern edge of the boreal forest and provides a definition of major forest structural zones that is consistent with these limits. The development of an initial global map of these zones and a review of the pattern of change in the boreal forest–tundra transition over the past 12 000 years are a particularly valuable background synthesis for understanding the interactive roles of climate and wildfire in shaping the northern boundary of the boreal forest. The climatic events immediately following a fire can promote or retard the tree population density (see Zasada, Sharik and Nygren, Chapter 3) and eventually reduce the reproductive capacity of the forest. The climate–fire interaction is a positive feedback relation in which trends for increasing or decreasing forest density following a fire are amplified.

Pastor & Mladenoff (Chapter 8) also find an abundance of complex system dynamics in the floristically simple southern boreal forest transition. A concluding comment to one of their sections, 'because it is evolutionarily very young, it is entirely possible that the boreal–northern hardwood *forest* landscape does not obtain a stable state even on very large scales because of underlying chaotic behavior arising from multiple feedback loops. . . and from the stochastic influences of climate. . .' could, by naming a different transition, be just as well a summary from Sirois' chapter. Pastor & Mladenoff focus on a range of feedback relations in the southern boreal forest to northern hardwood forest transition. They point out that the relatively low tree species diversity that characterizes the boreal forests is composed of species with a wide range of attributes that are important to ecosystem function (see also Nikolov & Helmisaari, Chapter 2). The low diversity tends to couple functional roles to the population dynamics of one or two species. There exist a number of interaction pathways among nutrient cycling, forage quality for animals, fuel build up, life history attributes of species, and plants and animals that can lead to a situation in which different types of forests can occupy similar sites.

The theme of complex system dynamics underlies the first two chapters in this section; the third chapter, by Glebov & Korzukhin (Chapter 9), echoes a chorus. In this chapter, the processes of bog formation or paludification within the boreal forest biome are studied as an internal transition between forest and bog. Glebov & Korzukhin find that bog-forming processes are amplified as paludification proceeds, and that the reverse of this process is also amplifying once it is initiated. Forest and bog are presented as two of the possible states on the boreal landscape. One can become the other through a web of linked processes that are strongly mediated by climate. Stratigraphic data on bog history as well as a review of observations throughout the boreal forest are analyzed to support this view. As a closing section, a nonlinear model of bog formation is developed and this model is analyzed for conditions under which one can expect the formation of bog, flooded forest, or forest.

The richness of pattern and the relation of this pattern to potential dynamic processes catalyze a need to better understand the larger scale patterns of the boreal forest. Ranson & Williams (Chapter 10) provide a background for those interested in understanding the use of remote sensing technology in boreal forest studies. Their discussion includes optical remote sensing (using solar radiation that is reflected, absorbed or transmitted at a particular wavelength) and microwave remote sensing. The review of the considerations in using remote sensing techniques provides a valuable introduction to what is a daunting literature for many plant ecologists. The use of remote sensing is particularly promising in increasing the understanding of energy interactions of soil and vegetation, in linking energy relations and ecosystem processes, and in quantifying forest change in space and time.

The final chapter of this section, by Solomon (Chapter 11), ties together the long-term dynamics and large-scale patterns of the boreal forest in terms of the needs of a boreal forest simulator. This chapter gives an overview of the patterns and processes discussed in detail in the earlier chapters and also provides a transition (and challenge) for the development of a unified boreal forest model that is discussed in the following section. Solomon describes the large-scale patterns of change in the boreal forest in the past. It appears that the areal extent of boreal forest worldwide during the full-glacial (Ice Age) condition of 18 000 years ago was considerably less than today. Solomon sees the construction of a unified boreal forest simulator as a needed development to synthesize our understanding of present forests and the forests of the past, and to predict the changes in the boreal forests of the future. The need for the model-based synthesis is fueled to a great degree by the dynamic changes in the forests in the past 18 000 years. The boreal forests of the geological-time 'yesterday' were different from those of today. The forests of tomorrow may be different again. The patterns produced by the complex, nonlinear dynamics mentioned in the chapters of this section and others in this book point to the use of models to synthesize and advance our understanding of the forces producing the boreal landscape.

7 The transition between boreal forest and tundra

Luc Sirois

Introduction

The boreal forest and the Arctic tundra are two major biomes which cover a total of 30 million square kilometers in the Northern hemisphere. They are characterized by contrasting dominant life forms and productivity levels associated with the presence of trees in the boreal forest and their general absence in the Arctic tundra. The same general differences exist between forest and grassland but the ecological factors responsible for the position of the forest–grassland transition zone are likely to differ from those operating for the boreal forest–tundra transition zone. A causal explanation of the climate–vegetation pattern of the northern forest border is as yet obscured by the poorly known ecophysiology of the tree species (Oechel & Lawrence 1985). Despite the lack of knowledge of the exact mechanism involved, there is general agreement that thermal characteristics of the climate are of major importance in determining the northern edge of the boreal forest. Somewhere at the interface between boreal forest and tundra, tree species face severe climatic conditions, which reduce their growth and development. This induces changes at the plant population and community levels, which can be observed in the transition zone between these two biomes. Forests of this boundary area are likely to be sensitive to environmental changes and often display low regenerative capacity. Various types of ecological alteration, for example climatic changes and wildfires, take place over a wide range of areal dimensions and in different periods. These result in a temporal and spatial vegetation patterning typical of the northern forest border.

An ecological synthesis of the circumpolar boreal forest–tundra transition zone has not yet been produced. The diversity of tree species, the limited knowledge of vegetation history and of the ecological processes taking place over this large area, as well as linguistic barriers with Russian literature, make it difficult to attempt a unified treatment of this transition zone as a whole. This chapter will first show the specific differences in and structural similarities of different areas of the circumpolar boreal forest–tundra transition zone. This is achieved by a description of the tree species composition and by a first approximation of a map of the circumhemisperic transition zone. Secondly, it will illustrate what is believed to be climatically induced spatial pattern and temporal dynamics observed in the northern forest borders of a limited number of regions, namely North America, Fennoscandia and the Soviet Union. Finally, the potential role of the fire–climate interaction in the Holocene development of the forest–tundra transition zone is exemplified by a case from north-eastern North America.

Terminology

The great diversity of approaches and criteria which have been used for the characterization and the subdivision of the boreal forest–tundra transition zone has resulted in a non-unified terminology (Ahti, Hämet-Ahti & Jalas 1968; Hustich 1979; Tuhkanen 1984). Frequently used are floristic (Aleksandrova 1980; Andreev & Aleksandrova 1981), climatic (Tuhkanen 1980, 1984) and physiognomic–structural (Hustich 1966; Atkinson 1981; Payette 1983) criteria. These criteria are not reciprocally exclusive and the predominance of one or another is conducive to specific constraints and advantages. The approach chosen depends on the scope and on questions to be investigated. In this chapter, I will use predominantly physiognomic–structural criteria to define zones and patterns. Floristic criteria will be used to illustrate regional specificity, and climatic criteria will be integrated as part of the causative factors. So far, a worldwide-accepted terminology to designate zones and pattern is still lacking (Hustich 1979). Herein, the following terminology and definitions, largely inspired from Hustich (1966, 1979) and Payette (1983), will be used (Fig. 7.1).

Continuous forest limit: Up to this line, forest covers all the water-free landscape features, including upland and lowland (Payette 1983). This limit is believed to be coincident with the isopleth where climate allows sexual regeneration of trees on an annual basis (Hustich 1966). This is a major phytogeographical concept, which still needs to be warranted by time-series data.

Physiognomic forest limit: This corresponds to the limit where forest stands are widespread in mesic sites, regardless of their sexual reproductive capacity (Hustich 1966, 1979).

Forest limit: This is the northern limit of forest stands found in sheltered areas of the northern part of the forest–tundra (Payette 1983).

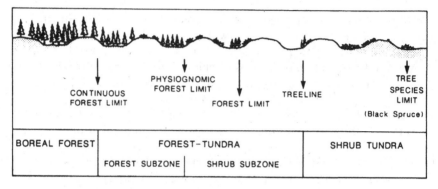

Fig. 7.1. Schematic representation of the major divisions of the boreal forest-tundra transition zone (Payette 1983; redrawn with the permission of the author.)

While the minimum dimensions of these stands have never been stated, they should correspond to a visible entity on a regular (approximate scale 1:50 000) aerial photograph. In this respect, most ecologists do not consider the northernmost so-called 'forests' consisting of Salicaceae species (Maycock & Matthews 1966; Aleksandrova 1980) as constituting the forest limit.

Treeline: This is the northern limit where forest species with an arborescent habit and reaching a minimum height of 5 m (Payette 1983) are found. This concept is totally different from Atkinson's treeline (Atkinson 1981) used in some works (Scott, Hansell & Fayle 1987), which actually corresponds to the continuous forest limit. It differs also from the treeline defined by Larsen (1965) as the zone where forest stands occupy 50% of the soil surface.

The above limits bound the major vegetation zones of the boreal forest–tundra transition zone as follows:

Taiga: Northern part of the boreal forest ranging from the northern limit of the closed crown coniferous forest to the southern limit of the forest–tundra. The northern limit of this zone corresponds to the economic forest-line *sensu* Hustich (1966, 1979), and to the continuous forest limit according to Payette (1983). In the taiga, the mesic forest stands are variously dense, but generally the open-crown lichen woodlands form a continuous cover on the uplands while feathermoss forests or forested bogs occupy the moister lowlands.

Forest–tundra: Vegetation zone lying between the continuous forest limit and the treeline. It is characterized by the patchiness of the forest cover in the landscape, the treeless areas being dominated by lichen and shrub species. The physiognomic forest limit divides the forest–tundra into the forest subzone in the south and the shrub subzone in the north

(Payette 1983) (Fig. 7.1). In the forest subzone the forest cover is widespread in the lowlands but quite sparse on drier upland sites. In the shrub subzone, where the climate has an obvious stressful effect on tree development, the forest stands generally become restricted to a few sheltered sites. Stunted growth forms are a common feature in such an environment; coniferous stands are mostly represented as krummholz.

Regional characterization of the boreal forest–tundra transition zone

The circumhemispheric belt between the southern limit of the taiga and the treeline shows a latitudinal vegetation zonation which is similar in many regions. In contrast, there is longitudinal variation in tree species composition and vegetation pattern; this variation should be considered to gain a global perception of the whole system.

Soviet Union

From Fennoscandia eastward to the Ural range, there is an increasing frequency of *Abies sibirica*, *Larix sibirica* and *Picea obovata* growing with *Pinus sylvestris* in the boreal forest–tundra transition zone. In the more continental climate east of the Ural range, *Larix sibirica* seems to be the most important tree species in the northern part of the transition zone while *Pinus sibirica* is more frequent further south. As the continentality of the climate increases toward eastern Siberia, *Larix dahurica* and *Pinus pumila* are frequent on upland sites (Tikhomirov 1971). These two species form the northernmost forests of the world, reaching lat. 72°N (Woodward 1987). In eastern Siberia, *Populus suaveolens*, *Chosenia macrolepis* and some *Betula* species occur in the valleys (Tikhomirov 1971) and go beyond the coniferous treeline, into the Arctic zone (Aleksandrova 1980). As in north-western North America, the range of most species drops southward near Soviet Beringia (Hustich 1966).

A number of vegetation maps of the Soviet Union (Lavrenko & Sochava 1954; Anonymous 1964) have been translated (Shabad 1965) or reproduced in English reports (Tikhomirov 1960; Hustich 1966; Bazilevich, Drozdov & Rodin 1971; Tuhkanen 1984). It appears that what is defined as the northern taiga zone by Soviet ecologists consists of a continuous, low-productivity forest frequently dominated by *Larix* and *Pinus*. Open-crown forests are a common feature in this area, which extends broadly from lat. 65° to 70°N (Fig. 7.2). It is structurally close to what we consider here as taiga. North of this zone is a strip of forest–tundra dominated by larches and pines and showing a typical patchy distribution of forest stands (Tikhomirov 1971), similar to that observed in the North American forest–tundra.

Fennoscandia

The most frequent tree species in the boreal forest–tundra transition zone of Fennoscandia are *Picea abies*, *Pinus sylvestris*, *Betula pubescens* and *Populus tremula* (Tuhkanen 1980). The deciduous species frequently

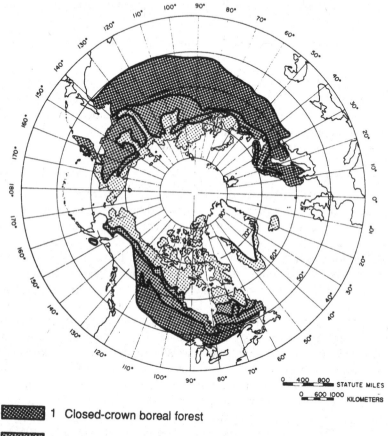

1 Closed-crown boreal forest

2 Taiga

3 Forest-tundra

4 Tundra

Fig. 7.2. Sketch map of the boreal forest-tundra continental vegetation zones. This map has been produced from information contained in maps from Anonymous (1964), Hustich (1966), Ahti, Hämet-Ahti & Jalas (1968), Krebs & Barry (1970), Tikhomirov (1971), Rowe (1972), Tukhanen (1980), Hämet-Ahti (1981), Olson, Watts & Allison (1983), Payette (1983) and Ritchie (1984).

extend north beyond the coniferous species, presumably in response to the damp oceanic climate (Hämet-Ahti 1963; Sjörs 1963; Ahti, Hämet-Ahti & Jalas 1968; Hämet-Ahti & Ahti 1969). The presence of south–north oriented highlands in Norway, along with a steep west–east oceanity gradient along the Atlantic coast, has a strong influence on the vegetational zonation of Fennoscandia. This is reflected in the maps of Sjörs (1963) and Ahti, Hämet-Ahti & Jalas (1968), both of which show an obvious longitudinal component in zonation. The northern boreal zone of Ahti, Hämet-Ahti & Jalas (1968) corresponds to the subarctic and boreomontane subzones in Sjörs' system (Sjörs 1963). This zone extends from the northern part of Fennoscandia southward into the highlands and seems to be structurally equivalent to what we have defined as taiga. The tree species component of the Fennoscandian lichen woodlands includes mainly *Pinus sylvestris* and to a lesser extent *Betula pubescens* (Oksanen & Ahti 1982). The orohemiarctic zone of Ahti, Hämet-Ahti & Jalas (1968) is named the Woodland–Tundra zone by Sjörs (1963). This zone, a narrow fringe between the northern boreal zone and the maritime tundra in the northernmost part of Fennoscandia (Fig. 7.2), consists basically of forest–tundra where *Betula pubescens* forms the treeline.

North America

In the North American boreal forest–tundra transition zone, the most frequent coniferous species are black spruce (*Picea mariana*), white spruce (*Picea glauca*) and larch (*Larix laricina*). West of Hudson Bay, black spruce and larch are most frequent in poorly drained sites while white spruce dominates in mesic sites. On the Quebec–Labrador peninsula and in Newfoundland, white spruce is restricted to areas influenced by the damp maritime climate of Hudson Bay and the Atlantic Ocean. Inland, black spruce forms almost monospecific forests in a large range of ecological conditions; larch is generally only a secondary species in such forests.

Jack pine (*Pinus banksiana*), fir (*Abies balsamea*) and sitka spruce (*Picea sitchensis*) are less widely distributed species in the transition zone. Jack pine is found in xeric habitats such as sand dunes (Desponts 1990) whereas fir is mainly observed in areas with a maritime–oceanic climate such as Labrador–Newfoundland (Lamb 1985). Sitka spruce is distributed on the Pacific shore, where it forms the treeline on the Alaska peninsula (Viereck 1979). Some deciduous tree species (*Populus balsamifera*, *P. tremuloides* and *Betula papyrifera*) form small populations, frequently associated with ecological disturbances such as fire or flooding, throughout the area.

Rowe's forest map of Canada (Rowe 1972), along with more regional studies (Rousseau 1952; Ritchie 1959; Viereck 1979; Payette 1983), allows us to depict the general zonation of northern North American vegetation. In Alaska, the treeline is situated at the southern Brook Range piedmont and drops southward from Kotzebue to Kodiak Island (Larsen 1989). The great difficulty in defining latitudinal zonation in this area is attributed to high elevation terrain situated where the forest–tundra transition zone would otherwise exist (Viereck 1979). The relatively flat topography of northern Canada is conducive to a simpler latitudinal zonation. Broadly speaking, the treeline is NW–SE oriented across Canada. It reaches its northernmost position in the Mackenzie River Delta area (lat. 68°N) (Black & Bliss 1978), drops south to James Bay (lat. 55°N) (Rowe 1972) and crosses the Quebec–Labrador peninsula at approximately 58°N (Rousseau 1952; Payette 1983). The area between the treeline and the continuous forest limit, the forest–tundra itself, increases in width from western to eastern continental Canada. This pattern can be partly explained by a southward rise in altitude from Ungava Bay to a point about 150 km north of the St Lawrence River (Hare & Ritchie 1972). Greater fluctuations in the July position of the Arctic air mass in eastern Canada (Bryson 1966) can also increase the width of the forest–tundra in this part of the country. South of the continuous forest limit is the large expanse of open crown forest of the taiga. The ground vegetation in the taiga west of Hudson Bay is generally dominated by shrubs and herbs, although some lichen woodlands are observed on upland sites (Rowe 1972). Lichen woodlands are far less extensive over this area than east of Hudson Bay (Rowe 1984), where they constitute the dominant forest type in the taiga and forest–tundra uplands (Larsen 1989).

Climate–vegetation relations

Elucidation of the relation between the northern vegetation boundaries, especially the treeline, and climatic indices has been a challenge to ecologists for decades. An ambitious attempt to relate climatic characteristics to northern phytogeographical regions is offered by Tuhkanen (1980, 1984). He suggests a system of climatic–phytogeographic regions based on a cubic model in which the axes are gradients of temperature sums, continentality–oceanity and aridity–humidity. The continentality–oceanity gradient, calculated by using the Conrad index, and the aridity–humidity gradient are related to the total precipitation during the growing season. They are not completely independent climatic indicators and appear to be associated mainly with the longitudinal pattern. Tuhkanen (1980) himself agrees that thermal characteristics of the climate, especially during the growing season, are responsible for or at least coincident with phytogeographical boundaries in northern latitudes.

Table 7.1. *Forest divisions and potential evapotranspiration on the Labrador–Ungava peninsula*

Division	Typical value of PE along boundaries (cm)	Dominant cover type
Tundra		Tundra
	30–32	
Forest–tundra		Tundra and lichen woodland intermingled
	36–37	
Open boreal woodland		Lichen woodland
	42–43	
Main boreal forest		Closed-crown forest

Source: Hare (1950). Reproduced with the permission of the American Geographical Society.

In this respect, the 10 °C isotherm of the warmest month of the year has been considered coincident with the treeline for many years. Although this line parallels the position of the treeline across northern Canada, it generally lies further north, especially in continental areas (Tuhkanen 1980). Nordenskjöld's line (Nordenskjöld & Mecking 1928, p. 73) integrates the temperature of the warmest (v) and coldest (k) month of the year to define a polar line situated where $v = g - 0.1k$. Nordenskjöld was not clearly affirmative in relating this line to the treeline, as some ecologists have done since (Larsen 1974; Tuhkanen 1980). Nevertheless, it has been shown that this line represents an improvement in fit compared with the 10 °C isotherm (Larsen 1974).

Another approach uses the temperature sum over a given threshold as an isopleth value coinciding with the treeline position. The 3 °C isopleth (sum of positive monthly mean temperature/12) shows a good fit, although some departures exist locally between real and predicted positions of the treeline (Tuhkanen 1980). Similar results were obtained by Young (1971) using the 35 °C isopleth calculated as the sum of monthly positive means over 0 °C.

The precipitation component of climate was not considered by Hare (1950) to be a limiting growth factor for trees in eastern Canada. Hare suggested that a relation existed between the major north-eastern American vegetational zones and given values of Thornthwaite's potential evapotranspiration (Tab. 7.1). The circumhemispheric forest–tundra transition zone occurs where potential evapotranspiration is approximately 34–35 cm (Larsen 1980).

Bryson (1966) stressed the striking correlation between the northern forest border and the modal July position of the front that separates the Arctic air mass from the Pacific air mass in Canada. This correlation fits

Table 7.2. *Representative values of various thermal parameters*

	Western Canada (100°W–120°W)		Labrador-Ungava (60°W–80°W)	
	Treeline	Closed crown forest	Treeline	Closed crown forest
Net radiation (kly)				
annual	18	28	19	31
thaw season[a]	19	29	22	31
Thornthwaite's PE (cm)	34	43	28	41
Duration of thaw season (days)	120	168	145	175

Note: [a] Period with mean daily temperature >0 °C.

Source: Hare & Ritchie (1972). Reproduced with the permission of the American Geographical Society.

better west of Hudson Bay compared with the Labrador–Ungava peninsula, which is also influenced by a rather high frequency of 'United States' and Atlantic air masses. Preliminary results by Krebs & Barry (1970) suggested that the relation between the modal July position of the Arctic air mass and the northern vegetation boundary also holds over Eurasia. These authors point out that the position of the July Arctic air front is not *per se* causative of the vegetational boundaries but influences the local energy budget during the growing season. In this connection, Hare & Ritchie (1972) suggested that northern vegetational zonation coincides with given values of several radiation parameters. Mean total and absorbed solar radiation isolines drop sharply southward over Canada by as much as 20° between 150° and 70° west, as do northern vegetation boundaries. The mean yearly and seasonal net radiation values observed at the treeline and in the closed crown boreal forest are fairly stable over the area (Table 7.2). Values of 22 and 30 kilolangleys for the forest–tundra and taiga respectively have been reported from the Soviet Union (Skorupskii & Shelyag-Sosonko 1982). In Alaska, the 16 kilolangley isopleth matches the treeline better than either the 10 °C July isotherm or Nordenskjöld's line (Hare & Ritchie 1972). In comparison, the west–east discrepancy between the value of potential evapotranspiration and the duration of the thaw season indicates that these are less accurate indicators of the vegetational boundaries than the radiative parameters (Hare & Ritchie 1972).

A more reductionist approach, that is at the organism level, could shed some light on climate–vegetation relation in the boreal forest–tundra transition zone. Several valuable ecophysiological studies looking at tree physiology in cold environments have been carried out at the alpine treeline, where low temperatures induce a negative carbon balance and hinder completion of the phenological cycle (reviewed by Tranquillini (1979)). In the Rockies, windy conditions during the winter are associated with high desiccation, abrasion and mortality in foliage of *Picea engelmanii* and *Abies lasiocarpa* which is unprotected by the snow pack (Hadley & Smith 1983, 1986, 1987). Winter water stress at the treeline should not be viewed as a universal phenomenon, however, since it has been suggested that *Abies balsamea* and *Picea mariana* do not experience it (Marchand & Chabot 1978). None the less, insufficient snow cover at the northern or alpine limit of several tree species results in stunted growth forms (Payette 1974) which are thought to have a higher photosynthetic to ligneous tissue ratio (Kulagin 1972).

In subarctic environments, the annual distribution of heat could be a more significant ecological factor than the mean annual temperature. For instance, it has been shown that white spruce growth is positively correlated with summer–fall temperature but inversely related to winter–spring temperature (Garfinkel & Brubaker 1980). However, work with black spruce at the treeline suggests that photosynthetic activity is more frequently limited by light intensity than by temperature during the growing season (Vowinckel, Oechel & Boll 1974). Some ecologists have argued that the problem of the treeline and the forest limit is one of population biology associated with the reproductive ecology of tree species (Marchand & Chabot 1978; Payette 1983). Black spruce seed viability (Elliot 1979*a*) and germinative capacity (Sirois 1988) appear to be related to the temperature gradient in the transition zone; experiments suggest that germination of this species is drastically reduced below 15 °C (Black & Bliss 1980). The threshold of 50% seed maturation in Scots pine requires 890 annual degree-days >5 °C, which is achieved at the treeline in only 2% of years (Henttonen *et al.* 1986). Because the minimum temperature sum for the development of a substantial number of flowering buds is approximately the same (Pohtila 1980), good seed-years occur very rarely in northern boreal and subarctic areas and a seed shortage is likely to occur at any given time.

In summary, one can find a certain degree of agreement between northern vegetation boundaries and many climatic indicators. However, none of them can yet be attributed with unequivocal causative significance (Carter & Prince 1981) because of insufficient ecophysiological knowledge about the major tree species. Nevertheless, at the formation level, thermic parameters such as radiative input and potential evapotranspiration show a global, but inaccurate, fit with major northern vegetational zones and give clues as to the nature of mechanisms operating at the limits. Moreover, the identification of a correspondence between the major boundaries and the modal position of the Arctic air front in summer and winter may constitute a first step towards the interpretation of historical movement of the treeline and other boundaries in terms of Holocene atmospheric circulation patterns. To develop this, further improvement of the existing general circulation models, along with increased knowledge of the physiology and reproductive biology of northern tree species, is needed.

Spatiotemporal patterns in the boreal forest–tundra transition zone

Vegetation patterning in the transition zone landscape is linked to the low resilience of forest communities to disturbances. Intensive logging has resulted in anthropogenic tundra expansion south of the treeline in the Soviet Union (Tikhomirov 1961; Tyrtikov 1978). Reindeer grazing of tree seedlings (Hustich 1966) and catastrophic herbivore insect outbreaks (Kallio & Lehtonen 1973; Sepälä & Rasta 1980; Lehtonen 1977) have also induced fragmentation of the forest in the transition zone of Fennoscandia. Decreased flooding frequency in recent decades is associated with a low rate of regeneration in white spruce and current shifts from forest to tundra communities near the treeline in the Mackenzie area (Pearce, McLennan & Cordes 1988). However, none of these factors has the global effect of climatic change and climate-fire interaction in determining the pattern of the forested landscape in the transition zone. Consequently, it is these two processes which will be focused on in the following discussion.

The climatic change during the Holocene (Lamb 1977) is thought to be the main forcing factor in expansion and recession of northernmost forests in the Soviet Union (Tikhomirov 1961, 1963; Bray 1971; Khotinskiy 1984), in Fennoscandia (Hustich 1958; Karlén 1976, 1983; Kullman 1983, 1985) and in North America (reviewed by Ritchie (1984, 1987a)). Climatic changes having an astronomical origin induce a global influence on the biosphere. The Milankovitch theory (Berger et al. 1984), based on the earth's axial tilt and orbital eccentricity and the timing of the perihelion, predicts the general climatic trend during the past 18 000 years as shown in Fig. 7.3. The seasonality of the climate increased sharply during the Holocene between 16 000 and 6000 BP, with a maximum amplitude around 9000 BP corresponding to a summer and winter radiation difference compared with present values of +8% and −8% respectively (Kutzbach & Guetter 1986). Between 16 000 and 9000 years BP, the estimated July temperature was 2.5 °C higher than at present. Winter and summer precipitation was lower than at present between 15 000 and 12 000 years BP. From 9000 BP to the present, solar radiation has decreased over the Northern hemisphere (Ritchie 1987a). The

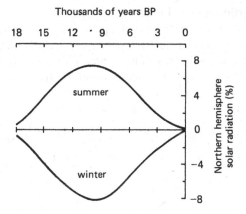

Fig. 7.3. Solar radiation departure from the present value in summer and in winter for the Northern hemisphere during the past 18 000 years according to the Milankovitch theory of climatic change. (Adapted from Kutzback & Guetter (1986); redrawn with the permission of the authors and the American Meteorological Society.

northern vegetation has reacted to the Holocene and recent climatic changes in various ways and with different lag periods, depending on the region.

Soviet Union

Holocene pollen and macrofossil records from the present taiga zone of the Soviet Union indicate afforestation beginning as early as 12 000 BP with arborescent birches and spruces. A general northward movement of the treeline between 8100 and 9200 BP is supported by several macrofossil trees found 100–200 km into the present tundra zone. These macrofossils consist of birches in western and eastern Siberia and of larches in the central Tamyr peninsula. Subsequent global cooling after 8000 BP resulted in a marked southward shift of the forest and transformation of northern taiga into forest–tundra in the western part of the country.

Another phase of forest progression occurred during the second half of the Atlantic period (6000–4600 BP). Spruce stumps from this epoch suggest that the tundra zone disappeared almost completely in the northern European USSR. Larch stump macrofossils substantiate forest encroachment 300, 100 and some tens of kilometers north of the present forest limit in the Yamal and Tamyr peninsulas and in the eastern maritime region respectively. Pollen and macrofossil evidence suggests a northern forest recession between 4600 and 4100 BP, a new advance between 4100 and 3200 BP and a degradation between 3200 and 2500 BP (Khotinskiy 1984). There are no literature reports of the response of forests to the Little Ice Age in the Soviet Union. The last forest retreat left only scattered groves, witness to the past warmer climate, in the most sheltered sites of the modern tundra (Khotinskiy 1984). Northern forest vegetation of the Soviet Union has responded positively to the global twentieth century warming trend documented by Jones, Wigley & Kelly (1982) and Jones, Wigley & Wright (1986). Northward migratorial trends have been observed in tree species, and forests are now found on patterned ground generated under a former tundra environment (Tikhomirov 1961, 1963). In places, the movement of forest onto previously treeless areas is reported to be as rapid as 700 m per year (Uspenskii 1963, *in* Bray 1971).

Fennoscandia

The high elevation of northern Fennoscandia has meant that forest limit movements associated with Holocene climatic change have occurred along both altitudinal and latitudinal gradients. Karlén (1976, 1983) and Kullman (1987a) showed that Scots pine forests grew above the present limit of coniferous forests during the whole interval between 8000 and 3000 BP, even when the post-glacial isostatic rebound of 120–160 m is deducted. Although the exact timing of tree establishment is uncertain, the numerous Scots pine subfossils found up to 70 km north of its present limit testify to the presence of pine populations and therefore more suitable growth conditions sometime between 6000 and 4000 BP (Eronen 1979; Eronen & Hyvärinen 1982). Pronounced cold periods marked by glacier expansions and a lowered treeline occurred between 7500 and 7300 BP and around 4500, 2800 and 2200 BP, as well as in the recent Little Ice Age (Karlén 1976) when the estimated summer temperature was 2 °C lower than the modern average (Kullman 1987b). Downward movement of the pine forest limit in the Fennoscandian highlands was not a regular process during the post-hypsithermal period, since ^{14}C datings of dead logs found above the present forest limit cluster around 6800, 6000, 4600, 1500 and 900 BP, which are thus interpreted as warm periods (Karlén 1983). However, Kullman (1987a) pointed out the possibility of interpreting wood remains from certain periods as indicating either a climate warm enough to allow tree growth or a climate suitable for dead wood production and its long-term preservation. The Little Ice Age period was marked by a large-scale pine forest decline associated with a pronounced regeneration gap and increased mortality among mature individuals in the highlands of Fennoscandia (Kullman 1987b,c). There is no evidence that the pine forest limit has moved upwards during the relatively warm twentieth century (Karlén 1976), although numerous single trees germinated up to 150 m above the forest limit during the 1920s and 1930s (Hustich 1958). It thus seems that only exceptionally warm climatic periods of relatively long duration (50–100 years) have been responsible for pine forest limit progression (Karlén 1976). In contrast, birch (*Betula pubescens*) in the northernmost part of Swedeh has reacted strongly to the recent climatic change (Sonesson & Hoogesteger 1983). It has expanded upwards by up to 50 m and increased markedly in population density at lower altitude in response to the twentieth century 0.5 °C rise in temperature and 13% increase in precipitation.

North America

Despite the relative geographical uniformity of that part of northern North America east of Alaska, marked west–east differences in deglaciation chronology and post-glacial vegetation history exist (Ritchie 1987*b*). In the Mackenzie Delta, melting of glacial ice was completed by 15 000 BP and spruces succeeded shrub tundra around 10 000–9000 BP. Thus afforestation in most of western Canada occurred in the context of a hypsithermal climate, although most pollen data suggest that the vegetation response to the warmest period actually occurred somewhat later (Ritchie, Cwynar & Spear 1983). In contrast, the last remnant of Laurentide ice sheet persisted until approximately 6500 years BP on the central Quebec–Labrador peninsula (Vincent 1989). Spruce migrated rapidly into this area (Richard 1979, 1981*a*; Richard, Larouche & Bouchard 1982) under deteriorating climatic conditions (Lamb 1985; Ritchie 1987*b*).

Paleodata from the forest–tundra transition zone of the Northwest Territories consistently suggest that earlier during the Holocene, forests either occupied positions north of the present forest limit or were more continuous into the transition zone than at present. On the Tuktoyaktuk peninsula (Northwest Territories), the following vegetation sequence has been suggested (Ritchie & Hare 1971):

12 900 to 11 600 years BP	dwarf birch tundra
11 600 to 8500	forest–tundra
8500 to 5500	closed-crown spruce–birch forest
5500 to 4000	tall shrub tundra
4000 to present	dwarf birch heath tundra.

The authors suggested that the shift from forest to tundra was associated with a southward displacement of the modal July position of the Arctic air front of 350 km compared with its location during the hypsithermal period. Other works in the same area, using pollen and macrofossil analysis, support the hypothesis of a retreat of the continuous forest limit during the second half of the Holocene period (Hyvärinen & Ritchie 1975), possibly 50–70 km south of its maximum extension (Spear 1983). Nichols (1974, p. 652) argues that several fluctuations of the continuous forest limit occurred in central Canada during the whole post-glacial period, with a maximum extension of approximately 250 km north of its present position between 5500 and 3500 BP. However, because no fossil logs have been found, the former presence of forests in this area remains to be proved (Payette & Gagnon 1979, 1985). East of Hudson Bay, early Holocene landscape vegetational dynamics associated with climatic changes have mostly been documented south of the treeline. Periods during the early Holocene in which spruce pollen was relatively more abundant than now can be bounded between 4200 and 3200 BP in the James Bay area (Richard 1979), 5200 and 4700 BP in central northern Quebec (Richard, Larouche & Bouchard 1982) and 4700 and 3500 BP in Labrador (Lamb 1985). The subsequent decrease in relative abundance of spruce pollen is clearly interpreted by Lamb (1985) as a signal of a reduction of forest cover in the forest–tundra transition zone following deteriorating climate and fire–climate interaction.

During the Little Ice Age, the early explorer Samuel Hearne drew a map suggesting that the treeline in the Keewatin district, west of Hudson Bay, was further south and west of its present position by up to 100 km (Ball 1986). The situation seems to have been quite different east of Hudson Bay where data suggest that the treeline moved only a short distance in response to the Little Ice Age climate (Payette & Gagnon 1979; Gagnon & Payette 1981). On the Quebec–Labrador peninsula, black spruce went some tens of kilometers north of the present treeline and the species limit is thought to represent the remnant of the forest limit during a warmer episode (Payette & Gagnon 1979), the dating of which is currently under investigation. As in Fennoscandia and the Soviet Union, the global warming trend of the twentieth century has induced local treeline expansion or increased seedling establishment near the forest limit in Alaska (Griggs 1937), in central Canada (Scott, Hansell & Fayle 1987) and in northern Quebec (Morin & Payette 1984; Payette & Filion 1985).

This review has illustrated the sensitivity of the northern forest border ecosystem to climatic change. A certain degree of circumhemispheric synchronicity can be observed in the vegetation response during the early post-glacial hypsithermal, and during the twentieth century warming (Fig. 7.4). Corresponding global northern forest retreat seems more difficult to warrant for lack of macrofossil evidence. Some climatic changes could be associated with the distribution of the total amount of heat received by the earth and have only a regional effect on the vegetational landscape. Predicted higher sensitivity to climatic warming of coniferous forests in maritime areas (Kauppi & Posch 1985) could also account for inter-regional differences in vegetation response. However, the data presently available are still too diffuse to corroborate this

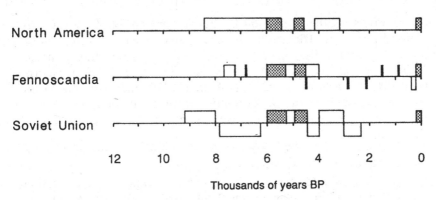

Fig. 7.4. Schematic representation of selected reports of tree population fluctuations in response to Holocene climatic forcing in the circumhemispheric boreal forest–tundra transition zone. Shaded section of the rectangles represents a time-coincident event; bold bars represent single data points. Upper scale represents forest progression; lower scale, forest regression.

possibility. On the other hand, several pointers suggest that disturbances, and forest fires in particular, could act in a synergic way with the climate to influence the vegetational landscape dynamics in the forest–tundra.

The role of fire in the holocene development of the boreal forest–tundra transition zone

Wildfires have been recognized as a driving ecological factor in boreal coniferous forests of the Soviet Union (Barney & Stocks 1983; Antonovski, Ter-Mikaelian & Furyaev, Chapter 14 of this volume), in northern Fennoscandia (Zackrisson 1977) and in North America (Heinselman 1981a). They play a major role in releasing nutrients locked up in dead and living organic matter (Van Cleve & Viereck 1981) and prescribe the start and end of vegetational successions (Dix & Swan 1971; Maikawa & Kershaw 1976; Black & Bliss 1978; Foster 1985; Morneau & Payette 1989). Wildfires control the age structure of forest stands (Van Wagner 1978; Cogbill 1985; Johnson & Van Wagner 1985; Sirois & Payette 1989), influence species composition and physiognomy of the vegetation (Johnson & Rowe 1977), and are a determinant of diversity and stability throughout the vegetational landscape (Heinselman 1981a; Romme 1982; Baker 1989). In steady-state conditions, fires act as a natural disturbing agent, triggering numerous mechanisms that lead to forest community renewal.

    In areas where climate has limiting effects on tree growth and development, numerous examples suggest that wildfires could have a much more drastic effect on the vegetation landscape development. Several pronounced fire-induced depletions of tree populations and shifts from forest to tundra communities have been reported in the transition zone of various regions including northern Sweden (Zackrisson 1977), Siberia (Kriuchkov (1968a,b) quoted by Wein (1975)) and northern North America (Hustich 1951a, 1966; Lutz 1956a; Ritchie 1962). It is suggested that this phenomenon, recurring over a long period of time, could have shaped the modern configuration of the forest–tundra.

Radiocarbon dating of charcoal found at the podsol surface of some treeless sites in the forest–tundra transition zone of Keewatin (Bryson, Irving & Larsen 1965) testifies to a 280 km fire-induced deforestation south of the modern treeline some 3500 years ago. Payette & Gagnon (1985) suggest that the modern forest–tundra of north-eastern North America is the end-product of a once densely populated coniferous zone that experienced long-term deforestation during the Holocene period. They provide datation of conifer charcoal from 116 predominantly treeless sites along a south–north transect in the forest–tundra east of Hudson Bay. The results show that a fire-induced deforestation process has been active in this area during the past 3000 years, particularly between 3000 and 2100 BP, 1800 and 1050 BP, 800 and 650 BP and 450 and 100 BP. The latitudinal position and age of the sampled charcoal suggests that there is no relation between the onset of deforestation and the distance from the present treeline. Intensive charcoal sampling in a much smaller region revealed fire-induced deforestation periods from 2350 to 2100 BP, 1650 to 1450 BP, 1350 to 1050 BP, 850 to 650 BP and 400 to 200 BP (Millet & Payette 1987). These periods fit well with those found over a large area by Payette & Gagnon (1985) and match to a considerable extent the cold climatic periods identified on the basis of sand dune development in the same general area (Filion 1984). Instead of a large-scale southward movement of the forest–tundra zone and treeline as reported in Keewatin (Bryson, Irving & Larsen 1965; Nichols 1974, 1975), the data provided by Payette and his associates suggest a progressive fragmentation of the forest cover in the forest–tundra transition zone during the past millennia. Deforestation did not occur at each fire event; periods of successful post-fire forest recovery have been identified between 3500 and 2700 BP, 2000 and 1600 BP (Gagnon & Payette 1981) and between 2100 and 1800 BP, 1050 and 800 BP and 650 and 450 BP (Payette & Gagnon 1985). Post-fire regeneration success hinges mainly on the regenerative potential of the tree species and on the coincidence of a fire and a climatic period more or less favorable to tree reproduction.

The effect of the climatic cooling trend of the Holocene on post-fire forest regeneration capacity can be assessed by the evaluation of fire impact on tree populations distributed along a south–north climatic

Table 7.3. *Percentage of upland (U) and lowland (L) sites in each post-fire density : pre-fire density ratio class*

Ratio class	BF[a]		FFTs		FFTn		SFT	
	U	L	U	L	U	L	U	L
0.00—0.25	21.6	7.8	45.8	25.0	44.0	8.0	92.9	12.5
0.26—0.50	15.7	7.8	20.8	37.5	16.0	6.0	5.4	19.6
0.51—1.00	15.7	11.8	16.7	25.0	18.0	16.0	0.0	16.1
1.01—2.00	21.6	35.3	12.5	10.4	18.0	38.0	0.0	30.4
>2.00	25.5	37.3	4.2	2.1	4.0	32.0	1.8	21.4

Note: [a] BF: upper boreal forest; FFTs: southern part of the forest subzone; FFTn: northern part of the forest subzone; SFT: shrub subzone.

gradient. The large fires of the 1950s, which burned over 5500 km^2 of land vegetation in representative parts of the upper boreal forest and the forest–tundra of northern Quebec, provide an opportunity to address the question of the role of fire in the Holocene development of forest–tundra at the demographic level. Black spruce forms almost monospecific forests in that area. Since seedling establishment of black spruce is very low after 20–25 post-fire years, owing to seed shortage (Morneau & Payette 1989; Sirois & Payette 1989) and increase in lichen cover (Brown & Mikola 1974; Cowles 1982), stand density appears to be determined by the relative success of seedling establishment during the early stage of stand development. Sirois & Payette (1991) evaluated tree population density immediately before and 30 years after 1950s fires in 410 sites in northern Quebec representative of four eco-regions distributed along a south–north climatic gradient. They found that low post-fire regeneration occurred in the uplands whereas the regeneration was high in the sheltered and more humid lowlands (Table 7.3). The long-term mainten-ance of closed-crown forest appears to be associated with a high seedling density in the initial stage of post-fire forest recovery. The impact of fire on tree population density in the uplands was most pronounced in the shrub subzone of the forest–tundra, where 93% of the krummholz stands experienced a minimum drop in population density of 75%. A pro-nounced decrease in population density was recorded in nearly half of the upland arborescent populations of the forest–tundra. Successive de-pletion of the tree population following several destructive fires appears to be the main deforestation process in the southern forest–tundra, whereas krummholz stands of the northern forest–tundra are frequently removed by a single fire event. More than one fifth of the upland samples from the upper boreal forest, where post-fire regeneration is generally successful, showed a minimum drop in tree population density of 75%.

Paleofire records (Desponts 1990) and the fire rotation period for the upper boreal forest (Payette *et al*. 1989*b*) suggests that 70 fire cycles have occurred in this area since it has been afforested (Sirois & Payette 1991). If each of those fires had had the same effect as the 1950s fires, then the boundary between the boreal forest and forest–tundra would be far south of its present position. The marked decrease in the tree density of a significant proportion of the northern boreal forest upland sites could be considered as an important step toward the southward expansion of the forest–tundra, several tens of kilometers into the boreal forest. This would suggest that the 1950s fires were particularly destructive. The fires occurred at a juncture between fifty years of a global warming trend, during which the fuel built up, and a cooling trend in the subsequent decade (Jones, Wigley & Kelly 1982; Jones, Wigley & Wright 1986) potentially hindering post-fire regeneration. The overall deforesting effect of the 1950s fires suggests that short-term (*c*. 10 years) climatic fluctuations could be of high biogeographical significance for long-term vegetation dynamics in the transition zone. Since low viable seed production at the treeline seems a general rule (Elliot 1979*a*; Black & Bliss 1980), it would appear justifiable to assume that cold climatic conditions in the forest–tundra are more limitative to tree sexual reproduction than to tree growth. With regard to the long-term maintenance capacity of black spruce populations in fire-free conditions near the treeline (Payette *et al*. 1985, 1989*a*), it is suggested that the present patchy vegetation pattern of the north-eastern north American forest–tundra is associated with catastrophic fires during climatic periods of the Holocene that were detrimental to tree reproduction (Fig. 7.5). Tree reproductive biology appears as a keystone of the forest–tundra ecology in this area. The global validation of this hypothesis requires further knowledge of fire ecology and reproductive characteristics of the tree species in the Fennoscandian and Siberian forest–tundra.

Concluding comment

Many important dimensions relating to the forest–tundra vegetational pattern have not been addressed in this chapter. These include, for example, permafrost distribution, nutrient cycling, light characteristics, moisture gradients and plant communities. Interested readers are referred to Van Cleve *et al*. (1986), Larsen (1989) and Bonan & Shugart (1989) for excellent treatment of these topics.

Despite regional variations due to species composition, the boreal forest–tundra transition zones of North America, Fennoscandia and the Soviet Union display some structural analogies in latitudinal zonation. This circumhemispheric belt has been characterized by instability at the

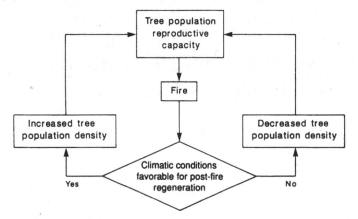

Fig. 7.5. Hypothetical effect of the fire–climate interaction on high-latitude forests. Other factors being equal, post-fire regenerative capacity is related to pre-fire stand density. Recurrent fires during climatic periods conducive to high reproductive capacity maintain a closed-crown canopy, as is the case in boreal forest. Recurrent fires when the climate is limitative to seed production and seedling establishment result in open woodlands and barelands. Post-fire seedbed conditions are very variable over short distances and have a considerable influence on tree reproduction success. This is why fire impact on subarctic tree density at the landscape level is highly variable.

community and landscape levels during the past millennia and appears especially responsive to climatic change as exemplified by several instances of progression or recession of forest populations reported in this chapter. In the context of a significant warming trend, current forest succession models predict pronounced qualitative community changes in the southern boreal forest which will bring it closer, in floristic terms, to the higher diversity, northern deciduous forest (Solomon 1986; Pastor & Post 1988). However, most of the productivity increase is expected to occur in the transition zone, where large expanses comprising low-density populations and bare ground previously covered by forest are available for afforestation. Community inertia and a lag response to the climatic warming trend (Davis & Botkin 1985; Solomon 1986; Payette *et al.* 1989*a*) might, however, slow the pace of this progression. Simulation and field studies are currently under way to investigate whether forest fires can act as a catalyst to increase the rate of vegetational change induced by the climatic trends in the boreal forest-tundra transition zone. I hypothesize that open lichen woodland subjected to recurrent fire offers one of the most promising opportunities to predict the mean and long-term effects of the global warming trend on the boreal zone.

8 The southern boreal–northern hardwood forest border

John Pastor and David J. Mladenoff

Introduction

In this chapter we will examine the nature of the southern boreal forest, explicitly its transition with the cool temperate deciduous forest that occurs to its south. Although this mixed deciduous and conifer forest occurs on portions of several continents in the northern hemisphere, it is most intact and most studied in North America as the northern hardwood and conifer biome (Braun 1950; Curtis 1959). Therefore, although we will describe this zone in other regions, we will concentrate on the North American expression of it. For convenience, we will also adopt the shortened common name 'northern hardwoods' for the cool temperate and largely deciduous forest to the south while recognizing that certain conifers are also important components of this forest.

The southern boreal forest contains extreme contrasts in ecosystem properties corresponding to the locations of different species within a mosaic landscape. These properties are accentuated at its boundary with the northern hardwoods and conifers, which are transitional to the deciduous forests to the south. Within these two biomes, each genus, and in some cases different species within each genus, affects ecosystem properties in different ways. Such properties include nutrient cycling and productivity, disturbance type and frequency, and suitability as an animal habitat. Consequently, there are strong feedbacks between succession, species dominance, resource limitation, disturbance regimes, and trophic structure (Fig. 8.1). These feedbacks may be stabilizing or destabilizing, the latter often causing cycles at various times and in various places. In addition, the two biomes share a common assemblage of early succes-sional species; how succession proceeds to a community of boreal conifers or northern hardwoods depends on the vagaries of site and regional conditions, such as soil type and climate, as well as on seed

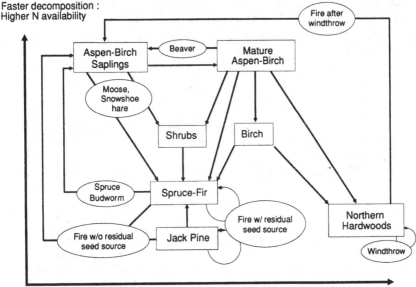

Fig. 8.1. Multiple successional pathways at the southern boreal–northern hardwood forest border and their interactions with nutrient cycles, climate, disturbance types, and herbivores.

dispersal and the influence of animals on species composition. Therefore, this landscape is an ideal setting for describing the relationships between population, community and ecosystem dynamics.

In this chapter, particular attention will be paid to the interaction between species characteristics and ecosystem properties. We will conclude with a discussion of how these interactions affect successional dynamics, spatial patterns, and the future of the southern boreal forest.

Location, origins, and general characteristics

The southern boreal forest abuts the northern hardwood forest in northeastern North America, southern Scandinavia and western European Russia – where it is known as the boreonemoral zone (Walter 1979) or mixed forest zone (Eyre 1968) – and in north-eastern China (Olson, Watts & Allison 1983), and northern Japan (Eyre 1968) (Table 8.1; see also Fig. 11.1). In Scandinavia and Russia, while the boreal forest remains relatively intact albeit highly managed, the region formerly occupied by northern hardwoods has been almost entirely agricultural for

Table 8.1. *Major tree species dominant in the boreal–northern hardwood transition in North America, Europe and Asia*

North America	Europe	Asia
Boreal		
Abies balsamea	*Pinus sylvestris*	*Abies nephrolepis*
Pinus banksiana	*Picea abies*	*Picea jezoensis*
Picea glauca	*Populus tremula*	*Populus davidiana*
Picea mariana	*Betula pubescens/*	*Betula platyphylla*
Populus tremuloides	*B. pendula*	*Betula ermanii*
Populus grandidentata	*Sorbus aucuparia*	
Betula papyrifera		
Northern hardwoods		
Acer saccharum	*Quercus robur*	*Acer mono*
Betula alleghaniensis	*Carpinus betulus*	*Quercus mongolica*
Tsuga canadensis	*Tilia cordata*	*Fraxinus mandshurica*
Pinus strobus	*Fagus sylvatica*	*Pinus koraiensis*
Tilia americana		*Tilia amurensis*
Fagus grandifolia		*Ulmus propinqua*

centuries. In northeastern Europe, the boreal–northern hardwood border is narrow and brief from Finland and south-east across European Russia, until the boreal zone itself abuts the steppe, as it does in central Canada. In both North America and Eurasia the southern boreal forest contacts deciduous forest in areas under maritime influence, and a direct boreal forest–steppe contact occurs in continental areas (Sukachev 1928; Eyre 1968; Walter 1979). The boundary between the two biomes in China is short and, unlike the other two regions, is largely an altitudinal boundary rather than a continental boundary (Yang & Wu 1987; Zhao 1987). Of the three areas, the boundary is most intact in North America, although much of the area is in an early-successional state because of timber harvesting and some farming during the past century (Curtis 1959). The boundary in North America runs eastward from northern Minnesota through northernmost Wisconsin and upper Michigan. Two arms extend north-west and north-east into Ontario around the east and west ends of Lake Superior, but these two arms do not connect over northernmost Lake Superior, where the northern hardwood species are not present, and only aspen and birch occur with spruce and fir (Little 1971). From the Lake Superior region the transition moves eastward through east-central Ontario and southern Quebec to northern Maine.

The natural vegetation of both the boreal and temperate deciduous zones of Europe is simpler than in North America, owing to reduced tree species diversity. Europe has not had a similar diversity of species as

Table 8.2. *Eastern North American and Eurasian conifers of boreal to cool temperate 'deciduous' zones*

Species	Zone
Eastern North America	
Pinus strobus	boreal/transition/n. hardwoods
P. resinosa	boreal/transition
P. banksiana	boreal/western transition
Picea glauca	boreal/transition
P. mariana	boreal/transition lowland
Abies balsamea	boreal/transition
Thuja occidentalis	boreal/transition lowland
Larix laricina	boreal/transition lowland
Tsuga canadensis	transition/n. hardwoods
Eurasia	
Pinus sylvestris	boreal/boreonemoral
Picea abies	boreal/boreonemoral

present in North America, particularly among the conifers *Pinus*, *Abies* and *Thuja*, since the Pleistocene (Table 8.2). In particular, the region lacks an analog to the cool-temperate eastern hemlock (*Tsuga canadensis*) of North America. It is only in far eastern Asia and northern Japan that this conifer diversity is again apparent (Sukachev 1928; Walter 1979). In addition, there is the appearance of greater simplicity and clearer divisions between zones because of the long history of human impact in the Eurasian region, which makes a discussion of the natural vegetation of the transition zone somewhat speculative. Because of its depauperate tree flora, in Europe the transition zone is more simply an intergrading of the boreal and more southern deciduous zone species. In eastern North America the transition zone includes this mixture plus additional species characteristic of this 'transition' zone itself, such as eastern hemlock, eastern white pine (*Pinus strobus*) and yellow birch (*Betula alleghaniensis*).

The congruence of the boreal–northern hardwood boundary with climatic isopleths has long been suspected (Marie-Victorin 1938; Raup 1941). Summers at the boundary are warm but short, winters are long and cold. The Holdridge Life-Zone Classification, based on hypothesized correspondence between vegetation and climate properties, predicts that the boundary should occur at a ratio of evapotranspiration to precipitation of less than 1.0 and a mean annual biotemperature of 6 °C (biotemperature is the annual average temperature with all readings of less than 0 °C set to 0 (Holdridge 1964; Emanuel, Shugart & Stevenson

1985*a*). This is somewhat north of the actual boundary in North America as mapped by Rowe (1972) and Olson, Watts & Allison (1983) (see Fig. 11.1). In eastern North America, the boundary corresponds more precisely with isotherms denoting an average of 2 days per year with temperatures over 32 °C, and a 120-day frost-free period (weather data from U.S. Department of Commerce 1968). Recently, Arris & Eagleson (1989) suggested that the northern limit of deciduous forest in North America corresponds to the −40 °C isotherm. Below this temperature, deciduous species cannot supercool sap to prevent ice from bursting xylem and phloem cells. The boundary also seems to correspond with synoptic weather patterns, particularly the winter position of Arctic frontal zones and the occurrence of July tropical air masses less than 10% of the time (Bryson 1966).

The correspondence with annual precipitation is weaker; for example, annual precipitation at the boundary in North America ranges from over 1 m along the Atlantic seaboard to half that in the continental interior. However, when precipitation falls to less than 30 cm annually, boreal forests intergrade with northern prairies as in Manitoba, Saskatchewan and the Soviet Union.

In North America, the southern boreal forest is centered largely on the Canadian Shield, a large craton of Pre-Cambrian granite, gneisses, and diabases. Atop the Shield is spread the debris of the Wisconsinan Glaciation, and virtually the whole of the boundary between the southern boreal and northern hardwood forests lies north of the southern extent of this glaciation. Consequently, the soils are relatively young and unweathered and derived mainly from igneous or metamorphic bedrock worked into till and outwash. Spodosols are the characteristic soil of the region, although in drier areas inceptisols and entisols with spodic characteristics are more common (Jenny 1980). In New England, topography and bedrock partially control the location of the boundary (Siccama 1974), but feedbacks between litter quality and soil nitrogen availability may also strongly affect local development and stability of northern hardwood and boreal conifer patches, as will be discussed below.

The origins of the current species assemblages of boreal and northern hardwood forests in North America have been reviewed by Webb, Cushing & Wright (1984) and Davis (1984). The current assemblage of spruce–fir–aspen–birch that is commonly considered boreal forest arose only 6000 years BP in North America (Webb, Cushing & Wright 1984). Prior to this, paper birch (*B. papyrifera*) was found south of the range of spruce (Webb, Cushing & Wright 1984). In eastern North America, spruce (*Picea* spp.) arrived near the current boreal–northern hardwood boundary approximately 10 000 years BP; sugar maple (*Acer saccharum*) arrived around 6000 years BP. Other species characteristic of northern hardwoods, such as beech (*Fagus grandifolia*), hemlock, and white pine, arrived near the current border along the Atlantic seaboard at about the same time as maple, but arrived much later (approximately 4000–3000 years BP) in the mid-continent (Davis 1984).

Species diversity and characteristics

The species diversity of the southern boreal–northern hardwood transition is surprisingly great. In a study of 103 stands across the southern boreal boundary in the Great Lakes region of North America, Maycock & Curtis (1960) list 42 tree species and over 300 shrub and herb species in both northern hardwood and boreal conifer forests. Of these, only 40 were confined mainly to the deciduous northern hardwood stands. The diversity of most stands in this region is greatest in the boreal–northern hardwood transition zone, but decreases with succession and increasing moisture (Auclair & Goff 1971).

Within the southern boreal forest of eastern North America, there is an amazing fidelity of species across stands. Stands across the Great Lakes region have 80% of their species in common; southern montane boreal stands from the Rocky Mountains or the Great Smoky Mountains have fewer species in common with the continental forests of the southern Canadian Shield (Curtis 1959).

Most herb species are circumpolar and common to both biomes, but tree genera, rather than individual species, are circumpolar and distinctly different between the biomes (Larsen 1980). However, at a finer scale across the landscape within a given region, boreal or northern hardwood stands segregate along environmental gradients, particularly moisture, in a way that obscures more global patterns (Maycock & Curtis 1960). In general, where the regional climate is warm enough to support northern hardwoods, boreal stands are confined to wet or cool microsites but creep out onto upland sites where the climate becomes colder. This local segregation along moisture and climatic gradients applies to individual species as well. For example, balsam fir (*Abies balsamea*) and white cedar (*Thuja occidentalis*) become more important with increasing soil moisture, while white pine and white spruce (*Picea glauca*) become less important. Sugar maple is most important on mesic sites and least important on xeric or hydric sites, and jack pine (*Pinus banksiana*), black spruce (*Picea mariana*), and hemlock have bimodal distributions in that they are most important at the extremes of the moisture gradient (Curtis 1959; Maycock & Curtis 1960). Beech co-occurs with sugar maple on mesic sites, but is more dominant on xeric sites in New England (Siccama 1974). Basswood (*Tillia americana*) replaces beech in the western Great Lakes region (Curtis 1959). Yellow birch occurs across the full moisture spectrum and also has the distinction of being the major hardwood found in otherwise nearly pure hemlock stands (Curtis 1959). Aspen (*Populus tremuloides*) and paper birch are successional species on better and poorer sites, respectively, in both boreal and northern hardwood forests.

This diversity of species would not be so important to the functioning of the boreal forest border were the species not also very different in traits important to ecosystem functioning (Table 8.3). Different genera, and in some cases different species within each genus, have different combinations of shade tolerance, litter quality, nutrient-stress tolerance, browse quality, browsing responses, and reproductive characteristics. It is through this diversity of functional characteristics that ecosystem properties in the southern boreal forest are intimately linked to population and community dynamics. An elaboration of this theme forms the main thread of the remainder of the chapter.

Functional diversity

The functional diversity of the southern boreal–northern hardwood border arises from several, species-related factors: (1) a diversity of tissue chemistry that partially determines litter decay and nutrient availability, fuel loadings, and browsing patterns by mammals; (2) the strong effect of each species on the availability of limiting resources, and the different responses of these species to resource limitation; (3) a diverse array of life history traits that partially determines colonization and persistence after disturbances. Aspects of nutrient cycling, fire regimes, and herbivory are also covered in other chapters within this volume. Here, we will concentrate on the contrasting nature of these processes in northern hardwoods and boreal conifers, particularly as they influence dynamics such as succession and spatial patterns in the landscape (see Fig. 8.1).

Nutrient cycling, resource use, and productivity

Nitrogen is the nutrient most limiting to production in both northern hardwood (Mitchell & Chandler 1939; Safford 1973; Nadelhoffer, Aber & Melillo 1983; Pastor *et al.* 1984) and boreal forests (Weetman 1968*a*; Stewart & Swan 1970; Van Cleve & Zasada 1976; Krause 1981; Mahendrappa & Salonius 1982; Van Cleve & Oliver 1982; Pastor *et al.* 1987). Because the cycling of nitrogen is intimately coupled with that of carbon and strongly influenced by species composition of these forests (Flanagan & Van Cleve 1983; Van Cleve *et al.* 1983*b*; Pastor *et al.* 1984; Chapin, Vitousek & Van Cleve 1986; Mladenoff 1987) it influences numerous other processes, such as herbivory and successional dynamics. Therefore, we shall concentrate on nitrogen cycling rather than a broad survey of the cycles of other, non-limiting nutrients.

The carbon and nitrogen cycles are reciprocally linked because the availability of soil nitrogen is a limiting factor to net carbon fixation in these forests, but the type of carbon compounds produced in leaf litter in turn affects nitrogen availability by limiting soil microbial activity (Melillo, Aber & Muratore 1982; Flanagan & Van Cleve 1983; McClaugherty *et al.* 1985; Chapin, Vitousek & Van Cleve 1986; Pastor *et al.* 1987). To the extent that species cycle these two elements differently, the ecosystem properties resulting from these linked cycles, such as productivity and herbivory, are altered whenever one species replaces another during succession or across environmental gradients (Van Cleve *et al.* 1983*b*; Pastor *et al.* 1984; Zak, Pregitzer & Host 1986; Mladenoff 1987).

An important difference between northern hardwoods and boreal conifers is in the decay rates and chemistry of their litters. The leaf litters of sugar maple and yellow birch have high nitrogen and low lignin contents, allowing them to be easily decomposed by soil microbes (Gosz, Likens & Bormann 1973; Melillo, Aber & Muratore 1982; McClaugherty *et al.* 1985). Consequently, soil nitrogen availability is usually high in northern hardwoods, ranging from 80 to 120 kg N per hectare per year (Nadelhoffer, Aber & Melillo 1983; Pastor *et al.* 1984; Mladenoff 1987; Zak, Host & Pregitzer 1989). This high nitrogen availability supports the high productivity of these forests, which ranges from 7 to 12 Mg ha^{-1} annually (Cannell 1982). It is possible that nitrogen is only marginally limiting to these forests, particularly in nearly pure stands of sugar maple (Zak, Host & Pregitzer 1989). However, admixtures of conifers such as hemlock or possibly white pine depress nitrogen availability locally (Mladenoff 1987), possibly creating a fine-grained mosaic of nitrogen cycling regimes even within an individual stand.

In contrast, the leaf litters of spruce and balsam fir have lower nitrogen and higher lignin contents (Melin 1930; Daubenmire & Prusso 1963; Hayes 1965; Lousier & Parkinson 1976, 1978; Flanagan & Van Cleve 1983; Fox & Van Cleve 1983; Moore 1984), making them less easily decomposable. Immobilization of nitrogen by microbes decomposing these leaf litters continues for three or more years (Moore 1984) in contrast to two years or less for sugar maple and yellow birch (Melillo, Aber & Muratore 1982); in fact, net immobilization can continue well into humus formation (Gordon & Van Cleve 1983). Consequently, net nitrogen availability is depressed in boreal conifer stands relative to that in northern hardwoods (Flanagan & Van Cleve 1983; Gordon 1983; Gordon & Van Cleve 1983; Van Cleve *et al.* 1983*b*). Nitrogen availability beneath boreal conifers is almost always below 40 kg ha^{-1} annually, approximately half that in northern hardwoods (Flanagan & Van Cleve 1983; Gordon 1983; Pastor *et al.* 1987); residence times of nitrogen in the forest floor are as high as 50–60 years with the residence times being

Table 8.3. *Major characteristics and ecosystem interactions of tree species*[a]

	Ecosystem properties				Interactions with mammals		
	Evergreen/ deciduous	Shade tolerance	N-stress tolerance	Litter quality	Food preference	Browse response	Cover preference
Common early-successional species							
Populus tremuloides	deciduous	low	moderate	high	high	sprout/decline	low
Prunus pensylvanica	deciduous	low	low	high	moderate	self-prune/decline	none
Betula papyrifera	deciduous	low	high	high	moderate	sprout	none
Boreal species							
Abies balsamea	evergreen	v. high	high	low	low	self-prune	moderate
Picea glauca/P. mariana	evergreen	moderate–high	high	low	none	none	high
Sorbus americana	deciduous	moderate	high	high	moderate	hedges-decline	none
Pinus banksiana	evergreen	v. low	high	low	low	none	moderate
Northern hardwoods							
Acer saccharum	deciduous	v. high	v. low	v. high	high	resprout/branch	none
Betula alleghaniensis	deciduous	high	low	high	high	resprout/branch	none
Tsuga canadensis	evergreen	v. high	high	low	high	dies	high
Pinus strobus	evergreen	low	high	low	low	branch	high
Tilia americana	deciduous	high	v. low	high	high	sprout	none
Fagus grandifolia	deciduous	v. high	moderate	moderate	mast high browse/low	sprout	none

	Reproductive traits					
	Reproductive mechanism	Dispersal mechanism	Seed shadow	Seed bank longevity	Seed bank recruitment	Germination & establishment requirements
Common early successional species						
Populus tremuloides	seed/sprout	wind	long/broad	seasonal	none	light/disturbed soil
Prunus pensylvanica	seed	bird	long/broad	v. high	episodic	light/disturbance
Betula papyrifera	seed/sprout	wind	moderate/long	moderate	high/episodic	light/disturbed soil
Boreal species						
Abies balsamea	seed	wind	moderate/broad	short (<3 yr)	low	moderate shade
Picea glauca/P. mariana	seed	wind	moderate	moderate	episodic	moderate light/disturbed soil
Sorbus americana	seed	bird	long/broad	high	episodic	light/disturbance
Pinus banksiana	seed	wind	moderate	long/on and off tree	high/episodic	high light/fire
Northern hardwoods						
Acer saccharum	seed/mod. sprouting after injury	wind	short	brief	low	established under canopy undisturbed seedbed
Betula alleghaniensis	seed/mod. sprouting after injury	wind	moderate/long	moderate	moderate/episodic	moderate light, disturbed/exposed soil, logs
Tsuga canadensis	seed	wind	moderate	brief	low	established under canopy/litter free seedbed (soil or logs)
Pinus strobus	seed	wind	—	—	—	—
Tilia americana	seed/sprout	—	short	—	—	—
Fagus grandifolia	seed/root sprouts w/out disturbance	vertebrates – birds & mammals	bimodal/patchy	short	low	established under canopy/undisturbed seedbed

Note: Boreal species: *Populus tremuloides* (aspen), *Abies balsamea* (balsam fir), *Picea glauca* (white spruce), *Sorbus americana* (mountain ash), *Prunus pensylvanica* (pin cherry), *Betula papyrifera* (paper birch), *Pinus banksiana* (jack pine).
Northern hardwood species: *Acer saccharum* (sugar maple), *Betula alleghaniensis* (yellow birch), *Tsuga canadensis* (eastern hemlock), *Pinus strobus* (white pine), *Tilia americana* (basswood), *Fagus grandifolia* (beech).

longest in black spruce ecosystems, in contrast to an average of only 6 years in northern hardwood stands (Van Cleve *et al.* 1983*b*). In addition to the low litter quality of boreal conifers, the cooler macro- and microclimates to which these species are confined also depress microbial activity and hence nitrogen availability (Timmer & Weetman 1969; Flanagan & Van Cleve 1983; Gordon 1983).

The low nitrogen availability is partly compensated for by a greater efficiency in the use of this element by conifers (Chapin, Vitousek & Van Cleve 1986). The efficiency is achieved by several mechanisms, such as lower amounts of nitrogen required to produce a given amount of tissue, an ability to photosynthesize with low leaf nitrogen concentrations (Mooney & Gulmon 1982; Sprugel 1989), and a low leaf turnover rate (Sprugel 1989). However, these adaptations also contribute to the low rates of nitrogen cycling because a high nutrient use efficiency means high C:N ratios and hence low decomposability of leaf litters; low leaf turnover rates mean slower rates of nitrogen return to the soil; and lignin, resins and other recalcitrant carbon compounds are produced partly to physically protect leaves during the three or more years that they are retained. Consequently, productivity of boreal conifer forests is lower than that of northern hardwoods, ranging from 3 to 8 Mg ha^{-1} annually (Cannell 1982). Productivity often declines with time in boreal conifers (Gordon 1983) and this decline may be related to the depression of nitrogen availability caused by the recalcitrant leaf litter (Weetman & Nykvist 1963; Heilman 1966; Viro 1967; Williams 1972; Miller *et al.* 1979; Pastor *et al.* 1987).

Curiously, nitrogen cycling in the early successional aspen and birch forests that are common to both northern hardwood and boreal conifer seres is intermediate between these two extremes (Gordon 1983; Van Cleve *et al.* 1983*b*; Pastor & Bockheim 1984). This is partly because of the intermediate lignin and nitrogen contents of aspen and birch litter (Melin 1930; Daubenmire & Prusso 1963; Gosz, Likens & Bormann 1973; Louiser & Parkinson 1976, 1978; Melillo, Aber & Muratore 1982; Flanagan & Van Cleve 1983; Berg, Ekbohm & McClaugherty 1985; McClaugherty *et al.* 1985), but also because of the high retention rates of nitrogen in perennial tissues once taken up by these species (Gordon 1983; Pastor & Bockheim 1984). Overall stand nutrient use efficiency can be very high in these early successional forests, particularly when aspen is underlain by shade-tolerant northern hardwoods that elevate nitrogen availability while cycling it in the shade where aspen cannot (Pastor & Bockheim 1984). Consequently, productivity often peaks during mid-succession in both northern hardwood and boreal conifer seres (Marks 1974; Gordon 1983; Van Cleve *et al.* 1983*b*, Pastor & Bockheim 1984).

Disturbance regimes

Fire and windthrow are the major disturbances on boreal forests and northern hardwoods, respectively. These processes have different sizes, distributions, frequencies, and intensities and are therefore major sources of structural and functional heterogeneity at the boreal–northern hardwood border.

The role of fire in structuring the boreal forest is well recognized and has recently been reviewed in Wein & MacLean (1983). Despite an abundance of research on the behavior of fire and the responses of different tree species to it, the factors that control fire occurrence and its effects at the southern boreal–northern hardwood boundary are still not well understood.

Weather is the major regional factor controlling the occurrence of fires, but which meteorological regimes are necessary for fire occurrence is not clear, and varies with stand type, topography, and fuel buildup. Drought, particularly in the spring, and prolonged hot weather have long been recognized as necessary for severe fires in the southern boreal forest (Spurr 1954; Haines & Sando 1969; Heinselman 1973; Swain 1973). This is partly supported by the increased fire recurrence intervals eastward from Minnesota to Maine and the Canadian Maritime Provinces in direct relation to increased precipitation and moister springs (Wein & Moore 1977; Heinselman 1981a; Cogbill 1985). The occurrence of warm dry weather appears to be caused by particular synoptic weather patterns. For example, between 1974 and 1980, large stationary high-pressure systems over north-central North America resulted in an average of 190 000 hectares burned by more than 700 fires annually in northwestern Ontario alone (Stocks & Street 1983).

However, not all forests burn during severe fire weather, and those that do burn do so in very different ways. Fire cycles in the mixed hardwood–conifer forests of northern Minnesota and Ontario vary greatly (Stearns 1949; Frissell 1973; Heinselman 1973, 1981a, Henry & Swan 1974; Grigal & Ohmann 1975; Swain 1980; Ohmann & Grigal 1981). Jack pine–spruce forests in this region have severe crown and/or surface fires of 10 000 acres approximately every 50 years while aspen–birch–fir forests have similar fires every 80 years (Heinselman 1981a). Forests dominated by white or red pines with open understorey have a more complex fire regime, with light surface fires every 20–40 years that keep the understorey open and severe crown fires at 150–180 year intervals that open the canopy (Frissell 1973; Heinselman 1973). These fires tend to be somewhat smaller than the more frequent fires in jack pine–spruce and aspen–birch–fir covertypes, covering approximately 1000–10 000 acres. Northern hardwood–hemlock forests have the longest fire cycle, greater than 300 years on the average, with an average size an order of magnitude less than that in the pine forests (Stearns 1949; Henry & Swan 1974; Heinselman 1981*a*).

Fire occurrences in the southern boreal forests have both deterministic and stochastic components (Heinselman 1981*b*). First, these return intervals are related to fuel flammability and persistence, which are controlled by the same chemical factors that control decomposition. Thus, jack pine and spruce, with their resinous and therefore recalcitrant and highly flammable litter that decomposes slowly but burns easily, have the shortest return intervals, while northern hardwoods with their rapidly decomposing litter with fewer resins have the longest return intervals. A slow nutrient cycling regime is thus directly related to a greater fuel accumulation and therefore a higher probability of fire return. However, by increasing nitrogen mineralization rates, fire may compensate for the depression of nitrogen availability caused by recalcitrant litters (Ohmann & Grigal 1979; Dyrness & Norum 1983; Viereck & Van Cleve 1986). Other deterministic factors controlling fire occurrence include the increased amounts of arboreal lichens (themselves fuels) and snags with stand age, thereby increasing the probability of large crown fires. However, lightning strikes, wind direction and speed, and 'spotting' due to sparks and flaming brands introduce randomness into the fire pattern (Heinselman 1981*b*), and stochastic fire cycle models are well corroborated for the southern boreal forest (Van Wagner 1978). In any case, once a major fire is started, it can spread to adjacent stands of any age or type, and in a severe drought year any stand can burn (Heinselman 1981*b*).

Clark (1988*b*) hypothesizes a synergistic interaction between climate cycles and fuel accumulation. Charcoal analyses of lake sediments in northern Minnesota suggest regional fire cycles that harmonize with multiples of a 22-year drought cycle, but which are particularly widespread during breakup of early successional stands when large amounts of fuelwood are returned to the forest floor in a few years. Clark (1988*b*) predicts that fires should occur with high probability every 33–44, 55–66, and 77–88 years, which corresponds to the timing of aspen breakup on sandy, loamy, and clay soils, respectively (Graham, Harrison & Westell 1963).

Recovery of the forest after fire depends on an interaction between fire size and intensity and the reproductive characteristics of the available species. For example, Swain (1980) found that fire regimes of 70 years in northern Minnesota will perpetuate aspen–birch forests if there are no firebreaks because the patch created is sufficiently large to prevent conifers from seeding in, but aspen and birch can reproduce from root suckers supplemented by wind-dispersed seeds. More complex topography contains firebreaks, protecting small patches of unburned conifers that can seed in surrounding areas. In addition, the serotinous cones of jack pine and black spruce and the buried and dormant but still viable seeds of pin cherry perpetuate their dominance after frequent fires over white and red pines and northern hardwoods, respectively (Fowells 1965; Marks 1974). In northern hardwoods, white pine was an important component of the presettlement forest, probably due to climatic conditions producing increased fire several hundred years ago (Curtis 1959).

In northeastern Minnesota, Ohmann & Grigal (1981) found that fire intensity and consequently the response of boreal forest differed with season of burn. A spring burn was less intense and destructive of the forest floor and its contents of reproductive structures, allowing more reproduction by vegetative propagation as well as the seed bank. A summer burn removed the forest floor down to mineral soil and favored seedling establishment by newly dispersed seed. The spring burn favored species that are capable of resprouting, such as aspen and birch, or those that are persistent in the dormant soil seed bank such as birch or *Prunus* species, and to some extent spruce. In addition, the few late-spring-seeding species such as aspen and red maple (*Acer rubrum*) would be at a favorable advantage in a mast-year burn, since their seeds either germinate or perish the year of dispersal. These species are among the most consistent seed producers of all the species in the boreal–northern hardwood region (Godman & Mattson 1976), perhaps because spring maturing of seeds avoids the effects of dry years.

In contrast to spring fires, severe summer burns kill most vegetative structures, as well as many dormant seeds. In this situation the only viable dormant seeds present are likely to be those of species that produce very long-lived seeds capable of migrating down into mineral soil deeply enough to escape destruction by fire, such as species of *Prunus* (Marks 1974) and shrubs such as red elder (*Sambucus pubens*) (Mladenoff 1990). At this depth, some soil disturbance is needed to bring the seeds up to the surface for germination. Because new seed of aspen would be destroyed, the fall-seeding species that benefit most from a mineral soil seedbed would be most favored, which would include birch and the conifers. In this case, those species with a better crop and better dispersal, and closer in proximity, would have an advantage.

Windthrow has been less intensively studied in boreal conifers and northern hardwoods than fire, but recent research indicates that it can be an important component of the disturbance regime, particularly in northern hardwood-dominated stands in the western portion of the transition zone (Bormann & Likens 1979b; Canham & Loucks 1984). Perhaps one reason why windthrow has been less recognized is its rarity and stochastic nature. Recurrence intervals for large, catastrophic windthrows are generally greater than 1000 years at the southern boreal–

northern hardwood border (Lorimer 1977; Canham & Loucks 1984; Frelich 1986). Along the North Atlantic seaboard, hurricanes can be a significant, although stochastic, disturbance superimposed on a fire regime with different recurrence intervals and patch sizes (Henry & Swan 1974).

In northern hardwoods, windthrow is important throughout the age of a stand. Since stand-initiating catastrophic storms have a mean return time of more than 1000 years (Canham & Loucks 1984), considerable tree replacement occurs within that interval. In the boreal forest, fires may return in intervals that are less than the age of the major dominants, thereby continually keeping the same species occupying the site, with little change in canopy dominants. In northern hardwoods the return time of major windstorms exceeds the lifespan of dominants by several times. In these forests canopy replacement by the creation and filling of small treefall gaps is important in compositional change. In many areas across the range of northern hardwoods an annual rate of new gap creation of 1% of the forest has been found, with 9–10% of the forest stands in gaps of varying age and with gap closure occurring by growth and expansion of subcanopy trees in approximately 20 years (Runkle 1982; Mladenoff 1985). This interaction of disturbance regimes in northern hardwoods at two different spatial and temporal scales and the variable stand structure that develops may be one factor in the greater number and variety of dominants that can persist in mature stands, as opposed to the boreal forest.

Locally, windthrow can be important in some southern boreal forests in montane areas. For example, windthrow and subsequent stress on exposed edges of openings is responsible for the wave-like pattern of stand history and structure in high-elevation balsam fir forests in New England and Japan (Sprugel 1976, 1984; Sprugel & Bormann 1981). In these forests, a patch opened by windthrow 'migrates' downwind because trees at the downwind side of the patch are increasingly exposed. Simultaneously, the forest recovers from the upwind side of the patch. This results in a unique stand structure of crescent-shaped patches aligned perpendicular to the prevailing winds with the oldest trees on the windward side of the patch.

Windthrow differs from fire in two important ways. First, wind deposits large amounts of debris on the forest floor, rather than removing debris. This should have a substantial effect on the cycling of nutrients, first because of the large pools of nutrients stored in boles and branches and second because these materials are recalcitrant to decomposition. For example, the opening of the canopy with the simultaneous return of large amounts of woody debris to the forest floor creates a unique spatial pattern of carbon and nitrogen cycles in high-elevation balsam fir forests

(Sprugel 1984). On a smaller scale, N mineralization increases in gaps of dominant sugar maple and hemlock, with the increase being greater in hemlock gaps than maple (Mladenoff 1987). A second major way in which windthrow differs from fire is that seed pools remain relatively intact after windthrow, whereas surface seed pools are either destroyed wholesale or, in the case of serotinous species, selectively activated after fire. As a consequence, species diversity in patches recovering from windthrow may be much greater than in patches recovering from fire, but at present neither this hypothesized pattern nor its consequences for other ecosystem properties has been tested.

In northern hardwoods, where windthrow rather than fire is most important, the potential successional sequence is more complex because of longer-lived species and greater diversity in dominant species' life histories. Without burning, northern hardwood stands that are logged or blown down regenerate from sprouts and advanced reproduction, primarily of sugar maple (Mladenoff & Howell 1980; Dunn, Guntenspergen & Dorney 1983). Sugar maple reproduction is ubiquitous in most northern hardwood stands. Its large seed, which allows penetration of the undisturbed forest litter by the radicle, and its shade tolerance favor it over its associates, which are either less tolerant or have particular seedbed requirements (Godman & Krefting 1960; Tubbs 1965). Similarly, Dunn, Guntenspergen & Dorney (1983) found that a mature northern hardwood–hemlock stand subjected to complete canopy removal from catastrophic windthrow regenerated directly from existing sugar maple saplings, new seedlings of yellow birch from remaining seed trees, and sprouts. An initial stage of pioneer species such as aspen or birch need not occur.

Locally, windthrow in northern hardwoods can be synergistic with fire. In the western Great Lakes region, severe drought can occur several times in 100 years during summers of persistent high-pressure systems that block north-eastward-moving moisture from the Gulf of Mexico (Lorimer & Gough 1982). If these conditions occur within a few years of a characteristic large-scale windthrow, a severe fire can occur in a former northern hardwood stand. In such a case the same species would be favored as after fire in boreal forests, depending on propagule availability and seasonality of burn. Similarly, following late nineteenth and early twentieth century logging, slash fires resulted in extensive stands of aspen and birch on former northern hardwood sites in the western Great Lakes region (Graham, Harrison & Westall 1963).

Animal habitat

Many boreal herbivores have their greatest densities at the southern border of the boreal forest, but prefer boreal communities over northern hardwoods when these elements are mixed in the landscape. Examples of such animals include moose (*Alces alces*), beaver (*Castor canadensis*), woodland caribou (*Rangifer tarandus*) and snowshoe hare (*Lepus americana*). In contrast, whitetail deer (*Odocoileus virginiana*) are more prevalent in northern hardwoods than in boreal communities.

Mammalian herbivores profoundly influence nutrient cycles and successional pathways in boreal forests by physically altering habitats and selectively foraging on certain plant species while avoiding others (Wolff 1980; Haukioja *et al.* 1983; Bryant & Kuropat 1980; Bryant, Chapin & Klein 1983; Bryant *et al.* 1985; Bryant & Chapin 1986; Bryant 1987; Naiman, Melillo & Hobbie 1986; Naiman, Johnston & Kelley 1988; Pastor *et al.* 1988; Johnston & Naiman 1990*a*). However, the full implication of these changes for both animal populations and landscape structure is not yet clear.

Beaver are the best example of a boreal herbivore that physically alters its habitat (Naiman, Johnston & Kelley 1988; Johnston & Naiman 1990*b*). Pre-settlement beaver populations in boreal North America are estimated at 10 million (Naiman, quoted in Newman 1985), and it was this resource that brought Europeans to boreal North America with all their subsequent modifications of the environment (MacKay 1967; Martin 1978; Newman 1985). In the boreal landscape of Voyageurs National Park in northern Minnesota, beaver ponds cover 13% of the land area, with an additional amount in heavily browsed zones around the ponds (Naiman, Johnston & Kelley 1988). Thus, beaver create a patch mosaic in the landscape, with the size and longevity of the ponds in direct proportion to topographic control and food supply (Johnston & Naiman 1990*b*). Both the aquatic and terrestrial patches cycle nutrients in ways not seen in adjacent forests and free-flowing streams (Naiman, Melillo & Hobbie 1986; Naiman, Johnston & Kelley 1988).

Selective foraging causes complex interactions between mammalian herbivores and ecosystem carbon and nitrogen cycles. First, boreal herbivores such as moose, beaver, and hare do not browse at random, but selectively browse early-successional hardwoods such as aspen and birch because these species contain lower concentrations of carbon-based, secondary compounds than do conifers (Bryant & Kuropat 1980) or produce these compounds only during juvenile phases (Bryant, Chapin & Klein 1983; Tahvanainen *et al.* 1985). Secondly, aspen and birch cease producing secondary compounds during maturity because they can grow rapidly enough to escape ungulate and hare browsing, at least under low to moderate levels (Coley, Bryant & Chapin 1985), but this rapid growth is possible only in soils of moderate to high N availability. In contrast, slow-growing conifers that are generally confined to soils with low N availability continue to produce carbon-based defenses because severe N limitation prevents rapid growth to escape browsing (Coley, Bryant & Chapin 1985; Bryant, Tuomi & Niemela 1988). Thirdly, defoliation can alter carbon–nitrogen balances in deciduous plants by inducing defense compound production, in part because of large carbon reserves in stems and large roots and in part because defoliation seems to cause fine root mortality and hence nitrogen stress (Tuomi *et al.* 1984; Bryant, Tuomi & Nierrela 1988). Finally, the same compounds that deter mammalian herbivory also decrease the rate of decomposition (Coldwell & DeLong 1950; Bryant & Kuropat 1980; Melillo, Aber & Muratore 1982; Flanagan & Van Cleve 1983; Bryant & Chapin 1986; Horner, Gosz & Cates 1988) for the simple reason that both ruminant digestion and decomposition are microbially mediated. Over time, selective browsing on early-successional hardwoods and avoidance of conifers such as spruce increases conifer dominance (Snyder & Janke 1976) and decreases litter quantity and quality and consequently soil nitrogen availability (Pastor *et al.* 1988). This in turn may induce further secondary compound production in residual hardwoods and shift competitive balance further toward conifers with low N demand (Bryant & Chapin 1986).

There are several processes that complicate this pattern further. First, some deciduous shrubs, such as hazel (*Corylus cornuta*), respond to browsing by developing into dense hedges that inhibit invasion by other species (Trottier 1978). Hazel litter decomposes easily and increases soil nutrient availability (Tappeiner & Alm 1975). Thus, development of browsing-induced hazel patches may stabilize food supply for ungulates, perhaps long enough to develop coincident patches of high soil N availability. Secondly, although not a preferred food, balsam fir is sometimes heavily browsed by moose in late winter when snow covers most other available shrubs or when heavy browsing has greatly reduced deciduous food supply (Peterson 1955). This may partially compensate for the competitive edge that conifers obtain when deciduous shrubs are heavily browsed. Finally, patches of unbrowsed species such as spruce or alder (*Alnus* spp.) are often required by hare or moose for winter cover, leading to gradients of decreasing browsing intensity away from cover patches (Hamilton, Drysdale & Euler 1980; Wolff 1980; J. P. Bryant, personal communication; J. Pastor & D. J. Mladenoff, personal observations). Thus, decisions that herbivores must make at different times and at different scales may cause development of spatial patterns in nutrient cycles across the landscape (Fig. 8.2).

Mammalian herbivores seem to have less effect on northern hardwoods than on boreal forests. This may be due in part to physiognomic differences between boreal and northern hardwood forests that result in different habitat suitabilities for herbivores in terms of both food and cover. The birch and aspen of the boreal forest have open canopies, which permit light to reach the forest floor, thus encouraging a well-developed understorey with abundant browse within reach of mammals. In contrast, mature northern hardwoods are characterized by a high closed canopy that casts a denser shade, resulting in a more open, sparse understorey, particularly in younger stands before gaps become frequent (Curtis 1959). The canopy becomes punctuated by small gaps as the stand continues to age (Runkle 1982; Mladenoff 1987) and it is in these gaps that browse supply is concentrated (Swift 1948; Stoeckeler, Strothmann & Krefting 1957; Frelich & Lorimer 1985). However, with large-scale disturbance such as occurred with logging in the past 100 years, deer populations reach levels not typical of the original northern hardwood landscape and can have a significant effect on both young and remaining old stands (Swift 1948; Graham 1954; Stoeckeler, Strothmann & Krefting 1957). In old growth forests, deer yard in eastern hemlock stands during winter because hemlock is a preferred food and snow is not as deep under the dense canopy (Dahlberg & Guettinger 1956; Anderson & Loucks 1979). Although the deer browse on both hemlock and sugar maple seedlings in these stands, the sugar maple are able to resprout and withstand browsing while hemlock generally dies; eventually, hemlock

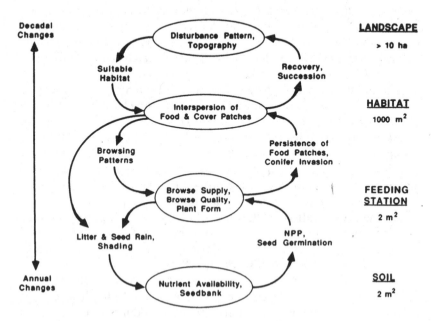

Fig. 8.2. Interactions between mammalian herbivores and habitat properties at nested scales.

forests in which deer yard could be converted to sugar maple forests (Frelich & Lorimer 1985; Alverson, Waller & Solheim 1988).

The role of insects in boreal forests is covered in detail elsewhere in this volume (Holling, Chapter 6). However, a few words may be in order here. First, deciduous, early-successional species may withstand defoliation better than conifers because their indeterminate growth habit allows them to produce new shoots the year of defoliation, whereas complete defoliation generally kills conifers (Kramer & Kozlowski 1979). Thus, spruce budworm (*Choristoneura fumiferana*) defoliations generally kill spruce and fir (Morris 1963), but tent caterpillar (*Malacosoma disstria*) defoliation of aspen merely slows growth for several years (Duncan & Hodson 1958). Killing the conifers causes a reversion to early successional stages (Morris 1963) while defoliation of aspen temporarily releases understorey balsam fir (Duncan & Hodson 1958). A combination of tent caterpillar and spruce budworm attacks can, over time, cause a stable limit cycle between balsam fir and aspen, whereby the caterpillar accelerates succession to balsam fir with the budworm following and resetting succession backwards (Ahlgren & Ahlgren 1984).

Succession, ecosystem processes, and spatial patterns in the southern boreal forest

An interesting facet of the southern boreal–northern hardwood border, with profound implications for ecosystem behavior, is the common assemblage of early-successional species that have different functional characteristics from the late-successional species in both biomes (Fig. 8.1, Table 8.3). Aspen, paper birch, and pin cherry are the major, shade-intolerant pioneer species both to northern hardwoods with or without a pine component (Gates 1912; Gleason 1924; Darlington 1930; Grant 1934; Nichols 1935; Eggler 1938; Kittredge 1938; Graham 1941; Braun 1950; Marks 1974; Fralish 1975) and to boreal conifers (Cooper 1913; Bergman & Stallard 1916; Lee 1924; Weaver & Clements 1929; Grant 1934; Nichols 1935; Kell 1938; Buell & Gordon 1945; Braun 1950; Oosting 1956; Larsen 1980). In contrast, jack pine is an early-successional conifer confined mainly to boreal regions. This common assemblage of early-successional species caused some confusion about whether hardwoods or conifers were the 'climax' forest of the region at the boundary between boreal and northern hardwoods in North America (e.g. Gleason 1924; cf. Weaver & Clements 1929; Braun 1950). This confusion was partly resolved by an adoption of an individualistic theory of plant interactions (Maycock & Curtis 1960; Larsen 1980) as well as a recognition of multiple pathways of succession depending on local conditions and episodic events (Graham, Harrison & Westell 1963; Fralish 1975; Larsen 1980). For example, boreal conifers are more frequent on cooler sites or at the extremes of the moisture continuum (Cooper 1913; Maycock & Curtis 1960; Larsen 1980) while northern hardwoods follow the early-successional species on the more mesic and warmer sites (Kittredge 1938; Maycock & Curtis 1960), with a strong component of white pine in rockier areas (Heinselman 1973). However, the vagaries of seed dispersal and establishment add a strong, seemingly stochastic, flavor to the entire process of succession. For example, many former northern hardwood sites in the western Great Lakes region, following logging and fire, are now occupied by stands colonized by aspen with spruce and fir becoming dominant (Curtis 1959; Maycock & Curtis 1960). Future climatic shifts may determine whether these sites are once again invaded by northern hardwoods or remain in boreal species with periodic renewal by disturbance. Brown & Curtis (1952) and Maycock & Curtis (1960) showed that stands within the boreal–northern hardwood border range from those typically dominated by each group separately, to stands containing mixtures of boreal and northern hardwood species. In this transition region where climate is ambivalent and highly stochastic in limiting these northern (boreal) and southern (northern hardwood) species, site conditions and microclimate take on greater importance in determining the balance of boreal vs. northern hardwood species within a stand. The contrasting moderating effects of Lake Superior along its north and south shores illustrate this effect. Along the north shore of the Lake, where the northern limit of the northern hardwood species is being approached, well-developed stands of sugar maple occur on high ridges overlooking the Lake, where the cold is ameliorated (Marschner 1974). Conversely, south of Lake Superior where regionally northern hardwood species are best developed, a band of boreal forest several kilometers wide exists in the cool local environment directly on the lake shore (Curtis 1959).

If site and stochastic events of establishment result in a stand of mixed boreal conifers and northern hardwoods, future compositional change is likely to be highly variable. However, if such events result in the development of a typical spruce–fir or northern hardwood stand, given similar climate, internal ecosystem feedbacks and the two systems' differing propensity for disturbance type may help to keep the site occupied by the original system (Siccama 1974; Pastor *et al.* 1984; Pastor & Broschart 1990). For example, Siccama (1974) hypothesizes that the cooler microclimate beneath closed spruce–fir canopies inhibits invasion by hardwoods, reinforcing a regional climatic discontinuity in the Green Mountains of Vermont. Similarly, the depression of nitrogen availability by conifer litter may inhibit invasion by more nitrogen-demanding hardwoods (Nadelhoffer, Aber & Melillo 1983; Pastor *et al.* 1984; Mladenoff 1987). Such positive feedbacks between tree species and limiting resources may reinforce successional trends toward one state and establish sharp boundaries between stands.

Therefore, although regionally there is a fairly even mix of boreal conifers and northern hardwoods, at a landscape scale boreal conifers do form 'climax' stands distinct from northern hardwoods, often separated from each other by sharp boundaries (Braun 1950; Rowe 1972). This has led Rowe (1972), for example, to classify the entire region in North America containing the boreal–northern hardwood border as the Great Lakes Conifer–Hardwood Forest, an entity distinct from either Northern Hardwoods or Boreal Conifers proper.

However, clearly it should not be surprising that initial stand composition and succession can vary considerably in this region of overlapping ecosystem types, where a great variety of northern and southern species reach the edge of their contiguous ranges for many reasons (Little 1971). Given the variety of events that control species composition and stand initiation in all regions, even similar sites will not necessarily result in the establishment of similar forests (McCune & Allen 1984a,b). The complex interactions between climate cycles, soil types, fuel decomposition and flammability, topography, animal populations, reproductive strategies, and disturbance intervals (Fig. 8.1) have led some to suggest that there is no succession in boreal forests in the linear sense in which it is applied to more temperate regions (Rowe 1961). The mosaic of forest types simply cycles among combinations of different species depending on whether factors that promote change occur synchronously and therefore amplify each other or whether they occur out of phase and thereby dampen each other (Heinselman 1981a; Wein & El-Bayoumi 1983). In the boreal–northern hardwood transition region, particularly where strong local elevational controls are lacking, succession is best characterized not as a fixed movement toward a chosen climax, but rather as change away from a stochastically determined beginning (Wein & El-Bayoumi 1983; Roberts 1987).

There are stochastic, linear, and cyclic components of succession in boreal forests (Wein & El-Bayoumi 1983). Stochastic aspects are small and frequent but unpredictable from the internal dynamics of the system, and cause no long-term change in species composition, although local changes in species dominance may result. Examples include lightning strikes on individual trees, browsing of individual stems, and random dispersal and germination of seeds. However, when these random processes occur over a large scale, then predictable and linear changes akin to succession in temperate regions occur. Thus after large fires succession proceeds through the classical shade-intolerant aspen–birch to shade-tolerant spruce–fir sequence, but the exact sequence depends on the availability of seeds in the seed bank or in the landscape, as discussed above.

Cyclic patterns in vegetation succession, as well as in animal populations and soil properties, are the most striking feature of the southern boreal forest. Cyclic fluctuations in boreal regions appear to occur when one species has a strong influence on availability of a limiting resource. The simplest example is the dense shading cast by conifers, inhibiting seedling establishment and eventually causing periodic fluctuations even in single-species populations, as shown elsewhere in this volume (Antonovski, Ter-Mikaelian & Furyaev, this volume, Chapter 14).

Another example is the potential stable limit cycle between birch and spruce caused by their opposite effects on, and subsequent responses to, nitrogen and light availabilities in modeled forests (Pastor *et al.* 1987). In these simulations, birch invades disturbed areas and increases nitrogen availability because of its easily decomposable litter; it is succeeded by spruce because spruce is more shade-tolerant but spruce in turn depresses nitrogen availability because of its recalcitrant litter. This leads to spruce dieback, opening of the canopy, increased decomposition and nitrogen availability, and re-invasion by birch. Thus, having a community composed of a few species with strong and opposite effects on limiting resources may cause limit cycles to develop.

The well-known cyclic nature of many boreal animal populations may be related to changes in plant availability or food quality during succession, the consequent spatial heterogeneity, and the feedbacks between browsing and individual-plant and whole-ecosysten responses (Gill 1972; Haukoija & Hakala 1975; Haukioja 1980; Rosenzweig & Abramsky 1980; Wolff 1980; Laine & Henttonen 1982; Haukoija *et al.* 1983). Furthermore, changes in browsing intensity during herbivore fluctuations may affect recovery of subpopulations of browse species and thereby amplify cycles in plant populations. In addition, Bryant, Tuomi & Niemela (1988) hypothesize that coevolution between hares and their browse species is related to the 10-year hare cycle.

Whereas boreal forests are composed of shade-intolerant hardwoods with easily decomposable litter (aspen, birch) and shade-tolerant conifers with recalcitrant litter (spruce, fir), northern hardwoods have a much less bimodal distribution of species characteristics (Table 8.3). Because no one species has an overriding effect on resource availability in northern hardwoods, and because they are longer-lived than boreal species, these forests may not have strong cyclic patterns of succession.

The important point to recognize here is that succession in both boreal and northern hardwood forests can begin with a common compositional stage after disturbance, but ecosystem properties diverge depending on the assemblage of succeeding species and their effects on resource availability. By strongly affecting resource availability, the succeeding community can cause further changes or it can be an assemblage of species that remains relatively stable in the absence of catastrophic, external disturbance. This leads to a fragmented patchwork of very contrasting ecosystems across the landscape.

As a result, the landscape does not seem to have a characteristic scale with a stable species composition, but rather may consist of a nested pattern of non-steady-state mosaics (Baker 1989). The disturbance environment appears to have several grains, with a large-scale grain causing some areas to burn or blow down more frequently than others, preventing a stable mosaic on a large scale; smaller-scale heterogeneities in soils, seed dispersal, and herbivore impacts prevent a steady state within disturbed patches. Baker (1989) theorizes that the southern boreal–northern hardwood landscape is highly fragmented because of the interactions between an external fire-patch regime driving fire occurrence and an internal fire-patch regime driving patterns of recovery, with propensities to burn changing during succession.

This fragmented pattern can arise not only from stochastic disturbances but also from the positive feedbacks in nutrient, herbivore, and fire cycles that cause divergent patterns of succession starting from initially similar conditions. Such divergences are characteristic of chaotic systems dominated by strong positive or negative feedbacks (May 1976). It is well known mathematically and supported to some extent empirically that biological systems that exhibit limit cycles can, when positive feedbacks are amplified, shift into chaotic behavior (May 1976). Interestingly, the lynx–hare cycle, perhaps the best known of all boreal forest limit cycles, appears to behave chaotically over the long term (Schaffer 1984, 1985; Schaffer & Kot 1986). While some argue that chaotic systems should eventually go extinct (Berryman & Millstein 1989), there are no general rules yet to predict how long this should take. Recall that the current assemblage of boreal and northern hardwood species arose only about 6000 years ago in the fossil record (Davis 1984; Webb, Cushing & Wright 1984); while this may seem long by human standards, it is only 30 generations or fewer for tree species that live 200 years or more. Thus, because it is evolutionarily very young, it is entirely possible that the boreal–northern hardwood landscape does not obtain a stable state even on very large scales because of underlying chaotic behavior arising from multiple feedback loops among plants animals, and the nitrogen cycle, from a common successional pathway to both boreal conifers and northern hardwoods, and from the stochastic influences of climate.

Climate change and the future of the southern boreal forest

Given the complex nature of the southern boreal forest at its border with northern hardwoods, it is extremely difficult to predict how it may respond to climate change. On one hand, synoptic weather patterns are partly responsible for the position of the boundary and some of its dynamics such as fire, but changes in these weather patterns are not predicted by the current generation of atmospheric circulation models. On the other hand, the fact that climate is neither optimal nor absolutely inhibiting to the existence of either boreal or northern hardwood species means that even small changes can greatly enhance or depress the growth of individual trees, depending on whether the change is toward or away from the species' climatic optimum.

Both the magnitude and rapidity of changes in climate are important factors to consider when forecasting changes in the southern boreal forest. If changes in temperature are as large, as forecast by some General Circulation Models (Manabe & Stouffer 1980), then most of the land currently occupied by boreal forest will be replaced by grassland in mid-continent areas and northern hardwoods in areas under maritime influence (Emanuel, Shugart & Stevenson 1985a). An important point here is that there is limited land area in the north to which boreal species may migrate. Even with less drastic changes in climate, the rapidity of the change may cause greatly restricted ranges of both northern hardwoods and boreal conifers because the species may not be able to disperse northward fast enough to keep up with shifts in climate zones (Davis & Zabinski 1990).

The movement of species northward in response to a warmer climate may have secondary effects that amplify those directly induced by climate change alone. For example, should sugar maple migrate into areas now currently occupied by spruce, nitrogen availability may be enhanced, further increasing productivity (Pastor & Post 1988). However, this can happen only on soils that retain adequate water to support the growth of maple; otherwise, the boreal forest will be replaced by an oak-pine savanna of much reduced biomass (Pastor & Post 1988).

Certain processes important to the structure of this landscape, such as fire, animal browsing, and insect attack, will almost certainly be altered in a warmer climate, and will in turn have their own secondary effects via other feedbacks. At this stage it is not possible to say with any degree of certainty what will happen.

In all likelihood what will happen is that the different species will respond individually to a changing climate, migrating in different directions and at different rates. New combinations of species will arise and old ones will disappear as occurred in the past (Davis 1984). Thus, any prediction of how the southern boreal forest will respond to a changing climate requires recognition of the different functional roles of the current species and the heterogeneity of soils and disturbance regimes in the landscape.

The research of this paper was supported by NSF grants BSR-8817665 and BRS-8906843.

9 Transitions between boreal forest and wetland

F. Z. Glebov and M. D. Korzukhin

In the boreal zone, precipitation exceeds evaporation and the forests are inclined to be paludal. There are two contradictory points of view regarding the nature of the transitions between boreal forest and bog. The first such view conceives of irreversible paludification of the forests due to the advancement of bog and self-paludification of the forests. The second surmises a dynamic equilibrium between forest and bog shown initially by the general lessening of the paludification process in recent times and secondly by the periodic afforestations of bogs and the depaludification of paludified forests.

The transitions between boreal forest and bog may be represented as a phytocenotic, continual series of ecosystems. Transitions among these ecosystems are reversible in time and space. A paludification series consists of automorphous forest → paludal forest → bogged forest → treed bog → open bog → regressive lake–bog complex; a series of depaludification is regressive lake–bog complex → secondary open bog or secondary treed bog → secondary bogged forest → automorphous forest. This work covers the latitudinal belt of taiga of the West Siberian Plain (Fig. 9.1) that consists of northern-, central-, southern- and sub-taiga subzones and is limited by permafrost in the north and soil salinity in the south.

Major works on the transitions between forest and bog

According to Sukachev (1914*a*), there are two opposite pathways of mire evolution. The first pathway involves intensification of moistness with peat accretion and an eventual transition to ombrogenic supply (Abolin 1914). This process results in the formation of a bog that covers a wide area and has a flat surface. The vegetation is like that described for raised bogs with pools and ridges. The second pathway is one of bog

Fig. 9.1. The taiga zone of the West Siberian Plain. Roman numerals indicate: I, subzone of northern taiga; II, subzone of central taiga; III, subzone of southern taiga; IV, subzone of subtaiga.

afforestation (Flerov 1899; Yurev 1926) and is typical of small, ombrogenic, high-domed sphagnum bogs. Self-drainage can occur in these bogs and a sphagnum pine forest then forms, resulting in a secondary bogged forest. It should be pointed out that such secondary bogged forests are distributed as widely as primary ones.

In Sukachev's (1914*a*) and Abolin's classic papers, secondary bogged forests are specified only in terms of descriptions of the vegetation. There are many stratigraphical data which indicate the existence of such forests. However, there are few works where conditions for the formation of secondary bogged forests derived from stratigraphical data are described in the light of our current scientific knowledge. Descriptions of specific series of phytocenotic successions in the north-western USSR, which include stages of secondary bogged forests originating during the oligo-trophic phase of mire complexes, are given by Galkina (1946, 1959).

Over the past hundred years, many workers have expressed widely differing opinions about the progressive and irreversible self-paludifi-cation of forests due to deforestation and self-paludification of soils in the taiga zone (Tanfilyev 1888; Graebner 1904; Hesselman 1907; Levitskii 1910; Haglund 1912; Cajander 1913; Kaks 1914; Sukachev 1914*a*; Tsin-zerling 1929; S. S. Archipov 1932; Dokturovski 1932; Sambuk 1932; Shennikov 1933; Pyatkov 1935; V. P. Williams 1939; Robinson, Hughes & Roberts 1949; Kampfmann 1980; P. D. Moore 1982; and others). Investigations providing evidence for the reversibility of forest paludifi-cation (Malmström 1932; Burenckov, Kotscheev & Malchevskya 1934; Melechov & Goldobina 1947; Tamm 1950; Kotscheev 1955; Huikari 1957; Lag 1959; Pyatetskii & Morozova 1962; Nickiforov, Melnickova & Kolymtsev 1981; Alaback 1982) did not start appearing until somewhat later and are less well known. The works of P'yavchenko (1953, 1954, 1956, 1963, 1979, 1980, 1985) are of special note and it is he who should be considered as the founder of the concept of the dynamic equilibrium existing between forest and mire. P'yavchenko believed that 'the paludi-fication process of forest soils ... is a contradictory one involving the unity and struggle of two opposite tendencies – paludification and depaludification. The fact why to this time the forests of our taiga zone have not been substituted for peat bogs but occupy a territory many times greater than mires should be explained by this contradiction.' (P'yav-chenko 1954, p. 278). Further on (*loc. cit.*, p. 286) he supposed that 'the processes involved in recent paludification of the forests ... are in the majority of cases of a temporary nature, arising and attenuating under the influence of factors of environment and vegetation'. P'yavchenko suggested that three quarters of boggy territory becomes thinly deposited peat bog and temporarily watered forest land. He also suggests that in recent times bog formation has on the whole slowed down because peat has already accumulated in most depressions, and because of water erosion and positive tectonics of the earth's surface in boreal regions. He also noted that south of the taiga zone mires are afforested, whereas in the forest-steppe they are subject to drought and cracked (P'yavchenko 1985).

In the West Siberian Plain, coverage of peat within the boreal zone is about 50%, reaching 70–80% in some locations. (This high degree of peat formation is taken into account by many investigators in the naming of the zone itself: for example, Gorodkov (1916*a*) distinguished cedar pine–boggy and urman–boggy subzones, and Orlov (1968) named the whole zone as a forest–mire.) In the landscapes of the West Siberian Plain, and in distinction from other regions of the boreal zone, both forests and mires have equally strong tendencies to replace each other. The recent geographical situation, as well as its future development, depends significantly on this pattern of interactions. Forest–mire transitions are especially diverse here, and the Plain is ideal for their study. A formulation of a theory for the dynamic equilibrium in forest–mire relations for the West Siberian Plain was developed by Glebov (1988) and ecological models representing this theory by Glebov & Korzukhin (1985, 1988, 1990).

Historically, the exceedingly high degree of peat formation over the vast area of the West Siberian Plain gives the illusion of progressive paludification, occurring over an extended period of time. It is in relation to the Plain that the total paludification concept in forests has been presented in numerous publications, since the beginning of the century. These involve scores of workers and have been written from various scientific positions: landscape science (Gordyagin 1901; Micheev & Dibtsov 1975); botanical geography and geobotany (Kuznetsov 1915*a,b*; Gorozhankina & Konstantinov 1978); peatland science (Baryshnikov 1929; Liss & Berezina 1981); foresty (Krilov 1961); soil science (Dranitsin 1914; Karavaeva 1982); geomorphology (Dranitsin 1914; Zemtsov 1976); hydrology (Nazarov *et al.* 1977); and paleoclimatology (Zhukov 1977*a*). We feel that it is appropriate to consider whether there is sufficient evidence to support the theory of total paludification of the boreal zone in the Plain.

Vast lowland areas of a high degree of paludification are found not only throughout West Siberia but also in other parts of the boreal zone, for example in northern Europe and in Canada, where the degree of paludification also reaches 50% and more (Moss 1953; Ruuhijärvi 1983). All highly paludified plains have the features typical of 'peat basins' reported by Nikonov (1955): (1) water balance favorable for paludification; (2) flat relief, promoting the stagnation of soil-groundwaters and preventing drainage, despite maturity of the river network; (3) abundant, widespread and relatively shallow reservoirs in which groundwater accumulates; and (4) a kettle form of the surface of Quaternary rocks which allows water to penetrate into paleodepressions.

Patterns of hydrology and climate are especially significant for peatland formation. It is here that one may reveal the unique causes of the degree of paludification in the lowlands of the Plain which distinguish it from other 'peat basins'. The hydrological regime of the Rivers Ob and Irtysh and their tributaries, with a system of backwaters preventing water flow from watersheds, is of major importance (Wendrov *et al.* 1966; Wendrov, Gluh & Malik 1967; Malik 1978). There is still much uncertainty, however, about the mechanism of climate influences. In the opinion of climatologists (Shwareva 1963; Mezentsev & Karnatsevich 1969; Zhukov & Potapova 1977), the high degree of paludification of the Plain is due to the water–heat balance of the area. Because of the insufficient heat input, total water input exceeds evaporation and excess humidity arises in the soil. This is supported to some extent by data from Canada: Nichols (1969) demonstrated through radiocarbon dating of basal layers that peatlands were formed primarily during cooler periods. Since a high degree of paludification causes a greater dissipation of thermal energy resources linked to excess evaporation, Zhukov (1977*b*) concludes that drainage amelioration will at the same time be heat amelioration. The poor heat conductivity of peat results in a considerable cooling of its superficial layer in conditions of cold or a continental climate, and in the case of drainage.

Zhukov's (1977*b*) study is concerned with southern subzones, in which the increased dissipation of heat through evaporation is expressed very weakly or not at all. In more northern subzones, freezing processes of drained peat deposits will be of considerable importance. Ivanov (1976) believes that if bog drainage occurs to the south of the permafrost zone, continentality will grow and the permafrost line will move south.

It should be noted that in Ireland, a country of blanket mires and domed bogs, there are 1250 mm of rainfall and 250 days with rain per year (Moore & Bellamy 1974). Thickness of peat deposits of 10–13 m is not uncommon here; the average peat deposit thickness for domed bogs is 7.5 m and for blanket mires 2.5 m (Barry 1969).

It is interesting that the degree of bog convexity (an indication of the activity of the bog formation process) decreases as the climate becomes less continental (Morrison 1955). In Norway, bog plant associations of dwarf shrubs and *Sphagnum fuscum* reach as far north as 66°; in the more continental European part of the USSR they reach only 53° (Eurola 1968). In northern Finland, intensive paludification of the forests and the formation of bogs in their place are explained by the maritime climate and nearby deposition of waterproof rocks on the earth's surface (Eurola *et al.* 1982). So it is seen that a humid, maritime climate such as that found in Ireland and Fennoscandia is most favorable for the bog formation process, whereas a continental climate, with relatively hot summers, and which is drier (500–800 mm of rainfall per year, 400–500 mm of that falling during the warm period of the year), as occurs in the taiga zone of West Siberia, is less favorable.

Elements of the concept of dynamic equilibrium are present in some works, which proceed to adopt quite a different concept.

While not denying that endogenous conversion of podzolic soils into boggy soils exists, Nemtchinov (1957) considered that this conversion occurs so slowly that boggy soils should for all intents and purposes be regarded as stable. An element of this concept may be observed in theories that suggest that paludification can result from devastation of forest stands by fires or fellings, and reduced evapotranspiration (Tanfilyev 1888; Ototskii 1906; Cajander 1913; Wisotzkii 1925). Logically, depaludification could occur if forests were left to regenerate (Malyanov 1939; Melechov & Goldobina 1947; P'yavchenko 1953; Kotscheev 1955). The role of bogged forest in drying soils has been demonstrated in studies by Heikurainen (1967). He found that both clear felling and selective harvesting of forest stands over peatlands resulted in an increase in the water level. This increase is less with clear felling than it is for selective harvesting. This phenomenon is also documented in Päivänen (1982).

The idea that processes which prevent paludification take place in automorphous forests is gaining support. For example, investigations carried out in the Ural region throughout the subzone of middle taiga between 1964 and 1969 are documented in a book, *Soswa Priobye* (1975), detailing the concept of irreversible paludification. The concluding chapter of the book *Natural Regimes of the Middle Taiga of West Siberia* (1977), which was written after the organization of the field station base in this region and summarized results of its work, says: 'Our observations show that dark-needle ecosystems play the part of a powerful 'pump' which constantly pumps out excess water coming from bogs . . .' (p. 291).

The concept of irreversible paludification, which flourished early in the 1970s owing for the most part to Neustadt's (1971) paper 'Paludification of the West Siberian Plain is a world natural phenomenon', became less categorical in the 1980s in the mind of this same author: he recognized that the process of bog advancement into forests was repeatedly weakened (Neustadt & Malik 1980). Irregularities of bog advancement into forests are noted in the monograph by Liss & Berezina (1981).

The authors above measured the slowing of the bog formation process in thousands of years; Dolgushin (1972) writes about fluctuations measured in hundred of years. According to his data, from the ninth to the fifteenth century the bog formation process was slow, and some bogs became covered with forests. In another paper (Dolgushin 1973) he states that within the subzone of middle taiga, rains make the soil in the paludified forest warmer, which accelerates summer defrosting, supplying the trees with more O_2 and nutrient substances, and thus increasing tree increment. One may draw the conclusion that rains favor the afforestation of bogs, but not swamp forests.

Finally, we shall consider direct evidence for the dynamic equilibrium between forest and mire formation processes.

On the basis of investigations in the north of the taiga zone, where the bog formation process is considered to be the most aggressive, Ivanov & Shumkova (1967) drew a conclusion about the stability of present-day borderlines of bogs and forest belts localized along rivers. For bogs to begin to advance into riverside forest belts, the available water balance must be disturbed.

According to the data of Lvov (1976), there are three latitudinal belts in the Tomsk region: northern (processes of paludification of dry land are developing); intermediate (bogging and drying of bogs are in balance); and southern (most of the bogs are becoming dry and covered with forest).

Orlov, who at first supported the concept of total paludification of the taiga zone (Orlov 1963), has more recently come to the conclusion, based on analyses of aerial photographs and comparisons with data on contemporary tectonics, that there are areas that are paludified and some that are depaludified (Orlov 1968, 1975).

In the south of the taiga zone the general balance of the transition between forest and bog tends toward depaludification. Most fens of the Barabinsk lowland become dry on the surface (Kuzmina 1949; Wagina 1982). Comparison of forest service data for 1888 with data from following years showed a considerable reduction in bog areas owing to an increase in forests (Taran 1964). The conclusion that in the south of West Siberia depaludification is presently taking place is made on the basis of stratigraphical analyses of peat deposits in a border belt of secondary bogged forests within fens of the Ob–Irtysh watershed (Platonov 1963; P'yavchenko 1965; Glebov & Alexandrova 1973).

Regional transitions between forest and mire

Subzone patterns

Moving from the subzone of middle taiga toward southern taiga, an improvement in trophic conditions of phytocenoses, soils and peat deposits parallels a sharp rise in the total degree of paludification of treed bogs, and secondary bogged forests start to play a significant role (Tables 9.1 and 9.2). In middle taiga, low-moor deposits are practically absent, whereas in southern taiga they are found predominantly over treed bogs; in both cases, high-moor deposits prevail over open bogs. In both subzones the thickness of low-moor deposits is less than that of high-moor deposits; thickness of deposits in general is less in treed bogs than in open bogs.

Table 9.1. *Classification of vegetation in the central and southern parts of the West Siberian Plain*

Vegetation type	Area 100 km²	%
1. *Subzone of middle taiga*	4918	100
1.1. Automorphous forests	1828	37
1.2. Hydromorphic vegetation	3090	63
1.2.1. Paludified forests (pine forests)	345	7
1.2.2. Primary bogged forests	894	18
1.2.2.1. Spruce and cedar pine forests	492	10
1.2.2.2. Pine forests	401	8
1.2.3. Treed (from pine forests) domed bogs	772	16
1.2.4. Open domed bogs	1079	22
2. *Subzone of southern taiga*	3732	100
2.1. Automorphous forests	1944	52
2.2. Hydromorphic vegetation	1788	48
2.2.1. Paludified forests	137	4
2.2.1.1. Spruce and cedar pine forests	53	1
2.2.1.2. Pine forests	84	2
2.2.2. Primary and secondary bogged forests	270	7
2.2.2.1. Forests of spruce, cedar pine and birch	120	3
2.2.2.2. Pine forests	150	4
2.2.3. Treed (from pine forests) domed bogs	798	21
2.2.4. Open bogs	583	16
2.2.4.1. Domed bogs	564	15
2.2.4.2. Transitional bogs	19	—
3. *Subzone of subtaiga*	1374	100
3.1. Automorphous forests	634	46
3.2. Hydromorphic vegetation	299	22
3.2.1. Primary and secondary bogged (birch) forests	52	4
3.2.2. Treed bogs	88	6
3.2.2.1. Domed bogs (from pine forests)	15	1
3.2.2.2. Transitional bogs and fens (pine, birch)	73	5
3.2.3. Open fens	159	12
3.3. Agricultural lands	441	32

Source: Glebov (1988).

It is seen that in these two subzones, a rise in trophic content of vegetation and soil toward the south, an increase in afforestation of bogs, and a retardation of the peat accumulation rate are directly related. This direct dependence confirms the conclusion drawn on the basis of ecological investigations that a rise in soil trophic content and climatic warming intensify forest formation processes and weaken those of mire formation. In the subtaiga subzone, however, this pattern breaks down: in hydro-

Table 9.2. *Peat deposits in bogs of different subzones of the West Siberian Plain*

S = area (above line, 100 m²; below line, %); h = thickness of peat deposit (m).

Subzone		Treed bogs						Open bogs					
	Vegetation type (from Table 9.1)	Type of peat deposit						Type of peat deposit					
		fen		mixed and transitional		raised		fen		mixed and transitional		raised	
		S	h	S	h	S	h	Vegetation type (from Table 9.1)	S	h	S	h	S	h
Middle taiga	1.2.3.	—	—	$\frac{50}{7}$	1.5	$\frac{663}{93}$	1.5	1.2.4.	—	—	$\frac{47}{5}$	1.6	$\frac{968}{95}$	1.7
Southern taiga	2.2.3.	$\frac{318}{42}$	1.4	$\frac{166}{22}$	2.0	$\frac{280}{36}$	2.2	2.2.4.1	$\frac{127}{23}$	2.2	$\frac{57}{11}$	3.0	$\frac{354}{66}$	3.1
								2.2.4.2.	$\frac{13}{70}$	1.5	$\frac{6}{30}$	1.1	—	—
Subtaiga	3.2.2.1.	56	1.2	—	—	15	1.4	3.2.3.	153	1.2	—	—	—	—

Source: Glebov (1988).

morphic parts of the landscape, open bogs are of absolute dominance, as in middle taiga. Soil salinity in the subtaiga subzone prevents bog afforestation.

The greatest thickness of peat deposit is in southern taiga where processes of bog afforestation are most intensive. This indicates that such thickness results from past climatic conditions.

Topology patterns

Bogs in the West Siberian Plain were instrumental in smoothing an original mesorelief (Archipov *et al.* 1970; Zemtsov 1976). This is not, however, evidence of recent paludification of the forests. The degree of hydromorphism in an ecosystem depends on the relative height of its soil surfaces in relation to neighboring ecosystems: paludified forests are located, as expected, in depressions.

The extent of influence of a bog on adjacent dry land is determined by the gradient of the mineral 'shelf' at the bottom of the peatland. The less this gradient is, the stronger will be the influence, exhibited by a greater width of the paludified forest belt adjoining the bog.

West Siberian bogs developed in conditions of a more continental climate than that in Europe, and that is why borders between ridges and hollows in a hummock–ridge complex are clear-cut and narrow; ridges are more elevated than hollows. Bog profiles are less domed, and abundance and area of secondary lakes significantly greater; this is evidence of the 'senescence' of West Siberian bogs and therefore of the loss of their aggressiveness.

The stratigraphical analysis of the peat deposits showed that secondary bogged forests have developed within the southern taiga subzone and even in the most bogged parts of the subzone of middle taiga, evidence of the recent bog afforestation.

Since blanket mires similar in relief to those found under conditions of the maritime climate of western Europe are not found over the Plain, the high degree of paludification of the Plain must be caused by not only recent but also past topo-hydrological conditions, as stated by Shumilova (1969).

Key ecosystem mechanisms of the transitions between forest and mire

The main reason for the possible existence of forest stands over bogs and fluctuations in their performance

Generalizing the results of investigations of a topo-ecological profile (Fig. 9.2) that was run through a domed high bog complex which had a

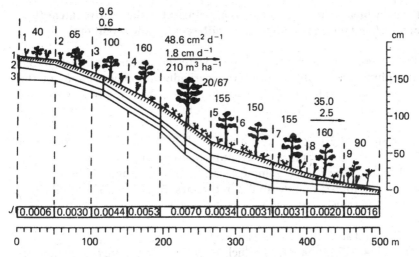

Fig. 9.2. Topo-ecological profile of domed bog. Averaged levels of swamp waters over the vegetative period 1969–74: 1, maximum; 2, average; 3, minimum; J, gradient of the bog surface. Horizontal lines with an arrow above the bog surface indicate places of the introduction of isotope tracers; figures above them indicate the rate of horizontal filtration (upper figure) and 'unit' water discharges (lower figure). Figures above the bog surface: lower figures are number of sample areas, upper figures are growing stocks in cubic metres per hectare. Vertical lines in the peat deposit are observation wells where water levels were measured.

homogeneous stratigraphy of the peat deposit over the whole area, similar physical and chemical characteristics and was covered in even-aged pine forest stands with different growing stocks in various parts of the profile, the following may be said. The possibility of the existence of trees over a bog, and the performance of the growing stock, depends on the aeration of the root layer that is achieved due to the simultaneous or separate realization of two factors: (1) variability of the water level regime during the vegetative period; and (2) high flowage of water. An index incorporating a combination of numerical values of the flowage (i.e. rate of horizontal filtration estimated by means of the radioisotope method) and variability of the level regime was used to calculate a 'unit' of water discharge in the root layer of the peat deposit. This discharge was estimated as the product of the horizontal filtration rate and the difference between minimum water level averaged over many years and the level during the experiment (measured by using a radioisotope).

Primary biological productivity as a coefficient of the relative intensity of forest and mire formation under different trophic conditions

Three sample areas in bogged forests were chosen for this investigation: their phytocenoses and soils differed markedly in trophic content, but the growing stock value of each was similar and high for bog conditions $(160-210 \, m^3 \, ha^{-1})$; the quantity of absolute dry woody phytomass was also similar $(165-172 \, Mt \, ha^{-1})$. The latter is of extreme importance for ascertaining the role of the forest stand in the biological turnover of forest–mire ecosystems of different trophic content.

The three areas were: sample area 1 in the oligotrophic bilberry–bog–moss–pine forest; sample area 2 in the mesotrophic marsh-trefoil–bog–moss–pine forest; and sample area 3 in the eutrophic grass–marsh–cedar pine forest.

The total amount of living phytomass in all the forest types is about the same $(174-177 \, Mt \, ha^{-1})$, of which 94–97% is in the trees. The above-ground phytomass greatly exceeds that of the below-ground phytomass and ranges from 69 to 73% of the total phytomass. In contrast, the amount of below-ground phytomass of the herb–dwarf shrub layer (consisting mainly of bog plants) is greater than the above-ground phytomass and ranges from 54 to 70% of the total phytomass. Substance quantity in the moss cover is inversely proportional to the trophic content of the soil. The greatest quantity, by mass, of chemical elements is in the eutrophic cedar pine forest; this decreases in the mesotrophic pine forest and falls again in the oligotrophic pine forest. Between eighty and ninety percent, by mass, of the elements are tied up in the trees; this is lowest in the oligotrophic pine forest (Fig. 9.3).

The same patterns are typical of both annual primary phytomass production and the biological turnover of chemical substances. These can be calculated from gross primary production, litterfall, and net primary production (the difference between the first two components) and are seen to decrease as trophic content drops. In the eutrophic and mesotrophic situations, phytomass increase and turnover of elements occur predominantly in the trees, whereas in the oligotrophic situation they prevail in the bog part of the phytocenosis (Fig. 9.4).

In general, phytocenoses of bogged forests are composed of trees, dwarf shrubs, hygrophilous herbs, and mosses. Peat formation is due mainly to the decomposition of below-ground parts of plants that have died.

Hygrophytes are represented mainly by herbaceous plants, whose below-ground part exceeds that above-ground, and also by mosses, in which the above-ground decay is small and growth is in stem tips with simultaneous dying off of the below-ground parts. Hygrophytes thus

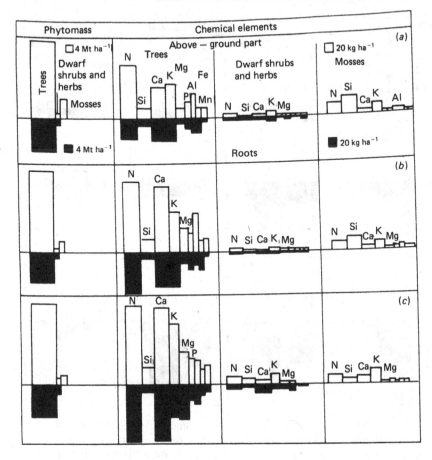

Fig. 9.3. Total phytomass amount and its chemical element content.
(a) Oligotrophic dwarf shrub–bog–moss–sphagnum–pine forest; (b)
mesotrophic trefoil–bog–moss–pine forest; (c) eutrophic grass–
marsh–cedar pine forest.

retard the processes of peat decomposition; they disperse elements into
the vertical profile of the peat deposit, suppress biological turnover, and
intensify peat formation.

The above-ground part of trees is much greater than that below-
ground. The ash content of wood is insignificant, and trees are long-lived:
the role of the trunk and branches in biological turnover and peat
formation is not large. However, an annual input to the soil of cation
elements and nitrogen from fast-mineralizing leaf-fall intensifies the
process of peat decomposition, concentrates nitrogen and ash substances
in the superficial layer of the peat deposit, stimulates biological turnover,
and reduces peat accumulation rate.

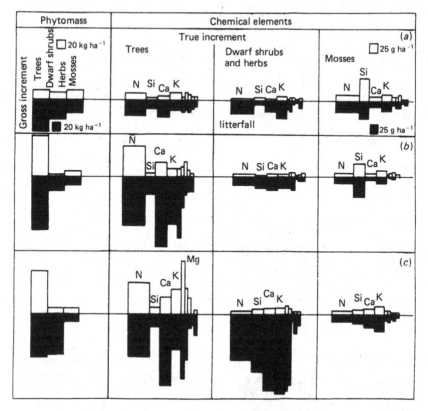

Fig. 9.4. Annual phytomass production and biological turnover of chemical elements. (*a*) Oligotrophic dwarf shrub–bog–moss–sphagnum–pine forest; (*b*) mesotrophic trefoil–bog–moss–pine forest; (*c*) eutrophic grass–marsh–cedar pine forest.

Trees stimulate the forest formation process, and hygrophytes stimulate the mire formation process. In a situation of relatively high nutrient content, a stand of bogged forest consists of species that are enriched in cations and produce much leaf litter (dark-needle species, birch): these stimulate biological turnover and accelerate the rate of peat decomposition. An increase in soil nutrient content thus favors the forest formation process and inhibits that of mire formation. Nutrient conditions act in a similar way to the decrease in heat supply as one moves to more northern latitudes.

Stages of forest bogging under different trophic conditions

In the mire formation process, there is a very slow accumulation (fractions of a millimeter per year) of peat and a gradual retardation of the succession rate in forests, while the ecosystem moves from forest to bog.

The rate of forest bogging is strongly affected by trophic conditions. The decline in available nutrient content in the course of peat deposit growth results in the replacement of more nutrient-demanding dark-coniferous species first, then of birch and then of pine. In oligotrophic situations, replacement of forest by bog occurs rapidly. In a stand of paludified forest, generations that appear before and during bogging (Fig. 9.5) are morphologically clearly distinguished and the thickness of the peat layer is not great in such situations. The paludified forest, which does not pass through a stage of primary bogged forest with insignificant peat accumulation, is replaced by a pine–dwarf shrub–sphagnum treed bog, where trees are not edificators. In high-nutrient or eutrophic situations, the forest environment persists for a much longer time, and paludified forest is replaced with a primary bogged forest that accumulates a thick peat deposit.

The process of the conversion of paludal forest into primary bogged forest is different depending on different available nutrient contents. Two extreme types of paludification will be considered here. The first type is ombrogenic as well as underflood with water from advancing high bog complexes. Paludification has two stages: exogenesis and endogenesis (see Fig. 9.8). At the exogenous stage, which takes place in the paludal forest, the main pioneers in bogging are sphagnum mosses. Bogging starts at the border with bogged complexes in depressions and begins with the outcropping of hollows grown over with hyperhygrophilous mosses. If the soil surface is a little raised, then bogging occurs as a slow advancement of a ridge of hygro- and meso-hygrophilous mosses. The height of such a ridge is 40–50 cm; its width can be up to 20 m. In the case of more elevated soil surfaces and in locations further away from the bog, the ridge divides into separate sphagnum cushions, which arise in microdepressions. These are invaded simultaneously by sphagnums and dwarf shrubs. The ground litter depth increases, and a process of illuvial horizon formation takes place, preventing water leakage. Because living, labile organisms are the main components, this process may easily be reversed by climatic fluctuations. The duration of the cycles of such paludification–depaludification is no more than 200–300 years. Annual height increment in sphagnum mosses reaches 10 cm and more (Tyuremnov 1949). Accordingly, the rate of reversible bogging of dry lands is very great: meters per year. Available data on this are rare but include those of Sparro (1924), who noted a rate of 11 m yr^{-1}, and Lvov (1976), who noted one of 1 m yr^{-1}. The degree of decomposition in the layer of organic matter decreases from the surface downward, but by this stage even lower layers are not yet peat, and tree roots are still found in the mineral soil.

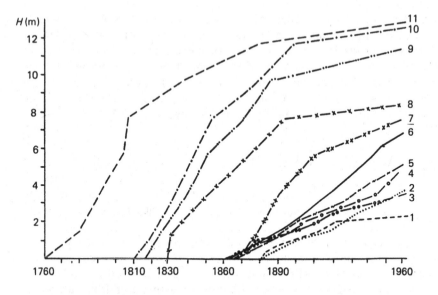

Fig. 9.5. A course of height growth of model trees in the paludified oligotrophic pine forest. This is an example of very fast paludification. Model trees 8–11 belong to the first generation, which arose prior to paludification, and model trees 1–7 to the second generation, which arose after paludification. The paludification period was between 1860 and 1890.

The endogenous stage, which takes place in the bogged forest, starts after the following events have occurred in some favorable sites.

(1) The formation of the illuvial gleyed horizon has come to an end, as a result of which a new, higher water level, independent of groundwater flow and vadose water level, has appeared above it.

(2) An organic layer of dead peat-forming plants, the lower part of which is true peat, has reached a critical thickness (Δh) (as specified by Ivanov (1975)) that is equal to the average amplitude of fluctuation over many years of a new water level typical of a 'peatgenous' horizon. (Far from all plants, including hygrophytes, form peat because after dying they decompose above the soil surface, as evidenced by the botanical analysis of peat.) In the lower part of this layer humidity is high whereas in the upper part it is low, the result of which is the incomplete decomposition of organic matter, i.e. peat accumulation.

(3) Separation of tree root systems from the mineral soil has occurred. Together with the appreciable increase in the hygrophilous part of the phytocenosis, this has resulted in a weakening of the forest stand as well as a diminution in size and density of trees, which are already unable to prevent peat accumulation by their litterfall.

As a result, the peat accumulation becomes irreversible under usual current fluctuations of climate. The system has already passed the stage at which a continuous external action, i.e. watering, is necessary for its existence.

The second type of forest paludification is quite rare, occurring mainly in river valleys, and characterizes eutrophic paludification particularly under the action of pressure from influent groundwaters. The process runs without the participation of sphagnous mosses; peat accretion proceeds because of the dying off of tree root systems and herbaceous hygrophytes. A well-decomposed woody and woody–grass–bog peat is formed. With pressure from the groundwater supply, the range of water level fluctuations is small, and consequently the peatgenous layer is very thin (up to 15–20 cm). The oxygen that is necessary for its formation is supplied with the influent waters. Reduction of the peatgenous layer's thickness is also caused by the large phytomass of high-ash litterfall, which is completely and rapidly decomposed by microorganisms, reducing acidity and stimulating microbiological processes. (The fact that as the trophic content of bogs rises, the thickness of the peatgenous horizon falls is noted on the basis of the analysis of extensive literature material by Tyuremnov (1949).) As with the oligotrophic type of paludification, the formation of a gleyed horizon takes place in the soil, but the head of water prevents complete gleization. It is impossible to distinguish exo- and endogenous stages in such paludification.

A number of intermediate types of paludification lie between these two extremes.

There are some quantitative characteristics for describing the transition between exogenesis and endogenesis. Peat increment is a result of two differently oriented processes, i.e. accumulation and decomposition of dead phytomass. During exogenesis, both processes proceed rapidly and spasmodically: a paludified area can become completely depaludified. During endogenesis, paludification is slow, continuous and stable; the peat accumulation can be checked, but does not stop (under usual climatic fluctuations). For this reason, the period of time that is necessary for the formation of the peatgenous horizon can be determined only approximately, bearing in mind that it is caused by an average rate of peat accumulation in the Holocene of roughly $0.5 \, \text{mm} \, \text{y}^{-1}$. The minimum thickness of the peatgenous horizon in bogged forest is 40 cm (root systems of trees have already passed through the mineral soil and are now located in peat). Accordingly, the time taken for horizon formation is 800 years. During this time the gleyed horizon would also be formed. Gradients of the soil surface in paludified and bogged forests are 0.01–0.02. A simple calculation shows that the rate of overpeating of such gradients under advancement of bogs is 3–6 cm yr^{-1}, i.e. during endogenesis it is two orders of magnitude lower than during exogenesis (cm yr^{-1} as against m yr^{-1}). In 800 years a borderline of bogged forest will advance only 25–50 m.

The transitions between forest and mire in the Holocene

Botanical, spore–pollen and radiocarbon analyses of natural peat outcrops at the River Ob (Fig. 9.6) and of peat pits in ridge–hummock complexes with pools (Fig. 9.7) within the subzones of middle taiga (Glebov et al. 1980) and southern taiga (Glebov et al. 1978) have revealed a number of long-term endogenous successions in bog development that are caused by mire formation processes oriented toward hydromorphism. However, as a background to these endogenous successions, 'inverse' variations in botanical composition are evidence of ridge expansion and reduction in the surface of lakes coinciding with xerothermic periods of the Holocene: toward the end of AT/beginning of SB and the end of SB/beginning of SA (Fig. 9.7).

Such warmings have been detected by many workers using a variety of methods: by Koshkarova (1986) on the basis of carpology investigations of peat bogs; by Kovalev, Klenov & Arslanov (1972) and Korsunov & Vedrova (1980, 1982) on evidence from soil science; and by C. A. Archipov & Votah (1980), Nickolskya (1982) and Firsov et al. (1982) from spore–pollen analysis.

During xerothermic periods, there was continued disturbance and afforestation of peatlands and depaludification of paludified forests that was especially intensive at the AT–SB boundary. This is evidenced by patterns in the formation of a 'border' horizon in European peatlands (with intervals in peat accumulation during its existence of up to 1500 years (Neustadt 1965)), an exceedingly high degree of peat decomposition in this horizon, the large size of tree stumps buried here and very large (c. 0.5 cm) annual ring width in some of these tree stumps. Tree rings of this width are not observable in forest stands of present-day drained domed bogs (Sukachev 1914b).

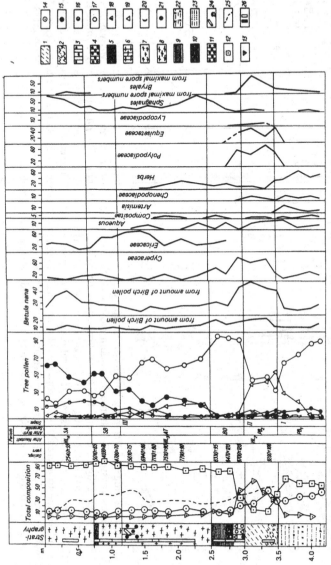

Fig. 9.6. Stratigraphy and spore-pollen spectrum of a natural peat outcrop 'Lukashkin Yar'. Low fen peat: 1, hypnum; 2, grass–hypnum; 3, woody–fern; 4, woody (dark-needle species and birch). Transitional peat: 5, woody (all woody species of Siberia) – cotton grass; 6, woody–sphagnum. Raised bog peat: 7, *Angustifolium*; 8, *Fuscum*; 9, *Eriophorum vaginatum*; 10, pine–*Eriophorum*; 11, pine. Totals: 12, the sum of pollen of woody species; 13, the sum of herbaceous pollen; 14, the sum of spores. Pollen: 15, pine (*Pinus sylvestris*); 16, cedar pine (*Pinus sibirica*); 17, birch; 18, fir (*Abies*); 19, spruce; 20, larch; 21, alder; 22, loam with horsetail; 23, loam; 24, tree stumps and trunks; 25, the degree of peat decomposition; 26, a point of sampling for [14]C.

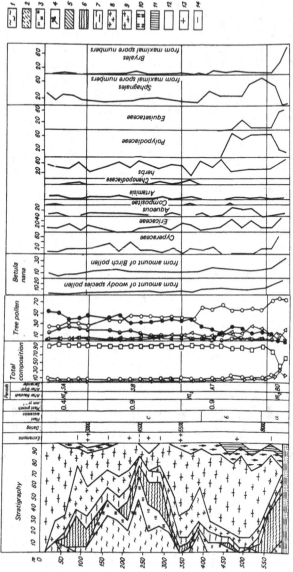

Fig. 9.7. Peat pit in a ridge of the ridge-hummock complex in the subzone of middle taïga. Peat-forming plants: 1, hummock sphagnous mosses; 2, hummock hypnum mosses; 3, *Scheuchzeria palustris*, *Carex limosa*; 4, *Menyanthes trefoliata*; 5, hummock sedges, except for *Carex limosa*; 6, *Eriophorum vaginatum*; 7, *Sphagnum magellanicum*; 8, *S. angustifolium*; 9, *S. fuscum*; 10, bog dwarf shrubs; 11, wood (mainly pine); 12, fiber, a species relation of which was not identified. Extreme points on the stratigraphical diagram of the peat deposit: 13, increase of content of hygro- and hydrophyte residues due to decrease of mesophytes; 14, increase of content of mesophyte residues due to decrease of hygro- and hydrophytes. Other standard abbreviations as in Fig. 9.4. Plant successesion: a, paludified cotton grass–sphagnum–pine forest; b, pine–dwarf shrub–sphagnum bog; c, ridge–hummock complex.

Dynamical model of forest–bog relations

Before presenting our own construction, we would like to mention two dynamical models dealing with peat accumulation in vertically distributed systems. Jones & Gore (1978) have subdivided the whole peat layer into 1 cm elementary sublayers, and have produced obvious balance equations for dead organic matter dynamics (m_i is ith-layer biomass):

$$\dot{m}_1 = Q - \lambda_1 m_1, \quad \dot{m}_i = -\lambda_i m_i \quad (i \geq 2). \tag{9.1}$$

Here, Q is dead organic matter influx, taken as constant, and λ_i are species-specific decomposition coefficients, depending on height due to the oxygen concentration gradient. Experimental evidence showed that

$$\lambda_i = a - ib, \tag{9.2}$$

so that below a certain height, $h \lesssim a/b$, conditions are completely anaerobic and decomposition rate becomes equal to zero. Nevertheless, the authors did not give any information about the origin of the peat formation layer and restricted themselves by detailed numerical application of the model to two 2500-year-old bogs in England. In addition, they did not mention gradual plant succession at the bog surface, which would lead to changes in Q and λ_i.

In the same volume, the model by Clymo (1978) involves the compression of peat under its own weight and considers decomposition in more detail.

The basis of our approach consists of (1) a proposed structure of the peat formation layer (Fig. 9.8(a)–(c)) that results in a two-phase process of bogging, and (2) taking into account the feedback between evolution of ecological parameters of the peat formation layer, state of vegetation above the bog surface, and the rate of peat accumulation.

We intend to concentrate mainly on ombrogenic bogs, and propose below a simplified and qualitative formalization of the considered system. The full variant can be found in Glebov and Korzukhin (1985, 1988, 1990).

Below, we shall denote as T the trophic mineral content of peat; W, oxygen concentration; λ, specific dead organic matter decomposition rate; Q, total litterfall; h, absolute height of dead organic matter layer above the mineral soil; and Δh, height of peat-forming layer. We also need the biomasses of hygrophilous, m_{hyg}, and woody, m_{tre}, species.

The central ecological mechanism which rules all evolution is connected with the specific form of $W(h)$ (Fig. 9.8(d)). This variable strongly affects the rate of dead organic matter decomposition, $\lambda = \lambda[W(h)]$ (Fig. 9.8(e)). In our first-approximation model, it is enough to operate by the variables, averaged over height, that is,

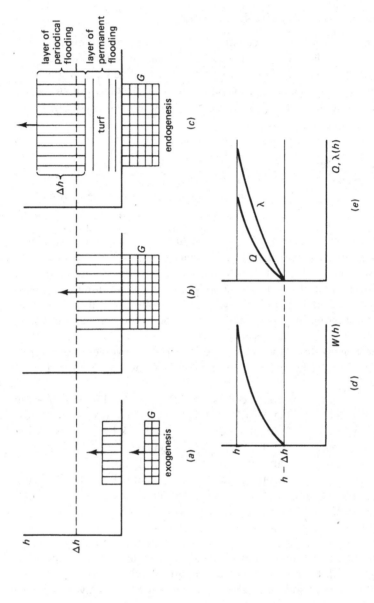

Fig. 9.8. Genesis of peat formation layer. (a), (b), (c) Sequential stages of its evolution in the course of bogging. G, thickness of gley horizon. (d), (e) Height profiles of W, Q and λ.

$$W = \frac{1}{h} \int_0^h W(z) \, dz \qquad \text{when } h < \Delta h; \qquad (9.3a)$$

$$W = \frac{1}{\Delta h} \int_{h-\Delta h}^h W(z) \, dz \qquad \text{when } h \geq \Delta h, \qquad (9.3b)$$

and the same for Q and λ.

It should be noted that if we intend to consider long-term forest–bog dynamics (with characteristic timescales $\tau_{fb} \approx 10^3$ years), we may average all fluctuations in the vegetation state emerging due to cover disturbances which induce demutational successions (with characteristic timescales $\tau_{dem} \approx 10^2$ years). Under this approximation, we may consider different vegetation type biomasses as some given functions of ecological conditions, in our case of W and T:

$$m_{hyg} = m_{hyg}(W, T), m_{tre} = m_{tre}(W, T). \qquad (9.4)$$

These functions can be identified from some successional models or from vegetation classifications.

For as long as the peat formation layer is underdeveloped, $h < \Delta h$, height change is given by the equation:

$$\dot{h} = Q(m_{hyg}, m_{tre}) - \lambda(W, T, m_{hyg}, m_{tre})h. \qquad (9.5)$$

where λ is the average decomposition rate for given m_{hyg} and m_{tre}, and Q and λ are measured in height units.

After origination of the complete peat formation layer, $h \geq \Delta h$, height dynamics is given by

$$\dot{h} = Q(m_{hyg}, m_{tre}) - \lambda(W, T, m_{hyg}, m_{tre}) \Delta h. \qquad (9.6)$$

The fundamental difference between these two equations is essential: the first has stable equilibrium

$$h^\circ(W, T) = \frac{Q[m_{hyg}(W, T), m_{tre}(W, T)]}{\lambda[W, T, m_{hyg}(W, T), m_{tre}(W, T)]} \qquad (9.7)$$

whereas the second has not.

Our model can correctly describe the adduced scheme of bogging if we suppose that h° rises when W falls (or that Q falls more slowly than λ). Under this proposition, h° can reach the critical value $h^\circ = \Delta h$, when the equilibrium vanishes and the system moves into the region of irreversible peat accumulation. This usually ceases in the late phases of the bogging process, when Q becomes small because of poor productivity.

Ecological data, and the considerations cited above concerning forest–bog system evolution, lead to the following picture of long-term system dynamics, illustrated by Fig. 9.9.

Let us denote as Γ the critical curve

$$f(W, T, \Delta h) = 0 \qquad (9.8)$$

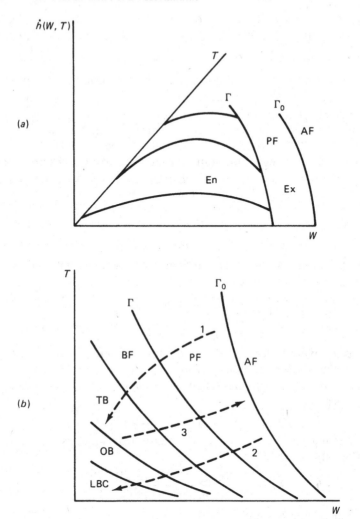

Fig. 9.9. (*a*) Qualitative dependence of peat accumulation rate, $\dot{h}$, on W, T. Line Γ is given by Eqn (9.8). Ex denotes zone of exogenesis in the course of initial bogging. En denotes zone of exogenesis: spontaneous growth of peat mass. Automorphous forest (AF) exists on line $Γ_0$, paludal forest (PF) between $Γ_0$ and Γ. (BF, bogged forest; TB, treed bog; OB, open bog; LBC, regressive lake-bog complex.) Maximum $\dot{h}$ is reached under intermediate values of W, T. (*b*) Disposition of forest–bog ecosystems on {W, T}, plane. Inclination of all border lines corresponds to the influence of W, T on the bogging process: increasing of W, T promotes the formation of forest. Numbers indicate: 1, 2, course of bogging beginning from good and poor trophic situations; 3, course of bog afforestation under W increasing as a result of drainage.

which originates from Eqn (9.6) under $\dot{h} = 0$. This function describes some line at the plane $\{W, T\}$ and divides exo- and endogenous regimes of system dynamics (Fig. 9.9(a),(b)). At the right hand of Γ, Eqn (9.5) works; at the left, Eqn (9.6).

Figure 9.9(a) plays an auxiliary role and shows the behavior of the peat accumulation rate, h, as a function of W, T.

Figure 9.9(b) represents an ordination diagram in $\{W, T\}$ coordinates, with addition of system evolution trajectories for different situations. Ordination ought to be fulfilled on the basis of functions (9.4), and the trajectories obtained from Eqn (9.6).

Our field experience allows us to relate all ecosystems with $h \le \Delta h$ to the class denoted above as paludal forest (the decisive character being plant root disposition in the mineral soil).

If needed, the described model can easily be connected with concrete numerical data, and expanded by means of addition of new variables (e.g. thickness of gleyed horizon, subdivision of h into a number of sublayers, etc.).

Concluding comment

We have described the major ecological mechanisms involved in the long-term process of bogging and peat formation, and the corresponding changes in the plant communities, for wetlands in the taiga zone of the West Siberian Plain. The discussion is developed at three spatiotemporal levels.

The first level is a regional one, in which transitions between adjacent and separate ecosystems of forest and mire appear. The second level is a mechanistic one that reveals intra-ecosystem relations of trees and hygrophytes. Trees stimulate the forest formation process, intensify biological turnover and retard peat accumulation. Hygrophytes stimulate the mire formation process, distributing ash and nitrogen vertically through the peatland profile, inhibiting the biological turnover and intensifying peat accumulation. Particular features of these processes under different trophic conditions are described.

The third level represents transitions between forests and mires in the Holocene. There were periods during BO and at the joins of AT/SB and of SB/SA when the climate was very dry (or there were positive tectonic movements of the earth's surface), as a result of which mire afforestation occurred. This was especially intensive at the join of AT and SB, when it was accompanied by the destruction of peat deposits.

The process of bogging has two different phases – initial bogging (exogenous stage) and late bogging (endogenous stage) – which differ principally in their ecological mechanisms and reaction to climatic changes. A concise qualitative formalization of the forest–bog system in the form of a dynamical model is presented. We support the idea of the existence of a balance between forest and bog in the boreal zone, which can shift depending on climatic conditions.

10 Remote sensing technology for forest ecosystem analysis

K. Jon Ranson and Darrel L. Williams

Introduction

Investigators worldwide are fundamentally concerned about vegetation changes within forest ecosystems across spatial and temporal scales. The nature and extent of the impacts of these changes, as well as the feedbacks to global climate, are being addressed through modeling the interactions of the vegetation, soil, and energy components of the forest ecosystem. Mathematical models of the dynamics of forest succession and soil processes, combined with observations of forest ecosystems, can provide much of the insight required to comprehend these processes and construct initial cause-and-effect relations. In practice, model testing is complicated by the difficulties of making necessary observations at the required scales. Observations of ecological factors, such as the successional stage of landscape units, at regional scales and over periods of decades or longer are particularly difficult. To effectively utilize the modeling strategies described elsewhere in this book, large amounts of information are required concerning the scope and state of boreal forest ecosystems. Remote sensing technology provides the only feasible means of acquiring this information, repeatedly, with a synoptic view of the landscape. In addition, the incorporation of remotely sensed data into models of forest ecosystem dynamics can be used to characterize northern–boreal forest ecosystems, especially with regard to the interpretation of landscape patterns and processes at local and regional scales.

The launch, by the National Aeronautics and Space Administration (NASA), of the Landsat multi-spectral scanner (MSS) in 1972 ushered in a new era for terrestrial ecologists by providing a synoptic view of the global landscape over an extended period of time. With remote sensing technology (measurements and models) and a 19-year base of satellite observations, the potential exists to forge direct links between ecological hypotheses formulated at various scales. It is likely that remote sensing can play a strong role as forest ecosystem models progress from local to regional and, ultimately, to global perspectives (Eagleson 1986; Moore *et al.* 1986; Baker & Wilson 1987).

In this chapter we have endeavored to provide background information about some of the remote sensing technology applicable to forest ecosystem analysis so that readers may gain an understanding of how to utilize this technology in their work. We discuss sensor systems that operate in the optical and microwave energy regimes, as well as the types of information that can be derived from various sensor systems. We also examine the role of radiative transfer modeling for understanding, interpreting, and interfacing remote sensing measurements with ecosystem models. In addition, the uses of satellite data for examining the temporal and spatial patterns of forest ecosystems are discussed. Finally, we describe NASA's future Earth Observing System (EOS) and the implications for studying global ecosystems.

Background

The interactions of electromagnetic radiation with forest canopies produce an integrated response related to biomass, structure, physiological processes, and vigor, as well as to the characteristics of the underlying soil. For example, the magnitude of solar radiation scattered from plant canopies has been related to phytomass and photosynthetic activity (Myers 1983; Goward 1985; Goward, Tucker & Dye 1985). This knowledge has been applied to data collected by earth-viewing satellite sensor systems to obtain consistent, timely, and reliable information with regard to the distribution, vigor, and seasonality of vegetation biomes at regional, continental, and global scales (Justice *et al.* 1985; Tucker, Townshend & Goff 1985; Goward *et al.* 1987; Goward 1989). These relatively recent successes in monitoring and assessing the dynamics of vegetation communities have been the direct result of fundamental research and intensive field measurement programs involving the collection of *in situ* spectral reflectance data of earth surface materials. One recent program was the First International Satellite Land Surface Climatology Program Field Experiment, or FIFE. During the growing seasons of 1987 and 1989, vegetation, soil, meteorological, and remote sensing measurements were made simultaneously over a grassland biome in Kansas, USA (Hall *et al.* 1989). A second such experiment (BOREAS), which will focus on the boreal forest ecosystem, is now being planned for 1993–4.

Typically, remote sensing devices designed for terrestrial applications utilize energy in the reflective (300 nm to 3000 nm), emissive or thermal

(3000 nm to 15 000 nm), and microwave or radar (1 mm to 0.8 m) wavelengths of the electromagnetic (EM) spectrum. These sensor systems are designed to detect the energy returned from a surface, generally within narrow portions or bands of the spectrum. The energy from the surface may be sensed passively, using the sun as the illumination source, as is the case for the Landsat MSS. Alternatively, active sensor systems utilize radar or laser technology to directly illuminate the surface and record the returned energy. In addition, sensor systems may have different spatial resolutions depending on such factors as sensor altitude, optics, and data rate and volume considerations. In the following sections we focus on the concepts and utility of passive optical sensors and active radar sensor systems and their roles in analysis of forest ecosystems.

Optical remote sensing

Solar radiation incident upon a forest canopy may be reflected, absorbed, or transmitted by leaf tissues, cells, and pigments. The interrelation among these three components of the radiation balance can be expressed as follows:

$$I_\lambda = R_\lambda + A_\lambda + T_\lambda, \tag{10.1}$$

where I_λ denotes incident energy at a particular wavelength, λ; R_λ, A_λ, and T_λ denote energy reflected, absorbed, and transmitted at that wavelength. When working with data acquired in the optical portion of the EM spectrum (i.e. 300 nm to 3000 nm), it is often more useful to think of this interrelation in terms of reflectance, where

$$\rho_\lambda = 1 - (\alpha_\lambda + \tau_\lambda), \tag{10.2}$$

(i.e. the proportion of reflected energy at a particular wavelength (ρ_λ) is equal to the incident energy reduced by the proportion of energy that is either absorbed (α) or transmitted (τ)).

Figure 10.1 is a representative plot of the complete optical properties of a green leaf, illustrating the partitioning of energy into reflectance, transmittance, and absorptance as a function of wavelength. One can see that, in general, changes in leaf transmittance parallel changes in leaf reflectance, but transmittance is usually slightly lower in magnitude (Myers 1983; Hodanova 1985). The dominant factors that control leaf reflectance as a function of wavelength region are shown in Fig. 10.2.

Visible region

In the visible portion of the spectrum (400–700 nm), leaf pigments such as chlorophyll a and b, carotenoids (carotenes and xanthophylls), and anthocyanins are the dominant factors controlling the magnitude of reflectance. Most of the incident energy is absorbed by these pigments in

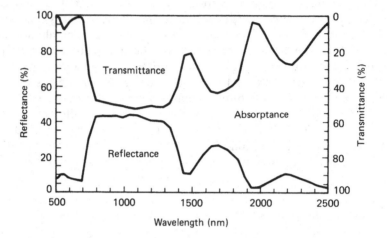

Fig. 10.1. Representative plot of the complete optical properties of a green leaf, illustrating the partitioning of energy into reflectance, transmittance, and absorptance.

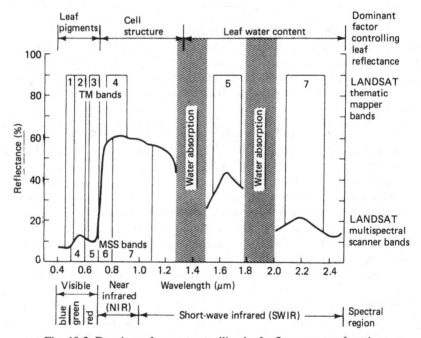

Fig. 10.2. Dominant factors controlling leaf reflectance as a function of wavelength region. Courtesy of NASA/Jet Propulsion Laboratory.

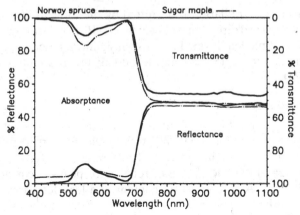

Fig. 10.3. Comparison of needle vs. broadleaf optical property data for Norway spruce and sugar maple. Note the greater proportion of absorption of Norway spruce needles across all wavelengths.

the leaves; the remaining energy is reflected and transmitted in nearly equal amounts. Because of the large amount of absorption, canopy reflectance in the visible portion of the spectrum is due, predominantly, to radiation scattered in the uppermost portion of the canopy.

Near infrared region

The internal structure of leaves, such as the arrangement of the cells of the leaf mesophyll, controls the level of reflectance in the near infrared (IR) region (720–1300 nm). In general, about half of the incident energy is reflected, nearly half is transmitted, and typically less than 10% is absorbed in the near IR region, but the exact partitioning of energy into ρ, τ, and α will vary as a function of species. Figure 10.3 permits comparison of mean reflectance, transmittance, and absorptance characteristics for a deciduous, broadleaf species (sugar maple, *Acer saccharum*) versus a needleleaf, evergreen species (Norway spruce, *Picea abies*) (Williams 1991). Close inspection of Fig. 10.3 will reveal that the percentage absorptance is greater at all wavelengths for the spruce needles, whereas transmittance is greater at all wavelengths for leaves of sugar maple. In the near IR region, both reflectance and absorptance are higher for Norway spruce than for sugar maple; conversely, transmittance is greater throughout the near IR for a broadleaf species than for spruce needles.

In summary, incoming solar irradiance in the near IR wavelength region can penetrate (transmit) down through and reflect back up through multiple layers of leaves within the vegetative canopy. This results in increased values of near IR reflectance relative to the visible wavelengths because radiation transmitted by the top leaf layer can be scattered in an upward direction by lower layers. Both measurements and modeling of multiple leaf layers have shown significant increases in near IR reflectance as more leaf layers are added, up to about six layers (Colwell 1974).

Middle infrared region

The moisture content of leaves controls the magnitude of reflectance in the middle IR region (1300–2600 nm), with much of the incident energy being absorbed by water in the leaf. The strongest water absorption bands occur near 1400 and 1900 nm; reflectance peaks occur at about 1600 and 2200 nm. As the moisture content of a leaf decreases, reflectance in the middle IR region increases, and the difference in magnitude between reflectance peaks (i.e. 1600 and 2200 nm) and valleys (i.e. 1400 and 1900 nm) decreases (Goward 1985).

Given the effects of leaf pigments, cell structure, and water content on the optical properties of a leaf, and the interrelation of these variables with leaf age and plant health, one can assess plant canopy vigor, leaf area index, and total green biomass by examining leaf and canopy reflectance characteristics. The middle IR portion of the spectrum is useful for assessing relative photosynthetic activity because of the strong relations between leaf moisture content, stomatal conductance, respiration and, therefore, the rate and efficiency of photosynthesis.

Use of reflectance for assessing canopy attributes

Reflected light from the surface provides the basic source of information for remote sensing systems sensitive to visible and near IR wavelengths. In addition to foliar optical properties, the amount of light, or radiance, perceived by a sensor is a function of numerous canopy variables. Canopy attributes such as leaf area index (LAI), biomass, canopy cover, three-dimensional structure, foliage angle distribution, and soil or background reflectance determine the amount and spectral distribution of sunlight reflected from a canopy. Consider, for example, the significant differences in the magnitude of spectral reflectance which were observed at the needle (leaf), branch, and canopy level for Norway spruce (Fig. 10.4) (Williams 1991). In comparing the spectral plots shown in Fig. 10.4, the most obvious feature of change is the dramatic drop in the magnitude of reflectance along the near IR plateau in going from the needle to the branch to the canopy level. A similar trend is evident across the visible wavelength region, particularly around the 'green peak' located at c. 550 nm. Williams (1989) summarized four significant factors which were believed to be contributing to the reduction in reflectance observed at the branch and canopy level relative to the needle level: (1) shadowing, especially in the visible wavelength region; (2) absorption of radiation by twigs, branches, bark, and litter; (3) changes in the indices of refraction of the needles within the canopy as a function of needle age (Westman & Price 1988), resulting in enhanced scattering of the light; and (4) the clustering or packing of needles on the branches which may enhance the trapping of radiation (Gates & Benedict 1963; W. K. Smith & Carter 1988). It should also be noted that other scene variables such as sun angle, topography, meteorological conditions, and atmospheric characteristics may also affect the measured reflectance.

Satellite-borne optical sensors do not measure reflectance directly, but measure radiance instead. The basic measurement equation for the spectral radiance from a scene can be written as (J. A. Smith 1984):

$$L_\lambda(\mu_o, \phi_o; \mu_r, \phi_r) = \frac{1}{\pi} \iint fr_\lambda(\mu_o, \phi_o; \mu_r, \phi_r; p)E_{total,\lambda}(\mu, \phi)\, d\mu\, d\phi, \quad (10.3)$$

where

L = spectral radiance at wavelength band λ;

μ_o, μ_r = cosines of the angles of incidence and reflection, respectively;

ϕ_o, ϕ_r = azimuth angles of incidence and reflection, respectively;

fr = bidirectional reflectance distribution function;

p = a set of biophysical parameters for the target;

E_{total} = total of direct and diffuse solar irradiance at the surface.

The bidirectional reflectance distribution function (BRDF) is an intrinsic property of the surface and thus not affected by variations in the magnitude of solar irradiance. It is the BRDF that contains information about the various canopy parameters in a scene. Earth remote sensing

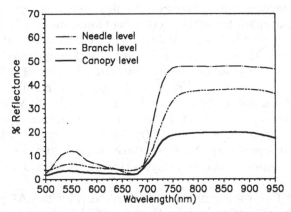

Fig. 10.4. Plots of mean spectral reflectance data obtained at the needle (leaf), branch, and canopy level for Norway spruce.

satellites acquire radiance of a scene at a single incidence or viewing angle, so there is insufficient information to characterize the BRDF. Fortunately, good results have been obtained by relating radiance measurements to vegetation canopy characteristics.

The usefulness of global satellite radiance observations for assessing changes in the amount and distribution of green standing vegetation has been demonstrated (Justice 1986) and a relation between these vegetation changes and variations in atmospheric CO_2 has been shown (Tucker et al. 1986). A few studies have attempted to relate processes such as canopy photosynthesis and resistance (e.g. Tucker & Sellers 1986) and net primary productivity (e.g. Goward et al. 1987) to these satellite data, thus providing a potential tool for understanding carbon flows in terrestrial ecosystems.

Even though these results are encouraging for large scale vegetation monitoring, there are still problems to solve before these techniques can be used with confidence in the boreal zone. For example, consider the use of the Normalized Difference Vegetation Index (NDVI), which exploits the absorption of red light (due to chlorophyll) and the high reflectance of near IR light by green leaves (Tucker 1979). Given the radiance (L) of a sensor in these two bands, NDVI is simply:

$$NDVI = \frac{L_{\text{Near-IR}} - L_{\text{Red}}}{L_{\text{Near-IR}} + L_{\text{Red}}}. \qquad (10.4)$$

For agricultural crops and grasslands, NDVI increases with increasing leaf area or biomass (Prince & Tucker 1986; Aswar, Myneni & Kanemasu 1989). However, for northern conifer forests the trend may be reversed, suggesting that there are fundamental differences between the optical properties of conifers and other vegetation types. A number of researchers have independently measured branch and/or canopy level spectral reflectance characteristics for a variety of forest species. They all found spruce to have the lowest near IR spectral reflectance of any species that was measured. Norman & Jarvis (1974) found nearly total absorptance of incoming solar irradiance near the bottom of a Sitka spruce plantation. In addition, an inspection of canopy level reflectance data provided by Kleman (1986), F. G. Hall et al. (1990) and Williams (1991) indicates that spruce seems to violate the widely accepted concept that near IR reflectance should increase as a function of increasing LAI. Data collected independently by these three investigators for three different species of spruce indicate a decrease in near IR reflectance with increasing LAI.

An important factor affecting the relation of NDVI with LAI is the reflectance of the underlying surface. Huete (1989) has shown that the effects of bare soil reflectance introduce errors in the neighborhood of

20%. Ranson & Daughtry (1986), however, demonstrated that a snow background may actually improve phytomass or LAI estimates with NDVI.

These results indicate that the boreal zone may present difficulties when trying to infer levels of biomass, LAI, or productivity by using general vegetation indices such as Eqn (10.4). However, the use of remote-sensing-derived information in concert with models of forest processes may produce useful information, especially over large areas. For example, the FOREST-BGC (BioGeochemical Cycles) model of Running & Coughlan (1988) can implement satellite-derived estimates of LAI to drive calculations of carbon, water, and nitrogen cycles through a forest ecosystem on a regional basis. Their model has also been used to interpret the seasonal trends of satellite-derived weekly global vegetation indices. High correlations were found with annual photosynthesis, transpiration, and net primary production (Running & Nemani 1988). In addition, Sellers (1986, 1987) has demonstrated through radiative transfer modeling how a simple ratio of red and near IR band reflectance (which is closely related to NDVI) is a nearly linear indicator of photosynthetically active radiation (PAR), minimum canopy resistance, and photosynthetic capacity. These are static or dynamic variables included in most forest ecosystem models (Shugart 1984). The ability to derive these parameters for a spatially explicit model (e.g. ZELIG (T. M. Smith & Urban 1988)) will enable the comparison of model and remote sensing image statistics over large areas.

New developments in optical remote sensing

Remote sensing technology is constantly improving and new instrument systems are beginning to show promise for forest ecosystem research. One such instrument is the Advanced Solid-state Array Sensor (ASAS) developed by NASA. This sensor is an airborne, off-nadir-pointing, imaging spectrometer that measures radiance from terrestrial targets (Irons *et al.* 1991). ASAS can acquire images of a target at seven viewing angles (e.g. 15°, 30°, 45° forward-looking; 0° (nadir); and 15°, 30°, 45° aft-looking) across twenty-nine 15 nm wide wavelength bands. Figure 10.5 illustrates a sequence of images acquired over a northern–boreal transitional forest area in Voyageurs National Park in northern Minnesota, USA. The changes in image tone and brightness are caused by differences in the solar illumination and viewing geometries. Because these types of changes in scene reflectance are related to canopy structure and species composition, these data can provide additional information about the forest canopy in comparison to nadir looking sensors. For example, Fig. 10.6 compares ASAS-acquired radiance data for two forest types: black spruce (Fig. 10.6(*a*)) and aspen (Fig. 10.6(*b*)). These data were acquired

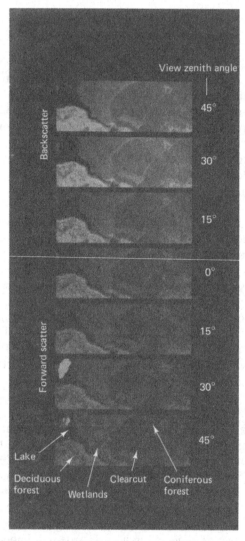

Fig. 10.5. Sequence of ASAS images acquired over a northern–boreal transitional forest area in Voyageurs National Park, Minnesota, USA, on 26 June 1988. Solar zenith angle was 34°. (Courtesy of Biospheric Sciences Branch, NASA/Goddard Space Flight Center.)

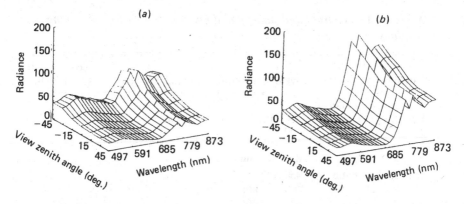

Fig. 10.6. Comparison of ASAS-acquired radiance data for two forest types: (*a*) black spruce; (*b*) aspen.

with the aircraft flying parallel to the principal plane of the sun. The plots show the differences in radiance from 497 nm to 880 nm at the seven view angles described above. The differences in the magnitude of radiance between the aspen and spruce canopies are related to the optical properties of the leaves, whereas differences in the slopes of the response surfaces are related to canopy structure (e.g. needles vs. leaves, branching angles, heights, and density). Research is currently being conducted to utilize these types of data to estimate canopy structure (density and height distributions), albedo, and intercepted PAR.

Another new sensor system being evaluated by NASA is the Airborne Visible, Near-infrared Imaging Spectrometer (AVIRIS) (Vane 1988). This instrument has two hundred and twenty-four 9.8 nm wide spectral bands covering the 400 nm to 2400 nm portion of the spectrum. Preliminary studies have shown that these fine spectral resolution data may yield information about canopy lignin and nitrogen status (see, for example, D. L. Peterson *et al*. 1988). These and other innovative sensor systems are being developed and tested for possible flight on the Earth Observing System (EOS) (NASA 1984) scheduled for launch in the late 1990s. Key aspects of the EOS mission are discussed near the end of this chapter.

Microwave remote sensing

The current array of earth remote sensing satellites (see Table 10.1) are designed primarily to utilize optical (reflective and thermal) technology. High frequency of cloud cover and low sun angles limit data quality and availability over boreal forests. Because radar systems are active and utilize longer wavelengths, it is possible to 'see' through cloud cover and acquire data independent of sun angle. Active radar satellite systems

Table 10.1. *Spectral bands and spatial resolution for earth-observing satellites*

Satellite	Wavelength bands	Spatial resolution	Years of data (as of 1990)
Passive sensor systems			
Landsat MSS	500–600 nm	80 m	18
	600–700 nm		
	700–800 nm		
	800–1100 nm		
Landsat Thematic	450–520 nm	30 m	8
Mapper	520–600 nm		
	630–690 nm		
	760–900 nm		
	1550–1750 nm		
	2080–2350 nm		
	10400–12500 nm	120 m	
SPOT-1	500–590 nm	20 m	4
	610–680 nm		
	790–890 nm		
	510–730 nm	10 m	
AVHRR	550–680 nm		
	730–1100 nm		
	3500–3900 nm		
	10500–11500 nm		
	11500–12500 nm	1100 m	11
Active sensor systems			
Seasat	23 cm	20 m	<1
SIR-A	23 cm	20 m	<1
SIR-B	23 cm	20 m	<1
SIR-C[a]	3 cm	15–90 m	
	6 cm		
	23 cm		
ERS-1[b]	6 cm	30–100 m	
JERS-1[c]	23 cm	18 m	
RADARSAT[d]	6 cm	10–100 m	

Note: [a] Scheduled for future launch.
[b] European Space Agency Remote Sensing Satellite.
[c] Japanese Earth Resources Satellite.
[d] Canadian radar satellite.

have only recently been deployed for earth observations and then for only limited missions of short duration. The first of these was the Synthetic Aperture Radar (SAR) on board the Seasat satellite launched in 1978. This sensor was designed primarily to monitor ocean waves, but land areas of North America, western Europe, and the northern polar regions were covered during the three-month mission. Two other short-duration radar missions, Shuttle Imaging Radar (SIR) -A and -B, were flown aboard the space shuttle in 1981 and 1984, respectively, providing new information about the use of radar for studying the vegetation resources of the earth. During the next decade a series of satellite-borne radar missions are planned by the European Space Agency, the United States, Canada, and Japan. Table 10.1 provides additional information about these satellites.

Radar measurements

The power received at a radar is a function of scene parameters and the characteristics of the radar system. The radar equation can be written as (Ulaby, Moore & Fung 1982):

$$\bar{P}_r = \frac{\lambda^2}{(4\pi)^3} \int_a \frac{P_t G^2 \sigma^0}{R^4}\, dS, \tag{10.5}$$

where

P_r, P_t are received and transmitted power (watts), respectively;
λ = wavelength in meters;
G = power gain;
σ^0 = average value of scattering cross-section;
R = distance in meters from radar to target;
A = area illuminated by radar;
S = surface, where the surface integral is taken over a.

The scattering cross section, σ^0, provides the link between the physical properties of the forest scene and the radar observations analogous to the BRDF for optical data. All of the parameters on the right side of Eqn (10.5), except σ^0, are controlled by the design of the radar system, thus making it relatively easy to determine σ^0. The characteristics of the scene that control σ^0 are primarily surface geometry and moisture content. Surface geometry can be described as roughness, slope, and vertical or horizontal inhomogeneity, which combine to describe the canopy structure. The parameter σ^0 is sensitive to the moisture content of vegetation and soil because the electrical properties of these materials, as expressed by their dielectric constant, ε, changes with moisture content. The ε value of water is around 77, whereas ε is less than three for freshly fallen snow, dry soil, or wood. Thus σ^0 will increase as the moisture content in vegetation or soil increases.

Radar wavelength and polarization

Remote sensing devices are sensitive to vegetation structure at the scale of the wavelength. Thus, optical devices sense canopy structure on the

Table 10.2. *Names, frequencies, and wavelengths for earth resources remote sensing radar systems*

Name	Frequency (Ghz)	Wavelength (m)
P-band	0.44	0.68
L-band	1.30	0.23
C-band	5.0	0.06
X-band	10.00	0.03

order of micrometers. Radar devices, however, with wavelengths in the centimeter to decimeter range, sense the structure of leaves, stems, and even boles. This can provide additional information about the forest canopy volume as well as the underlying surface. Common wavelengths used for current earth science radar instruments include 3 cm (X-band), 6 cm (C-band) and 23 cm (L-band). An additional wavelength of 68 cm (P-band) is currently under study as part of the NASA/JPL (Jet Propulsion Laboratory) airborne SAR system. Table 10.2 lists the band-names, wavelengths, and frequencies for these types of radars. The shorter wavelength of X-band radar results in direct backscatter from foliage and is sensitive to leaf size, foliage density, and crown shape. Multiple interactions between vegetation and soil can occur with C-band radar. The L-band provides direct backscatter from the soil surface, as well as interactions with boles and large branches. P-band radar should interact primarily with the ground and boles, with a good deal of penetration into the surface under dry conditions. Because these longer radar wavelengths penetrate deeper into the canopy they can provide information about the forest canopy volume and underlying surface. The differential scattering of microwaves in different frequencies by a forest is illustrated in Fig. 10.7.

Radar systems can also be designed to transmit and receive energy at different polarizations, making them sensitive to the orientation of canopy scatterers. Transmit or receive polarizations may be oriented to either vertical (V) or horizontal (H). Thus up to four different polarization combinations can be utilized: VV (transmit Vertical, receive Vertical), HH, VH, or HV. When taken together, these four polarization combinations can be used to produce polarization signatures, which can provide additional information about the structure and composition of a forest (Durden, Van Zyl & Zebker 1989).

These differences in the scattering behavior of radar from forests can be seen as varying levels of brightness on an image (a brighter return indicates higher backscatter). For example, Fig. 10.8 presents an image

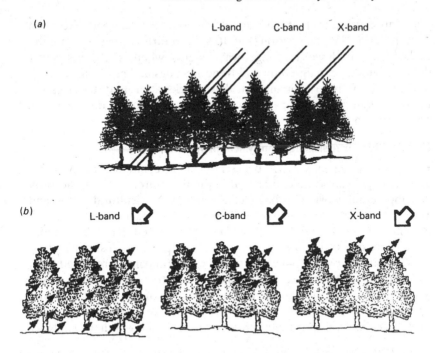

Fig. 10.7. (*a*) The different wavelengths of radar sensors interact with different parts of a forest canopy. (*b*) The penetration depth of different wavelength radars results in backscattering from different levels within the canopy. X-band = 3 cm; C-band = 6 cm; L-band = 23 cm.

combining data from three frequencies (C-, L-, and P-band) simultaneously acquired by an airborne SAR. The imaged area covers a portion of International Paper's Northern Experimental Forest near Bangor, Maine, USA (Ranson & Smith 1990). The forests are primarily spruce, hemlock, and mixed hardwoods (aspen–birch). In addition, there are several clearcuts and bogs in the area. Within the image, roads, grassy areas, bogs and forest clearings appear dark. Mixed hardwood stands appear dark gray, whereas the extensive hemlock and spruce stands are lighter and more mottled in appearance. In addition, the absence of backscattering from water bodies causes them to appear black on all three images.

Mean returned power in HH and VV polarizations are shown for several scene categories in Fig. 10.9 (*a*) and (*b*) respectively. The higher radar return from the forest stands for HH polarization (Fig. 10.9(*a*)) at longer wavelengths (i.e. P- and L-band) suggests that the radar beam is penetrating the canopy, interacting with the boles and the ground, and then backscattering. The lower power returned at HH from the bogs and grassy areas indicates that the radar beam was either absorbed or forward scattered. For VV polarization (Fig. 10.9(b)) returned power data for the forest stands are lower for longer wavelengths, whereas returned power data are similar in magnitude and trend for bogs and grassy areas. The task remaining is to capitalize on these backscattering differences between ecosystem classes as a source of information for inferring ecosystem parameters.

Use of radar for assessing biophysical properties

Relations among microwave backscatter and vegetation canopy parameters have been well documented in the literature. Much of the early developmental work has been conducted on agricultural crops and grasses (e.g. Brakke *et al.* 1981; Ulaby *et al.* 1984). In a study of small balsam fir trees, Daughtry & Ranson (1986) showed that C-band radar backscatter cross section was sensitive to LAI up to 2.5. Westman & Paris (1987) studied a pygmy pine and cypress forest (maximum reported LAI, 2.5) and found that the sum of C-band backscatter cross sections in VV and VH polarizations may be a good indicator of forest leaf area and phytomass. These results suggest that further studies should be conducted in a natural forest environment to determine whether the higher

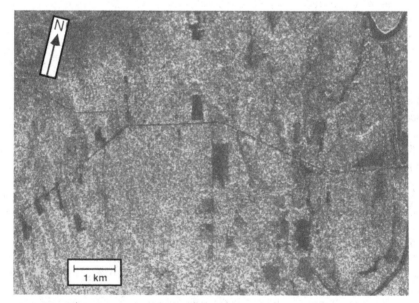

Fig. 10.8. Synthetic Aperture Radar images of a forest area in Maine, USA. Dark areas represent clearcuts, bogs or grassy meadows. Lighter areas depict forest cover.

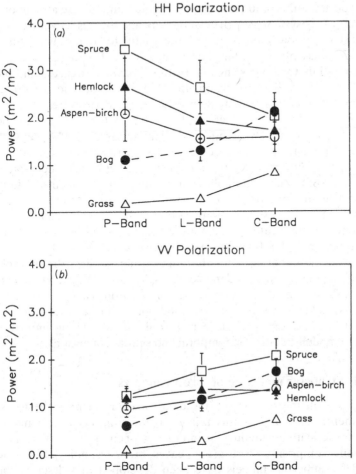

Fig. 10.9. Variation of SAR returned power for five ecosystem classes: (*a*) HH polarization; (*b*) VV polarization.

LAI and phytomass typical of northern forests can be successfully inferred from radar multiple frequency data. Forest biomass and various canopy structure parameters were found to be highly correlated with L-band SAR data, especially with HV polarization (Sader 1987). Riom & LeToan (1981) found this frequency and polarization to be highly correlated with pine tree heights. In addition, there is evidence that soil moisture and surface characteristics can be inferred from backscatter cross-section data (Ulaby, Batlivala & Dobson 1978; Dobson & Ulaby 1986).

Currently, radar systems with multiple polarization capabilities are being studied as a means to improve the detection of forest characteristics such as species composition and canopy architecture. These are important drivers of ecosystem pattern development and there is some evidence to suggest that radar data may be quite useful for determining these factors. For example, using Shuttle Imaging Radar-B data (L-band and HH polarization, three view angles), Hoffer, Lozano-Garcia & Gillespie (1986) have shown good discrimination between pine and deciduous swampland. Similar results may be possible for northern forest conifers and spruce bogs. In addition, Zebker, Van Zyl & Held (1987) found good contrast between conifer stands and clearcut areas in Maine with multiple-polarized, aircraft SAR data. The sensitivity of L-band scatterometer data to discriminate among conifer species has been demonstrated by Sieber (1985) using multipolarization angle measurements of individual spruce and pine trees. Further study is required to extend these findings to natural forest conditions, although the 10–20 m spatial resolution of current and future space-borne radars should be adequate to resolve typical gaps left by fallen mature trees.

Based on the experiences of these and other studies, the multiple frequencies, polarizations, and incidence angles utilized by radar systems may provide information about forest composition, canopy structure, phytomass, LAI, and soil moisture conditions. Further evaluation of radar data should also provide a new tool for developing improved ecosystem models in both the temporal and spatial dimensions.

Radiative transfer modeling of forest canopies

The key to utilizing remote sensing data for characterizing the forest environment, especially as this relates to inferring ecosystem model parameters, is in understanding the interaction of energy with the various components comprising a forest (e.g. leaves, branches, boles). Since the early 1970s numerous models have been developed that describe the interaction of radiation (both optical and microwave) with vegetation canopies.

The optical reflectance models of J. A. Smith & Oliver (1972) and Suits (1972) are well-known examples that consider a vegetation canopy as uniform horizontal layers of infinite extent. More recent developments have produced radiative transfer models that consider canopies with row structure (Norman & Welles 1983; Suits 1983) and true three-dimensional structure (Kimes & Kirchner 1982). Li & Strahler (1985, 1986) developed a geometric optics reflectance model for coniferous forests using cone-shaped tree crowns. Recent developments in ray tracing techniques show promise for characterizing the scattering of radiation from a leaf based on cellular structure and composition (Bone, Lee & Norman 1985; Brakke & Smith 1987).

A number of models have been developed for the microwave region. Attema & Ulaby (1978) calculated backscatter from a vegetation canopy modeled as a cloud of water droplets. Since that time models have become more detailed and include the shape and size distributions of leaves, branches, and stems (Ulaby, Moore & Fung 1982; Eom & Fung 1986).

Canopy reflectance models have been developed to account for variations in many of these parameters, and such models provide a convenient method for understanding both the energy interactions within plant canopies and the expected signal impinging on a remote sensing device. The radiative transfer mechanisms within the soil–forest complex can be modeled by using a hierarchical set of models that treat individual canopy components, larger tree units, collections of trees (i.e. stands), and finally overall terrain models. The radiative flux calculations can be used in three ways. First, they determine the energy interactions (i.e. reflectance, transmittance, absorption) within the soil or canopy layers. Spectral dependencies of individual leaf or needle components may also be potentially relatable to physiologically induced changes in leaf morphology or biochemistry. Secondly, the calculations at various spatial scales can guide the development of biophysical inference algorithms for relating remotely sensed radiance to canopy attributes. Thirdly, radiative transfer modeling may be used to examine fundamental scattering properties of forest canopies.

For example, remote sensing studies have revealed that spruce canopies, in contrast to deciduous species, absorb greater amounts of near IR radiation as canopy density increases (D. L. Williams 1989). Leaf scattering and canopy reflectance models can be used to systematically evaluate at what level (i.e. leaf, shoot, or tree) this increased IR absorption occurs. Thus, new information with regard to unique morphological and/or physiological adaptations made by some forest species to enhance their survival under certain environmental conditions can be gained. This may be extremely important for forest ecologists to draw upon in developing truly representative forest succession models.

Generally, canopy reflectance models have been used to study the bidirectional reflectance, thermal properties, or microwave backscatter of earth surface features (Fung & Ulaby 1984; J. A. Smith 1984). Successful inversion of models to estimate canopy parameters has been demonstrated. For example, Goel and co-workers showed that biomass, leaf area index, leaf angle distribution, and canopy cover for agricultural crops could be inferred from bidirectional reflectance (e.g. Goel, Strebel & Thompson 1984). Li & Strahler (1985) were also successful in estimating coniferous forest crown density and tree height. This suggests a coupling of remote sensing measurements with forest succession models that incorporate leaf area, biomass, and tree height (e.g. FORET model of Shugart & West (1980)). To capitalize on this potential, a physically based approach for understanding the sensitivity of ecosystems to changes in climate and anthropogenic factors is required.

Temporal and spatial analyses

It is the structure of the individual-tree and gap models which permits the biology and physiology to be linked across a large range of space and time scales, from the individual-tree to the landscape level and from yearly to decadal time scales. In principle, hypotheses at one scale (e.g. individual-tree response to nutrient level) can be incorporated into these models so that the implications at the ecosystem level can be examined. Conversely, observations of ecosystem dynamics at regional levels can be used to test the hypotheses which constitute the forest growth models. That is, do the forest dynamics model's predictions correspond to observation? Satellite data provide a means for examining change across the forested landscape. Change over time can be assessed by comparing images acquired at different dates. New methods for examining spatial pattern variability in forests are emerging (see, for example, O'Neill *et al.* 1990). The following section provides examples of both types of analysis.

Changes in forest composition over time

In practice, hypothesis testing among scales is complicated by the inability to make observations at the required scales, i.e. particularly at the regional level, where observations of ecosystem parameters such as productivity and species composition are impracticable. However, the advent of satellite platforms, advances in the automatic processing of digital image data, developments in digital computation, and improved understanding of light interaction with vegetation canopies permits reliable and affordable remote sensing observations of vegetation state and dynamics at local, regional, and even global scales.

For example, a forested landscape observed from a satellite remote sensing platform can be characterized in terms of successional states. Each 'patch' in the landscape, or associated group of picture elements (pixels) in the image data, is cycling between these states. For example, in a boreal forest area disturbed by fire, aspen, jack pine, and spruce are present. Initially, aspen dominates the stand and develops a closed canopy. As the aspen stand matures and individual aspen trees die, the conifers begin to be visible from above. The stand assumes the appearance of a mixed conifer–deciduous stand. Finally, the conifers dominate and eventually a disturbance event such as fire or clearing is required to return to the 'ground state' of the life cycle. Because the species in each of these ecosystem states have different leaf optical properties, morphologies, and shadowing geometries, etc., each reflects light in a distinct way. Thus, each state forms a 'spectral ecostate' and can be identified with remote sensing data. Each pixel group representing a patch in the landscape can then be classified into one of the ecostates by using spatially registered, multi-year satellite data and standard image processing algorithms.

Hall *et al.* (1987) working in the Superior National Forest in Minnesota, USA, used multi-date satellite data to examine changes occurring in the forest over a ten-year period. From the calculated proportions of the areas changing to different successional stages, they estimated transition probabilities between ecostates. Given the observed transition probability matrix and the observed initial state, the forest landscape dynamics were modeled by a Markov chain and future states of the landscape projected. The data (transition matrix) can also be used to validate and improve patch growth models. Furthermore, Hall, Strebel & Sellers (1988) suggest that this approach is also useful for mapping and monitoring net primary productivity and the time rate of change of ecosystems.

Spatial analysis with fractal dimensions

As environmental and anthropogenic factors alter the distribution of species and phytomass across the landscape the structure of forest patches in terms of number, size, and shape also changes. A quantifiable indicator of these characteristics, suggested by Mandelbrot (1977, 1983), relates the areas of patches to perimeters or size distributions. This quantity was named the fractal dimension by Mandelbrot and has since been used to examine the spatial characteristics of a number of natural phenomena.

Generally, for perimeter–area analysis a relation of the form:

$$P \propto c\sqrt{A}^{D_\mathrm{p}} \qquad (10.6)$$

can be used to determine the fractal dimension, D_p, over a landscape given a set of patch area (A) and perimeter (P) data from digitized maps or satellite images. The parameter c is related to shape (e.g. $c = 4$ for squares and $2\sqrt{\pi}$ for circles). While many other indicators of spatial patterns are available (O'Neill *et al.* 1990), the fractal dimension is a convenient one with simple interpretation and is easily calculated from remote sensing imagery (Lovejoy 1982). As a general rule, landscapes with regular shaped patches (e.g. circles and rectangles) will yield values of D_p close to 1.0, whereas landscapes with irregularly shaped patches will have D_p values approaching 2.0. This provides a convenient interpretation of spatial patterns since forest patches within agricultural or logged areas yield lower fractal dimensions than forest patches found in wilderness areas (Krummel *et al.* 1987; O'Neill *et al.* 1990).

Variation in fractal dimension with scale is an indicator of the occurrence of various scale-dependent processes operating in the spatial domains giving rise to the observed fractal dimension changes. Analysis of patch size distribution employs the power-law assumption of Korcak's law:

$$Nr(A > a) = Fa^{-B}, \tag{10.7}$$

where $Nr(A > a)$ is the number of polygons (or patches) of size A above a, and F and B are two positive constants. Mandelbrot showed that for simple models the fractal dimension of the cumulative perimeter of all polygons, D, was equivalent to one half of B when using $\sqrt{a}$ to approximate the diameter, or characteristic length, of a given polygon. For the case of simple scaling fractals, $1 \le D_p \le D \le 2$ must hold (Cahalan & Joseph 1988).

If log–log plots of the data produce straight lines then the polygons across the distribution of sizes are considered to be self-similar. That is, there is no difference, in a statistical sense, between a large polygon and a small polygon enlarged to the appropriate size. An abrupt change in slope, then, suggests that different forces are operating at different scales.

Remote sensing image data can be easily analyzed to provide data for fractal analysis. One needs only to develop a classification scheme to identify polygons of interest and calculate areas, perimeters, and frequencies. Cahalan & Joseph (1988), using Landsat data, showed these relationships to be useful for examining clouds for coherence in spatial structure. Using image data from the French satellite, SPOT, Ranson & Smith (1991) showed that patterns of North American forests have lower fractal dimensions for managed forests and higher fractal dimensions for wilderness forests. These concepts may prove useful for the analysis of ecosystem dynamics such as discussed by Pastor & Mladenoff in Chapter 8 of this book.

Spatial analyses of regional-scale imagery may also serve as a guide to the stratification of study areas into regions subject to similar, underlying biological or man-induced processes. Using different remote sensing platforms with various resolutions allows one to examine a region under varying spatial ranges. These concepts may also provide insight into ecosystem change on a global scale when applied to global data sets such as those to be provided by the Earth Observing System.

The Earth Observing System

In the light of growing concerns with anthropogenically induced changes at the global ecosystem level, scientists are recognizing the need for repetitive data sets over the entire surface of the earth (NASA 1986). The Earth Observing System (EOS) is a multidisciplinary mission planned to provide the observational and data handling capabilities needed to study the earth as a system, with emphasis on understanding the global-scale processes that operate at or near the planet's surface. The instruments now under consideration for flight on the EOS platforms will observe at wavelengths ranging from ultraviolet to microwaves, allowing, for the first time, near-simultaneous, global-scale measurement of many environmentally significant parameters. The combined set of instruments that are finally selected for flight on the EOS observatories are expected to have capabilities in the areas of surface imaging (land and sea), tropospheric and stratospheric temperature and moisture sounding, including clouds and precipitation, atmospheric dynamics and chemistry, physical oceanography, particles and fields, and solid earth processes.

A unique aspect of the EOS concept is that the mission is viewed as an earth science information system. Accordingly, an EOS Data and Information System (EOSDIS) is being developed to provide for data production, validation, archiving, documentation, and dissemination to a wide community of scientists representing varied disciplinary and interdisciplinary interests.

Current planning calls for the launch of two series of NASA observatories, EOS-A and EOS-B, with three observatories in each series. The target launch date for the first observatory in the EOS-A series is late 1997. The first observatory of the EOS-B series will be launched two-and-a-half years after the launch of EOS-A. The observatories in each series will be replaced at five-year intervals. Through the sequencing of launches, it will be possible to establish a consistent time series of earth science measurements lasting over fifteen years, as may be required to detect changes in many climatically significant variables.

The NASA EOS observatories will be complemented by a National Oceanic and Atmospheric Administration (NOAA) observatory, two European Space Agency (ESA) observatories, and a Japanese observatory, all carrying earth science research instruments in sun-synchronous polar orbits. The NASA observatories will orbit at an altitude of 705 km with afternoon equatorial crossing times, and the ESA observatories will have morning equatorial crossing times. There are also plans to fly a few EOS research instruments as attached payloads on Space Station Freedom.

In addition to the observatories described above, there are plans for a related platform dedicated to the Synthetic Aperture Radar (SAR) being developed by the Jet Propulsion Laboratory. The EOS SAR will operate in the L-, C-, and X- microwave bands with multiple polarization capabilities.

The broad scope of scientific issues to be addressed in the EOS era provides a rich opportunity for international scientific cooperation. Implementation of such a program, aimed at understanding the earth as a globally interacting system, will require coordination of operations, data exchange, and analysis of measurements, not only from orbiting remote sensing instruments, but from *in situ* measurements as well. There will also have to be laboratory efforts and, crucially, there have to be interdisciplinary modeling studies which draw upon the data from all these sources to provide the understanding that is necessary to address the issues of global change. The EOS program is taking the necessary steps in all these areas. It is noteworthy that in addition to the flight hardware discussed above, NASA has selected a large set of major interdisciplinary studies related to developing a better understanding of the climatic, hydrologic, and biogeochemical cycles of our earth.

A more thorough description of the EOS mission, as well as some of the key instruments under consideration for flight on EOS platforms, is provided by Covault (1989) and by Ormsby & Soffen (1989).

Summary and comment

In this chapter we have outlined some of the capabilities and potential of using remote sensing as a tool for monitoring and modeling forest ecosystems. To summarize, the roles of remote sensing measurements and related modeling in forest ecosystem analysis are to: (1) provide fundamental understanding of energy interactions within the vegetation–soil complex; (2) provide a link between the energy environment and ecosystem processes; and (3) provide observations for quantifying changes in the forest ecosystem. The scope of the problem ranges from characterizing the interception of photosynthetically active radiation by leaves and constituent cells, to the integrated terrain media scattering of radiation perceived by satellite sensors. The key is to use that information as a driver of ecosystem processes, and to interpret changes in the forest ecosystem at varying temporal and spatial scales. Thus, the approach is necessarily integral and consists of interfacing forest ecosystem and radiative transfer process models, and acquiring supporting measurements at laboratory, field, aircraft, and satellite levels.

Satellite observations of terrestrial vegetation are available globally at a resolution of 80 m for selected areas dating from the launch of Landsat 1 in July 1972. Daily, global coverage of the terrestrial landscape at a resolution of 1.1 km has been acquired by the NOAA satellites since 1976. The future holds much promise with the advent of the Earth Observing System, as well as European, Canadian, and Japanese earth remote sensing satellites. The task before us is to capitalize upon the expertise of scientists in the many disciplines required to understand ecosystem processes at local to regional scales.

11 The nature and distribution of past, present and future boreal forests: lessons for a research and modeling agenda

Allen M. Solomon

Introduction

The objective of this chapter is to define the properties of a unified boreal forest model. Models are constructed for specific purposes, which themselves guide the selection of, for example, which factors to include, and the level of abstraction and realism. In the case of the unified boreal forest model, certain tasks are obvious and have strongly influenced model construction. Fundamentally we require a model that will permit us to examine long-term forest ecological behavior under a wide range of stable and chronically or catastrophically shifting conditions. However, we can describe more specific tasks for which a boreal simulator is needed.

Most importantly, we lack a vehicle for learning how realistically boreal forests can be defined as abstract ecosystems. Thus, we may wish to develop a mathematical model that can predict forest behavior over both time and space, based on quantitative relations between growth processes and forces intrinsic and extrinsic to boreal forest stands. Construction of the model would simply reflect in part the desire to formulate in testable ways our hypotheses on the importance of various processes that affect the long-term behavior of forest stands. Iterative model construction, comparison of its predictions with field data, further model modification, and so on, allows us to understand what the real forests 'know' concerning the processes of interest that the model does not, and to correct the model accordingly. The model may also serve for exploring long-term ramifications of current temporal or spatial vegetation patterns, which are not otherwise intuitively obvious. Many questions derived from scientific curiosity alone require examination of patterns and processes over decades or even centuries.

A second set of questions involves learning how boreal forests functioned in the past. The behavior of forests over the several generations which occur during a millennium involves intrinsically interesting phenomena, which cannot be measured in mere lifetimes of scientists. Boreal forests have grown under full-glacial climates occurring during 90% of the time over the past two million years, indicating that study of their 'normal' circumpolar configuration may be as instructive as studies of their modern, but atypical, configuration. Boreal forests experienced conditions of relatively rapid warming between 16 000 and 6000 years ago. As they migrated across large regions of Eurasia and North America, the genetic makeup of species may have changed, as warmth-adapted ecotypes were favored (Solomon & Tharp 1985). At times, migration speed may have lagged the new availability of growth sites, leaving empty niches and allowing migration speed rather than climate to force, and be reflected by, vegetation measured in paleoecological records (e.g. Davis 1983).

Indeed, these records can provide very precise and occasionally quantitative estimates of stability and change in prehistoric community properties. However, the records contain no direct information on the causes of this stability or change. In contrast, the model described above must be constructed from either known or hypothesized cause-and-effect relations, allowing one to test hypotheses concerning the causes of recorded prehistoric boreal forest behavior.

A third set of questions is concerned with predicting the future of boreal forests. We wish to predict the response of boreal forests to future global changes in climate and atmospheric chemistry. The long debate concerning the role of various ecosystems as carbon sources or sinks (e.g. Woodwell *et al.* 1978; Broecker *et al.* 1979; Baes, Bjorkstrom & Mulholland 1985) will not be solved in the foreseeable future, but is beginning to focus on the boreal regions (e.g. D'Arrigo, Jacoby & Fung 1987; Tans, Fung & Takahashi 1990). Given a detailed scenario of climate change, the model could provide accurate estimates of the response of above-ground and below-ground components of circumpolar boreal ecosystems in terms of annually or decadally changing carbon storage.

Additional uncertainties about future biodiversity and ecosystem–human interactions in high-latitude regions, which may change with changing climate, could be reduced with a realistic boreal forest simulator. For example, most of the world's migratory wildfowl depend upon wetland habitats in high latitudes for nesting. Changing distributions of forests and shifting vegetation-related soil moisture dynamics would have severe consequences for wildfowl reproduction. Similarly profound consequences are likely for high-latitude aboriginal populations that still live directly on the natural resources provided by present distributions of organisms and their associated biotic communities.

Another set of concerns regarding effects of future climate change on boreal forests involves the forest products industry. Approximately three quarters of the global softwood growing stock is in boreal forests (US Department of Agriculture 1982). The forest industry depends upon an uninterrupted supply of harvested timber. If the forests were to undergo widespread mortality of mature trees because of climate change, a great supply of timber would be followed in quick succession by very small stocks. If tree establishment failed consistently, growing stocks would decline regularly. The long-term consequences for growing stock availability of slight but chronic changes in tree establishment, forest growth and mortality must be known now, if appropriate management strategies that can assure the continuity of wood supply are to be implemented.

These potential needs and uses for a dynamic boreal forest simulation model must be balanced with the nature of the boreal forest system. What abiotic and biotic factors and relationships define the forest now, and defined it in the past, and which ones will be important in the future? How tightly linked are boreal species and boreal conditions, and how flexible are the ecological relationships? The answers to these questions will define the properties of a dynamic boreal forest simulation model. Their implications for model form and function may be discerned in part by examining the nature of today's boreal forest systems. Additional implications can be derived from analysis of past boreal forests and the hypothesized forces that may have shaped them. Finally, we can reverse the cause-and-effect question, and explore the forces that are likely to shape boreal forests of the future, hypothesizing the nature of the forest that could result.

Boreal forests of today

Depending on definition, the circumpolar boreal forest may be the most simply defined forest biome of the globe. While not free of controversy, classifications of boreal forests rest upon a commonly perceived image of homogeneous arboreal stands, dominated by conifers (*Abies*, *Picea*, *Pinus*, *Larix*) during later stages of succession, and by arboreal members of the birch (*Betula*, *Alnus*) and willow (*Salix*, *Populus*) families in early successional stages. Although there are no circumpolar tree species, many shrub species, and an even greater number of moss and lichen species, are distributed on all continents in which the zone occurs, with endemism being comparatively rare. The northern border of the forest is bounded by treeless tundra; this border may be defined as that point where closed-canopy forest terminates, or where full-grown individuals or groves of trees no longer appear on the tundra (my preference), or that point where even stunted trees no longer grow (see Sirois, Chapter 7, this volume, for additional discussion). The southern boundary of the boreal forest is often defined as the point where thermophilous deciduous trees or prairie steppe vegetation is encountered. Again, boreal forest can be defined by the presence either of continuous forest, or of fragmented elements mixed with non-boreal species (the definition assumed in this paper), or as that point where only non-boreal species grow.

The environmental controls and ecological processes which characterize boreal forests have been the subject of several books (e.g. Larsen 1980; Wein & MacLean 1983; Tuhkanen 1984; Van Cleve *et al.* 1986) and recently were thoroughly reviewed (Bonan & Shugart 1989). In addition, other chapters of this volume deal with that subject in detail. Rather than abstract this information, I will review a few salient points of relevance to model form and function.

Hare & Ritchie (1972) emphasize that species are more or less interchangeable, because boreal vegetation types (including non-forest communities) are more directly defined in terms of community structure (physiognomy), life forms of species, horizontal spacing and vertical layering. Specifically, they argue that the geometry attainable by whatever species happen to occur defines differing characteristics of the atmosphere through their influence upon aerodynamic roughness, albedo, evaporativity and snow retention capability. Conversely, they point out that climatic characteristics of atmospheric heat (solar radiation, net radiative heating, duration of growing season) permit only a very few geometric configurations of vegetation at any point on the boreal landscape. Probably because of this tight linkage between forest structure and thermal properties of the atmosphere, the aspect presented by all boreal forests is of a monotonous vegetation and an impoverished flora (Tuhkanen 1984).

The use of structure rather than composition to classify boreal forests is plausible in the light of the strong climatic and edaphic factors which control the nature of the vegetation. Most boreal classifications cite the climatic constraints on this northernmost forest biome. These include cold, snowy winters in which daily temperatures do not reach the freezing point for at least a month, and short, warm summers with only the four summer months each exceeding 10 °C (Ritchie 1987*a*). Precipitation is not often a consideration in boreal forests, although none receive more than moderate amounts and many receive low precipitation.

The boreal temperature regimes result in a latitudinal distribution of boreal forests (Fig. 11.1). The primacy of heat as a limiting factor is also obvious in the subdivisions of boreal forests into latitudinal

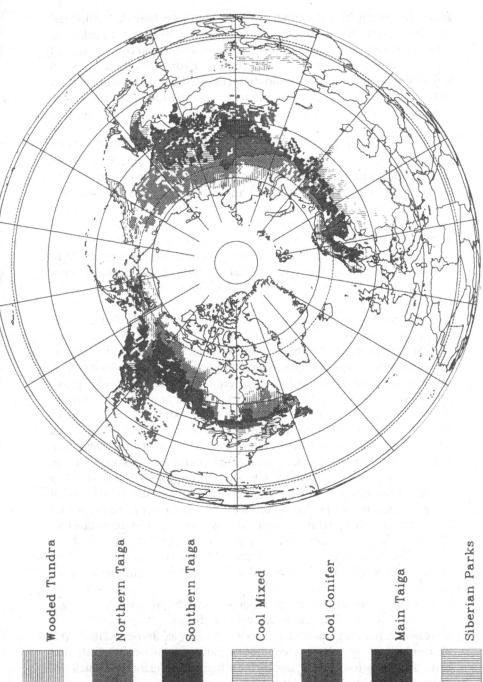

Wooded Tundra

Northern Taiga

Southern Taiga

Cool Mixed

Cool Conifer

Main Taiga

Siberian Parks

Fig. 11.1. Geographic distribution of boreal forest types (from Olson, Watts & Allison (1983)).

zones (northern boreal, middle boreal, southern boreal, hemiboreal of Hämet-Ahti (1981) being largely equivalent to the northern maritime taiga, northern taiga, southern taiga and cool hardwood conifer types of Olson, Watts & Allison (1983) and to similar zones of others). Several longitudinal subdivisions commonly recognized within the circumpolar belt of boreal forest also appear to be temperature-controlled. For example, much of the North Asian but little of the North American boreal forest is dominated by exceedingly cold winter temperatures ($<-40\,°C$), widespread permafrost, and deciduous conifers (*Larix* species). Strong continentality of central and western Siberia, reflected by extreme winter cold and summer warmth, contrasts the minor temperature differences between winter and summer produced by maritime influences in western Scandinavia and Alaska. Maritime influences which moderate winter *and* summer temperatures and add constant humidity appear responsible for birch woodlands along western and eastern edges of Eurasia (Hämet-Ahti & Ahti 1969), low-growing thickets of *Pinus pumila* which cover thousands of square kilometers in northeast Asia (Hämet-Ahti 1981), and heathlands in place of forests along most oceanic coastal areas.

As humidity and reduced seasonality can supersede the fundamental control by temperature to produce coastal heathlands, edaphic factors too can overcome the thermal control of boreal forest qualities. This is most commonly expressed in soil moisture regimes which foster development of unforested bogs and mires where forest would otherwise grow. Olson, Watts & Allison (1983) map 115×10^7 ha of northern, main and southern taiga, which is accompanied by approximately 9×10^7 ha of bogs and mires. Gore (1983) maps mire cover exceeding 10% of the area of the boreal forest of North America, and much of that in western Eurasia. The commencement and maintenance of bogs depends on climatic variables, particularly low evapotranspiration rates and low growing season temperatures, although size, shape, depth and rate of increment of specific bogs is determined by local water table properties (see Glebov & Korzukhin, Chapter 9, this volume). The importance of soil water, in combination with temperature, in regulating very localized vegetation composition and structure is formalized in the model by Bonan (Chapter 4, this volume), and applied by Bonan, Shugart & Urban (1990).

Similarly, thermal control of vegetation can be overridden by large-scale disturbance. For example, wildfire frequency and intensity is increased in part by climatic variables which enhance dryness of fuels. It is also enhanced by substrate conditions (some climate-controlled) that favor accumulation of fuels, and by the structure of the forests, which will in turn determine both fuel availability and its wetness. If mosses and lichens are able to colonize burns and form continuous mats, tree establishment may be precluded for many decades, despite a thermal regime which permits tree growth. Large-scale insect attacks can also open the forest canopy for long periods of time, as sunlight resulting from defoliation of trees allows the growth of mosses and lichens, which inhibit the establishment of tree seedlings.

These considerations suggest a boreal forest model that could be largely empirical and non-dynamic, and still be capable of predicting the geography and structure of modern boreal forests. Radiation measures, or even seasonal temperatures, could be employed to predict structural features that determine standing crop, sunlight and groundcover required for tree reproduction, and so on. The latitudinal distribution of boreal forests has been reproduced quite accurately by a combination of continentality (i.e. seasonality), heat sums, and evaporation (Tuhkanen 1984). Marine influences, which reduce boreal forest stature, can be defined as constants, through proximity to oceans. Soil moisture status and permafrost depth and cover, which determine presence of non-forested bogs and muskegs, could be similarly treated. Within these extrinsic relationships, one could apply correlations of disturbance regime with climate to estimate frequency and intensity of wildfire and insect attack, and hence the distribution of stand ages at which forest succession is set back in time. Static regressions could even be used to characterize the rather complex interactions of closed forest, permafrost, fire, moss and lichen cover, and moisture status of soil surfaces described by Bonan (Chapter 4, this volume). Indeed, the unified boreal forest simulator described in this book is capable of reproducing these static properties of boreal forests. However, the simulator is also capable of considerably more, allowing it to be applied to both the past and the future.

Forests of the past

An instructive exercise in gleaning characteristics of the circumpolar boreal forest is to examine its geographic distribution in the past, particularly during full-glacial time when global climate was significantly different from that today. Unfortunately, there is no globally comprehensive map of full-glacial boreal forest distribution. A map exists of full-glacial Holdridge Life Zones, classified from output of a general circulation model (GCM) of the atmosphere (National Geographic Society 1989). However, that mapped distribution of boreal forest types is considerably different from distributions suggested by measurements at specific locations (e.g. Davis et al. 1975; Huntley & Webb 1988) or estimated with other models in specific regions (e.g. Overpeck & Bartlein

1989). It is possible that either the GCM or the Life Zone Classification (or both) are too inaccurate to describe the actual distributions.

Guetter & Kutzbach (1990) modified the Köppen classification to generate a modern vegetation map on which boreal forest types (Df, Dw) were correctly placed where boreal forest grows today. Their full-glacial simulation maps much of the Df and Dw classes further south, and the boreal classes cover only about half of the area now occupied by those classes. Even this cover may be an exaggeration of full-glacial boreal forest distribution, as much of the Df and Dw type is placed in central Europe and the Mediterranean where the few maps published indicate steppe vegetation. The winter-wet, summer-dry climate probably responsible for the growth of full-glacial steppe in this region is not included among the classes of Guetter & Kutzbach's (1990) classification.

CLIMAP members (1981) published albedo maps derived from vegetation samples and other considerations, for use in GCMs of full-glacial climatic conditions. A very small albedo value (10–14%) is characteristic of both modern rain forest and boreal forest (map 9a, CLIMAP 1981). That same albedo class covers only a very small portion of the full-glacial landscape at latitudes greater than 30° (map 9b, CLIMAP 1981); this portion is almost entirely in the eastern United States, with none in Europe and central Asia.

Thus it appears most likely that little boreal forest existed during at least the last full-glacial period, and probably during the other full-glacial periods that compose most of the past two million years. This is somewhat surprising considering that the lower temperatures associated with conditions of continental glaciation probably produced a mid-latitude thermal regime much more compatible with boreal forest requirements than today's regime (e.g. Bryson & Wendland 1967; Moran 1972; Williams, Barry & Washington 1974). One might hypothesize that the generally drier conditions revealed by most full-glacial climate reconstructions reduced the amount of full-glacial boreal forest to a small fraction of its present areal coverage. This may be an adequate explanation for the apparent absence of full-glacial boreal forests in Eurasia. However, examination of boreal forest composition in North America during that time indicates that additional extrinsic variables controlled the boreal forest.

E. Lucy Braun (1950) adhered to a concept of full-glacial vegetation in eastern North America as a simple southerly displacement of today's ecosystems, with the same or similar species composition. The landscape contained a few 'refugia', mainly on the Ozark Plateau and southern Appalachian Mountains, in which certain thermophilous species were able to survive until the onset of modern conditions. This view is still expounded by some workers, albeit with different refugia (e.g. Delcourt

& Delcourt 1987). Yet, it has been apparent for many years that species composition and geographic distribution during full-glacial time varied considerably from that of today.

Particularly instructive are the compilations and syntheses of eastern North American pollen data by Webb (Solomon & Webb 1985; Webb 1985, 1987; Jacobson, Webb & Grimm 1987). These data reveal boreal forests of greater variety than those found today. At the maximum of continental glaciation, 18 000 years ago, boreal forests, composed of spruce and accompanied by very little of the now ubiquitous pine, covered much of the unglaciated central United States. The presence of large proportions of sedge pollen in the spruce forests strongly suggests that this region was covered by open spruce woodland rather than closed-canopy forest, because the spruce–sedge pollen combination is confined today to open forest-tundra in the western Ungava Peninsula, north of James Bay in eastern Canada (Jacobson, Webb & Grimm 1987, Plate 2). Examination of western Ungava climatic conditions, which coincide with spruce woodland distributions today, reveals more moisture, a higher humidity, and less extreme temperatures in both winter and summer seasons than elsewhere in North America. However, these conditions may be only partly responsible for the full-glacial boreal forests.

To the south and east of the full-glacial spruce woodland in the United States, boreal forest was dominated by pine species, with large areas of the south-east also containing hickory and some oak. This transverse orientation of the two boreal forest types (spruce to the north-west, pine to the south-east) remained largely intact until about 14 000 years ago, when spruce woodland began to assume a latitudinal distribution, moving eastward close to the continental ice sheets. By 12 000 years ago, pine had migrated north along the eastern seaboard, to reach the ice sheets in eastern Canada. Sometime between 12 000 and 10 000 years ago, the spruce woodland disappeared and the pine moved westward, adjacent to retreating ice sheets, producing (for the first time in perhaps 100 000 years!) the essential spruce–pine composition and closed-canopy character of modern North American boreal forests.

The spruce woodland and pine–hickory forests were not the only boreal forest anomalies during glacial conditions. Particularly apparent in late-glacial time was the widespread presence in more southerly boreal forests of admixtures of thermophilous hardwoods with spruce and pine, although these forests are very rare today. Bryson & Wendland (1967) speculated that the mixed forests could have resulted from reduced seasonal temperature extremes, allowing boreal species, which are limited by summer high temperatures, to mix with thermophilous hardwoods, which are limited by winter low temperatures. Subsequent analysis of paleoecological indicators of seasonality (Moran 1972), and of astronomical forcing of seasonality revealed by deep sea cores (e.g. Kutzbach & Guetter 1986), have confirmed that the range of temperature between summer and winter was small in late-glacial time (e.g. 12 000 to 10 000 years ago), and increased thereafter. Solomon (1982) and Solomon & Shugart (1984) tested the effects of reduced seasonality on late-glacial boreal forest composition by applying late-glacial temperature scenarios to a stand simulator (FORENA) which was constrained by moisture and temperature. By simulating forest growth on the several exposures and soil types which would have contributed pollen to a nearby pollen record, they found that seasonality could account for the unique admixtures.

In retrospect, I believe that the simulations were demonstrating that the boreal and thermophilous hardwood mixtures required rough topography in which the unlike species could grow adjacent to each other but in separate communities. The topographic requirement was met at only a few sites, which recorded the unique assemblage, and hence reduced seasonality is not a sufficient cause to produce these forests. Simulations (A. M. Solomon, unpublished) with slightly differing climate variables lead to mixtures of the hardwoods with *Pinus resinosa*, *Abies fraseri*, and *Picea rubens*. These last three are southern boreal species, which have very limited areal extent today, but the *Picea rubens* in particular could very well have dominated full-glacial forests and coexisted close to the hardwoods. The narrow geographic ranges of the three boreal species suggest that they have little genetic variability today.

Loss of genetic variability during rapid climate change suggests another potential source of the unique boreal forests of full- and late-glacial time. Individuals or isolated populations that were adapted to grow vigorously in cool summers would be at a competitive disadvantage, and could be eliminated from the population, when climate warmed quickly at the end of the last glacial. Thus, hardwood populations that could previously grow with boreal spruce and pine would lose their thermophobic members, reducing the temperature limits within which the species as a whole could operate. Solomon & Tharp (1985) demonstrated by simulations with the FORENA forest stand model that a heat-sum tolerance range in red oak greater by only 6% than that found today would allow white spruce and red oak to coexist on the same site.

Glacial boreal forests operated in an atmospheric CO_2 concentration different from that found today. Recent analyses of CO_2 bubbles trapped in the ancient ice of polar ice sheets reveal that full-glacial atmospheric CO_2 concentrations hovered around 200 ppm (parts per million by volume), often reaching 180 ppm. This contrasts with the 280 ppm of pre-industrial time, and the 350 ppm now found in the atmosphere. At 100 ppm, the partial pressure of CO_2 becomes so low that plants have difficulty taking in CO_2 through their stomata, that is, they undergo CO_2 starvation. Solomon (1984) suggested that low atmospheric CO_2 concentrations could be responsible for the absence from most North American full-glacial forests of nutrient-demanding tree species, such as hemlock and beech. CO_2 may also have played a role in defining the nature of the full-glacial boreal forests. Low CO_2 concentration may have been directly responsible for the open character of central spruce forests (south-eastern pine forests could also have been woodlands, but no indicators of structure are available). However, low CO_2 concentration could have had a more indirect role in segregating pine and spruce, in that pine species do not demonstrate changes in water use efficiency (drought tolerance) with changing CO_2 concentration (P. J. Kramer, pers. comm., April 1984), while tolerance of spruce to drought may have decreased with low CO_2, thereby causing it to be excluded from the drier coastal plain and nearby areas of the south-east.

Thus, modern boreal forest structure and composition may be of very recent origin, and may be the result of extrinsic forcing variables very different from those that constrained the forests over most of the past two million years. What those forces may have been is still obviously a matter for hypothesis and speculation. Most importantly, both the extrinsic forces and the biotic responses of species were probably different in the past from those found within the modern boreal forests. A unified boreal forest simulator must therefore treat species separately and independently, allowing species combinations not necessarily common on today's landscape. If species are defined only by their current associations and geography, many of the patterns and processes characteristic of past boreal forests will not be available for examination by the model. In addition, independent treatment of species must include the capability to model individualistic responses of species to certain extrinsic variables that are currently geographically constant, such as atmospheric CO_2 concentrations. These requirements exceed by a considerable margin the characteristics of models that follow from spatial considerations, derived from the preceding analysis of present geography of boreal forests and their climate and edaphic factors.

Boreal forests of the future

Consideration of boreal forests of the past necessarily began with the measured characteristics of those forests, and proceeded to the hypothesis of causes for the patterns which can be tested with a unified boreal forest simulator. Consideration of boreal forests of the future must begin with the likely causes of change and stability, and must proceed to hypotheses of forest responses, revealed by a unified boreal forest simulator. The requirements of a model for predicting future forest dynamics are not the same as those of models for understanding past environments, or modern relationships. Consider the changing extrinsic forces envisioned during the next one or two centuries.

The most obvious change is that expected from rising concentrations of radiatively active gases (greenhouse gases) caused by activities of humans, especially through burning of fossil fuels. At current rates of increase, atmospheric CO_2 concentration could double by the year 2030 from its preindustrial level of 280 ppm. Coincidentally, world population will also double by 2030 at current rates of increase. The doubling of CO_2 concentration is merely a reference point and, as with population increase, one can expect chronic increases in greenhouse gas concentrations to continue for the foreseeable future, barring an agreement to severely reduce greenhouse gas emissions among industrialized nations, which depend on fossil fuel consumption for their present living standard, and Third World nations, which posses no means to industrialize except large coal reserves.

The issue of greatest concern in greenhouse gas increases is their effect on climate, and of climate on managed and unmanaged biota of the earth. Prediction of climatic response to increasing greenhouse gases can be made for subcontinental regions with GCMs (e.g. Hansen *et al.* 1984; Manabe & Wetherald 1987; Schlesinger & Mitchell 1987). In addition, local and regional climate can be examined by substituting specific time periods from the instrumental weather record which are thought to represent analogs for future climate (e.g. Wigley, Jones & Kelley 1986; Jaeger 1990; Pitovranov & Jaeger 1990).

Despite differences among GCMs, and between GCMs and climate analogs, several patterns emerge. The greatest temperature change is expected to be in polar regions, and the least in equatorial areas. A warming of 1.5–4.5 °C worldwide produces a warming of 5–10 °C in high latitudes (Jaeger 1988). In addition, winter temperatures are expected to increase more than summer temperatures, effectively reducing the seasonal temperature range, and increasing growing season length disproportionately in comparison with summer temperature increases. Precipitation projections are much more contentious and ambiguous primarily because the GCMs operate on spatial scales too coarse to include the topographic and atmospheric properties responsible for precipitation in most places. However, warming is expected to produce a general increase in intensity of the hydrological cycle although there is no indication of whether the increased evaporation will equal or exceed the increased precipitation. Increased droughtiness over continental interiors and increased precipitation nearer to coastal regions has been simulated with GCMs (e.g. Mitchell 1983; Manabe & Wetherald 1987).

A geographical implication for boreal forests of one GCM (Mitchell 1983) is shown in Figure 11.2 (Solomon & Leemans 1990). This figure presents the geographical range of boreal forest life zones defined by Holdridge (1947), based on annual heat sums and annual precipitation in the climate data base of the International Institute for Applied Systems Analysis (Leemans & Cramer 1990). Approximately 25% of today's boreal forest would have to die out, giving way to steppe and deciduous forest, if boreal boundaries are controlled by annual heat sums and precipitation as assumed by the Holdridge life zone system. Particularly large losses of forest are projected in Russia, just west of the Ural

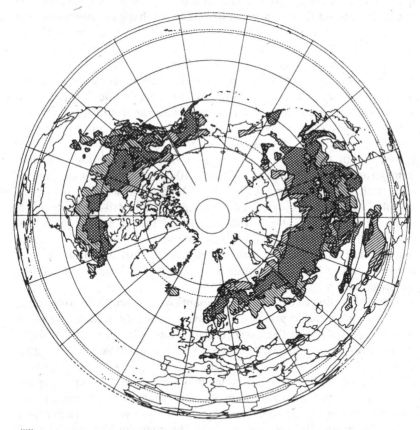

▨ Boreal Forest, Doubled CO$_2$

▨ Boreal Forest, Modern

Fig. 11.2. Geographic distribution of boreal forest life zones (Holdridge 1947), based on modern climate (Leemans & Cramer 1990) and on climate of a doubling of CO$_2$ (Mitchell 1983).

Mountains, and in North America, east of the Rocky Mountains. Notably, this projection suggests that less than half of the boreal forest lost to the south would be replaced with new forests to the north and on the Himalayan Plateau. This projection does not imply that the areas in which boreal forest life zones are unchanged (crosshatched in Figure 11.2) would necessarily support stable vegetation; the presence of ecotypes within populations suggests increased mortality among thermophobic members of ecotypic populations at any point in their geographic ranges.

The important question which modeling must answer concerns the speed of climate change, not its magnitude. We saw that the boreal forest was quite plastic in the past, responding to changing environmental variables with changes in species associations, structural properties of stands, and areas occupied, including its disappearance from, and reoccupation of, large regions of the world. Unlike the Pleistocene climate changes, which apparently took place over thousands of years, however, the future climate change of about the same magnitude is expected to take 100 years or less. We once calculated (Solomon et al. 1984, pp. 13–14), from age-at-reproduction of northern hardwoods and boreal softwoods, and from seed transport distances reported in the scientific literature, that maximum potential rates of migration of wind-transported seeds may be 20–30 km per 100 years (half that amount for trees with animal-transported seeds), which is consistent with observations from paleoecological evidence in the field (10–45 km per 100 years (Davis 1983)). We further calculated that if atmospheric CO_2 doubled in 100 years, the Mitchell climate scenario used in Figure 11.2 requires that heat-sum isotherms migrate northward at 400–600 km per 100 years, an order of magnitude more rapidly than the calculated and measured tree migration rates.

This decoupling of climate determinants from species and community distributions may also not be new, having probably occurred at many times and places during the last several millennia (Davis et al. 1986; Webb 1986). However, the speed of change may be so rapid that problems other than the obvious inability of trees to migrate quickly arise. The time required to complete tree life cycles may be critical. The most rapidly maturing trees require a minimum of 5 or 10 years to reach reproductive maturity. Slow-growing species require 50 or more years to reach reproductive maturity after seeds germinate. The question of concern here is whether the life cycles of trees can be completed before the environmental limits to tree growth are exceeded by climate changes. The climate resulting from a doubling of CO_2 could kill trees before they reach reproductive maturity. Even if reproductive maturity is reached, the seeds they produce will belong to species that cannot grow where the seeds fall. Trees whose seedlings can now survive belong to adults which the climate may prohibit from growing there in 50–100 years; adults which can grow there in 50 or 100 years have seedlings which may be unable to survive the climatic conditions that prevail now.

The response by boreal forests may involve the extinction of slow-growing species and the selection of species that complete their life cycles very quickly, that is, early-successional trees and shrubs. In addition, the ability to reach reproductive maturity in the shortest time requires that the early successional tree species grow in the sun, that is, in the absence of closed-canopy forests. In essence, this suggests an open landscape of scattered trees which form woodlands or savannas, and a concomitant absence of the dense forests we know today.

These responses imply that only direct effects of climate change will impact boreal trees and communities. Yet, other environmental components are also expected to change drastically. For example, atmospheric chemistry is likely to be much different in 100 years from that at present. Differential responses of tree species to atmospheric CO_2 concentrations were discussed above. A current hypothesis in northwestern Europe is that atmospheric contaminant concentrations may be serving both as growth inhibitors (sulfates) and as fertilizers (nitrogen compounds, CO_2). Hari, Raunemma & Hautojarvi (1986) examined effects of several contaminants in the absence of any projected climate change. They used measured relationships of tree growth to atmospheric toxicants and nutrients, along with scenarios of atmospheric chemistry changes, to project a gradual increase in Finnish boreal forest growth for a few decades, followed by permanent chronic loss in growth. The model they used suggested that negative impacts of soil acidification would eventually overcome all other effects from atmospheric contaminants.

Forest stand simulations applied to combined effects of warming and CO_2 (Solomon & West 1987; Solomon 1988) suggest that the negative effects of climate change on growth will be several times more important than the positive influence of carbon fertilization. Yet, it is also clear that the stand model used was inappropriate for predicting future tree or stand responses to CO_2 in that simulated effects of carbon fertilization on growth, mortality, reproduction and competitive relationships among species were based on only a few long-term studies of two non-boreal tree species. Indeed, until carbon fertilization can be studied in functioning forest stands, it seems unlikely that any simulation model can be trusted to define future forest response to this variable.

Atmospheric CO_2 concentration is not the only component of atmospheric chemistry in boreal forests likely to change in the future. Consider Figure 11.3 (Andersson 1991), which shows the global distribution of soils vulnerable to acidic precipitation, and the present and future

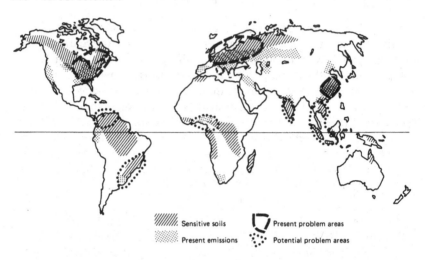

Fig. 11.3. Geographic distribution of soils vulnerable to acidic precipitation, and of present and potential future high concentrations of
acids in the atmosphere (Andersson 1991).

distribution of large amounts of acidic gases in the atmosphere. Note the
large regions of vulnerable boreal forests in both North America and
Europe. Andersson indicates that these areas would escape effects of the
dense populations which generate high regional concentrations of the
gases. However, the future geography of the gases is predicated on the
assumption that current geographic growth trends of human populations
will continue. Yet, if warming occurs as expected, and if human populations continue expanding, much of today's boreal forest could be
suitable for occupation by dense populations of humans.

In addition to the development of large new urban areas in today's
boreal forest regions, human needs for food are also likely to change the
landscape. The limit to tillage agriculture appears to coincide with the
southern boreal forest boundary. Thus, Fig. 11.2 may also be used to
suggest a new limit for growth of crops. However, a doubling of human
population will certainly require far more new agricultural land than is
indicated in Fig. 11.2, even if we assume that (1) all current agricultural
land remains productive; (2) all tillable land now used to feed livestock is
put into crops used for direct consumption by humans; and (3) livestock is
raised on non-tillable soils, including those of boreal regions. The
resulting increase in non-forested pastureland for animal husbandry and
intensity of farming of all arable soils will not only reduce boreal forest
growth, but will also remove corridors of natural vegetation suitable for
use by migrating tree populations. Hence, the land use changes will
probably magnify effects of rapid climate change on boreal forests,

reducing their capability to respond and enhancing probabilities of local extirpations of species and communites.

What is the nature of a unified boreal forest model which can predict forest responses to these environmental forces of the future, and which is not already specified by model needs for study of the present and the past? The singular problem faced by future boreal forests involves the high rate of environmental change. The resulting potential problems related to timing of life cycles suggest a model with the capability to treat life cycle stages independently, and to deal with the relationships of different stages to structural characteristics of stands, especially the sunlight regime. The fact that the life cycle stage most vulnerable to environmental variation is establishment further suggests that establishment should be treated mechanistically in the model. The obvious role of species migration in future boreal forests is also critical, indicating a model in which propagule transport can be simulated over large regions, once again as a mechanistic process. The additional importance of changing land use also points toward a model which can treat regional landscapes, as well as individual points.

The other problem not encountered in assessments of present and past boreal forests is the effect of changing atmospheric chemistry. The current hypotheses concerning effects of atmospheric contaminants as nutrients as well as toxicants should be incorporated into the model structure for testing on current landscapes as well as for projecting onto future ones.

The foregoing analysis of past, present and future environmental forces and ecological responses defines requirements of a unified boreal forest model. The chapters on detailed characteristics of boreal forest functioning (i.e. Chapter 3 by Zasada, Sharik & Nygren, Chapter 6 by Holling, Chapter 7 by Sirois, Chapter 8 by Pastor & Mladenoff, and Chapter 17 by Duinker, Sallnäs & Nilsson) suggest other requirements as well. No one model or approach can meet these requirements, and none will in the future, at least in the absence of new information on ecological effects of atmospheric contaminants, relationships of carbon storage and release to climate, and so on. Yet, the capability to model boreal forest dynamics at both the stand level and the landscape level, revealed in the chapters by Leemans (Chapter 16) and Korzukhin & Antonovski (Chapter 13), indicates that the unified boreal forest model is nearly complete, already useful for many of the tasks it must perform, and a very great advance over modeling capabilities of just a few years ago.

Computer models for synthesis of pattern and process in the boreal forest

Introduction

Rik Leemans

Models for simulating different aspects of vegetation dynamics have become increasingly popular during recent decades. Initially, mathematical modeling was only accessible to well-trained biomathematicians, but with the increasing availability of small and faster computers and with the development of modern software, it has been applied by more traditionally trained ecologists and foresters. Recently, many papers that present different models and applications within ecology have been published (e.g. Emanuel *et al.* 1984; van Tongeren & Prentice 1986; Running & Coughlan 1988; Tilman 1988; Costanza, Sklar & White 1990; Keane, Arno & Brown 1990).

Simulation models can help in the understanding and management of ecosystems. Such models are usually the only tool available for translating a collection of hypotheses for ecological processes into a testable representation of how the whole ecosystem functions. Simulation models can be used not only to evaluate hypotheses generated by field studies and ecological experiments, but also for situations where the more traditional ecological approach is less applicable, for example for studies that span several research generations, such as the study of processes involved in forest succession and gap-phase replacement of individual trees within a stand (Watt 1947). Ecological hypothesis-testing by experiment and field studies for such long-term and large-scale processes is almost inevitably incomplete and must be supplemented by simulation experiments.

Simulation models consist of a collection of hypotheses, most often in equation form. These hypotheses define how the major parts of the model change over time (Swartzman & Kaluzny 1987). The models that we are most concerned with in this section are those that include mathematical descriptions that control the various processes of forest dynamics.

Models based on such principles are mechanistic, in contrast to empirical models where a relationship between the process and controlling variables is established without explicitly considering the underlying mechanisms. Despite this section being limited to the more mechanistic models, we still encompass a diverse array of approaches. Different aims and theoretical viewpoints define the model structure and type used. All authors discuss not only their model structure, but also the underlying theory.

Models for forest dynamics range from simple to complex, from simulation of single tree growth to simulation of whole ecosystems and landscapes, and encompass temporal scales that may range from days to centuries. The most uncomplicated models, conceptually, are highly concrete, yield-oriented empirical models that are often intended as management tools. These models often depend on huge databases and are usually only applicable to those areas for which data are explicitly available (Duinker, Sallnäs & Nilsson, Chapter 17). Some other conceptually uncomplicated models are highly abstract and require few data: Horn (1975b), for example, presented a Markovian model for forest succession, based on observed transition probabilities for tree by tree replacement. Noble & Slatyer (1980) use a scheme of vital attributes of tree species to determine vegetation response after recurrent fires in Australian forest. None of these approaches involves modeling ecological processes directly.

Between these two extremes are somewhat more complex models that explicitly simulate ecophysiological processes connected with tree growth and interaction. The most popular are often direct descendants of the forest succession model JABOWA (Botkin, Janak & Wallis 1972; Shugart 1984), but other models have also been successfully developed (e.g. FOREST by Ek & Monserud (1974)). Dale, Doyle & Shugart (1985) reviewed the different types of forest model. They distinguished 'forest growth models' and 'community dynamics models'. The first type is generally applied to managed, even-aged forests with few species and is closely tied to empirical site-specific data. Most models developed by foresters for use in their management practices fall into this category. The latter type is used to study the effects on natural forest development of feedbacks between a mixed-species assemblage and its physical environment.

In this section, several community dynamics models are presented and discussed together with a large array of specific applications. Emphasis is directed toward recent developments with the community-dynamics approach to modeling the boreal forest. Examples are presented from North America (Bonan, Chapter 15, and Duinker et al., Chapter 17), Europe (Leemans, Chapter 16, and Duinker et al., Chapter 17) and the

Soviet Union (Antonovski, Ter-Mikaelian & Furyaev, Chapter 14, and Korzukhin & Antonovski, Chapter 13). This section does not attempt to be complete in dealing with all possible topics concerning models of the boreal forest, but merely attempts to give a report on the status of current research. Emphasis is directed toward the natural processes defining and influencing forest dynamics. Management applications are only briefly summarized as they fall outside the scope of this book.

Shugart & Prentice (Chapter 12) introduce modeling of forest dynamics by presenting a history of the development of individual-tree-based models, combined with a concise review of the other important model categories presently available. They then discuss possible approaches for expanding such generally accepted stand models toward larger spatial scales, ranging from landscapes to biomes and eventually the whole terrestrial biosphere. They conclude by stating that such developments become increasingly important for understanding the effects of the processes acting on different scales and influencing individual trees, forest stands, landscapes and biomes. Such increased understanding is necessary for a better assessment of forest response to large environmental changes, such as climate change.

Korzukhin & Antonovski (Chapter 13) take individual-tree-based models as a starting point. They discuss their strengths and weaknesses and then elaborate toward mathematical models generated from sets of differential equations. Such dynamical systems are capable of simulating the dynamics of age distributions within even-aged and uneven-aged stands. This is particularly feasible for the boreal forest, with its relatively few tree species. Although few complete age distributions from long-term forest dynamics are available, they successfully demonstrate that this approach can be linked to different disturbance regimes, especially if the disturbance reoccurs periodically. Finally they combine their model with a simple model for moss dynamics, comparative to that of Bonan & Korzukhin (1989). Within the whole boreal forest ecosystem the interaction between the moss layer and tree layer is important for its overall structure and, especially, successful regeneration of tree species (Bonan & Shugart 1989).

Antonovski, Ter-Mikaelian & Furyaev (Chapter 14) review the approaches available for modeling fire dynamics within the boreal forest. They conclude that the spatially explicit, long-term models for fire dynamics are among those poorest developed and that such models are necessary to mimic the effects of fire over large areas. To fill this gap they present their research directed at the creation of such a model. They concentrate their test simulations on western Siberia, where an extensive mosaic of different-aged forest patches exists, all of which have regenerated after wildfires. A large part of this area has been studied intensively by Furyaev & Kireev (1979). This study is exclusively summarized and presented in this chapter for the non-Russian-speaking audience. Their simulations clearly demonstrate the differences in the fire-dynamics models developed for forest stands and the more spatially detailed models such as their own.

Bonan (Chapter 15) reviews the different causal environmental processes that determine the mosaic of different forest types within the boreal forest. He quantifies the complex relations between these processes in a model that simulates soil moisture and permafrost regimes together with moss dynamics from easily obtainable topographic, soil, climatic and forest structure data. He has linked this model to a traditional gap model (Shugart 1984; Bonan 1988c) and a nutrient cycling model (Pastor & Post 1985) and demonstrates the interrelations between forest dynamics and environmental processes for different forest stands in North America by comparing some important aspects of forest structure with data from existing stands. Using sensitivity analysis, he then assesses some of the possible changes in biomass within his earlier simulated stands. This application provides a good example of the capabilities of such complex combined models.

Leemans (Chapter 16) describes a simulation model for the dynamics of a boreal forest stand. The model uses a similar approach to simulating a forest stand as the earlier, traditional gap models (as used by Bonan, Chapter 15). However, the parameterization of regeneration, growth and mortality are modified to mimic the specific dynamics of boreal forests (Leemans 1989). To demonstrate the more accurate simulation capacity of this model over the traditional gap models for certain applications, he compares the improved model with a traditional model for a simulation of a well-documented forest stand at the southern edge of the boreal forest in Sweden. Using both summary stand statistics and the characteristic size distribution of all species involved, he shows that the improved model mimics the dynamics of boreal forest more accurately. Although this chapter only describes the simulation of trees and their interactions, possible linkages with potential models of environmental processes are discussed in detail.

Duinker, Sallnäs & Nilsson (Chapter 17) concentrate on the use of stand models for forest management rather than ecological applications. They review the basic concepts in forest management and forest inventories and give two examples of management models used for long-term forest responses to management strategies and future timber assessment. They conclude that models based on only site-specific information can lead to erratic management strategies when extrapolated for forest growth during and after large environmental changes. Under such circumstances, the more general ecological stand simulators show a more realistic response, but are not yet well adapted to management purposes.

Although all chapters in principle deal with a different part of the spectrum of model theory and applications, there are, of course, overlaps. Such overlap gives an indication of the interrelations between models, processes, ideas and approaches on different temporal and spatial scales.

12 Individual-tree-based models of forest dynamics and their application in global change research

Herman H. Shugart and I. Colin Prentice

Introduction

The problem of predicting ecological responses to global environmental change has generated a rich array of scientific challenges for ecosystem ecologists. The International Geosphere–Biosphere Program (IGBP), chartered by the International Council of Scientific Unions in 1986, has as its objective:

'... to describe and understand the interactive physical, chemical, and biological processes that regulate the total earth system, the unique environment that it provides for life, the changes that are occurring in this system, and the manner in which they are influenced by human activities.' (International Council of Scientific Unions 1986.)

IGBP is one of the most exciting scientific opportunities of our time. The intellectual challenge is ultimately to unify ecological and geophysical sciences at the global scale: a large undertaking in which ecosystem ecology must play a central part.

One of the major tools for this multidisciplinary undertaking will be ecosystem simulation models implemented on high-speed computers. Global dynamics models exist for the atmosphere and oceans as physical systems, but not yet for ecosystems and their interactions with the atmosphere and oceans. This book represents the first attempt to develop a unified model of a major biome (the boreal forest) at the global scale. The model described in the chapters by Bonan (Chapter 15) and Leemans (Chapter 16) belongs to a well-established class of models that simulate the dynamics of forests by projecting the change in size of individual trees over time, through explicit representations of the interactions between arrays of individuals and their local resource environment. The consideration of individual trees in models to project global ecosystem dynamics may at first seem paradoxical. However, it is the interactions between individual trees and their local environment that determine processes such as gap-phase regeneration, which in turn crucially affect the composition, structure and texture of forests over large areas. These processes can be effectively simulated by individual-based models. Also, data on the size distributions of individual trees are relatively easy to obtain, provide an important source of data for testing dynamic models, and have historically been used as the starting point for understanding large-scale forest processes.

Historical background

The problems of understanding forests

Trees dominate the structure and function of ecosystems in which they occur. Yet trees are difficult subjects for the classical ecological techniques of demography and growth analysis. The high mortality rates of small trees imply a need to study many individuals to estimate sapling death rates. Yet adult trees have the potential to live for such long periods of time (centuries to millennia) that direct demographic measurements are impossible. The precise scaling up of what is known of the physiology of tree tissues to the level of the whole plant has proven to be a daunting problem: it is still difficult to predict with mechanistic physiological models either annual or decadal tree growth expected under particular environmental conditions, even for important commercial tree species growing in plantations (presumably the most straightforward case). These features conspire to make trees, and forests, awkward experimental objects. It is therefore essential to use indirect methods to infer the processes that produce observed patterns in forests.

Harper (1977) summarized the problems of tree demography in two sentences bridging adjacent paragraphs:

'In a sense a tree has to master all trades – to be successful in a variety of life stages and to meet the hazards of each layer of the vegetation that it penetrates.

The study of trees is a study of short cuts; the long life and large size of trees makes many of the conventional methods of plant biology impossible or unrealistic.' (Harper 1977, p. 600.)

This first statement implies that a full understanding of tree demography may ultimately require a synthesis, at some level, of much of terrestrial population ecology. The second statement mentions the 'short cuts' that are more or less mandated by the biology of trees. One of the principal 'short cuts' in forest biology is the extensive use of statistical data on the diameters of trees. Tree size distributions integrate aspects of forest dynamics on a timescale commensurate with tree lifetimes, and to some extent provide a surrogate for observational data that would take several human lifetimes to collect.

Tree diameter distributions

In 1898, DeLiocourt noted that when the density of stems in a mixed-aged forest was plotted in equal diameter intervals, the resulting curve had the property that the number of stems in each interval was in constant proportion with the next size interval. This exponential decline of numbers in diameter categories (Meyer 1952) has been used as a basis for forest management (Spurr 1952; Knuchel 1953; Leak 1964; Sammi 1969) in which forests are harvested to maintain a 'balanced' condition (Assmann 1970; Harper 1977) of numbers among size classes. It was also noted by foresters designing harvesting schedules for regional forests that, when the unimodal diameter distributions typical of even-aged forests (variously described using normal distributions (Hough 1932), the Gram–Charlier series (Osborne & Schumacher 1935), Pearson distributions (McGee & Della-Bianca 1967; Lenhart & Clutter 1971) and Weibell distributions (Bailey & Dell 1973)) were summed across ages, the resulting curve strongly resembled that expected for a mixed-aged forest in both shape and magnitude (Assmann 1970).

This relation, between diameter distributions in forests of mixed ages and the sum of the diameter distributions of a mixture of even-aged forests, does not necessarily imply that the diameter of trees could be used as a surrogate for age in demographic studies. On the contrary, there is ample evidence that size of trees and age are often only weakly related in natural forests (Harper 1977; Veblen 1986). Nevertheless, the diameter distributions of 'natural' forests tend to be relatively similar even when the forests themselves are quite different.

Since diameter distributions for forests that have not been recently disturbed by natural or human-caused events typically decline as a negative exponential curve, the curves appear relatively straight with a logarithmic transformation of the tree numbers. But this linearity is usually most pronounced in the smaller size classes. In many forests there is a tendency for the curves to have a 'plateau' or a change in slope in the larger size classes. These changes in shape have been given a variety of interpretations in attempts to understand the processes that produce both the general pattern and the patterns observed for a given forest.

In the case of diameter distributions drawn from large or wide area samples, the smaller trees may not necessarily be associated with the larger trees. Thus, the negative exponential curve associated with a given landscape could arise from a summation of curves of a non-exponential form associated with smaller pieces of the forest landscape. Leak (1973) and Goff & West (1975) postulated that the expected shape of a diameter distribution for small plots of land with no patches of even-aged forests should produce a 'rotated sigmoid' distribution curve with a steep negative slope for the smallest size classes, a flatter slope or plateau in the intermediate size classes associated with the lower mortality or faster growth rate of trees emerging into the canopy, and a steep decline in the numbers of extremely large individuals associated with the higher mortality of senescent, large trees.

This rotated sigmoid form has been reported for some North American deciduous forests (e.g. Schmelz & Lindsey 1965; Leak 1973) and for forests averaged over regions (e.g. Harcombe & Marks 1978). In old-growth forest in Europe, similarly, a dip in the diameter distributions in the size classes immediately smaller than the largest trees has been reported (Leibundgut 1982; Koop & Hilgen 1987; Leemans & Prentice 1987). This dip has been attributed to increased growth of trees attaining canopy position (Mitscherlich 1970; Koop & Hilgen 1987), causing trees to grow rapidly out of these size classes.

The size of the forest and the spatial scale of heterogeneity or homogeneity can confound the interpretation of diameter distributions. One might expect the patterns caused by tree to tree interactions to be more obvious at smaller scales. Diameter distributions of larger areas should reflect both small-scale processes (such as competition between trees of different sizes on the same small plots) and large-scale processes (such as the pattern or frequency of the disturbance regime).

Individual-tree-scale processes and forest simulation models

Tree–environment feedback

Interpretation of diameter frequency data requires at least an indirect consideration of the biology of the individual trees and an understanding of the interactions between trees and their environment.

Individual trees grow large enough to alter their own microenvironment and that of subordinate or neighboring trees. The species, shapes and sizes of trees in a forest have a direct influence on the local forest environment. The environment, in turn, determines the performance of different species, shapes and sizes of trees. Thus, there is a feedback from the canopy tree to the local microenvironment and subsequently to the seedling and sapling regeneration that may become the canopy of the future.

This type of feedback is most easily quantified for the light environment. The leaf area profile and canopy geometry determine the amount and quality of light at the forest floor (Anderson 1971; Cowan 1986).

Trees also alter their local environment with respect to the proportion of throughfall (Zinke 1962; Helvey & Patric 1965), soil moisture (Swift, Heal & Anderson 1979), rates of litterfall, litter decomposition and mineralization affecting soil nutrient availability (Zinke 1962), and other factors.

The importance of individual-tree-scale interactions in understanding the analysis of forest structure at all scales implies that these interactions must be considered in any attempt to project the dynamic response of forest ecosystems to global environmental change.

Individual-based models represent these interactions explicitly and form the basis of a 'special theory of forest dynamics' (Shugart 1984) that helps to explain and predict the response of forests to natural and unnatural disturbance and environmental change (Prentice 1986a).

Individual-based forest simulation models

Computer models that simulate the dynamics of a forest by following the fates of each individual tree were developed in the mid-1960s. The earliest was by Newnham (1964) and many others followed. They all involved using a computer to update a map of the sizes and positions of each tree in a forest (Shugart 1984; Prentice et al. 1990). Such models have been used increasingly as the computer power available to ecologists has increased (Munro 1974; Shugart & West 1980).

Huston, DeAngelis & Post (1988) pointed out that one major advantage of such models is that two assumptions implicit in the conventional approach to modeling population dynamics are rendered unnecessary. These assumptions are:

1. that unique features of individuals are sufficiently unimportant that individuals can be assumed to be identical; and
2. that the population is perfectly mixed so that local spatial interactions can be ignored.

These assumptions are particularly inappropriate for trees, which are sessile and vary greatly in size during their life span.

A class of individual-tree-based models that has been widely used in ecological (as opposed to forestry) applications is the so-called 'gap' model family (Shugart & West 1980; Shugart 1984). The global boreal forest model presented in this book is derived historically from this class of models; we will therefore discuss gap models in more detail.

Gap models

Individual-tree-based forest gap models simulate the establishment, growth and mortality of trees, usually on an annual time-step and on a plot of defined size (c. 0.1 ha). This approach has been applied to a wide variety of forested systems around the world (Fig. 12.1). Many of the models indicated in Figure 12.1 are variants of the FORET model (Shugart & West 1977). Shugart (1984) provides more detailed discussions of these models, their derivation and their testing.

The FORET model was originally derived from the JABOWA model (Botkin, Janak & Wallis 1972) which, in turn, is an ecological version of the earlier forestry models based on individual tree dynamics. The JABOWA model is, however, not spatially explicit as were the earlier forestry models, and features relatively simple protocols to estimate the model parameters for as many species as are needed to simulate the forests of a particular region. There is a considerable body of information on the performance of individual trees (growth rates, establishment requirements, shade and nutrient-stress tolerance, height–diameter relations and so on) that can be used directly in estimating the parameters of gap models, and various approximations have been developed to estimate values of less easily available parameters, for example leaf area allometry; the approximations are acceptable given that gap models aim for greater generality and less precision than most forestry models (Prentice & Helmisaari 1990).

Although the models differ in their inclusion of processes which may be important in the dynamics of particular sites or regions (e.g. hurricane disturbance, flooding), all forest gap models, including the model described in this book, share a common set of characteristics as follows.

1. *Model structure*
Each tree is modeled as a unique entity with respect to the processes of establishment, growth and mortality. Thus, the models have appropriate information to allow computation of species- and size-specific demographic effects.

2. *Spatial scaling*
Each individual on the plot influences and is influenced by the growth of all others on the plot. In other words, there is a formal assumption of horizontal homogeneity inside the simulated plot. This makes the size of the simulated plot critical. The plot area corresponds approximately to the zone of influence of a single individual of maximum size (Shugart & West 1979). This definition allows an individual growing on the plot to achieve maximum size while at the same time allowing the death of a large individual to significantly influence the resource environment on the plot. The magnitude of this influence for light competition depends on the relation between the crown size of the largest individuals and the plot size (Prentice & Leemans 1990). In boreal forests at high latitudes the light

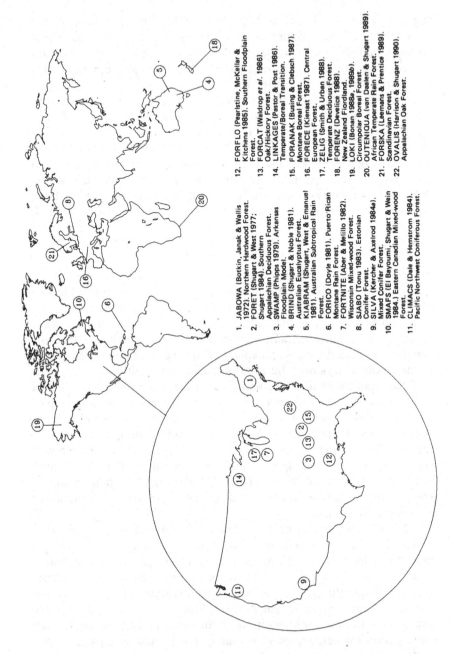

1. JABOWA (Botkin, Janak & Wallis 1972). Northern Hardwood Forest.
2. FORET (Shugart & West 1977; Shugart 1984). Southern Appalachian Deciduous Forest.
3. SWAMP (Phipps 1979). Arkansas Floodplain Model.
4. BRIND (Shugart & Noble 1981). Australian Eucalyptus Forest.
5. KIABRAM (Shugart, West & Emanuel 1981). Australian Subtropical Rain Forest.
6. FORICO (Doyle 1981). Puerto Rican Montane Rain Forest.
7. FORTNITE (Aber & Melillo 1982). Wisconsin Mixed-wood Forest.
8. SJABO (Tonu 1983). Estonian Conifer Forest.
9. SILVA (Kercher & Axelrod 1984). Mixed Conifer Forest.
10. SMAFS (El Bayoumi, Shugart & Wein 1984.) Eastern Canadian Mixed-wood Forest.
11. CLIMACS (Dale & Hemstrom 1984). Pacific Northwest Coniferous Forest.

12. FORFLO (Pearlstine, McKellar & Kitchens 1985). Southern Floodplain Forest.
13. FORCAT (Waldrop et al. 1986). Oak/Hickory Forest.
14. LINKAGES (Pastor & Post 1986). Temperate/Boreal Transition.
15. FORANAK (Busing & Clebsch 1987). Montane Boreal Forest.
16. FORECE (Kienast 1987). Central European Forest.
17. ZELIG (Smith & Urban 1988). Temperate Deciduous Forest.
18. FORENZ (Develice 1988). New Zealand Fiordland.
19. LOKI (Bonan 1988a, 1989b). Circumpolar Boreal Forest.
20. OUTENIQUA (van Daalen & Shugart 1989). African Temperate Rain Forest.
21. FORSKA (Leemans & Prentice 1989). Scandinavian Forest.
22. OVALIS (Harrison & Shugart 1990). Appalachian Oak Forest.

Fig. 12.1. Published individual-organism-based gap models.

angles are relatively flat, even at the height of the growing season, and the individual trees are typically narrow-crowned; these features mean that an individual tree's shading potential is relatively small, with important consequences for the forest dynamics (Bonan, this volume, Chapter 15; Leemans, this volume, Chapter 16).

Although horizontal homogeneity on the plot is assumed, the vertical structure of the canopy is modeled more explicitly. The sizes of individuals (height and leaf area, which are related allometrically or dynamically to diameter) are used to construct a vertical leaf area profile. Using a light extinction equation, the vertical profile of available light is then calculated so that the light environment for each individual can be defined. Satisfactory simulations of boreal forest dynamics have been obtained with a simple representation of individual tree crowns in which leaf area is uniformly distributed between the tree height and the crown base (Leemans & Prentice 1987; Prentice & Leemans 1990).

3. Environmental constraints and resource competition

All gap models simulate response of individual trees to the vertical profile of light availability on the plot. Other resources and environmental constraints are incorporated with varying degrees of realism in different models; these other constraints include soil moisture, fertility (either as a generic soil variable, or as an explicit nitrogen cycle), temperature, and disturbances such as fires, hurricanes, floods and windthrow.

The responses of trees are modeled on the principle that a tree of a given species has a certain potential diameter increment, survivorship or sapling establishment rate under optimal conditions, and that this potential is reduced by applying a series of dimensionless multipliers (normally taking values between 0 and 1), one for each type of constraint, derived from response curves. Species are often categorized into a limited number of functional classes (Fig. 12.2(a)–(d)), for example shade-tolerant versus shade-intolerant, and generic response curves assigned to all members of a class.

Competition depends on the relative performance of different trees under the current environmental conditions on the plot. These conditions may be influenced by the trees themselves (e.g. a tree's leaf area influences light available beneath it), or may be considered as extrinsic (e.g. air temperature). Competition for light is asymmetric and exploitative (a tree at a given height absorbs light and reduces the resource available to trees at lower positions in the canopy). In most gap models competition for below-ground resources (water and nutrients) is symmetric and non-exploitative (each tree experiences a resource level common to the plot, and this resource is not depleted in use, although some models now include more realistic formulations of below-ground

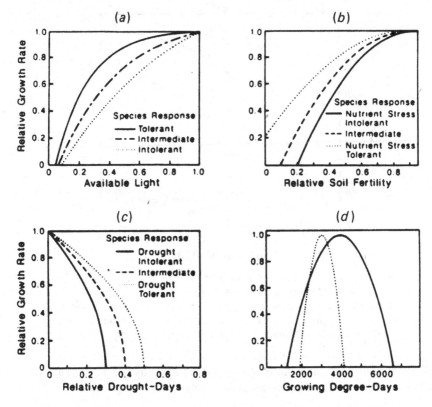

Fig. 12.2. Response functions used to modify annual diameter growth of trees for the environmental constraints of (a) available light, (b) soil fertility, (c) soil moisture, and (d) temperature. Most species can be assigned to response categories for light, nutrients and moisture; temperature responses are based on geographic ranges of species.

competition). Competitive ability depends strictly on the context of the modeled gap: the tree that has the best performance – relative only to other trees on the plot – is the most successful. Competitive success thus depends on the environmental conditions on the plot, which species are present, and the relative sizes of the trees; each of these varies through time in a realistic manner.

4. Growth
The growth of an individual is calculated using a function that is species-specific and predicts, under optimal conditions, an expected growth increment given a tree's current dimensions (Fig. 12.3). Actual growth is

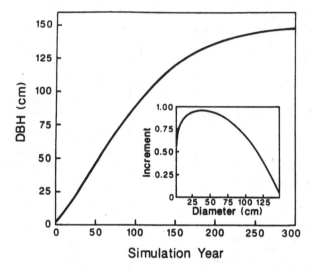

Fig. 12.3. Annual diameter growth as simulated in gap models: (*a*) optimal growth through time, and (*b*) (inset) transformed to predict potential annual diameter increment based on current diameter. Potential increment is multiplied by environmental response functions associated with environmental limitation to growth.

then computed via environmental response functions as described above, and the realized increment – an annual ring – is added to the tree.

5. *Mortality*

The death of individuals is modeled as a stochastic process. Most gap models have two components of mortality: age-related and stress-induced. The age-related component applies equally to all individuals of a species and depends on the maximum expected longevity for the species; this age is often of the order of 300 years, yielding an annual mortality rate of around 1–2%. Stress is deemed to occur in any year when a tree fails to reach a certain minimum diameter increment or growth efficiency (typically 10% of optimal). Stressed individuals are subjected to an elevated mortality rate.

Waring (1983) has discussed the use of growth efficiency (the ratio of stem volume increment to current leaf area) as an index of tree vigor, and presented evidence showing an inverse relationship between growth efficiency and the risk of succumbing to a pest outbreak. Prentice & Leemans (1990) found that the use of growth efficiency rather than diameter increment as an index of stress allowed an improved simulation of the structure of a boreal forest in which the understorey was populated by slow-growing shade-tolerant saplings.

6. Establishment

Several authors (van der Pijl 1972; Whitmore 1975; Grubb 1977; Bazzaz & Pickett 1980) have discussed species attributes that are important in differentiating the regeneration success of various trees. The complexity of the regeneration process in trees and its stochastic nature makes it nearly impossible to predict the success of an individual tree seedling. Most gap models are designed to treat regeneration in trees from a pragmatic view that the factors influencing the establishment of seedlings can be usefully grouped in broad classes (Kozlowski 1971a,b; van der Pijl 1972; Grubb 1977; Denslow 1980). Tree establishment and regeneration are stochastic, with maximum potential establishment rates constrained by the same environmental factors that modify tree growth. In each simulation year a pool of potential recruits is filtered through the environmental context of the plot, and a few new individuals are established. The relatively small number of individuals on any one plot combined with the stochastic nature of the demographic processes means that different plots can develop in somewhat different ways through accidents of mortality and recruitment, and these differences can be amplified during development. Thus, tests and applications of gap models focus on the average behavior of an ensemble of plots.

Gap models include three key features for a dynamic description of vegetation pattern: (1) responses of individual plants to the environment; (2) how individuals modify their environment; and (3) how accidents of mortality and establishment are amplified through the nonlinear processes of plant–environment feedbacks. The models are hierarchical in the sense that higher-level patterns (i.e. patterns at the population, community and ecosystem levels of organization) are predicted by integrating individual plant responses to environmental constraints. Applications of gap models illustrate the 'emergence' of higher-level patterns that are thus shown to be predictable, but not trivially so, from plant-level processes (Prentice & Leemans 1990).

Gap models are well suited for examining vegetation response to changing environmental conditions because the expression of plant response to the environment is not limited to reproducing present-day vegetation patterns. The individual-based approach also provides a link between more detailed ecophysiological models and the larger-scale expression of plant responses.

General issues in model scaling

Scale incompatibilities and 'stiff system' problems

Current discussions of biospheric dynamics and global ecology (Risser 1986) come at a time when there is renewed interest in time and space scales in ecological systems. An appreciation of scale is a prerequisite to unifying the dynamics of ecosystems on the land surface. It becomes particularly important to understand how particular causal factors came to be predominant at particular scales of observation, as discussed in several reviews (Delcourt, Delcourt & Webb 1983; Pickett & White 1985; Prentice 1986*a*,*b*). Hierarchy theory (Allen & Starr 1982; Allen & Hoekstra 1984; O'Neill *et al.* 1986; Urban, O'Neill & Shugart 1987) provides a theoretical underpinning for these studies.

Coupling models at different scales is a difficult problem. In many fields using dynamic models, so-called 'stiff system' problems, in which the time constants for different processes span several orders of magnitude, can exceed the capacity of modern computers. Ecosystem modeling can easily produce intractably stiff systems of equations, for example if fast processes that influence production and decomposition (biophysical responses of microbial growth and soil chemical kinetics, leaf surfaces with response times of seconds to minutes) are coupled with slow processes (tree mortality, soil organic matter turnover in a single dynamic model). Stiff system problems are typically solved by identifying subsystems of 'fast' and 'slow' variables that can be evaluated separately. In ecological modeling this same procedure is applied in an *ad hoc* manner in the choice of processes included in or excluded from a given model.

The problems of integration across scales become even more crucial when the system of interest becomes global and is expanded to include interactions of the biosphere with the atmosphere and oceans. Fig. 12.4 was developed by a NASA (National Aeronautics and Space Administration, USA) conference on the topic of scale considerations in interfacing climate models with ecological models (Rosenzweig & Dickinson 1986). This diagram was developed partly in response to climate modelers' interest in having dynamic models for the response of vegetative canopies at a space scale (*c.* 100×100 km^2) that greatly exceeds that of physiologically based models (indeed, this is on the outer fringe of the spatial domains of even the larger-scale ecological models), while being on the timescale of many physiological models (minutes to hours). Global change research is thus increasingly posing modeling problems for ecologists that are in unfamiliar parts of the space and time domain.

'Scaling up': principles and problems

Common to many of these problems is the idea of 'scaling up'. We would like to communicate, in some quantitative and relatively direct fashion, the outcome of a model at a relatively fine space- or timescale (corresponding to our understanding of relevant processes) to a coarser scale required for large-scale evaluations or in order to effect a coupling to larger-scale processes. In many physical systems scaling up can be done in a straightforward fashion because one can demonstrate analytically that certain terms of the equations can be ignored at larger scales of time or space. An excellent example is provided by the relation between the Penman–Monteith equation for evapotranspiration, in which the water vapor flux can be largely controlled by the canopy conductance (which depends on the canopy leaf area index and the stomatal conductance of the leaves), and coarser-scale evapotranspiration models such as the Priestley–Taylor equation in which the flux is determined by net radiation and temperature. It can be shown that the degree of dependence of evapotranspiration on surface characteristics is reduced as the spatial scale of measurement increases, the dependence being related to a 'coupling coefficient' that can be computed from the aerodynamic roughness of vegetation (G. S. Campbell 1977; Jarvis & McNaughton 1986).

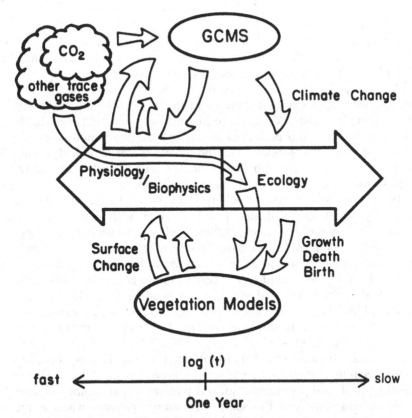

Fig. 12.4. Links among atmosphere–biosphere modeling efforts (by H.H. Shugart from Rosenzweig & Dickinson (1986)).

Unfortunately, the problems of scaling up from more complex models cannot usually be resolved so elegantly, but similar principles can often be applied in a more approximate way. Three important classes of problem include:

1. *Numerical problems*
Propagation of error, non-constancy of conditions assumed to be constant over relatively short (or small) measurement intervals, or rounding errors when solving equations with small time constants over long intervals can all cause numerical instability and unreliability when attempting to scale up, for example when scaling up mechanistic physiological models to large space and time scales. Gap models resolve this type of problem, with respect to tree physiology, by making use of simplified, robust phenomenological equations to represent physiological processes.

2. *Initial condition sensitivity and chaotic behavior*
It is now well established that many (perhaps most) systems of nonlinear differential equations show 'chaotic' behavior that is effectively unpredictable, in that slightly different starting points diverge to quite different trajectories. For example, starting with two sets of conditions describing the state of the atmosphere that are so similar as to be indistinguishable by measurement, the equations describing the fluid motion of the atmosphere have the property that the predicted atmospheric dynamics diverge over time. This insight makes the long-term prediction of weather appear to be impossible, except in a statistical sense.

Individual-based forest models typically show chaotic dynamics at relatively small spatial scales (commensurate with individual tree crowns). Because of the strong positive feedback between light competition and growth (big trees have an advantage that is progressively amplified through time) and the abrupt local changes that attend the death of large trees, slight differences in initial conditions and in the pattern of establishment and mortality simulated for a particular plot are amplified into large differences in the trajectory of succession on that plot. Thus, gap models should not be expected to predict the long-term dynamics of a small plot of forest, although the average of many simulations may predict the behavior of a larger area. This aspect of gap models is quite realistic; there are good reasons to expect that replicate small plots of land, even under ideal conditions of environmental similarity, would not show the same successional pattern, whereas the average behavior of an ensemble of plots is likely to be more consistent (Whittaker & Levin 1977).

3. *Transmutation across scales*

O'Neill *et al.* (1986) used this term to describe the tendency for appropriate representations of processes to be changed when viewed from a different point of reference. Fig. 12.5 (from data in Pastor & Post (1986)) shows the proportion of trees in different strata and in stands of different ages limited in their growth by light, water, temperature and nutrients. The result was obtained by a gap model from an open (bare ground) condition to a relatively mature forest. An interesting feature of this result is with respect to Liebig's so-called 'Law of the Minimum' when applied across scales. Liebig's law states that the most limiting factor will control growth. If one were to apply the law to individual plants (as could

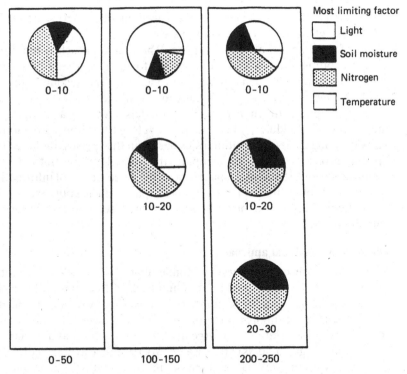

Fig. 12.5. Most limiting factor in a simulated successional sequence as a function of tree height and time. The simulated successional sequence is based on results from an individual-tree-based model of northern hardwood forest. The simulation is initiated at year 0. The sections of each circle are allocated according to the proportion of individual trees having the indicated factor as the most limiting factor to tree growth. Numbers under circles indicate height range. Figure redrawn from Pastor & Post (1986).

be done with the data in Fig. 12.5), the dynamics at the level of the entire canopy would not follow the law because some parts of the canopy would consist of trees limited by one factor and some parts would consist of trees limited by other factors. The point here is not to defend Liebig's law or to speculate at what level (if any) it should be applied. The point is that no such 'law' can work both at the individual level and at the canopy level. Indeed, any number of rules for allocation or optimization have this property. An optimal solution for the shape, function or cost or a part of a system can rarely be expected to conform to the shape, function or cost of the part when the entire system is optimized by the same criteria.

Scaling up ecosystem models for global application

We consider three approaches to the questions of how to scale up the results of individual-tree-based ecosystem models for large-scale evaluations of the impacts of global change. The first approach is the direct one, exploiting today's computer power to apply such models more extensively than has been attempted in the past. The second approach involves constructing computationally much simpler models incorporating key features of the individual-based models, either by a 'mapping' of the output of individual-based models into sets of transition probabilities, or by deriving a set of transition rules based on the species' biology. The third approach consists of mapping the model output into a low-dimensional environmental 'space' and using a geographical information system to display the results in physical space. Each approach has its distinctive strengths and they may all be pursued, separately or in various combinations.

Direct computational approach

Gap models are computationally demanding (1) because each plot is represented by explicit simulation of hundreds of trees, and (2) because of the need to simulate large numbers of plots for any one location. But with the coming of supercomputers, we can now realistically aim to carry out gap model simulations for large numbers of locations and scenarios.

In addition, gap models can be scaled up to simulate average landscape dynamics without any large increase in computational demands by applying a Monte Carlo approach (Prentice *et al.* 1990). Traditionally, gap model simulations have been performed with a series of environmentally identical replicate plots and the results averaged to produce a consistent estimate of stand dynamics. In the Monte Carlo approach, the variability among replicates is deliberately confounded with other aspects of landscape variability: environmental constraints (such as differences in slope, aspect and soil water capacity) and time since disturbance. Thus, instead of being initially identical, the plots are assigned fixed values of environmental variables to represent a statistical sample of the landscape. The disturbance regime is modeled as an additional stochastic process, destroying or modifying the vegetation on a plot with a probability that increases with time since the previous disturbance.

This modeling protocol allows for the 'emergence' of landscape-scale properties from tree- and plot-scale mechanisms. Unlike the dynamics of single plots, the dynamics of the simulated landscape in a constant environment tends towards equilibrium, typically within a few hundred years (I. C. Prentice, M. T. Sykes & W. Cramer, unpublished). Landscape simulation by this method is a useful tool for evaluating the generalized reponse of regional forests to scenarios of global climatic change.

Yet even with supercomputers, it is a tall order computationally to model a geographic array of landscapes across an entire biome. Also, the simple Monte Carlo procedure does not provide a direct way to incorporate large-scale spatial processes that are important in the boreal forest, such as wildfire, pest outbreaks and long-distance migrations of species in response to climatic change. Various simplifications and short cuts will therefore also be needed for global change applications, especially if large-scale processes are to be incorporated by embedding the landscapes in a geographic grid (Prentice *et al.* 1990).

One possible simplification includes increasing the computational efficiency of gap models. Fulton (1990) has shown that this is possible by treating individual trees not as unique entities, but grouped into structural classes. This approximation preserves the key features of gap models, including the interactions of growth and demography, while losing little in precision and gaining by two orders of magnitude in computation speed (M. Fulton, personal communication). Still further simplification is possible by abandoning mechanistic computation of plot dynamics and instead using Markov models as approximate analogs, incorporating generalized information on species replacements derived from gap model experiments or from simplified theory, as we discuss next.

Markov analog approach

Markov models of plant community dynamics are constructed by determining the probability that the vegetation on a prescribed area will be in some other vegetation type after a given interval. Thus, they require a scheme for classifying the vegetation into identifiable categories.

The manner in which the vegetation states are classified has varied among applications. Horn (1975*a,b*, 1976) used the species of canopy tree to define the states of a model developed for a temperate deciduous forest. The time-step of this model was the average generation time of the canopy trees. Waggoner & Stephens (1971) categorized the forest types according to the most abundant species (in terms of individual trees over 12 cm diameter at breast height (DBH)) on 0.01 ha plots and applied a Markov model over uniform time intervals. A spatial Markov model has been developed to simulate the dynamics of heathland vegetation (van Tongeren & Prentice 1986) and a transect Markov model has been implemented to simulate the vegetation dynamics of coastal dunes (Shugart, Bonan & Rastetter 1987).

Multivariate state classification schemes can also be used. For example, a small plot of land can be characterized by both the species of the largest individual and the number of individuals (e.g. highly stocked white oak dominated type, understocked loblolly pine stands, etc.). Hool (1966) used this approach in developing a Markov model of stand change over a large area.

In a Markov model, the number of model parameters is a function of the square of the number of states (or categories in the model). Thus, in the development of a Markov model, one is forced to trade off between the increased resolution in being able to enumerate many different system states and the parameter estimation problems that attend this greater resolution. Relatively uncommon transitions from one state to another need to be estimated with equivalent precision to that of the other more common transitions. This feature creates a need to observe the frequency of occurrence of rate transitions between states and implies large remeasurement data sets.

An alternative to direct measurement is to develop theoretical constructs that allow the estimation of the model parameters on some other basis. For example, Horn (1975a,b, 1976) somewhat simplistically assumed that the proportion of trees of a given species found growing below a canopy tree indicated the transition probabilities. Noble & Slatyer (1978, 1980) developed the 'vital attributes' concept, which uses regeneration, response to disturbance and longevity of plants to determine the parameters of a Markov model. Such devices make it possible to develop large Markov models (e.g. Kessell 1976, 1979a,b; Cattelino et al. 1979; Potter, Kessell & Cattelino 1979; Kessell & Potter 1980). Theories on Markov model structure and parameters based on biological attributes of species, as in the vital attributes approach, can be designed to be compatible with individual-organism-based simulators (Ian Noble, personal communication). For global applications, one 'brute force' method for applying results from an individual-tree-based simulation model would be to use the output from the model to estimate the parameters for a Markov model. A second, more subtle, approach would be to interpret the individual-tree-based model parameters in the context of a 'vital attributes' modeling approach to obtain a Markovian representation of the species mixture. Since both these approaches produce a Markov model, good practice would probably involve applying both techniques and then checking the resulting models for consistency. A Markov model, so derived, could then be used to simulate the expected regional dynamics. As Shugart, Crow & Hett (1973) pointed out, one can also represent a first-order Markov model as a system of linear ordinary differential equations, if the interest is in the average behavior of the vegetation expected for a large area.

Such Markovian models (or their linear differential equation equivalent) would, however, necessarily be simplifications of any individual-based model on which they were based, and as such they might not have the ability to reproduce the total range of dynamics in the base model. In particular, nonlinear dynamics such as hysteretic responses to environmental change or multiple stable equilibria would not be found in this class of models.

The vital attributes approach of Noble & Slatyer (1978, 1980) has historical antecedents in Humboldt (1807), Grisebach (1838) and particularly Warming (1909) and Raunkiaer (1934). It is philosophically allied with Gleason's (1927) emphasis on the biological properties of available plants as determinants of succession. The use of vital attributes to predict temporal patterns is analogous to the use of physiognomic characteristics to predict spatial patterns in vegetation (e.g. Raunkiaer 1934; Box 1981), an idea which inspires a third methodology for scaling up individual-based models.

Gradient substitution approach

The most obvious means of encoding information on large- or global-scale patterns of forests is a map. 'Natural vegetation' maps (Kuchler 1978) can be used to determine the potential vegetation of a region and, given a familiarity with a particular region, provide a baseline against which to evaluate changes in the vegetation. This conceptually simple approach has been used to indicate the potential effects of global climate change. For example, Emanuel, Shugart & Stevenson (1985a) used data from 7000 weather stations to extrapolate an expected world vegetation map based on the Holdridge (1967) life-zone prediction algorithm. Emanuel, Shugart & Stevenson (1985a) then modified the 7000 weather records according to the changes in temperatures predicted for each location by the Manabe & Stouffer (1980) General Circulation Model-simulated climate response to a doubling of CO_2.

There can be striking similarities between the ecological systems at different points in space that have similar geological and climatic features. This observation permeates the nineteenth century publications of the great German plant geographers from Von Humboldt through Drude and Graebner to Warming and Schimper, and forms the basis of several geographers' algorithms for mapping climatic types and potential vegetation (e.g. Koppen 1931; Thornthwaite 1948; Troll 1948; Holdridge 1967; Walter 1971; Box 1981).

In the same fashion that one might apply a gap model along an environmental gradient by supplying a range of site conditions and observing the degree to which the vegetation simulated by the model matches that found at equivalent points along the gradient, one can simulate the expected vegetation for a set of environmental conditions as specified in a geographer's algorithm. Given sufficient environmental data in a geographic information system, the resulting nomogram can be used to produce maps of expected vegetation in the form of simulated abundance of different tree taxa. The principal difference between this approach and the traditional geographers' approach is the former's ability to predict not only the equilibrium vegetation, but the transient or successional vegetation types as well.

Conclusions

We have attempted to provide a context for the use of simulation models based on individual-tree-scale as a basis for large-scale to global evaluations. As is clear from the other chapters in this book, there are important phenomena in the boreal forest that are, because of their spatial and temporal scales, not easily incorporated into such models. There are also important phenomena that probably should be included in future developments of models such as the one documented in this book. There must be continuing development of modeling and observational work in the boreal forest. The greatest value of having a base model of the boreal forests of the Earth is as a benchmark for this future work.

Several large issues must be addressed in future research. Important among these are:

1. An increased understanding of the transition between boreal forests and other biomes and the associated problem of understanding multiple life-form competition along complex environmental gradients.

2. Development of ecological studies and a hierarchy of consistent models with nested space- and timescales to allow interpretation of strongly scale-disconnected phenonena in the boreal forest. Two important problems in this area involve the coupling of plant physiology at a 'mechanistic' scale with models simulating community dynamics, and the inclusion of large-scale spatial phenomena (particularly wildfires, insect outbreaks and species migration) in ecological models.

3. Model testing, in the present case and against paleoecological data. The models discussed in this book have already been tested extensively, but continued iteration between model tests, field experimentation and data collection, and model improvement is important and should be highly productive.

We hope that a model of an entire biome, developed through co-operative efforts of a community of scientists, will prove as valuable as a focus for discussion as it is as a predictive tool for evaluating forest response to global environmental change.

13 Population-level models of forest dynamics

M. D. Korzukhin and M. Ya. Antonovski

13.1. Introduction

It has been a long-standing tradition in mathematical ecology to use dynamical equations in the modeling of population dynamics. These equations are intended to describe the ecological mechanisms ruling the observed dynamics of a system, and radically excel formalized descriptions such as regression formulae, which feature widely in the modeling literature, including that for forest modeling.

The various forms of forest (or stand) dynamics models can be subdivided, for our purposes, into three types (for a full review and details see Shugart (1984)):

(a) Models with zero-level averaging over a tree population. They are based on the dynamics of individual trees with individual coordinates: spatial models in the exact sense of the word (Newnham & Smith 1964; Ek & Monserud 1974; Mitchell 1969; West 1987; Gurtzev & Korzukhin 1988).

(b) Models with an intermediate degree of averaging. They operate by considering individual trees but without their spatial location (Botkin, Janak & Wallis 1972; Shugart & West 1977; Shugart 1984; Huston & Smith 1987; Leemans & Prentice 1989), and form the family of so called 'gap models' (see Shugart & Prentice, this volume, Chapter 12).

(c) Models with the maximum degree of averaging. They operate by considering groups of trees–parts of a whole population – chosen depending on the particular task (usual subdivisions are by size and/or by age) (Ek & Monserud 1979; Kapur 1982; Korzukhin, Sedych & Ter-Mikaelian 1987, 1988; Tait 1988; Korzukhin, Antonovski & Matskiavichus 1989; Prentice *et al.* 1989; Kohyama 1989).

Usually, different modeling techniques are more effective for solving one type of task and less effective for others. Approach (a) is suitable for even-aged stands, for detailed investigation of stand structure, and for testing of different models of competition. As a rule, it demands much input information and is too bulky for parameter estimation by means of comparison of the model and empirical behavior. Intermediate case (b) is the most appropriate when the stand under consideration is multi-aged and composed of many (say, three or more) species. It is important to note that whereas models of types (a) and (b) are purely imitative, models of type (c) sometimes allow analytical approaches because they have a form of dynamical equations.

Approach (c), a review and examples of which are given in this chapter, has been somewhat moved aside by the individual-tree technique. However, we believe that this classical approach initiated by Lotka and Volterra, is complementary to those of (a) and (b), and has not exhausted its potential in theoretical and applicative tasks (for a contrasting viewpoint see Huston & Smith (1987) and Huston, DeAngelis & Post (1988).

There are some fields of forest dynamics where dynamical equations are indispensable. For example, the dynamics of grass and moss populations as part of the forest ecosystem cannot really be described at the individual-plant level; the same is true for all non-living components of forest ecosystems, such as dead organic material, nutrients, etc. In many other fields, dynamical equations need not be used but may be appropriate and helpful. For example, the inevitable stochastic element in population dynamics when an individual-tree approach is adopted leads to difficulties in parameter estimation: the majority of calculative methods demand determination of parameter derivatives that are almost impossible to find in a stochastic system or can be achieved only with low accuracy. Dynamical equations, however, are ideally suited for the calculation of the small trajectory modifications under small parameter variations that are needed for finding the derivatives.

It is clear from these general modeling considerations that this technique can be most successfully applied for comparatively simple population systems. The technique results in numerical solving, effective parameter estimation and – the most attractive theoretical aim – possible analytical results. Further, many boreal forest ecosystems are composed of a small number of tree and above-ground plant species, so the use of dynamical equations in this context can be of applied interest. In contrast, dynamical models for multi-species tropical and deciduous forests consist of many equations and parameters, so the deterministic behavior of tree number trajectories becomes so complex that it is near-stochastic. As a result, the advantages of a simple description by means of a dynamical equation technique over an individual-tree approach cease to exist. Besides the aim of the review and development of population level models in forest dynamics, we have attempted to illustrate how, in our opinion, the approach to modeling should develop. At present, modeling appears to be more akin to art than science with each model being more or less specific, and each modeler choosing his own empirical rules of advancement. The view we advocate can be expressed as 'Try to use simplicity before sinking into complexities'. For example, if the requirement is to analyze the behavior of the overall population-level characters only these, if possible, should be used as state variables, without size or age distributions, individuals, etc. If single-level, horizontal crowns give reasonable results, these should not be changed to vertically distributed ones. If a single-variable equation of trajectory of tree growth is required as output, use a single-variable equation of growth until it gives us what we want. The model should be made more complex only if it does not provide the required output. Many examples of unnecessarily complicated approaches are found in the forest and ecology modeling literature.

In general, much modeling is performed on a 'trial and error' basis. Usually, the ecological mechanisms and the level of detail that is to be included in order to achieve the required information on a population or a forest system are not known. As yet, no attempts have been made to answer this basic question, although the whole issue is central to any modeling. Only when the necessary and sufficient relationships are known can one affirm that property A is explained by property B, that is, B is the reason for A's origin. Only a model constructed on this basis gives 'explanation' and can produce accurate and reliable results; all others are incomplete to some degree.

Forest modeling is currently in an active phase of its development. Both the expectations and the resources involved are high, and we believe that a certain degree of methodological accuracy should be sought in order to optimize output and to make the results more reliable.

13.2. Major divisions of forest modeling

The problems of the formulation of adequate forest models are in principle similar for individual-tree and dynamical equation techniques. In order to construct any model of a stand one should describe, for multi-aged populations: (1) the free growth of one tree; (2) the mechanism of competition between trees; (3) the dependence of the viability of one tree upon available resources and environmental factors; and (4) conditions necessary for seed viability. Of these points (1) and (2) are commonly required for individual-tree and dynamical equation techniques; (3) and (4) can be embodied in either deterministic or stochastic forms, thus giving the two discussed approaches to forest modeling.

Given these major divisions, in various forms and details, the modeler should be able, in principle, to pick out a model with desirable properties. However, at present we are some way from this ideal situation; there are still relatively few basic models and these have not been efficiently tested to allow them to be used in such a way.

13.2.1. Tree growth

The most physiologically correct and apparently most widespread approach to deriving tree growth equations involves the use of the concept of carbon balance (Richards 1959; Pienaar & Turnbull 1973; Aikman & Watkinson 1980; Bichele, Moldau & Ross 1980; Jarvis, Edwards & Talbot 1981; Waring & Schlesinger 1985; Mäkela 1986; West 1987; and many others). The simplest approach uses only one variable, m (total tree biomass), so that the balance takes the form

$$\dot{m} = S_L(m) \cdot A(R) - Re(m) - De(m). \tag{13.1}$$

Here $\dot{m}$ is the biomass growth rate, S_L is total leaf area, A is specific assimilation rate, R is resources of photosynthesis, Re is respiration, and De is decay losses.

In spite of the obviousness of Eqn (13.1), it is useful to note that when we consider long-term (e.g. one year) carbon balance, the difference $S_iA - Re$ cannot be less than zero. In fact, the balancing of Eqn (13.1) does not really represent the balance of tree biomass, but the balance of tree free assimilates (plus decay): free assimilates $= S_iA - Re$ when $S_iA > Re$ and $= 0$ when $S_iA < Re$.

Since Eqn (13.1) uses only one variable for the growth description of a whole tree, S_L, Re and De are expressed as functions of m. It is common for more than one variable to be introduced (Agren & Axellson 1980; Kull & Kull 1984; Oja 1985; West 1987; Mäkela 1988; and others). In such cases, the authors are immediately confronted with the problem of assimilate distribution. Empirical difficulties are obvious, but theoretical approaches, which should be based on the optimization theory of growth, have not been satisfactorily developed.

In this area, the work by Rachko (1979) is of particular note. In this work, the roots, stem and leaves of a tree are treated as independent variables, and the tree is assumed to be growing with the requirement of maximization of biomass increment during one year. The resulting description of photosynthesis is fairly realistic. Among recent works, we can mention Mäkela (1988) who considered optimal tree height growth, and Korzukhin (1985) who introduced competition into the optimal growth task. The work by Korzukhin & Ter-Mikaelian (1987), which included tree defense, is discussed in Section 13.2.2.

Thus, we believe that as the first step in a population level approach a growth model description should take the form of a one-variable growth curve (e.g. $D(t)$ – diameter or $m(t)$), and we recommend the use of the approach shown in Eqn (13.1) for this purpose.

Models differ in the forms of $A = A(R)$, $Re = Re(m)$ and $De = De(m)$. Here we shall not consider the vast literature on corresponding models, but note only that the simplest and most widespread is the hyperbolic function $A(R_I)$, where R_I is intensity of light. However, one needs to take care in the application of this classical light curve to a whole tree growing over a long period: the mechanism of photosynthetic adaptation can make the function $A(R_I)$ almost linear (Tcelniker 1978). Respiration of growth is proportional to $S_L A$ and can be taken into consideration by a coefficient before the term $S_L A$ in Eqn (13.1). In the simplest case, other components of respiration are taken to be proportional to m (Bertalanffy 1957; Richards 1959; Pienaar & Turnbull 1973; Aikman & Watkinson 1980; and others). In more complex cases, some elements of the respiration structure are introduced (Bichele, Moldau & Ross 1980; Mäkela 1986). $De(m)$ is usually taken as being proportional to m.

Some authors (Botkin, Janak & Wallis 1972; Reed 1980; Shugart 1984; Leemans & Prentice 1989; and others) have used a semi-empirical analogue of the balance equation (13.1):

$$\dot{m} = S_L A \cdot (1 - m/m^{max}). \tag{13.2}$$

Owing to Eqn (13.1) the value of m^{max} (maximum tree biomass (the root of $\dot{m} = S_L(m)A(R) - Re(m) - De(m) = 0)$) depends on the assimilation rate A. For example, A and m^{max} are less in poor conditions. On the other hand, owing to Eqn (13.2), m^{max} is independent of A, and only the growth rate depends on A.

13.2.2. Tree viability

The problem of tree viability modeling consists (a) of defining the concrete form of function

$$Y(x), \tag{13.3}$$

that is, the probability of survival of one tree over, one may suppose, one year, and (b) of defining the list of arguments, x, of the function.

Following the usual subdivision (Semevsky & Semenov 1982; Tait 1988), viability factors are divided into two large groups. The first of them, denoted as w, is connected with stochastical, often catastrophic, natural disturbances like fire and drought (Waring & Schlesinger 1985) so $w = \{$frequencies and magnitudes of disturbances, climatic parameters, etc.$\}$. These factors are the arguments of the density-independent component of viability, $W(w)$. The second group, denoted as v, arises from

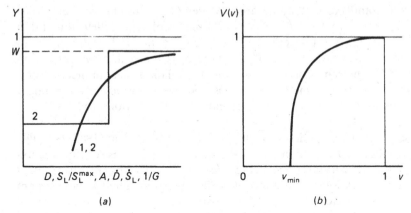

Fig. 13.1. Qualitative behavior and arguments of viability function Y for a single tree. (*a*) Empirical (1) and model (1), (2) forms following field data and models (see text for references and denotations). W is viability of mature, maximum-defensed tree. (*b*) Qualitative behavior of density-dependent viability component, V, on argument $v = A/A_{max}$; V is recommended for Base model 1 (§3.3) and used in Base model 2 (§ 4.3); $v_{min} = A_{min}/A_{max}$, where A_{min} is maintenance respiration of leaves or a whole tree.

tree–tree permanent interrelations in the community, and gives arguments of the density-dependent component of viability, $V(v)$, where $v = \{$all tree state variables and density$\}$. So we can represent x as $X = \{W, V\}$ and write

$$Y(w, v) = W(w) \cdot V(v), \qquad (13.4)$$

Below, we shall discuss the problem of $V(v)$ dependence only, which has been investigated much less than $W(w)$.

Several attempts have been made to establish the form of V on the basis of experimental data. In most cases, arguments of V used have been dimensional characteristics of the tree, its growth rate, and a parameter describing the pressure of competition. The dependence of V on these arguments is usually quite marked (Moser 1972; Monserud 1976; Glover & Hool 1979; Hamilton 1980; Buchman, Pederson & Walters 1983; Blagovidov 1984) as may be seen in Fig. 13.1(*a*). Glover & Hool (1979), Hamilton (1980), and Buchman, Pederson & Walters (1983) used the exponential function

$$V(v) = 1/[1 + e^{-\varphi(v)}],$$

where v is a set of arguments, different for each of the cited papers ($v = D, G, \dot{D}, \dot{D}/D$, competition index), and φ a linear function; here and below D is average tree diameter, N is tree number, and $G = (D^2 N)/4$

is the stand cross-sectional area. Moser (1972) found from empirical observations that $V = 1 - \exp(-aD)$, and Kohayma (1989) found that $V = 1 - aG$. For a description of empirical data, K. J. Mitchell (1969) used the argument $v = S_c/S_c^{max}$, where S_c is the real crown projection and S_c^{max} the same for a free-growing tree. This argument means that a suppressed tree 'remembers' its possible size if its growth is no longer suppressed. The same effect is proposed by Ek & Monserud (1974).

So far, the most profound approach to the problem has been the theoretical construction of Korzukhin & Ter-Mikaelian (1987) (see also Antonovski, Korzukhin & Ter-Mikaelian (1987)). Tree life function 'viability' is considered as being one of a number of activities (growth and proliferation in these models). The tree is assumed to be a system which distributes resources (free assimilates in these models) among the activities, and of high priority is the maximization of total seed production. This is proportional to accumulated leaf area and tree longevity, which in turn are determined by the amount of assimilates directed to defense. This distribution task was formalized and its solution showed that the share of assimilates directed to defense decreases monotonically with age. A trajectory of optimal growth $m(t)$ and realistic age behavior of seed production was also obtained: this was equal to zero until a certain age and thereafter increased monotonically. The results of such an analysis for a free-growing tree and for a population of even-aged trees were qualitatively close. Despite being sound from the biological point of view, this approach does not lend itself to practical applications.

More practical modeling realizations of empirical observations can be divided into discrete and smooth, the first type being appropriate to individual-tree level models and the second to all kinds of models. The preferred argument of V is usually D because of its relatively easy field measurement.

The application of a step-like function V was initiated by Newnham & Smith (1964) who used $V = 1$ when $\dot{D} >$ const and $V = 0$ otherwise. This was used by Aikman & Watkinson (1980) taking $\dot{S}$ instead of $\dot{D}$. Mäkela & Hari (1986) took $V = 1$ when $\dot{S} > 0$ and $V = 1 -$ const $\dot{S}_L/S_L$ otherwise. In a gap-modeling technique (Botkin, Janak & Wallis 1972; Shugart 1984; and many others), $V = 10^{-2/T}$ when $\dot{D} > 1$ mm yr^{-1} and $V = 10^{-0.2}$ otherwise, where T is the lifespan of the species. K. J. Mitchell (1975) has taken $V = 1$ when $NH^2 <$ const and $V =$ value sustaining $NH^2 =$ const otherwise (H is average tree height). Tait's (1988) model is the only one that uses density and growth rate simultaneously as viability arguments: $V =$ const $- N\dot{S}_L$. In West's (1987) model a tree dies when total assimilation becomes less than respiration of the leaves, $S_LA - r_Lm_L < 0$, that is equivalent to $A <$ const, because $S_L \approx m_L$.

The smoothed forms of V are also varied. Korzukhin, Sedych & Ter-Mikaelian (1987, 1988) applied in their quantitative model

$$V = (R_I/R_I^{max})^\varphi,$$

where R_I is the light available for a tree, and a very similar argument was used by K. J. Mitchell (1969) (cited above). Zemlis & Shvirta (1987) took $V = 1 - GN$, and Cherkashin (1980) took $V = v^\alpha \exp(-1/v)$, where $v = \dot{D}/D$. In their population-level model (see §13.4.3), Antonovski *et al.* (1989) and Korzukhin, Antonovski & Matskiavichus (1989) used $V = 1 - \text{const}\,(1 - v)^\beta$, where $v = A/A_{max}$ is the relative assimilation rate of a tree (Fig. 13.1b). Pukkala (1987) applied $V = 1 - G/t$ as seedling viability. The first case of a smooth viability curve $V(D)$ in gap modeling was presented by Leemans & Prentice (1989).

To conclude this short review, we note that the way forward as far as definition of viability is concerned must lie in the use of evolutionary optimization principles (Pianka 1978; Semevsky & Semenov 1982). As yet, their application has not led to algorithms for the definition of V which are reliable enough to be used in a forest dynamics model (see Korzukhin & Ter-Mikaelian (1987) and Antonovski, Korzukhin & Ter-Mikaelian (1987)).

13.2.3. Competition

Combined resource utilization leads to interaction between trees and unifies individuals in a stand, so competition can be considered as one of the two central phenomena that determine the dynamical behavior of a forest (the other is interaction connected with reproduction).

In essence, the problem consists of defining tree density-dependence of two functions: assimilation rate in Eqn (13.1), $A = A[R(N)]$, and viability in Eqn (13.3), $V = V[v(N)]$. Density can be the average number of trees per unit area, in population-level and local approaches, or several neighboring trees around the central tree in the individual-level approach.

As in the cases of tree growth and viability, there are two ways of introducing density, namely formal and mechanistic approaches. The first is widely represented in forestry modeling and involves the formal introduction of 'indexes' of competition, *CI*. When the individual-tree level is considered, *CI* are artificial functions of diameter, height, cross-sectional area, and biomass increments of neighbors of the tree in question (Newnham & Smith 1964; K. J.Mitchell 1969; Ek & Monserud 1974; Diggle 1976; Britton 1982; Smith & Bell 1983; Cennel, Rothery & Ford 1984) or of local density, that is number of trees (Plotnikov 1979), or of quite formal parameters of the central tree or its neighbors (Laessle 1965; Fires 1974; Lorimer 1983). When *CI* have been chosen, the correlation between the diameter increment of the central tree and *CI* is usually calculated, with the aim of testing the competition model.

When description is at the population level, that is dynamical equations are used, *CI* are some function of mean population variables such as $\dot{D}/D$ (Cherkashin 1980), *GN* (Zemlis & Shvirta 1987), H^2N (K. J. Mitchell 1975), *N* (Chjan & Chjao 1985), σ_L (Tait 1988), or *G* (Kohyama 1989).

The above approach is rather formal: in particular, *CI* cannot be used as an argument of *A* in Eqn (13.1) if one wants to interpret *A* in physiological terms.

The other, mechanistic, approach deals with the physical resources, *R*, available to a tree and tries to calculate their dependence upon density (taken again in two senses), that is to define the form of function

$$R(N, r), \tag{13.5}$$

where *r* are some dimensional characteristics of neighboring trees, or trees of the whole population. From general considerations it is clear that the results will depend on the type of resource, tree morphology, and spatial pattern of the trees.

Our understanding of competition for light is considerably greater than that for soil resources. Usually, a uniform distribution of tree foliage is assumed (a representation of the producing layer) that immediately leads to the application of the Lambert-Beer penetration law (Botkin, Janak & Wallis 1972, plus all gap modelers; also Ross 1975; Mäkela & Hari 1986; Oker-Blöm 1986). This law states that the amount of light at height level *h* is given by

$$R_I(h) = R_{I0} \exp\left[-\gamma \sigma^+(h)\right], \tag{13.6}$$

where γ is the extinction coefficient after transmission through one leaf layer, which can depend on *h*, $\sigma^+(h)$ denotes the amount of foliage layer above level *h*, and R_{I0} is the initial light flux.

However, the *sancta simplicita* of Eqn (13.6) is lost after the first refinement of the uniform model. Foliage in a stand is organized in tree crowns. Let us consider the question of light available to an average tree in a population (Korzukhin & Ter-Mikaelian 1982). Each tree has a horizontal monolayer crown with area S_L, and *n(h)* is the number of trees with height *h*. It is important that stem bases are distributed randomly over the plane, a requirement of Poisson's law. The extinction coefficient after one foliage screen transmission is equal to γ. By means of a geometrical probability technique, it was shown that under these conditions the classical formula (13.6) is true, where $\sigma^+(h) = \int_h^\infty S_L n(z)\, dz$. The lack of a Poisson distribution, however, violates Eqn (13.6): this law is necessary for the fulfillment of Eqn (13.6).

Another example is provided by the same system but with, say, two-layered crowns. Each layer is close enough to the other to avoid intersections with other crowns and has area S_c, so $S_L = 2S_c$. Thus, the whole-tree absorption coefficient becomes equal to $\Gamma = 1 - (1 - \gamma)^2$. Following our approach, $R_I(h)$ is given again by (13.6) but in the form

$$R_I(h) = R_{I0} \exp\left[\Gamma \sigma_c^+(h)\right],$$

where σ_c^+ is the area of crown projections above level h, which in our case consists of $\frac{1}{2}\sigma^+(h)$. Finally, we obtain

$$R_I(h) = R_{I0} \exp\left[-\tfrac{1}{2}\Gamma \sigma^+(h)\right],$$

that is the analog of Eqn (13.6) with effective extinction coefficient $\gamma' = [1 - (1 - \gamma)^2]/2$, a new violation of the uniform distribution model. Obviously k-layered crowns will give $\gamma' = (1 - \gamma)^k]/k$. The physical reason for this violation is clear: being organized in crowns, foliage screens have lost their freedom to move independently of each other, and only the whole crown (k screens together) has reserved this ability.

Several such grouping foliage models have been developed for crowns and stands (for review and references see Oker-Blöm (1986)). If the possibility of crown intersections is to be taken into account, the analytical calculation of $R_I(h)$ becomes almost impossible.

Although root competition is just as important in forest dynamics as competition for light, no one model for the former has been developed that corresponds to that for light. Most approaches do not use soil resources as explicit variables, although some do. As a rule, authors determine some 'qualitatively true' functions of total biomasses of competing trees (moss and grass if needed), represented as $B(M)$; it is usually assumed that root competition has a suppressive effect, so $dB/dM < 0$. These functions are then used as multipliers to assimilation rate function A (e.g. McMurtrie & Wolf (1983) and all gap modelers). The most recent example of this (Shugart 1984) is a multiplier for the individual-tree growth equation

$$B(M) = 1 - M/M^{\mathrm{max}},$$

where M^{max} is determined from field observations.

We can offer here a more realistic approach to the modeling of root competition, based once again on geometrical probabilities (Korzukhin 1986). Let us consider a population of identical screens, each having a thin, plane root system with area S_R. All are disposed at one level as is often the case for the boreal forest zone, so the process of competition is two-dimensional. Root systems are distributed according to Poisson's law over the plane, and from a unit area of the medium an amount R_{S0} of resources (water, oxygen, etc.) is available. These resources are divided equally among all root systems which overlap at a given point; that is, from area S_{R1}. which is not overlapped by other roots, the tree obtains an amount $R_{S1} = S_{R1}R_{S0}$ of resources; from area S_{R2}, which is overlapped by one neighbor, it obtains $R_{S2} = S_{R2}R_{S0}/2$, etc. So, the total resource available to a particular tree is

$$\sum_{i=1}^{\infty} \frac{1}{i} S_{Ri}R_{S0}, \ \Sigma \, S_{Ri} = S_R.$$

We are interested in the average amount of resource

$$R_S = \sum_{i=1} \bar{w}_i R_{S0}.$$

$\bar{w}_i$ are average areas of the ith overlapping points. It can be shown that under Poisson's law $S_{Ri}/S_R = \lambda^{i-1} e^{-\lambda}/(i-1)!$, where $\lambda = S_R N$, that is average coverage of unit area (root area index). Finally we have

$$R_S = \frac{R_{S0}}{N} [1 - \exp(-S_R N)]. \tag{13.7}$$

Magnitude R can now be used as an argument in the assimilation rate function, together with light: $A = A(R_I, R_S)$. One variant of the function was developed by Gurtzev & Korzukhin (1988) and applied in an individual-tree model of a linear pine stand; it was shown that taking into account the process of root competition improved appreciably the quality of growth description.

13.3. Even-aged monospecific stands

13.3.1. Introduction

If initially we do not take into consideration the problems connected with the migration and establishment of seeds, even-aged stand modeling consists of the same basic elements as modeling of multi-aged and multi-specific stands. In both cases one must be able to formalize the processes of competition of individual tree growth (including, perhaps, changes in morphology) and of tree mortality. However, the field data relating to even-aged stands are much more accurate and numerous than those for multi-aged ones.

So, it seems reasonable to adjust a population-level model on the basis of even-aged behavior first of all, and if this is successful, progress to multi-aged behavior. Surprisingly, this apparently obvious way of model development has not been accomplished until now; even- and multi-aged stand models have developed independently.

13.3.2. Empirical behavior of even-aged stands

The required data can be taken from numerous observations at permanent plots. We shall restrict ourselves to the level of description of a stand which uses only the average characteristics of a tree, but not tree distributions. In this case, the system is described by the following variables: tree biomass m, diameter D, height H, leaf area S_L, and seed production p (per year). The population variables are: tree number N, total biomass $M = mN$, leaf area $\sigma = S_L N$ (or close to σ is cross-sectional area G), and total seed production $P = pN$.

For the purposes of model development it is necessary to know the behavior of these variables under variation of initial density $N(0)$ and ecological (site quality) parameters φ. A brief summary of this behavior is given in Korzukhin (1986) and in McFadden & Oliver (1988).

Time behavior

Individual-tree variables have a simple form of monotonic sigmoidal functions; population variables (except N) are nonmonotonic and go through a maximum (Fig. 13.2). After crown closure, a specific system invariant arises, connected with the maximum amount of leaves, σ^{max}, which can be achieved in given ecological conditions:

$$\sigma(t) = S_L(t)N(t) \simeq \sigma^{max}. \tag{13.8}$$

Density behavior

Individual-tree variables m, p and $D[t, N(0)]$ taken at any given moment decrease monotonically with the increase in initial density $N(0)$ (although $H[t, N(0)]$ is sometimes not monotonic). Maximum differences are observed somewhere in the middle of the set of trajectories for m (Fig. 13.3).

If one considers tree number dynamics under $N(0)$, one can observe the effect of 'forgetting' of initial conditions (Fig. 13.3), that is

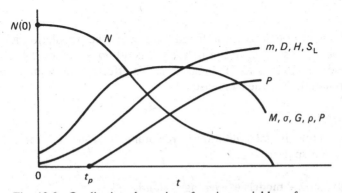

Fig. 13.2. Qualitative dynamics of major variables of even-aged systems; curves with the same type of behavior are unified; $\rho = m\,N^{3/2}$ (see text for definitions).

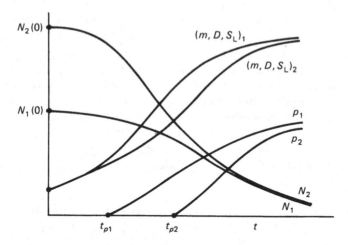

Fig. 13.3. A set of even-aged system trajectories obtained under various initial densities (see text for definitions).

$N[t, N_2(0)]/N[t, N_1(0)] \to 1$ for all $N_{1,2}(0)$. This 'drawing together' effect means that tree mortality depends upon density (and increases as density goes up).

Population variables M, P, G and $\sigma(t, N(0)]$ increase at first with increased $N(0)$. Owing to somewhat sparse data (*Results of Experimental* ... 1964; Hirano & Kira 1965; Buzykin 1970; Redko 1978), their maximum values M^{max}, etc. begin to decrease under very large $N(0)$ (Fig. 13.4). We call this effect 'overcrowding'. The time at which these variables reach their maximum values, $t_{M,G,o}^{max}[N(0)]$, decreases with increasing $N(0)$ (Fig. 13.4).

Site quality

These effects are obvious for individual trees. At the population level, tree numbers are smaller in the better site conditions (under the same initial conditions) (Fig. 13.5). In spite of this effect, total biomass and other population variables increase.

Tolerance effects

Under the same initial conditions, tree number is greater for more shade-tolerant species (Fig. 13.5).

13.3.3. Dynamical models of even-aged stands

Surprisingly, there are no even-aged stand models which are able to describe all properties of the system enumerated in Section 13.2.2; there

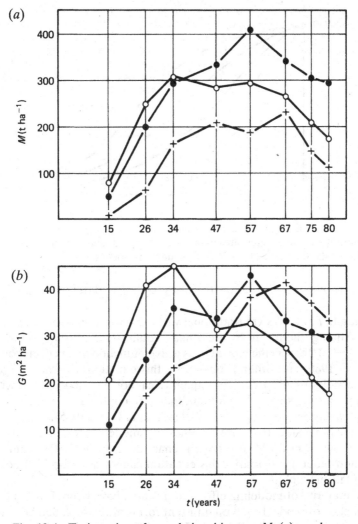

Fig. 13.4. Trajectories of population biomass M (a), and cross sectional area G (b), (which is proportional to σ) under initial densities $N_1(0) = 1000$ (pluses), $N_2(0) = 2400$ (filled circles) and $N_3(0) = 10\,000$ (open circles); data from permanent plots with *Pinus sylvestris* in Moscow region (*Results of Experimental* ... 1964). 'Overcrowding' effect consists of $M_3^{\max} < M_2^{\max}$ under $N_3(0) > N_2(0)$, whereas all $G^{\max}$ are approximately equal, so only $t_{G_i}^{\max}$ changes.

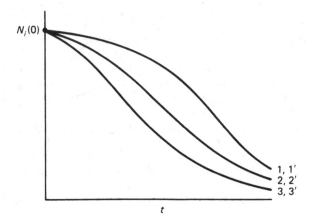

Fig. 13.5. Given the equal initial numbers of trees, N_i (0), N_i (t) falls rapidly for ameliorated site conditions ($1 \to 2 \to 3$) and with decreasing shade tolerance ($1' \to 2' \to 3'$) (qualitative behavior).

exist only some formalized constructions relating to different parts of the whole picture.

A number of works are traditionally devoted to tree number behavior only. Among them, there are linear equations $\dot{N} = -d(t)N$ where $d(t) = 1 - V(t)$ is specific mortality as a given function of age (Terskov & Terskova 1980; and others); obviously, this equation does not 'forget' initial conditions $N(0)$ and consequently cannot describe the 'drawing together' of trajectories. In addition, there are nonlinear 'Volterra-type' equations, e.g. $\dot{N} = -aN[1 - (b/N)^\lambda]$ (Chjan & Chjao 1985).

On the other side, there are 'productivity' models, operating with one variable M: $\dot{M} = aM^\alpha - bm$ (e.g. Pienaar & Turnbull 1973; Budyko 1977; and others); obviously, this equation cannot describe $M(t)$ going through its maximum.

The majority of modeling efforts in this area have been devoted to a special class of model based on different forms of the so-called '3/2 law' invariants (Reineke 1933; White & Harper 1970; Hozumi 1980; Lonsdale & Watkinson 1983; Zeide 1987; and many others):

$$\rho = yN^\alpha = \text{const.} \tag{13.9}$$

where $y = S_L$, m, D, or H. It is likely that the possible population level 'invariant' is interrelation (13.8), so Eqn (13.9) is, in fact, Eqn (13.8) rewritten in terms of other variables.

If our aim is to model the whole scenario adduced in Section 13.3.2, interrelation (13.9) appears to be very restricted. Firstly, it cannot be related to the part of the trajectory $M(t)$ which lies behind M^{max}. If we put $y = m$ (classical variant of Yoda *et al.* (1963)), then $\rho = mN^\alpha =$

$MN^{\alpha-1} = $ const; because $\dot{N} < 0$ this leads to $\dot{M} > 0$, because $(\alpha - 1)$ is greater than 0, so that Eqn (13.9) can be true only before M^{max}, where $\dot{M} = 0$. Secondly, it cannot take into consideration the important dependence of the trajectories on initial density $N(0)$, as this value is not 'remembered'.

The most developed approach of Hozumi (1980) uses, besides invariant (13.9), non-autonomous dynamical equations for individual-tree growth, and looks artificial.

All types of dynamical analysis of even-aged stands based on the 3/2 law give us a 'semi-model', which is intended to substitute the formulation and analysis of a complete nonlinear system of dynamical equations aimed at modeling the whole trajectory of the system. It seems to us that in this area there is a certain 'magic' attached to simple formulae, and that researchers have now extracted from them all that was possible.

13.3.4. Base model 1 of even-aged stands

Below we use the simplest model elements of growth, viability and competition for the composition of even-aged model 1. The aim of this model is to describe the maximum dynamical properties of even-aged stands enumerated above.

Calculations using growth equation (13.1) for free-growing trees and trees experiencing competiton show that it works well when one simply takes

$$S_L(m) = am^{\alpha}, \tag{13.10}$$

and when

$$Re(m) + De(m) = cm \tag{13.11}$$

(see, for example, papers in Richards 1959; Fires 1974; Gurtzev & Korzukhin 1988). Interrelation (13.10) can be derived from two well-established ties: $S_L \approx D^{\beta}$, $1.5 \leq \beta \leq 2$ (e.g. Mohler, Marks & Sprugel 1978; Mäkela 1986), and $m \sim D^{\delta}$, $2.5 \leq 3$ (numerous yield tables). As a result we obtain Eqn (13.10) with $0.5 \leq \alpha \leq 0.8$.

Following (13.4), total viability $Y = WV$. In the construction of V, we are working from the basis of the following propositions: (1) it should be based on a resources approach; arguments such as $\dot{D}$ are considered to be indirect reflexes of the real viability mechanism, which is in essence the use of tree resources for defense and repair; (2) we will abandon the 'memory' arguments such as $m(t)/m^{max}(t)$ for the sake of simplicity, since they demand a second growth variable (for $m^{max}(t)$ for example); (3) because photosynthesis is the central process of tree resource production we suppose that assimilation rate A should be the argument of V in Eqn

(13.4) (normalized for suitability):

$$v = A/A_{max}.$$

Following (13.1) and (13.5), A depends on N through R, that is $A = A[R(N, m)]$; when $N = 0$, $R = R^{max}$, $A = A_{max}$, $v = 1$, and $V = 1$, we obtain a free-growing tree.

Finally, we have base model 1 (specific death rate is equal to $1 - Y$):

$$\begin{cases} \dot{N} = \{1 - WV[A(R(N, m))/A_{max}]\}N, \\ \dot{m} = S_L(m)A[R(N, m)] - cm, \end{cases} \tag{13.12}$$

where $S_L(m)$ is given by Eqn (13.10), $W < 1$, $V < 1$, A is given by any model of photosynthesis, and R is given by the model of competition.

Below we adduce a simplified analytical example of the use of Eqn (13.12). Again, we consider the competition for vertically directed light among a population of Poisson-distributed trees with horizontal crowns, thinly spread over height. This gives (Korzukhin & Ter-Mikaelian 1982):

$$R_I(N, m) = R_{I0} \exp\left[-\tfrac{1}{2}\gamma S_L(m)N\right]. \tag{13.13}$$

Consider the situation when competitive interaction is weak, that is magnitude $x = \gamma S_L(m)N \ll 1$, and undertake Taylor's expansion of A near $x = 0$ and V near $v = 1$:

$$A(x) = A(0) - \left|\frac{dA}{dx}\right| x - \ldots,$$

$$V(v) = V(1) - \frac{dV}{\partial v}(1 - v) - \ldots,$$

where $A(0) = A_{max}$, $V(1) = 1$ and the argument of V is

$$v = A(x)/A_{max} = 1 - \frac{1}{A_{max}}\left|\frac{dA}{dx}\right| - \ldots.$$

Assuming that all first derivatives are not equal to zero, $|dA/dx| = a_1 \neq 0$, $|dV/dv| b_1 \neq 0$, we obtain a system of first approximation

$$\begin{cases} \dot{N} = \left[1 - W\left(1 - a_1 b_1 \frac{1}{A_{max}} x\right)\right] N, \\ \dot{m} = S_L(m)A_{max}\left(1 - \frac{a_1}{A_{max}} x\right) - cm, \end{cases} \tag{13.14}$$

Now suppose that density-independent mortality is absent, $W \equiv 1$, respiration and decay losses in m are negligible, and $a_1 x/A_{max} \ll 1$, that is growth is free. The result is an idealized system

$$\dot{N} = -rm^{\alpha}N^2; \quad \dot{m} = qm^{\alpha}$$

$(r = aa_1b_1/A_{max};\ q = aA_{max})$ and its solution under initial conditions $N(0),\ m(0) = m_0$:

$$m(t) = [(1 - \alpha)qt + m_0^{1-\alpha}]^{1/1-\alpha}; \tag{13.15a}$$

$$N(t) = \left\{\frac{1}{N_0} - \frac{m_0}{q} + \frac{r}{q}[(1 - \alpha)qt + m_0^{1-\alpha}]^{1/1-\alpha}\right\}^{-1}. \tag{13.15b}$$

It is clear that solution (13.15a) quickly forgets initial condition m_0 (mass of seed), so the formulae are simplified:

$$m(t) = [(1 - \alpha)qt]^{1/1-\alpha},\ N(t) = \left\{\frac{1}{N_0} + \frac{r}{q}[(1 - \alpha)qt]^{1/1-\alpha}\right\}^{-1} \tag{13.16}$$

We also need a generalized population variable of the form $F = m^\beta N$, which, from (13.16), is equal to

$$F(t) = \frac{r_1 t^{\beta/1 - \alpha}}{1/N_0 + r_2 t^{1/1-\alpha}} \tag{13.17}$$

$(r_1' = [(1 - \alpha)q]^{\beta/1-\alpha},\ r_2 = r_2 = r(1 - \alpha)^{1/1-\alpha}q^{\alpha/1-\alpha}).$

F represents M, σ and G when β is taken as needed, and under $\beta < 1$ it goes through a maximum when $t = t_F^{max}$:

$$F_{max} = N_0^{1-\beta}A_{max}^{2\beta}f_1(\xi), \tag{13.18}$$

$$t_F^{max} = \frac{1}{N_0^{1-\alpha}}\overline{A_{max}^{2\alpha-1}}f_2(\xi) \tag{13.19}$$

where we have included in f_1 and f_2 all dependencies upon the rest of the parameters, which are not of interest here: $\xi = \{a, a_1, b_1, \gamma\}$.

Equations (13.15)–(13.19) correctly describe many properties of an even-aged system:

1. From (13.15b), density $N[t, N(0)]$ forgets initial conditions $N(0)$ and converges to a 'magistral' trajectory (Fig. 13.3):

$$N(t) = \frac{1}{a_1b_1\gamma a^{1/1-\alpha}A_{max}^{2\alpha-1/1-\alpha}}\frac{1}{t^{1/1-\alpha}}$$

2. When site conditions ameliorate, A_{max} increases and $N(t)$ diminishes at any given t (Fig. 13.5) if one supposes $\alpha > 0.5$ (a fairly realistic condition).

3. If we consider more shade-intolerant species, then the reaction of V to diminishing of A becomes sharper, that is b_1 (and consequently r) increases and leads again to $N(t)$ diminishing (Fig. 13.5).

4. When N_0 increases, F_{max} increases also (for $\beta < 1$), that corresponds to the behavior of $\sigma^{max}(N_0$ and $G^{max}(N_0)$ (Fig. 13.4) under a relatively small $N(0)$.

5. Magnitude $t^{max}(N_0)$ (Fig. 13.4) also decreases under better site conditions (when $\alpha > 0.5$).

Taking into account respiration and decay in the tree growth equation, $\dot{m} = qm^\alpha - cm$ will make $M(t)$ go through a maximum because $m(t)$ becomes finite but $N(t) \to 0$, and addition of competition will obviously give a bunch of trajectories $m[t, N(0)]$.

Finally, the only dynamical effect that cannot be given by base model 1 (13.12) is the 'overcrowding' (Fig. 13.4) under large $N(0)$; it can be shown that this effect demands at least three dynamical variables.

One can easily see that the form of system (13.14) and all results do not, in fact, necessarily depend on the resource under competition: they are true both for light, and for soil resources. Bearing in mind the proposition about weak interaction, one can take Taylor's expansion of Eqn (13.7) and repeat the same calculations with $x = qS_R(m) \ll 1$.

The topic of numerical applications of system (13.12) is worthy of a separate paper. It seems to us that the above considerations show clearly the ability of base model 1 to be used for natural forest modeling.

13.4. Multi-aged stands

In the array of successively complicated elements in the field of forest dynamics, multi-aged monospecific stands are the simplest elements, ones which can be related to real natural forest. The only process that needs to be added here in comparison with an even-aged population is the origin of seedlings. This gives us a usual population demographical system with a complete collection of dynamical processes.

It is necessary, then, to undertake a model description of the combined dynamics of the set of age cohorts which represent the whole population. There are some variants of the mathematical embodiment of the population which are distinguished by age and consequently by tree size and we shall review them briefly.

13.4.1. Construction of demographical models

There are a number of similar ways to formalize a population's age dynamics. The first approach was illustrated by Von Foerster (1959) who considered $n(t, \tau)$ (the number of individuals with given age τ at time t) and processes of birth and death, ignoring growth. This corresponds to the model

$$\frac{\partial n}{\partial t} + \frac{\partial n}{\partial \tau} = -d(t, \tau)n, \qquad n(t, 0) = \int p(t, z)n(t, z)\,dz. \quad (13.20)$$

Here d and p are specific death and birth rates; note that they are independent of population density n, that is the model is linear. This simple approach has now been exhausted from the mathematical point of view (Sinko & Streifer 1967; Poluectov 1974) and is of no serious interest for forest dynamics. Extensions of model (13.20) have been developed in various directions. Gurtin & MacCamy (1979) introduced density-dependent mortality $d = d(\tau, N)$ where $N = \int n\, d\tau$ and obtained analytical results for partial cases of $d(N)$. Sinko & Streifer (1967) considered a two-dimensional system, which operates by combined age and size distribution $n = n(t, \tau, m)$ where m is any quantitative characteristic of the individual. Behavior of n in the most general case is derived by Kolmogorov's or Fokker-Plank's equation

$$\frac{\partial n}{\partial t} + \frac{\partial n}{\partial \tau} + \frac{\partial}{\partial m}[g(t, \tau, m, n)\, n] = -d(t, \tau, m, n)n \qquad (13.21)$$

with corresponding boundary conditions; here $g = \dot{m}$ is the growth rate of the organism. Competitive and other density-dependent aspects are taken into account by means of argument n in g and d. In this case, Eqn (13.21) is very complex for analytical consideration; Sinko & Streifer (1967) have examined a partial case when Eqn (13.21) is linear with respect to n and m, that is $g = g(t, \tau)$ and $d = d(t, \tau)$.

Another and more popular approach was proposoed by Leslie (1945), whose well-known matrix technique is the discrete analogue of the continuous-time model in Eqn (13.20) and is more suitable for solving by computer. Leslie dealt with age distribution $n(t, \tau)$. It is convenient for us to write out his model in a 'cohort' form and with a generalized variant:

$$\begin{cases} n(t + 1, \tau + 1) = Y(t, \tau, \langle n \rangle)n(t, \tau), & \tau = 1, \ldots, T - 1 \\ n(t, 1) = \Sigma\, \rho(t, \tau, \langle n \rangle)n(t, \tau) \end{cases} \qquad (13.22)$$

where t and τ are discrete, Y and p are viability and birth rate, and $\langle n \rangle$ is a generalized vector argument, $\langle n \rangle = n(t, 1), n(t, 2), \ldots, n(t, T)$ which describes density effects.

Leslie (1945) and many others have used this model with $Y, p = Y, p(\tau)$ only, that is, the simplest linear variant which enables powerful matrix analysis. For example, Enright & Ogden (1979) presented detailed numerical analysis of one- and multi-species rain forest population systems by means of a linear case of model (13.22) with empirically determined birth and death rates. It is clear that in the case of forest dynamics linear equations can be true either for low densities or for short time periods until argument $\langle n \rangle \simeq$ const.

Among recent works we can note a two-species age-distributed model, an application of system (13.22), presented in Korzukhin, Sedych & Ter-Mikaelian (1987, 1988). This was applied to 200-year post-fire successional dynamics in West Siberia. Growth curves for both species (birch and Siberian pine) were fixed, so only age number dynamics were analyzed, that is, behavior of magnitudes $n_1(t, \tau)$ and $n_2(t, \tau)$. Crowns were horizontal and light competition interaction was directed from higher to lower trees only, as in the gap-model approach. The dynamics of two age packages ('waves') observed in the field were observed numerically (see also §13.4.2.).

In forest dynamics applications the Lefkovitch (1965) approach is more popular; this offered the same technique for size distribution analysis (sizes of trees are measured much more easily than their ages). If one breaks the size axis m into Q intervals

$$(\mu_0, \mu_1)(\mu_1, \mu_2) \ldots (\mu_{Q-1}, \mu_Q) \tag{13.23}$$

and takes all trees whose sizes belong to interval i, that is $\mu_{i-1} \leq m \leq \mu_i$, then the system dynamics will be represented by the following scheme:

$$
\begin{array}{c}
\text{death} \qquad\qquad \text{death} \\
\underset{\text{Birth}}{\xrightarrow{\hspace{1cm}}} \quad n(t, 1) \overset{d_1}{\nearrow} \qquad n(t, 2) \overset{d_1}{\nearrow} \\
\underset{\text{Birth}}{\xrightarrow{\hspace{1cm}}} \quad n(t+1, 1) \qquad n(t+1, 2)
\end{array}
\tag{13.24}
$$

where $n(t, i)$ is tree number in the ith size interval, g_i is growth and d_i is death rate. It is proposed, for the sake of simplicity, that changes in sizes are small and occur only between neighboring classes. Dynamical equations, being discrete analogs of differential equations (13.21), are easily written from the balance scheme (13.24):

$$
\left\{
\begin{aligned}
n(t+1, i) &= n(t, i) + g(t, i-1, \langle n \rangle)n(t, i-1) \\
&\quad - g(t, i, \langle n \rangle)n(t, i) - d(t, i, \langle n \rangle)n(t, i), \quad i = 2, \ldots, Q; \\
n(t+1, 1) &= n(t, 1) + \sum_i p(t, i, \langle n \rangle)n(t, i) - g(t, 1, \langle n \rangle)n(t, 1) \\
&\quad - d(t, 1, \langle n \rangle)n(t, 1).
\end{aligned}
\right.
\tag{13.25}
$$

Here p is size-specific fecundity, argument $\langle n \rangle$ is analogous to that used in Eqn (13.22), and dependence of g, d, and p upon i means their dependence on size. In these equations, the magnitudes of size intervals $\Delta \mu i$ are considered to be included in functions g, d, and p.

The most frequently used variant of the highly generalized model (13.25) consists of taking birth, growth and death rates as depending only upon size (class number i). Buongiorna & Bruce (1980) have applied this model to the task of productivity maximization in this linear variant, when $g = g(i)$. Hartshorn (1975) has used this type of model for two tropical tree species, and Dyrenkov & Gorovaya (1980) for one spruce species. The value and restrictions of the linear approach were noted above. Kapur (1982) undertook an important extension of the task by introducing $g = g(N)$, that is, a rough reflection of density-dependent factors. The aim was again to maximize total stand productivity. A nonlinear model by Ek & Monserud (1979) treated the same object – a six-species forest in Wisconsin – as was analyzed by their individual-tree model (Ek & Monserud 1974). This is the only modeling work where such an interesting comparison was made. Unfortunately, the authors have not adduced the explicit form of their dynamical equations. A complex nonlinear model was formulated and applied to a seven-species boreal forest in South Siberia by Cherkashin (1980). Recent work of Kohyama (1989) is devoted to the analysis of tree diameter distributions dynamics, $n(D, t)$, in a rain forest population system which was close to an even-aged one; the model is presented by nonlinear system (13.25) with empirically determined death and growth rates as functions of G.

A major weakness in the described matrix and continuous-time multi-aged models is the rather formal realization of individual-tree growth and competition mechanisms. From this point of view, individual-tree models are better developed.

A third way to investigate age and size dynamics is by means of a combination of Leslie's (1945) and Lefkovitch's (1965) matrix technique, which can be made by means of straight generalization of our even-aged base model 1 in Eqn (13.12). In other words, a multi-aged model can be constructed simply as the sum of one-cohort equations (13.12). This method was proposed and realized by Antonovski, Korzukhin & Matskiavichus (1989) and Korzukhin, Antonovski & Matskiavichus (1989). This gives the system to which we shall refer as base model 2:

$$\begin{cases} n(t+1, 1) = \Sigma\rho(t, \tau, \langle n\rangle, \langle m\rangle)n(t, \tau) + f, \\ n(t+1, \tau+1) = Y(T, \tau, \langle n\rangle, \langle m\rangle)n(t, \tau), \\ m(t+1, \tau+1) = m(t, \tau) + g(t, \tau, \langle n\rangle, \langle m\rangle), \qquad \tau = 1, \ldots, T-1. \end{cases} \qquad (13.26)$$

Boundary conditions for the growth equation are $m(t, 1) = m_1$. Variable m, indeed, can be substituted by arbitrary size characteristics of a tree. Argument $\langle m\rangle$ has the same meaning as $\langle n\rangle$. Note that Leslie's system (Eqn (13.22)) is the partial case of Eqn (13.26); in order to obtain Eqn (13.22), one would simply not take into consideration all equations for $m(t, \tau)$, that is, one would suppose that growth curve $m(\tau)$ is set.

We have added one new element in base model 2 compared with Eqns (13.22) and (13.25), namely addendum f in the equation for first age class. This element describes seed influx in the system due to the background of seed that usually exists in forested areas and introduces a spatial aspect to the system analysis which is absent in the standard form of Leslie's and

Lefkovitch's models. It is interesting to note that gap models, on the contrary, use only the background of seeds and neglect seed production inside the considered area itself.

Let us briefly discuss the comparative potential of size and improved age cohort approaches, that is, systems (13.25) and (13.26). Major differences are the following. System (13.26) is purely deterministic, giving one trajectory of biomass and tree number for a given cohort, so that this even-aged sub-system at each moment in time is described by only two magnitudes: $m(t + k, k)$ and $n(t + k, k)$; k is the age of the cohort born at moment t. In contrast, Eqn (13.25) describes spreading of sizes (and numbers, correspondingly) about the m-axis, even for one cohort. In reality, after k steps each cohort will be partly presented in all size classes from the first to $(k + 1)$th. So certain stochastic mechanisms are contained in Eqn (13.25); in its strictest sense, as it deals with size distribution, it is analogous to a Markov chain.

The two systems also have similarities. Both give us size distribution, model (13.25) by the definition of $n(t, i)$ and model (13.26) after simple summation over a given size interval (Eqn (13.23)):

$$n(t, i) = \sum_{\tau} n(t, \tau), \qquad \tau_i \leq \tau \leq \tau_{i+1},$$

where τ_i and τ_{i+1} are the ages whose tree size m_i belongs to the ith interval, $\mu_i < m_i < \mu_{i+1}$. Both systems also give us age distribution, model (13.26) by the definition of $n(t, \tau)$ and model (13.25) by means of a principally clear but rather sophisticated procedure of watching the fate of each age cohort that spreads in the set of size cohorts. Let us examine a simple example. Consider the fate of a cohort born at time t, with number of trees $n(t, 1)$. At time $t + k$ there will be $n(t + k, k + 1) = g_1 g_2$ $\ldots g_k n(t, 1)$ trees in size class $k + 1$, and $n(t + k, k) = g_1 g_2 \ldots$ $g_{k-1}(1 - g_k - d_k)n(t, 1)$ trees in size class k, etc. Summing up all these numbers gives us the total number of trees of age $\tau = k$:

$$n(t + k, \tau) = \sum_{i} n(t + k, i).$$

It seems to us that base model 2 has some advantages over Eqn (13.25). For the latter, in the case of several species, a common set of size intervals (Eqn (13.23)) makes the species' growth resolution different; a particular set makes the whole system almost inoperable (e.g. if we need, as is often the case, to compare species heights for defining competitive relations). This approach is disquieting when one has to deal with a number of tree species with noticeably different growth rates and, moreover, try to include a description of, say, grass and shrub growth.

System (13.26) does not have any of these disadvantages. The size-class approach can, of course, be useful for one-species even-aged tasks aimed at analyzing size distribution dynamics.

In the next section we will show the simulation abilities of base model 2 using an example of the modeling of non-stationary age distribution behavior.

13.4.2. Empirical age dynamics in simple forests

First of all, we will describe the non-stationary age distribution behavior of a generalized coniferous species. Such a situation originates after a severe catastrophic disturbance, which entirely obliterates the initial 'maternal' stand and provides zero initial conditions

$$n(t = 0, \tau) \equiv 0$$

This situation is typical of major fires in taiga forests, total phytophagy defoliation (after which the trees quickly wilt and die), hurricanes, and, of course, cutting by man. The subsequent successions have been repeatedly described and analyzed in the literature (Semetchkin 1970; Leak 1975; Hett & Loucks 1976; Larson & Oliver 1979; Oliver 1981; and others).

Further, the intensity of seed influx f will be the central ecological parameter. Clearly, at sufficiently low f values, the area will be inhabited by a population with a monotonic age distribution $\partial n/\partial \tau < 0$, which attains equilibrium during one generation time. This case is typical for habitats with poor soil and climatic conditions.

In comparatively better conditions, and when f is large enough, pioneer individuals capture the area and exhaust the resources (light, soil oxygen, nitrogen, etc.), and the seedlings of the next ages die off owing to competitive suppression. The result is a 'package' of older trees and a zero-gap at that part of the age distribution which corresponds to younger trees (Fig. 13.6), a picture that has been described repeatedly (Zubarev 1965; Semetchkin 1970; Kazimirov 1971; Leak 1975; Sprugel 1976; Oliver 1978, 1981; Francline & Waring 1979; Larson & Oliver 1979; Glebov & Kobyakov 1984). For a full review of different types of non-stationary age distributions, see Katayeva & Korzukhin (1987). In all cases shown in Fig. 13.6, we have the 'running wave' or several waves along the τ-axis.

The subsequent dynamics may go one of two ways: (a) after one or several damping waves the age distribution attains equilibrium; or (b) the system enters an oscillating regime which can be either fully or quasi-periodic. For a full review of field observations, see Katayeva & Korzukhin (1987). Here we will list only some typical cases.

A prolonged (100–200 years) endogenous periodic cycle was reported for deciduous forests in the far eastern USSR (Kolesnikov 1956; Vasiliev

& Kolesnikov 1962; Rozenberg, Manko & Vasiliev 1972), for fir forests in the Ural region (Smolonogov 1970; and others), for beech Crimea forests (Sukachev & Poplavskaya 1927), and for balsam fir and eastern hemlock forests in the Great Lakes region (Hett & Loucks 1976). An endogenous

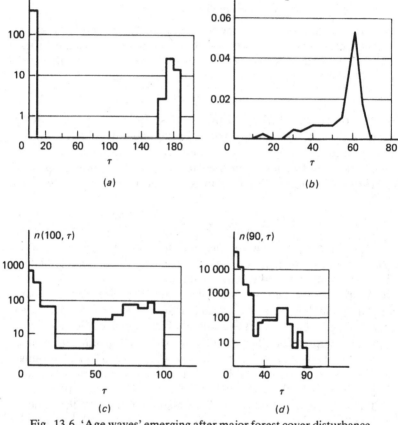

Fig. 13.6. 'Age waves' emerging after major forest cover disturbance in sufficiently dense stands. $n(t, \tau)$ is the number of trees in age interval $\Delta \tau = 1$ under age τ; t is moment of observation (equal to time since disturbance). Cases of strong suppression of juveniles by mature canopy: (*a*) post-fire succession, *Pinus sibirica*, East Siberia (Semetchkin 1970); (*b*) post-cutting succession, sum of oak, maple, and birch, New England (Oliver 1978). Cases of weak suppression: (*c*) post-fire succession, sum of spruce, *Pinus sibirica*, and fir, West Siberia (Korzukhin, Sedych & Ter-Mikaelian 1988); (*d*) post-fire succession, mixed oak forest, Appalachian Mountain (Ross, Sharik & Smith 1982).

periodic cycle of 60–80 years was reported for *Abies* forests in New England (Sprugel 1976; Reiners & Lang 1979; Foster & Reiners 1983; Moloney 1986) and for *Abies* forests in Japan (Oshima *et al.* 1958).

Unfortunately, only a few papers (e.g. Hett & Loucks 1976; Reiners & Lang 1979; Kozin 1982) give information about age distributions. Of great importance are the data about the size of area which is occupied by a single-phase stand: this area is usually between 0.02 and 1 ha (Oshima *et al.* 1958; Foster & Reiners 1983; Moloney 1986).

A further set of works is dedicated to spatial waves in pure dark-needled coniferous stands (Oshima *et al.* 1958; Sprugel 1976; Reiners & Lang 1979; Foster & Reiners 1983; Moloney 1986). If the system is observed in a single 'point' – reported above to be between 0.02 and 1 ha – it exhibits periodicity of age distributions of trees. Spatial waves are generated by synchronization of phases in different points owing to strong, unidirected winds. When there are no winds the phases of development become different.

The next section is devoted to a modeling analysis of periodic phenomena.

13.4.3. One-species oscillation models

First, we offer two simplified qualitative models (for details see Korzukhin (1980), Antonovski, Korzukhin & Matskiavichus (1989) and Korzukhin, Antonovski & Matskiavichus (1989)), which are some particular cases of the general multi-aged base model in Eqn (13.26).

The first system is differential and describes number dynamics in a system with three age classes (n_i are their tree numbers):

$$\begin{cases} \dot{n}_1 = V(\gamma\sigma)f - n_i; \\ \dot{n}_2 = n_1 - n_2; \\ \dot{n}_3 = n_2 - n_3. \end{cases} \tag{13.27}$$

Here we ignore the mortality in class 2 and consider the seed background as the only source of regeneration: its intensity is f and offspring survival is V, depending on total leaf area, $\sigma = S_2 n_2 + S_3 n_3$ ($S_1 n_1$ is neglected), and γ is a generalized interaction parameter (for example, light extinction coefficient). By using trivial qualitative equilibrium point analysis, it is easy to show that system (13.27) can realize sustainable oscillations. For example, when $V = \exp(-\gamma\sigma)$, this happens when

$$f \geq \frac{8}{\gamma(S_3 - 2S_2)} \exp\left[8(S_2 + S_3)/(S_3 - 2S_2)\right].$$

The discrete analog of system (13.27)

$$\begin{cases} n_1(t+1) = V(\gamma\sigma)f \\ n_2(t+1) = n_1(t) \\ \cdots\cdots\cdots\cdots \\ n_T(t+1) = n_{T-1}(t) \end{cases}$$ (13.28)

where $\sigma(t) = \Sigma\, S(i)n_i(t)$, with stepwise viability function and linear law of tree growth

$$V(\gamma\sigma) = \begin{cases} 1 \text{ when } \sigma < \sigma^* \\ 0 \text{ otherwise} \end{cases}; \quad S(i) = a(i-1)$$

also realizes sustainable age oscillations. For example, when $af > \sigma^*/(T-1)$, there is a regime with period

$$\theta = T + K; \quad K = \left[\frac{\sqrt{8T-7}+1}{2}\right];$$

$[z]$ denotes the integer part of z. The solution of (13.28) with zero initial conditions has the form of periodically running waves with K non-zero age classes in a package:

$$n_i(t) = (0, \ldots, 0) \to (f, 0, \ldots, 0) \to (f, f, 0, \ldots, 0) \to$$
$$\to (0, \ldots, 0, f, \ldots, f, 0, \ldots, 0) \to \ldots \to (0, \ldots, 0, f) \to (0, \ldots, 0).$$

The numerical and realistic oscillation model (see the source works cited) originates from the general form of base model 2 (Eqn (13.26)). Crown morphology was the most simple: crowns were horizontal, and again with areas (Eqn (13.10))

$$S_L(t, \tau) = \alpha m^\alpha(t, \tau)$$ (13.29)

(a and α are parameters); decay and respiration are given by Eqn (13.11). Assimilation was proportional to normalized light intensity

$$A(R_I) = bR_I(t, \tau)/R_{I0}, \quad A(R_{I0}) = A_{max} = b,$$ (13.30)

where R_I is given by (Korzukhin & Ter-Mikaelian (1982); *cf.* Eqn (13.13)):

$$R_I(t, \tau) = R_{I0} \sin \varphi \exp\left[1/2\gamma S_L(t, \tau)n(t, \tau)\right.$$
$$\left. - \gamma \sum_{\mu=\tau+1} S_L(t, \mu)n(t, \mu)\right];$$

φ is the angle of the sun. Within total viability, $Y = WV$, the density-independent component was constant:

$$W(\tau) = W_0 < 1,$$ (13.31)

and the density-dependent one had the form

$$V(v) = \begin{cases} 1 - q\left[\dfrac{1-v}{1-u}\right]^\beta & \text{when } v = \dfrac{A(R_I)}{A_{max}} \geq u \\ 0 & \text{otherwise} \end{cases}$$ (13.32)

($q \leq 1$, $u < 1$, and β are parameters), that realizes the possibility to put smooth (small β) and near-stepwise (large β) viability functions (Fig. 13.1). Only the background seed quantity was considered: as in Eqn (13.28), $p(\tau) = 0$.

Finally, we dealt with the system

$$
\begin{cases}
n(t+1, 1) = f \\
n(t+1, \tau+1) = W_0 V(v_\tau) \\
m(t+1, 1) = m_0 \qquad \tau = 1, \dots, T-1 \\
m(t+1, \tau+1) = m(t, \tau) + S_L(m_\tau) A(R_\tau) - cm(t, \tau)
\end{cases}
\tag{13.33}
$$

$$[m_\tau = m(t, \tau), R_\tau = R(t, \tau), v_\tau = A(R_\tau)/A_{\max}]$$

where V, S, A, and R are the functions defined above.

Parameter definition

Species lifespan T was set to 200 years. The value of α in Eqn (13.29) most commonly lies in the interval $0.5 \leq \alpha \leq 0.8$; we took $\alpha = 0.7$. At the end of the lifespan of coniferous species, the ratio $a_1 = m_L^{\max}/m_{\max} = 0.05$ (Larcher 1975), where m_L is leaf biomass. The mass–surface transition factor for needles was taken as $a_2 = m_L/s_L = 0.15$ [kg fresh mass m^{-2}]. After taking $m_{\max} = 1000$, we can calculate that $a = a_1 m_{\max}^{1-\alpha}/a_2 = 2.65$; this means $S_L^{\max} = 334\ m^2$ and $m_L^{\max} = 50$ kg. R_{I0} can be adopted as 1; φ equaled 0.76 [rad]. A typical value of b in Eqn (13.30) is in the order of 10^{-1} [kg fresh mass m^{-2} yr^{-1}]; we took $b = 0.15$. The parameter c of unified respiration and decay losses can be found from the condition $m(t) \to m_{\max}$ when $t \to \infty$ and competition is absent; this gives $am_{\max}^\alpha b \sin \varphi - cm_{\max} = 0$ or $c = 0.034$ [kg^{-1} yr]. W in Eqn (13.31) was equal to 0.98053 (by the age $T = 200$, 2% of trees are left); q equaled 0.5, that is, a moderate jump in V at $z = u$ was allowed.

The values of γ, f, β and u were varied in the search for oscillations. In this way, we experimented with generalized coniferous tree species having plausible parameter values from measurements and a number of free parameters for the searching of oscillations. Among the latter $\gamma \approx 0.3$–0.8, f usually varies from some hundreds to thousands, u is near the light compensation point for the whole tree, that is $u \approx 0.1$–0.3, and β measures the plasticity of tree response to shortage of resources, a value which is unknown. The initial conditions comprise zero densities, $n(0, \tau) = 0$ and $m_0 = m(0, 1) = 10^{-3}$ [kg]; $m(0, \tau) = 0$ for $\tau \geq 2$. After many numerical experiments we have concentrated on the value $\beta = 60$, which relates to all the adduced results. The system calculated up to a $t^{\max}$ of 2000 years.

All dynamical regimes in three-dimensional parametric space (γ, f, u) are given in Table 13.1, where the period of oscillations, θ, when they

Table 13.1. *Oscillation period θ* (γ, f, u) *and the type of dynamical behavior in Eqn* (13.33)

	$u = 0.1$			$u = 0.2$		
γ	0.2	0.4	0.6	0.2	0.4	0.6
f						
500	SE	SE	SE	SE	RO, 86	RO, 86
1000	SE	SE	RO, 85	RO, 86	RO, 82	QO, 43–71
2000	SE	RO, 82	RO, 82	RO, 83	QO, 52–68	QO, 49–66

Note: SE = stable, RO = regular, and QO = quasi-regular oscillations.

exist, is also shown. When the oscillations are quasi-regular, we adduce the observed interval of θ. It can be seen that low values of γ, f, and u promote a stable behavior; as their values increase the system becomes unstable and finally falls into a stochastic regime.

The behavior of some important characteristics as functions of γ for parameter values $f = 1000$, $u = 0.15$ is presented in Fig. 13.7. Up to $\gamma \simeq 0.2$, the system attains stable equilibrium. Near $\gamma = 0.22$ it becomes unstable, and a stable periodic regime arises (Fig. 13.8). Figure 13.7 shows the dependence on γ of oscillation amplitude n^{max} $(t, 100)$ and n^{min} (t, τ^{min}), where τ^{min} is the minimum age when the maximum number is at $\tau^{max} = 100$ (see also Fig. 13.8 where curves $n(t, \tau)$ are pictured at the time when the maximum goes through age 100). Note (Fig. 13.7) the high stability of the oscillation period, $80 \le \theta \le 88$, and the value of n^{max} whereas n^{min} is greatly reduced.

Leaf area index

The time behavior of leaf area index, $\sigma(t) = \Sigma_{\tau} S(t, \tau) n(t, \tau)$, shows the appearance of expected dynamical invariant (13.8): beyond $t \approx 50$–100, σ assumes a relatively constant value (the relative variations $\Delta\sigma/\sigma^{average} \simeq 0.15$) which is the maximum possible value for the given site and species parameters. This behavior corresponds well with the situations observed in even- and multi-aged stands. At $\gamma = 0$ (Fig. 13.7) (free-growing trees) σ^{max} is very large and equals 115, a totally unrealistic value for natural populations. However, $\sigma(\gamma)$ then falls quickly, attaining reasonable values by the time γ is only 0.1.

Ecophysiological invariant

Among the effects discovered the greatest interest was aroused by the appearance of some parametric invariant. In our previous works (Antonovski, Korzukhin & Matskiavichus 1989; Korzukhin, Antonovski & Matskiavichus 1989) we have found it occasionally, but now we are able

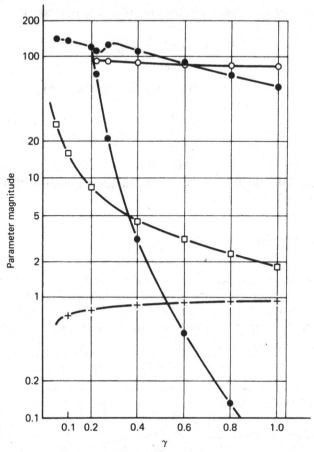

Fig. 13.7. Some quantities of system (13.33) with various values of the parameters and extinction coefficient γ. Two curves (filled circles) arising after γ 0.22 show amplitude of the younger wave; for meanings of $n^{\max} = n(t, 100)$ and $n^{\min} = n(t, \tau^{\min})$ see Fig. 13.8. Open circles, oscillation period θ (γ); squares, leaf area index δ (γ) in $T = 2000$; pluses, value of J (13.35).

to develop some theoretical base. Let us put the following question: what is the interrelation between maximum leaf area index, $\sigma^{\max}$, which a stand develops, extinction coefficient, γ, and leaf (or tree) viability? It is clear that when γ is large, under fixed leaf shade tolerance a stand can sustain a fewer number of layers than when γ is small. Physiologists know well that the lowest leaf layer in a stand exists only when its respiration losses are less than net assimilation. Below, we try to express this rule in model terms.

Let us consider the ideal situation when viability function V is step-wise:

$$V(R_I) = \begin{cases} 1, & \text{when } R_I/R_{I0} > u \\ 0 & \text{otherwise;} \end{cases}$$

that is, a tree with height h dies when for relative light supply the following condition is true

$$R_I(h)/R_{I0} \le u. \tag{13.34}$$

Further, when a stand develops σ leaf layers, the last layer receives

$$\exp(-\gamma\sigma)$$

share of initial light. So, the condition of existence of exactly σ^{max} layers, obtained from Eqn (13.34), is

$$\exp(-\gamma\sigma^{max}) = u.$$

It is suitable to introduce the function

$$J(\sigma, \gamma, u) = u \exp(\gamma\sigma), \tag{13.35}$$

which can be named as an 'ecophysiological invariant'. It connects three fundamental magnitudes relating to stand functioning. Among them, u can be identified with a compensation point on the leaf light curve measured for 24-hour leaf carbon balance (normalized to maximum light).

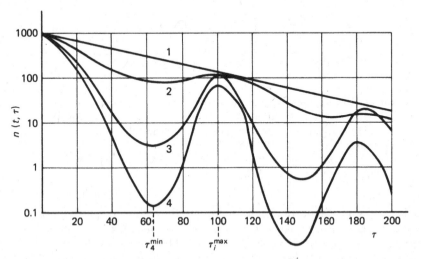

Fig. 13.8. Age distributions of tree numbers $n(t,\tau)$: solution of system (13.33) under various γ. 1, stable distribution for $\gamma = 0$, $n(\tau) = 1000 \exp(-0.0197\tau)$; 2, 3, 4, stable oscillations for $\gamma = 0.226, 0.4, 0.8$; curves are taken in time moment t when maximums were going through age $= 100$; see text for parameter values.

Table 13.2. *Dependence of leaf area index σ (first number) and invariant J (second number) on system parameters*

γ	$u = 0.1$			$u = 0.2$		
f	0.2	0.4	0.6	0.2	0.4	0.6
500	7.6 0.46	4.4 0.58	3.1 0.64	7.0 0.81	3.2 0.72	2.4 0.84
1000	8.9 0.59	4.8 0.68	3.4 0.77	6.3 0.71	3.4 0.78	2.6 0.95
2000	9.5 0.69	4.9 0.71	3.6 0.87	6.9 0.79	3.9 0.95	2.5 0.90

For our simplified viability model the equality

$$J(\sigma, \gamma, u) = J^{\max} = 1$$

should be valid. In reality, many factors deviate J from 1. Any tree has a finite probability of dying even if its relative light supply is greater than u. So, a stand cannot reach the value $\sigma^{\max}$, which corresponds to the stepwise form of V; and one will expect $\sigma < \sigma^{\max}$ and the value of J will be less than 1. Deviation of J from 1 is a measure of stand structure deviation from optimum structure. The concrete value of J is defined by the form of the V function and other system parameters. Returning to our model, we adduce the example of J calculation at $u = 0.15$ (Fig. 13.7) and for $u = 0.1$ and $u = 0.2$ (Table 13.2). Certain scattering and non-monotonic behavior of J can be explained by taking its values at $t = 2000$ and in an oscillation regime.

It seems to us that further development of this approach lies in calculations of ecophysiological invariant (13.35) for real stands, while of central interest is watching the degree of J stability with parameter variations and time, and analysis of its deviations from value $J^{\max} = 1$.

Table 13.2 also shows an approximate constancy of σ as a function of seed migration, f, under a given γ and u. Here we are witnessing a 'habitat-saturation' effect, which is observed in real ecosystems and explained, as above, by maximum light resource utilization. The effect was also studied by Korzukhin (1986).

Numerical experiments showed that a large curvature of viability function $V(v)$ Eqn (13.32) is essential for the appearance of the oscillations. The influence of q, u, and β can be seen from the value of the derivative at $v = u$: $V'(v)|_{v=u} = -q\beta/(1 - u)$. It is worth mentioning that the used value of $\beta = 60$ yields a 'near-step' function $V(v)$, which is close to purely stepwise viability functions widely used in forest models (see review in §13.2.2).

The dynamics of formation and passing of the first age wave is of great interest for forest ecology regardless of the question of oscillations. The

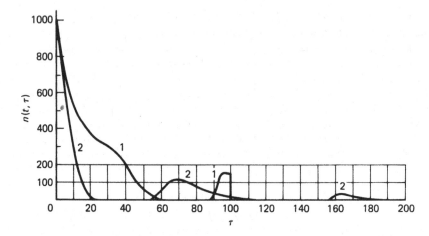

Fig. 13.9. Solution of system (13.33) under $\gamma = 0.4$ and $u = 0.2$: 1, first 'age package'; 2, stable oscillations.

first wave is much higher and narrower than the succeeding ones (Fig. 13.9). One can see that at $t = 100$, the system is on the edge of the next wave origin.

13.5. From stand to forest; addition of above-ground species

Up to this stage, we have discussed stand dynamics only, which is a serious contraction of real forest composition. However, although trees can dominate in certain senses, e.g. in terms of the proportion of live ecosystem biomass, essentially they can depend on other species. Among these, the above-ground plant species such as grasses, mosses and lichens are most important for boreal forests because they provide the boundary conditions for tree regeneration. In this aspect, trees as a life form exhibit an apparent weakness in their regeneration strategy: many trees suppress mosses and grasses and simultaneously their own seedlings. Sparse stands allow seedlings to grow but also promote their competitors. The interplay between these extremes can lead to interesting dynamical and life-strategy scenarios. Below, we describe some of the model embodiments of a simplified ecosystem consisting of one tree and one above-ground species as an example of a two-life-form community. Modeling activity in this field is very limited compared with tree systems only, so we can adduce only a few results.

13.5.1. Modeling of moss dynamics

Many dark-needled coniferous boreal forests are characterized by a noticeable moss organic layer on the forest floor. The thickness of this layer, taking into account both live and dead parts, is up to between 30 and 50 cm. The layer is an important structural component of a forest, controlling energy flow, nutrient cycling, water relations, and, through these, stand productivity and dynamics (Bonan & Shugart 1989). For example, soil temperature and depth of permafrost are directly related to the thickness of the layer. Another example is the dependence of the viability of different types of tree seeds on moss layer thickness (Fig. 13.10), which will be discussed below.

First, let us consider only live moss. The simplest formalization of its growth can be made in terms of carbon balance using one variable: green moss biomass μ [kg m^{-2}]:

$$\dot{\mu} = S_\mu A_\mu(R_\mu) - c\mu, \tag{13.36}$$

where respiration and decay are assumed to be proportional to μ, R_μ denotes mean light falling per unit area, A_μ is specific assimilation, and leaf area S_μ can be taken as proportional to μ ($S_\mu = a_\mu\mu$). The central idea of this model is that moss growth must be auto-restricted, that is, Eqn (13.36) must give $\mu \to \mu^{\max}$ in the absence of trees and other competitors. If we consider light extinction in a relatively thin moss layer with thickness h and a vertical profile of light $R_\mu(h)$, then, approximately,

$$R_\mu \simeq R(h/2) \simeq R_{\mu 0} \exp(-\gamma S_\mu \mu/2),$$

where $R_{\mu 0}$ is the light at the top of the layer. If we suppose that A_μ is proportional to R_μ ($A_\mu = bR_\mu$), this gives us a simple equation of moss layer growth:

$$\dot{\mu} = a_\mu\mu bR_{\mu 0} \exp(-\gamma a_\mu\mu/2) - c\mu = v \exp(-q\mu) - c\mu. \tag{13.37}$$

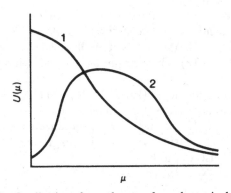

Fig. 13.10. Qualitative dependence of seed survival and germination, U, on moss biomass (or thickness) μ for boreal tree species. (1) Small seeds (birch, spruce); (2) large seeds, in particular *Pinus sibirica*, whose seed viability is maximal under a certain moss thickness (summarized in Katayeva & Korzukhin (1987)).

For sustainable existence of moss, v must be greater than c. This model coincides with that proposed for grasses by McMurtrie & Wolf (1983). A more realistic model was developed by Bonan & Korzukhin (1989) where the second moss variable v (dead moss biomass [kg m^{-2}]) and also the usual light curve for assimilation were taken into account:

$$\begin{cases} \dot\mu = S_\mu \dfrac{a_1(R_\mu - a_3)}{1 + a_2 R_\mu} - c_1\mu - c_2\mu; \\ \dot v \qquad\qquad\quad c_1\mu - c_3 v. \end{cases} \tag{13.38}$$

In this equation, c_1 is the specific decay rate of live moss, c_2 is its respiration losses, c_3 is the specific oxidation rate of dead moss, a_1 and a_2 are light curve parameters, and a_3 is the compensation point. The model was identified with the help of actual ecophysiological (Larcher 1975) and field (Van Cleve *et al.* 1983*b*) moss data.

13.5.2. Lichen–trees system modeling

The principal components for the construction of various models of the discussed type have been identified in previous parts of this chapter. For example, we may join the age-distributed base tree model 2 (Eqn (13.26)) with the simplest moss growth equation (Eqn (13.37)). In fact, this equation can be applied to any above-ground species because principles on which the equation is based are very general. For the sake of diversity, let us consider the lichen–trees situation that is described in detail by Sirois in Chapter 7 of this book. Among other properties of the system, there is a possibility of sustainable occupation of a given territory either by trees (represented by spruce) or by lichen; this depends on initial conditions of area settlement.

In the simplest case, lichen–tree interactions are expressed (a) in terms of dependence of incident light reaching lichen layer, $R_{\lambda 0}$, on tree leaf area σ: $R_{\lambda 0} = R_{\lambda 0}(\sigma)$, and (b) in terms of seedling viability dependence on lichen thickness, $U = U(\lambda)$. These arguments, λ and σ, interconnect tree and lichen dynamics.

Whereas the complete model (Eqns (13.26) and (13.37)) is difficult to analyze, some qualitative properties can be demonstrated on the basis of its simplified version. Let us take one generalized variable N (total number of trees) and consider only seed background. A very simple lichen–tree model turns out to be

$$\begin{cases} \dot N = U(\lambda)f - N \\ \dot\lambda = [v \exp(-kN - q\lambda) - c]\lambda, \end{cases} \tag{13.39}$$

where k is proportional to average single-tree leaf area multiplied by the extinction coefficient for trees. In this approximate description, N can be changed to total tree biomass M. The behavior of this system is typical of two-dimensional systems with 'competitive' interactions, which are encountered repeatedly in forest modeling (e.g. McMurtrie & Wolf 1983).

First consider the case of the suppression form of function $U(\lambda)$ only (Fig. 13.10). Let us take $U(\lambda) = 1 - \lambda/\lambda^*$ when $\lambda < \lambda^*$; otherwise $U(\lambda) = 0$ (λ^* is a border lichen thickness). Simple qualitative analysis of system (13.39) gives us four standard situations. By denoting $r = \ln(v/c)$, we have:

1. If $kf > r$ and $q\lambda^* > r$, we have one stable equilibrium $(f, 0)$, that is, trees outcompete lichen.

2. If $kf < r$ and $q\lambda^* > r$, we have one stable equilibrium (N^0, λ^0), where $N^0 < f$, $\lambda^0 < \lambda^*$, that is, lichen and trees coexist; N^0 and λ^0 can be found from the linear system that originates from Eqn (13.39) under $\dot{N}$, $\dot{\lambda} = 0$:

$$kN - q\lambda = r; \qquad N - (1 - \lambda/\lambda^*)f = 0.$$

3. If $kf > r$ and $q\lambda^* < r$, we have two stable equilibria $(f, 0)$ and $(0, r/q)$, which are realized depending on initial conditions $N(0)$, $\lambda(0)$.

4. If $kf < r$ and $q\lambda^* < r$, we have one stable equilibrium $(0, r/q)$, that is, lichen outcompetes trees.

The case of a non-monotonic function $U(\lambda)$ (Fig. 13.10) adds the possibility of the coexistence of two stable states: (N_1^0, λ^0) and $(N_2^0, 0)$.

Now let us imagine a movement along an environmental gradient from dry and warm conditions, which are favorable for tree growth and unfavorable for lichen (f is large, v is small), to wet and cold conditions where the reverse is true. In this case, system (13.39) will realize the consequences of the stable state moving from $1 \to 2 \to 4$ or $1 \to 3 \to 4$, that is, the transition from temperate forest without lichen to tundra without trees, passing between these extremes through the possibility of lichen–tree coexistence (boreal forest) (Fig. 13.11). There is absolutely no doubt that any generalization of this qualitative system (e.g. Eqns (13.26) and (13.38)) will retain this transition effect.

In a more realistic imitative, but not mathematical, model (Bonan & Korzukhin 1989) a very similar type of behavior was actually discovered in the field for tree–moss communities disposed along a cold–wet to hot–dry environmental gradient. The sites were typical of central Alaska and were dominated by black and white spruce, white birch and trembling aspen, plus sphagnum mosses. The two-compartment moss model (Eqn (13.38)) was used. Functions of seed germination suppression, $U_i(M)$ for each tree species i, were reconstructed from field data; here $M_\mu = \mu + v$ is the total moss biomass. The dynamics of the trees were described by using

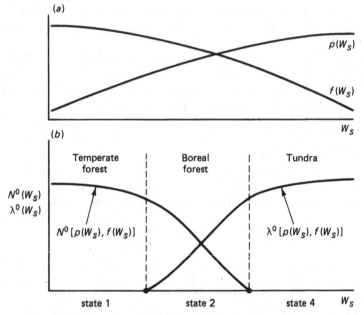

Fig. 13.11. Parametric evolution in lichen–tree system (13.39); if one considers p and f as monotonic functions of soil moisture W_S (a), then steady states of (13.39), $N^0(W_S)$, $\lambda^0(W_S)$, realize trajectories (b) which can be interpreted as different types of lichen–tree ecosystems disposed along an environmental gradient (a case of monotonic $U(\lambda)$, as a consequence of states: $1 \rightarrow 2 \rightarrow 4$). The conditions of realization of these states in system (13.39) are as follows.

State 1: $\ln [p(W_S/c)] < kf(W_S) < q^{\lambda^*}$.

State 2: $kf(W_S) < \ln [(p(W_S/c)] < q^{\lambda^*}$.

State 4: $kf(W_S) < q^{\lambda^*} < \ln [p(W_S/c)]$.

a common gap-modeling technique. Field observations along the gradient supported model results; see the cited work for details.

6. Concluding remarks

In this chapter, two aspects of forest modeling were touched upon. The first relates to the whole field of stand modeling and reviews the compulsory, elementary blocks which form the base of every type of model. Here, we presented the most contemporary methods of tree growth, inter-tree competition, and tree viability formalization.

The second modeling aspect concerned developing and running models of forest dynamics at the population level. These models are based on the use of certain dynamical equations derived with the help of mechanistic considerations. We demonstrated the researching abilities of these models by means of explanation of some central dynamical effects observed in even- and multi-aged forest stands. Among these effects, wave behavior of tree age distributions was the most demonstrative example of the abilities of population-level equations.

The authors feel deeply indebted to Erica Schwarz for her inestimable help in preparing the text for publication, and, of course, for her benevolent patience during numerous modifications of the paper. The authors would also like to express thankfulness to their colleagues in forest modeling for beneficial discussions.

Appendix

List of variables and parameters used

A	specific assimilation rate [kg m^{-2} yr^{-1}]
d	tree mortality ($=1-Y$) [yr^{-1}]
D	tree diameter at breast height [cm]
De	tree decay losses [kg yr^{-1}]
f	seed immigration rate [ha^{-1} yr^{-1}]
G	stand cross-sectional area ($=D^2N/4$) [m^2]
γ	light extinction coefficient [dimensionless]
H	tree height [m]
λ	lichen biomass [kg m^{-2}]
m	total tree biomass [kg of dry or fresh mass]
m_L	tree leaf biomass [kg]
M_μ	total moss biomass ($=\mu+\nu$) [kg m^{-2}]
μ	green moss biomass [kg m^{-2}]
$n(t,\tau)$	tree population age distribution [ha^{-1} yr^{-1}]
N	tree population density [ha^{-1}]
J	ecophysiological invariant [dimensionless]
ν	dead moss biomass [kg m^{-2}]
p	tree seed production [yr^{-1}]
P	population seed production [yr^{-1}]
R_I	light intensity [arbitrary units]
R_μ, R_λ	incident light on moss and lichen surface [arbitrary units]
$R_\mu(h)$	vertical profile of light in moss layer [arbitrary units]
R_S	amount of soil resource [arbitrary units]
Re	tree respiration losses [kg yr^{-1}]
S_L, S_R, S_C	tree leaf, root, and crown projection area [m^2]
S_μ	moss leaf area [m^2]
σ	population leaf area index [dimensionless]
σ_c	population crown projection area [dimensionless]

$\sigma^+(h)$	population leaf area index above height h [dimensionless]
t	time [yr]
τ	tree age [yr]
U	seed viability connected with moss and lichen effect [yr^{-1}]
V	density-dependent component of tree survival probability [yr^{-1}]
W	density-independent component of tree survival probability [yr^{-1}]
Y	tree survival probability $(=1-d)$ [yr^{-1}]

14 A spatial model of long-term forest fire dynamics and its applications to forests in western Siberia

M. Ya. Antonovski, M. T. Ter-Mikaelian and V. V. Furyaev

14.1. Introduction

In this chapter we will present our achievements in developing a spatial model of long-term forest fire dynamics. By 'long-term' we mean dynamics over hundreds or thousands of years rather than changes in forest patterns over one fire season. By the word 'spatial' we denote a model that describes the dynamics of a large, non-homogenous (from the ecological viewpoint) forested territory, taking into account interactions between adjacent landscape units.

Three types of spatial interaction are known to take place during long-term dynamics of boreal forests. These are the spread of fire, seed propagation and the spread of insects. We will consider only the first two types, for the following reasons.

(1) To us the methodologically correct way of developing a new model is to include new mechanisms and to verify the model at each stage of development. Because of the absence of spatial models describing forest dynamics on such time and space scales[1] it is desirable to decrease the number of new spatial interactions to be included in the model.

(2) According to the majority of forest researchers, wildfires are a dominant factor controlling formation and maintenance of boreal forest communities. In fact, present boreal forests represent a mosaic of different areas, each of post-fire origin. Thus the spread of fires seems to be the first type of spatial interaction to be included in the model. On the

---

[1] It should be emphasized that we are talking about models of long-term forest dynamics, taking into account the spatial interactions listed above, and not about the models that describe the processes of these interactions; the latter are widely represented in the literature.

other hand, taking into account the spread of insects without considering wildfires is not quite correct because the spread of insects inevitably leads to burning of damaged forested territory.

(3) Finally, seed propagation should be included in the model in order to describe post-fire invasion of burned territory. Verification of the simple spatial model (Ter-Mikaelian & Furyaev 1988) showed that ignoring seed propagation leads to qualitatively incorrect results of model runs.

In the early stage of our work on modeling forest fire dynamics we encountered a variety of opinions both on the general behavior of boreal forests and on different factors controlling fire processes; the following brief review partly reflects this variety.

14.2. The approaches available

For our purposes it is possible to classify models first as to whether they are locally or spatially distributed and second as to whether they are short-term or long-term. The following discussion does not pretend to be a complete review of existing models; it is intended to help us formulate the main questions to be answered. Since we are particularly interested in modeling long-term forest fire dynamics over large areas, we discuss only the models most closely related to our interests and purposes.

14.2.1. Short-term spatially distributed models

The models of this group simulate the pattern of a single forest fire. In all models a forest is considered to be a homogeneous surface conductor of fire; the rate of fire spread is assumed to be dependent on spread direction (thus the influence of wind is taken into account). Two approaches to modeling a fire process of this kind exist.

In Bajenov (1982), for a currently burning point on the surface, the location of points that will be ignited in the next moment is determined. The 'burning spot' is the conjunction of points currently burning and already burned, taken with their neighborhoods. Thus, a process of unfading fire spread is defined and the problem of localizing the 'burning spot' is investigated analytically. This approach seems to be of more theoretical than practical interest.

The technique used in another approach (Vorobiyov & Valendik 1978; Vorobiyov & Dorrer 1974; O'Regan, Kourtz & Nozaki 1976) is simulation of random fire spread on a grid with certain probabilities of fire spread from one cell to adjacent ones. (The partial case of this fire process (with probabilities of fire spread equal to 1) is a discrete analog of the process defined in Bajenov (1982).) In models using this approach a number of problems have been studied, for example calculation of mean size of a single fire; model tests coincide satisfactory with results of field experiments. For our purposes this approach is convenient for modeling fire processes during one fire season.

14.2.2. Long-term locally distributed models

Long-term locally distributed models can be conventionally subdivided into three groups. Let us discuss them in consecutive order.

Statistical models

Statistical models usually describe either distribution of fire intervals (i.e. period of time between two successive fires) or age structure of the studied area (Van Wagner 1978; Johnson 1979; Suffling 1983; Johnson & Van Wagner 1985); here and below by the term 'age structure' we denote not age structure of a stand but the overall age structure of a large forested area (i.e. the parts of an area occupied by forest of a certain age, the age being equal to the length of time since the last severe fire). In both cases either a negative exponential distribution or a Weibull distribution is used to describe observed data. The main features of these models are now described.

First, they are helpful for predicting fire occurrence within a concrete area, but they are of little use for describing spatial effects of forest fire dynamics such as the size of a single fire, the part of an area burned per year and so on. It is theoretically possible to expand these models and make them spatially distributed, for example to consider two-dimensional statistical distributions of fire intervals and size of area burned per year, or the like. However, estimation of the parameters of these distributions will cause a non-proportional increase in requirements for field data, so this way seems less appropriate.

Second, the parameters of distributions cannot usually be physically interpreted as this would restrict the possibilities of models of this kind and especially their application to predictions of future forest patterns.

Markovian models

Markovian models simulate dynamics of a single plot (which is considered to be ecologically homogenous) as a random trajectory of the Markov chain (Martell 1980; Kessell 1982; Korzukhin & Sedych 1983; Hall *et al*. 1987). The succession line represented in the plot is divided into successional stages. These stages compose the set of possible states of the Markov chain. The set of possible transitions with corresponding probabilities is defined within the set of states. The transition probabilities are assumed to be constant, i.e. they do not change with time. Fires correspond to transitions into stages with lower age; in particular, severe fires correspond to transition into a 'zero state', i.e. a state with age equal to 0 (completely burned area).[2]

The model developed in Kessell (1982) is the most complete. Kessell also raises, for the first time, a question about the necessity of including in the model the interaction between adjacent forest plots. Moreover, he proposes a concrete kind of interaction to be included in the model, namely seed propagation from one forest plot to adjacent ones. Unfortunately we have not seen his more recent papers but according to his earlier ones, the work on creating such a model was already in progress.

The description of a single plot used in Markovian models seems to us most reasonable at the moment; to us a more detailed description (e.g. including in a model the number of trees within the plot, their age structure, biomass, etc.) is premature. At the same time we again need to emphasize that correct simulation of spatial aspects of forest fire dynamics requires including in the model processes of spatial interaction. The kind of interaction proposed by Kessell is essential but not sufficient because the occurrence of major fires burning large forested areas is hardly probable in a model in which probability of burning is independent of what is happening in adjacent plots. In other words, fire spread simulation should be obligatorily added.

Gap models

Gap models simulate succession within a small plot (equal to 1/12 ha). A number of gap models (Shugart & Noble 1981; Kercher & Axelrod 1984b; Bonan 1988c) simulate the influence of fires on tree growth, mortality and reproduction. The merit of these models is that they describe the dynamics of a gap in detail. At the same time this merit is a shortcoming from the viewpoint of spatial modeling. Theoretically it is possible to simulate the dynamics of a large area as the mutual dynamics of a great number of interacting patches or plots. However, for the present state of computer development, running such a 'multi-gap model' would require a great deal of computer time. On the other hand, the output information from gap models seems to be redundant from the viewpoint of simulating dynamics of large forested areas. In our opinion, a more simple and aggregated model would be a more effective tool for simulating such dynamics, at least with the present state of the model art.

General remarks

A few general remarks on locally distributed long-term models should be made. Forest fire dynamics, as it is simulated in Markovian models,

---

[2] Let us remember that by the term age we denote the length of time since the last severe fire.

inevitably leads to settlement of a stable state. The state of a forest is considered to be stable if its age structure does not change over long periods of time. It is obvious that age structure is stable only when it is monotonically decreasing. In fact, the same assumption is made in statistical models dealing with a negative exponential distribution although it is validated only in the paper by Van Wagner (1978).[3] However, most age structures are not monotonically decreasing; they have at least one obvious global peak and a number of local ones (Heinselman 1973; Furyaev & Kireev 1979; Tande 1979; Suffling 1983). The hiatus in age structures after the year 1900 is usually explained through fire control. To us this assumption should be tested carefully because there is another reason for doubting the monotonically decreasing shape of an age structure, namely that there are irregular fires of high intensity that burn large forested areas and therefore cause peaks in age structure.

In fact, we have two alternatives. The first is to assume that forests are in general in a stable state (in the sense of stability of age structure); currently observed non-monotonic age structures are intermediate between two stable states of forest, the change of states being caused by external factors (increasing fire control). The second hypothesis is that the state of forest is generally unstable; this hypothesis involves another question, namely whether the instability of a forest is caused by climatic fluctuations or whether it can be explained by internal reasons (as a possible reason one could mention the accumulation of large amounts of fuel over large areas which is favorable for the occurrence of major fires). We will return to this problem later, because we consider it very important for the correct modeling of forest fire dynamics.

14.2.3. Long-term spatially distributed models

Unfortunately, applications within this group are the poorest of the three. In fact, only the model created by Marsden (1983) includes interaction between forest plots during dynamics, i.e. the probability of transition from a fire-burned plot to the next successional stage includes as a multiplier the ratio of plots at reproductive age to the total number of plots; thus he realizes the proposal made by Kessell (1982). The insufficiency of including only this type of spatial interaction was discussed above.

14.2.4. Review of background results

Let us now summarize the questions to be answered in the first place, which were formulated during the above brief review of existing approaches:

---

[3] Van Wagner considers forest areas to be large numbers of same-aged, equally sized stands, each having an annual probability of burning independent of age of stand; it follows that the age structure of the study area tends to be a negative exponential.

(1) Are present boreal forests in a stable state or is this state essentially unstable?
(2) To what extent is the size of wildfires controlled by fluctuations in climatic parameters? In other words, is it really necessary to include processes of spatial interaction between adjacent forest plots when modeling the spatial effects of forest fire dynamics or can the dynamics be explained solely in terms of fluctuations in climatic variables?

In Antonovski & Ter-Mikaelian (1987) and Ter-Mikaelian & Furyaev (1988), we presented the first version of a spatial model of forest fire dynamics. In this model a large forested area was simulated as a grid, each cell representing a forest stand. The Markovian model was used for simulating the dynamics of a single cell, with fire spread and seed propagation included as the interaction processes. The model was verified using data on long-term forest fire dynamics in North America (Heinselman 1973; Tande 1979). The following conclusions were drawn from the results of this testing of the model.

1. Boreal forests are not in a stable state but there is a stable fire regime in which during most years only a small area of forest is burned, with years of major fires occurring at irregular intervals.

2. There is a 'synchronization' effect of forest over a large area, presumably related to the accumulation of a similar, large amount of combustible material over the area. This can be shown by performing model runs with constant probabilities of forest fire maturity, which is a key parameter. The effect leads to the fire regime described above, even when the influence of climatic fluctuations is eliminated (i.e. key model parameters are constant). This proves the necessity of a spatial approach to modeling fire dynamics for large forested areas.

We consider these results to be a good background for further development of a spatial model of forest fire dynamics. The particular aim of this chapter is to include the influence of climate in this model.

14.3. Object of the study

In the model presented in this chapter we use a landscape approach to describe forest dynamics. The essence of this approach is the following. The area to be modeled is subdivided into ecologically homogenous plots, otherwise known as 'cells'. This is a usual procedure in geographical studies, with the rank of cells depending on the problem under consideration and the available data. In all cases an area is entirely covered by a grid of cells.

Each cell of a given type has its own set of ecological conditions (ecotype). Each type of cell is characterized by fire frequency and by only one type of post-fire successional dynamics (successional line). This last feature is a basic merit of the landscape approach because it allows the accurate prediction of the process of forest development after fire. For each successional line we consider only one sequence of successional stages (ignoring secondary successional lines). The duration of each successional stage in the absence of fire is considered to be constant and equal to the mean duration of the stages. Thus the complete area represents a mosaic of cells, each of which is characterized by the number of successional lines and the number of successional stages. According to our assumptions formulated above, a particular successional stage will correspond to a particular age of forest, which will be equal to the length of time since the last severe fire.[4]

Three types of spatial interaction between forest cells are known to take place during forest fire dynamics. These are the spread of fire from one burning cell to adjacent ones, seed propagation, and the spread of insects. We will consider only the first two types and shall now look at these in more detail.

First of all let us describe an idealized mechanism for the influence of fire on a grid. We are concerned only with severe fires, after which the vegetation of a cell is completely burned.

Let us assume that within a particular cell a source of fire occurs. If the vegetation in this cell is dry enough or, in other words, 'ready for burning' (i.e. in a state of fire maturity), then fire occurs and the vegetation in the cell is completely burned. If the vegetation of adjacent cells is not fire-mature then the fire stops on the margins of the burned cell; otherwise, fire is transferred to all adjacent fire-mature cells and the process continues. In this idealization, it is essential that the final pattern of the fire-burned area coincides with the conjunction of a few adjacent cells; this saves us from the necessity of considering the problem of changing sizes of cells during forest dynamics.

According to the mechanism described above, occurrence and spread of fire involve three events: the occurrence of a source of fire within a cell; the existence of fire-mature vegetation in this cell; and the spread of fire from the burning cell to adjacent ones. The most convenient parameters with which to characterize this process are the probability of the occurrence of a source of fire, the probability of fire maturity of a cell and the probability of the spread of fire from one cell to an adjacent one. As our goal is to describe long-term forest dynamics it seems reasonable

---

[4] A division of the successional line into stages is not of principal importance in the context of the current work. In fact, for us this division is nothing more than an aggregation of the data and a reduction in the number of model parameters.

to set the temporal step of this simulation at one year. A few comments on the parameters determining these probabilities should be made.

(1) We do not distinguish between natural and man-induced sources of fire; a soure of fire is simply caused by some or other external factor. The probability of the occurrence of a source of fire is independent of a cell's successional stage.

(2) The probability of fire maturity of a cell depends strongly on the successional stage of the cell and the climatic conditions of the current year.

(3) The probability of the spread of fire depends both on the successional stage of a cell (because the successional stage determines the chances of the fire being able to overcome various natural obstacles such as rivers, open areas, etc.) and on climatic conditions (most importantly direction and speed of wind).

(4) We assume all these probabilities to be independent of the location of a cell within the area under consideration.

Let us turn now to the mechanism for seed propagation. To include this in the model requires firstly the determination of the reproductive age of all species in the area under consideration and secondly, the determination of the probability distribution functions of the seed propagation distances (or at least mean values of these distances), again for all species. Knowledge of the probability distribution functions is necessary for calculating the number of seeds available to each fire-burned cell. This consideration clearly requires that all secondary successional lines be taken into account. In order to simplify the model we opted for using the following idealization of the seed propagation process.

For each successional line represented in the area to be modeled we define (1) the reproductive stage, and (2) the mean seed propagation distance averaged over all species. By reproductive stage we denote the successional stage by which all species represented in this line are assumed to have attained reproductive age, and therefore all seeds necessary for complete regeneration of this line are available. For each fire-burned cell one determines whether there exists a cell of the same type (i.e. belonging to the same successional line) with vegetation at a reproductive stage. If such a cell is available and the distance between this cell and the fire-burned cell is less than or equal to the seed propagation distance of the successional line to which both cells belong, then the fire-burned cell is assumed to be invaded and succession begins; otherwise the cell is assumed to be uninvaded.

14.4. Formulation of the model

There are two possible approaches to formulating the spatial model of forest dynamics described in the previous section. The first approach is to construct a system of differential equations that describe the dynamics in terms of those parts of the area occupied by forest at the same successional stage; these equations should be essentially nonlinear in order to describe spatial interactions between cells during the dynamics. The second approach is to construct a simulation model that produces random trajectories over a large area according to the mechanisms described above. We have adopted the second approach because it is more convenient in terms of including spatial interactions in the model.

Let us now turn to the formulation of the model. From the assumptions made above, it is seen that the parameters 'successional stage of cell' and 'age of cell' (i.e. the time since the last severe fire) are in fact equivalent. This means that if the successional stage of a cell is known, the age can easily be calculated, and vice versa. In the formulation of the model we will use only the parameter 'age of cell' in order to avoid possible confusion.

Consider a grid of size $L \times M$, where L is the number of rows and M the number of columns. The grid is a model pattern of a forested area; each cell of the grid represents a stand. In order to exclude possible marginal effects during dynamics the grid is assumed to be closed, i.e. cells $(i, 1)$ and $(1, j)$ are considered to be adjacent to cells (i, M) and (L, j) respectively, for $i = 1, \ldots, L, j = 1, \ldots, M$. The total number of successional lines represented in the grid is equal to N. Thus the state of each cell at time t is determined by the coordinates (n, k), where n is the number of the successional line to which this cell belongs, and k is the age of the cell.

Let R_n be the reproductive age of the nth successional line and D_n be the mean seed propagation distance of the nth line. This last parameter means that if cell (i, j) is at a reproductive age (i.e. its age is greater than or equal to R_n), the seeds from this cell can be transferred to all cells (i_1, j_1) that satisfy the condition

$$|i_1 - i| + |j_1 - j| \leq D_n. \tag{14.1}$$

Consider now the parameters controlling fire processes. Let Q be the probability of occurrence of a source of fire in one cell during one year. Let $P_{n,k}$ be the conditional probability that the vegetation of a cell whose state is (n, k) will burn should a source of fire occur within it. (From now on we will refer to this probability as the probability of fire maturity.) Let $V_{n,k}$ be the conditional probability that fire from a burning cell in a state (n, k) will spread to adjacent cells (i.e. the fire spread probability). We have omitted here the index t, indicating the dependence of these probabilities on time t, which is in fact dependence on climatic conditions. For details of inclusion of this dependence in the model, see Section 14.5. Finally, we assume the probabilities Q, P and V to be independent of the location of a cell within the grid.

The dynamics of the grid during one year (the temporal step of the model) are simulated in the following manner.

At the beginning of year t, cell (i, j) is in state (n, k). During year t the cell may be ignited; ignition may either occur within the area of the cell (the probability of this event is equal to Q) or it may be initiated from an adjacent cell that is already burning, i.e. $(i - 1, j)$, $(i + 1, j)$, $(i, j - 1)$, or $(i, j + 1)$. After being ignited the cell may burn with a probability $P_{n,k}$; if so the cell in its turn becomes a source of fire for adjacent unburned cells and the spread of fire may occur with a probability equal to $V_{n,k}$. Adjacent cells may then burn with probability P depending on their state, and so on.

If burned, a cell changes its state to $(n, 0)$; as mentioned above, only severe fires are taken into account. If seeds of the nth successional line are available (i.e. there exists at least one cell in the state $(n, s), s \geq R_n$, which satisfies condition (14.1)), the succession process will begin. The state of the cell will change again and at year $t + 1$ become $(n, 1)$. If no seeds are available, the cell will remain unoccupied in state $(n, 0)$.

If a cell is not burned, its age increases, with its state changing to $(n, k + 1)$.

There are a number of comments on differences between this model and the previous version presented in Antonovski & Ter-Mikaelian (1987), as follows.

1. In the previous version it was assumed that if a cell attained age T_n (the maximum longevity of the nth successional line) without burning, it self-destroyed and succession began again. In the current version we have abandoned this assumption, considering it to be unrealistic. At the same time, we have included a new parameter improving the quality of model verification, namely the comparison during model runs of the maximum ages of cells with those obtained from field data. For more details, see Section 14.6.

2. In the previous version, the process of fire spread was controlled by a single set of parameters P. It was assumed that fire would always spread from a burning cell to adjacent ones. In the current version we have divided the fire spread process into two parts. The first is the burning of a cell, controlled by parameter P, and the second is the spread of fire, controlled by parameter Q. The reasons for this were the following.

(a) The assumption that fire always spreads from a burning cell to adjacent ones is not correct. The fire spread process is evidently controlled not only by fire maturity of the forest but also by various obstacles (such as rivers, lakes, open areas, etc.), as well as by wind direction and speed.

(b) The use of probabilities of fire maturity in the current version was decided upon because these probabilities could easily be changed depending on climatic conditions. At the same time, the only way of estimating the accuracy of probabilities used in the previous version was to compare results of model runs with observed data; this made them virtually useless for investigating the effect of climatic fluctuations.

(c) Measurements of fire maturity carried out in a test region in western Siberia showed that the probability of forest fire maturity decreased markedly with the age of the forest. If this probability were the only parameter controlling the fire spread process, it would cause an irreversible aging of cells in the model. On the other hand, tests of the previous version of the model using North American forests showed that only when this probability increased with the age of forest did the results of model runs correspond to observed data.

These reasons forced us to consider two sets of 'fire probabilities' instead of the one used before. The following sections contain additional details on these probabilities.

14.5. Description of the test region

The region chosen for testing the model was a forested area in western Siberia. This region was studied for 20 years by one of the authors, who carried out numerous field experiments there. The results of this study are presented in Furyaev & Kireev (1979). Here we will give only a brief summary of those data important for model testing.

The area studied is situated on the Kas-Eniseyskaya plain in western Siberia at a latitude of 59°N and a longitude of 90°E, and covers an area of 165 000 ha. Four main successional lines are represented in the area. These are:

1. Fir-tree forests on loamy soils;
2. Spruce forests on humid loamy soils;
3. Spruce forests on damp peaty loamy soils;
4. Pine forests on sands.

In the following description we will use these numbers to refer to the different successional lines.

Table 14.1 shows the percentage of the total area occupied by forests of each successional line.

It is necessary to emphasize once more that successional lines are distinguished on the basis of a landscape approach (types of soils are

Table 14.1. *Percentage of area occupied by each successional line*

	Successional line			
	1	2	3	4
Area occupied (% of total)	11.1	34.4	25.2	29.3

Table 14.2. *Age margins of successional stages (years)*

No. of line	Successional stage							
	1	2	3	4	5	6	7	8
1	0–3	4–10	11–40	41–80	80–100	101–140	141–160	>160
2	0–3	4–10	11–40	41–60	61–120	121–160	161–180	>180
3	0–4	5–10	11–20	21–60	61–80	81–120	121–160	>160
4	0–1	2–15	16–40	41–80	81–120	121–160	161–180	>180

Table 14.3. *Maximum longevities of successional lines (years)*

	Number of line			
	1	2	3	4
Maximum longevity	250	300	300	320

Table 14.4. *Total stand biomass of each stage (m^3/ha)*

No. of line	Number of stage							
	1	2	3	4	5	6	7	8
1	0	25*	110	215	290	280	280	260
2	0	25*	105	172	182	240	280	260
3	0	10*	38*	54	140	148	205	240
4	0	47	82	188	280	300	—	—

Note: * Data marked with asterisks were not measured and have been calculated by using interpolation.

Table 14.5. *Critical values of FDI (thousands of mbar deg)*

No. of line	Number of stage							
	1	2	3	4	5	6	7	8
1	0	0.5	1.0	1.2	3.0	3.3	5.5	8.0
2	0	0.5	0.9	1.0	2.8	3.0	5.8	8.2
3	0	1.0	1.2	1.5	3.2	3.8	6.2	8.6
4	0	0.3	0.5	0.6	0.8	1.2	—	—

taken into account). This means that we consider the area occupied by each successional line to be the same over all trajectories of forest dynamics, which allows us to predict accurately the post-fire successional process for each cell.

All successional lines are divided into eight stages. The division was performed according to a scheme developed in Kolesnikov (1956) and Kolesnikov & Smolonogov (1960). For our purposes, complete names of these stages are unimportant so we shall not list them here. The important information we need from this division is the length of each stage together with corresponding characteristics. Table 14.2 shows the age margins for each stage; as mentioned above, primary successional lines only are considered.

Table 14.4 shows the average stand biomass for each stage within each successional line.

Table 14.5 presents data on the fire-danger index (FDI) used in Soviet forestry. This index is calculated as the sum of air temperature at 1300 hours (AT) multiplied by air humidity at 1300 hours (H) for each day over the period since the last 'rainy' day (i.e. day on which precipitation was greater than or equal to 3 mm):

$$FDI = \Sigma \, (AT \times H). \qquad (14.2)$$

In the study area, series of field experiments were carried out to estimate critical values of FDI. For each successional line and for each successional stage a certain cell was ignited daily. On the day on which ignition led to settlement of fire, the value of the FDI was calculated with the help of meteorological data. The value obtained was the critical value of the FDI at which the forest attained a state of fire maturity. A summary of these values is presented in Table 14.5.

The period from 1 May to 30 September (153 days) was taken as the fire-dangerous period. By the term 'fire-dangerous' we denote a period during which wildfire can occur. Thus with the help of Table 14.5, probabilities of fire maturity can be calculated for each fire-dangerous season. These probabilities are equal to the number of days during the

Table 14.6. *Fire years and percentage of area burned*

Year	Area burned (%)	Year	Area burned (%)	Year	Area burned (%)
1700	6	1830	18	1920	11
1787	5	1842	5	1921	14
1788	10	1860	44	1930	24
1793	16	1866	5	1931	8
1806	13	1870	84	1933	19
1808	15	1888	19	1946	17
1819	12	1891	7	1950	8
1820	5	1896	23	1952	12
1825	6	1909	5	1956	38
1829	5	1915	84	—	—

season which have an FDI higher than the appropriate critical value, divided by 153.

Let us now turn to fire dynamics of the area under consideration. One of the most important merits of the study area for model testing is that parts of its fire history have been reconstructed. To our knowledge, studies involved with reconstructing forest-fire history have been undertaken for three regions in North America and for the present study area in western Siberia. The results of these studies are presented in Heinselman (1973), Furyaev & Kireev (1979), Tande (1979), and Payette *et al.* (1989*b*). Payette *et al.* (1989*b*) is especially notable as it contains excellent detailed data on fires over the period 1894–1984 for a large area (about 54 000 km^2). Data on the fire history of the study area are not so detailed but cover a period from the beginning of the eighteenth century until 1970. Table 14.6 contains the dates of fire years together with the corresponding percentage of area burned.

The dynamics of the percentage of the area burned during each fire year are shown in Fig. 14.1*a*. From these data, it would be useful to calculate and present the corresponding distribution of percentage of area burned over the period; this distribution is of considerable value for model testing. Suppose that during a period of time T the number of years with fire recorded is equal to S. Let us denote S_{1-10} as the number of fire years during which the area burned was between 1% and 10% of the total area. By the same method we can define S_{11-20}, S_{21-30}, etc., with their sum being equal to S. The resulting set of quantities $S_{1-10/S}$, $S_{11-20/S}$, ... gives us a distribution of the percentage of area burned over the total period, constructed with a size step of 10%. This distribution for the area under consideration is presented in Fig. 14.1*b*.

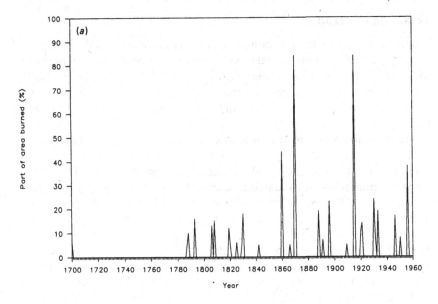

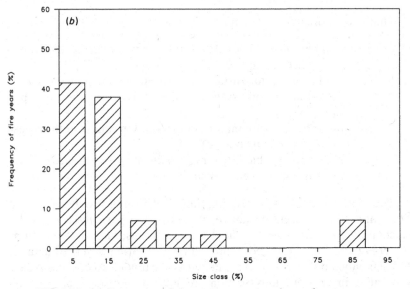

Fig. 14.1. (*a*) Dynamics of the percentage area burned per year on the Kas-Eniseyskaya plain. (*b*) Distribution of the percentage area burned per year (observed data).

14.6. Model setup

Verification of the model requires firstly estimation (or calculation if possible) of values of the model parameters and, secondly, formulation of the criteria for fitting the model to the observed data. First of all we will recap the model parameters. Before doing this there is one point worth noting. The model was formulated in terms of 'age of cell' although it was pointed out that the terms 'age of cell' and 'stage of cell' are in fact interchangeable. This is so in this case because we are considering only primary successional lines, and specification of successional stages is nothing more than division of these lines into segments of fixed lengths. Thus the model was formulated in terms of 'age of cell' simply to avoid additional explanations connected with the transition of cells from one successional stage to the next one. However, for verification purposes the use of 'stages' is very helpful because firstly it decreases the number of model parameters and secondly, data available are related to successional stages (and there are not enough detailed data to connect them directly with ages of cells). Therefore in this section, further description and use of parameters will be given in terms of 'stage of cell'. Values of parameters are assumed to be constant within each stage, so that for each stage mean values of the parameters (i.e. averaged over this stage) are used.

Thus the model parameters are:

N number of successional lines represented over forested area;

R_n reproductive stage of nth successional line;

D_n seed propagation distance for nth successional line;

Q probability of occurrence of fire source during one year per square unit;

$P_{n,k}$ probability of fire maturity of cell in kth stage of nth successional line during one year;

$V_{n,k}$ fire spread probability for cell in kth stage of nth successional line during one year per square unit.

The main parameters controlling the simulation of fire dynamics are Q, P and V. As described above, values of the parameter P can be calculated from the results of field measurements; values of the parameters Q and V can only be estimated by comparing observed data on forest fire dynamics with those simulated by the model. According to data presented in the previous section we have to estimate at least 33 parameters, namely the probability of the occurrence of a fire source Q, and the set of fire spread probabilities $V_{n,k}$, $n = 1, \ldots, 4, 1, \ldots, 8$. To begin verification of a model with 33 unknown parameters is a difficult task of questionable value. Therefore in order to decrease the number of

Table 14.7. *Characteristics of a generalized successional line*

Number of stage	Age margins of stage (years)	Total stand biomass ($m^3\,ha^{-1}$)	Critical values of FDI (10^3 mbar deg)
1	0–1	0.0	0.0
2	2–15	28.3	0.55
3	16–40	83.1	0.85
4	41–80	154.4	1.00
5	81–120	213.9	2.30
6	121–160	240.6	2.65
7	161–180	267.9	4.45
8	>180	266.8	6.10

parameters to be estimated we will consider one generalized successional line over the whole area. The characteristics of this line are presented in Table 14.7 and were obtained as weighted sums of the corresponding characteristics of the four successional lines represented in the test area (Tables 14.3–14.5) according to the percentage of the area occupied by each line (Table 14.1).

Since the number of successional lines N is equal to 1, from now on we will omit the first subscript n. The number of successional stages k varies from 1 to 8.

The forest area to be modeled was simulated as as grid of 25×25, with each cell representing 264 ha of test area. The reproductive stage R was set to 4; thus the mean reproductive age of the forest was 60 years. Seed propagation distance D was set to 2 (in cells), which means that maximum seed propagation distance was 4 km.

We must now consider the dependence of 'fire probabilities' on climatic fluctuations. Let us first turn to the probabilities of fire maturity P_k. As shown above, P_k can be calculated as the ratio of days with an FDI higher than the corresponding critical value to the total length of the fire-dangerous season. In order to study the dynamics of P_k, hydrometeorological data from the station nearest to the test area were considered. These data comprise daily measurements (four times a day) of the main hydrometeorological parameters for the period 1936–80. With the help of these data, the time series of P_k, $k = 2, \ldots, 8$ ($P_1 = 0$) was calculated. Figure 14.2a shows the time series of P_4 (which corresponds to a critical value of FDI of 1000 mbar deg; see Table 14.7). For $k = 2, \ldots, 8$, an autocorrelation function was constructed in order to check whether there existed a trend in values of P_k; no trend was found.

Our next step was to link P_k with some general climatic parameters.

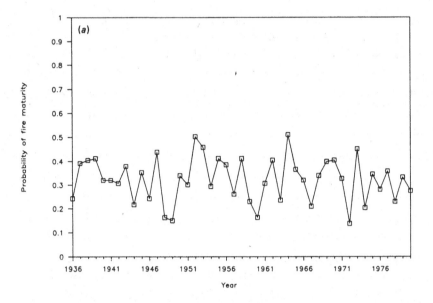

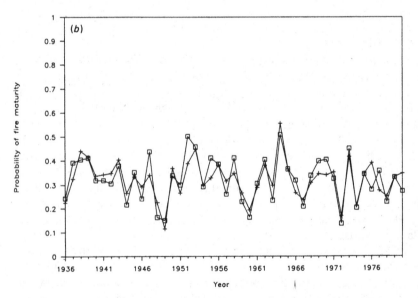

Fig. 14.2. (*a*) Dynamics of the probability of fire maturity for an FDI equal to 1000 mbar degrees. (*b*) Dynamics of calculated and estimated probability of fire maturity. Open squares, calculated curves; pluses, estimated curve.

Table 14.8. *Regression coefficients and multiple regression coefficients of Eqn (14.3)*

Number of stage (k)	Regression coefficients				Multiple regression coefficient
	a_{1k}	a_{2k}	a_{3k}	a_{4k}	
2	0.0511	0.00363	−0.00065	−0.0973	0.824
3	0.0491	0.00561	−0.00063	−0.2217	0.858
4	0.0509	0.00625	−0.00064	−0.2954	0.850
5	0.0326	0.00677	−0.00017	−0.3780	0.780
6	0.0232	0.00660	−0.00019	−0.2705	0.794
7	0.0109	0.00681	0.0000	−0.2354	0.798
8	0.0068	0.00493	0.0000	−0.1660	0.764

This was important for two reasons: firstly, to avoid detailed simulation of parameters AT and H during each fire-dangerous season and, secondly, to produce a tool for constructing various scenarios of possible climatic changes in terms of some general parameters. To achieve this, methods of linear regression were used. The climatic parameters initially used as predictors were:

(a) mean annual air temperature;
(b) mean seasonal air temperature;
(c) annual sum of precipitation;
(d) seasonal sum of precipitation;
(e) mean seasonal period between two successive rains;
(f) maximum seasonal period between two successive rains;
(g) minimum seasonal period between two successive rains.

By 'seasonal' we mean that the parameter is averaged over the fire season, which runs from 1 May until 30 September.

This list was reduced with the help of stepwise regression. Parameters selected as significant were (b), (d) and (f) for P_k, $k = 2, \ldots, 6$, and (b) and (f) for P_k, $k = 7, 8$. For the reduced list of parameters, coefficients of the following linear regression equation were estimated:

$$P_k = a_{1k}TR + a_{2k}MP + a_{3k}PR + a_{4k}, \qquad (14.3)$$

where TR is mean seasonal air temperature, MP is maximum seasonal period between two successive rains, and PR is the seasonal sum of precipitation.

Coefficients a_{jk}, $j = 1, \ldots, 4$, $k = 2, \ldots, 8$, and corresponding multiple regression coefficients are presented in Table 14.8.

An example of estimated values of P_4 is shown in Fig. 14.2b.

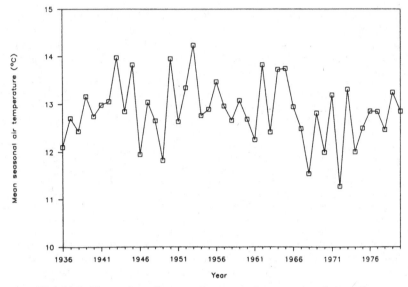

Fig. 14.3. Dynamics of mean air temperature averaged over fire season.

Since the problem of simulating probabilities of fire maturity P_k is reduced to simulating parameters *TR*, *PR* and *MP*, we studied dynamics of these parameters. Figures 14.3, 14.4 and 14.5 show the dynamics of *TR*, *PR* and *MP* respectively for the period 1936–80.

Table 14.9 contains a matrix of correlation coefficients between the three parameters. Low values of correlation coefficients allowed us to simulate the dynamics of *TR*, *MP* and *PR* independently. This result agrees with the conclusion made by Budyko & Izrael (1987) that there was no significant correlation between the dynamics of mean annual air temperature and annual sum of precipitation for western Siberia.

For each parameter an autocorrelation function was constructed; no trends were found. This meant that the time series of each parameter could be considered as a sequence of values of a randomly distributed variable. For each parameter a series of tests was carried out in order to fit a probability distribution which satisfactorily described sample data. Table 14.10 summarizes the results of these tests. It shouls be noted that parameters of normal distribution for *TR* and *PR* were estimated by using standard methods; parameters for Pearson's III type distribution were estimated with the help of scales of coordinates rectifying theoretical distribution curves. This method was developed by Kolosov in a series of works (see, for example, Kolosov & Liseev (1987)). For each parameter a Kolmogorov–Smirnov one-sample test against theoretical curves was carried out; corresponding probabilities are given in Table 14.10.

Thus, in summary, the probabilities of fire maturity P_k, $k = 1, \ldots, 8$ were simulated as a function (Eqn (14.3)) of mean seasonal air temperature (TR), seasonal sum of precipitation (PR) and maximum seaonal period between two successive rains (MP); coefficients of Eqn (14.3) are given in Table 14.8. At each step of the model, values of TR, PR and MP were generated independently as sample values from the corresponding probability distributions; parameters of these distributions (given in Table 14.10) were assumed to be constant over all model trajectories.

Let us turn now to probabilities of fire spread V_k, $k = 1, \ldots, 8$. Since values of V_k are unknown and should be estimated as a result of model runs, we assumed them to be constant over all trajectories to be modeled. This is equivalent to assuming that climatic conditions (e.g. wind-rose) in the test area which determine these probabilities are stable. For the initial assessment we assumed probabilities of fire spread to be proportional to total stand biomass of 1 ha of cell as given in Table 14.7. As mentioned earlier, the probability of fire spread depends on a fire's ability to overcome natural obstacles. This ability is determined by intensity which is related to stand biomass.

The final parameter to be considered is the probability Q of fire occurrence within one cell during one year. This parameter should also be estimated as a result of comparing model results with observed ones. An

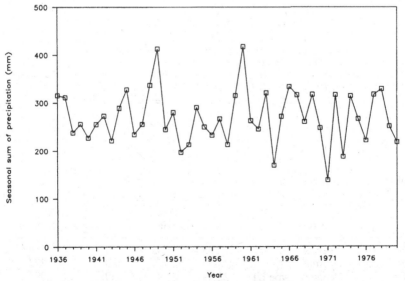

Fig. 14.4. Dynamics of seasonal sum of precipitation.

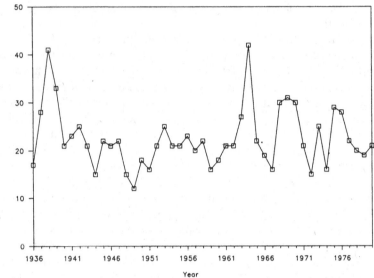

Fig. 14.5. Dynamics of maximum period between two successive rains.

Table 14.9. *Correlation matrix between TR, MP and PR*

	TR	*MP*	*PR*
TR	1.000	—	—
MP	0.136	1.000	—
PR	−0.424	−0.466	1.000

Table 14.10. *Probability distributions for TR, PR and MP*

Parameter	Type of distribution	Mean	Standard deviation	Skewness	K–S test
TR	Normal	12.85	0.652	0.0	0.668
PR	Normal	273.5	52.65	0.0	0.372
MP	Pearson's type III	22.49	6.29	1.0	0.379

initial value of 0.001, obtained in model runs for North American forests (see Antonovski & Ter-Mikaelian (1987)), was taken.

14.7. Model verification

Let us now discuss the criteria of model verification. Since our model produces random trajectories of forest fire dynamics, it seems pointless to directly compare a single trajectory with one reconstructed from field data. However, certain characteristics averaged over these trajectories are stable enough to be compared with corresponding characteristics of real forest fire dynamics. The distribution of percentage of area burned per year was taken as one such characteristic; the construction of this distribution was described in Section 5. This method was successfully used in Antonovski & Ter-Mikaelian (1987) and Ter-Mikaelian & Furyaev (1988) and will now be described briefly. Each model trajectory is divided into periods of length T years. For each period the distribution of percentage of area burned per year is plotted and compared with that constructed from field data; the same is done for the distribution constructed for the whole trajectory. In our model runs, T was taken to equal 300. Since our model is driven by a set of randomly generated parameters it is impossible to obtain a model trajectory for which the distribution of percentage of area burned per year, constructed for different periods T, coincides completely. On the other hand, we did not know whether the observed distribution constructed for the period 1700–1970 (see Fig. 14.1b) would coincide completely with an analogous distribution for another period, for example, 1400–1700, should data on fire dynamics during this period be available. Therefore coincidence between the observed distribution and the distribution for a single period T was taken as a criterion of model fitting. The distribution of the whole trajectory was followed in order to check whether its pattern was stable from the qualitative viewpoint.

As a second criterion, the mean number of years with fires per 100 years, obtained from the model, was compared with the actual value (the most recent value can be calculated from Table 14.6 and is equal to 10.7).

Finally, the mean age of forest in the last (8th) stage, obtained from the model, was compared with that observed in real forests. From Table 14.3 it can be calculated that the actual upper limit of forest age is approximately 300 years. Thus lower and upper age margins of the 8th successional stage are 181 and 300 years respectively, with a corresponding mean age of 240 years. This value was compared with the mean age of cells in the 8th stage as calculated by the model. Since in the model the 8th stage was assumed to be unlimited (see Section 14.4, p. 381), this comparison gives us a third criterion on the quality of model fitting to field data.

Using these criteria, numerous computer experiments on model runs were carried out. During these experiments, values of Q and V_k, $k = 1, \ldots, 8$ were varied in order to obtain the best correspondence between model results and observed data. The best results were obtained for the following values of Q and V_k:

$$Q = 0.0004, \qquad V_k = \{0, 0.47, 0.47, 0.5, 0.65, 0.725, 0.8, 0.8\} \qquad (14.4)$$

The distribution of percentage of area burned per year which coincided most with the observed one was obtained for the period 1200–1500; both distributions are shown in Fig. 14.6a. Figure 14.6b shows the dynamics of the percentage of area burned per year for the same period. The total length of the trajectory simulated was 3000 years. The mean number of years with fires per 100 years for set (14.4) was 9.5; the mean age of forest in the 8th stage was 261 years.

A few comments on the results of the model verification should be made. The first is concerned with the difference in modeled and observed distributions of percentage of area burned per year. As can be seen from Fig. 14.6a the percentage of years with fires belonging to the smallest size class (0–10%) is higher in the model runs than in real forests. This difference is most likely to be due to a shortage of data on small fires. It is clear that field reconstruction of all small fires during the past 270 years is impossible; Table 14.6 contains only years with fires large enough to burn at least 5% of the total area. In our model runs, even one cell being burned a year (which corresponds to 0.16% of the model area) contributes to the first size class of area burned per year, leading to a warped distribution compared with the reconstructed one.

The next point concerns estimated values of V_k. Our hypothesis regarding proportionality between probabilities of fire spread V_k and stand biomasses of corresponding stages is not completely valid, as illustrated by Fig. 14.7. One of the curves represents plotted values of set (14.4) and the other has been plotted proportional to stand biomass with a coefficient of 0.003. The difference is high for the first stages and tends to zero with increase in stage number. One possible explanation is that the biomasses presented in Table 14.7 do not include the lower layer of vegetation (mosses, grass, bushes, etc.). The contribution of this layer to total biomass is high for juvenile forests and low for mature forests. At the same time it is obvious that, being a conductor of fire, this layer increases the probability of fire spread.

The previous section describes an example of a model prediction of the effect of changing climatic conditions on forest fire dynamics. Because driving climatic parameters in the model are mean seasonal air tempera-

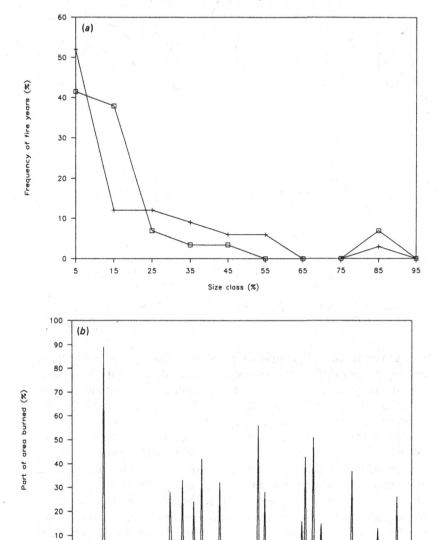

Fig. 14.6. (*a*) Distribution of the percentage area burned per year (observed (squares) and model data (pluses)). (*b*) Dynamics of the percentage area burned per year (model trajectory).

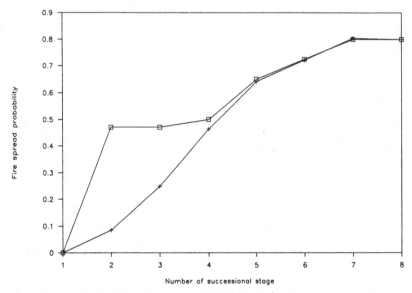

Fig. 14.7. Fire transition probabilities (model estimation (squares) and hypothesized (pluses)).

ture, seasonal sum of precipitation and maximum period between two successive rains during one season, it is possible to create an arbitrary scenario of their changes in time and simulate corresponding forest dynamics. Since mean seasonal air temperature is perhaps the most pertinent climatic parameter to investigate, we undertook a series of model runs for different values of this parameter. In these runs we varied only mean values of this variable, leaving its standard deviation and type of probability distribution unchanged. All other parameters were left unchanged (i.e. their values were set to those obtained in the process of model verification; see Section 14.5). It should be emphasized that we simulated forest fire dynamics under the new climatic scenario with stable parameters of the corresponding probability distributions and not the transitional period from the previous climatic scenario to the new one (although this would also have been possible). Figure 14.8*a–d* presents the results of these runs.

Figure 14.8*a* shows the dependence of the number of years with fires per 100 years on mean seasonal air temperature. It is obvious that higher air temperature causes greater fire maturity of forests which in turn leads to an increase in the number of fire years per 100 years. The shape of the curve showing an increase in mean size of area burned per year (Fig. 14.8*b*) is also expectable. With the help of these data one can easily calculate the corresponding fire rotation periods; fire rotation period is inversely proportional to mean size of area burned per year and rep- resents the period of time during which a particular area will be com- pletely burned.

In contrast, the results shown in Fig. 14.8c, are somewhat unexpected: in this figure we see the mean size of area burned per year with fire decreasing as mean seasonal air temperature increases. There is a possible explanation for this. We have seen that the probability of fire maturity decreases with the age of the forest. Decrease in mean seasonal air temperature leads to a decrease in the probability of fire maturity, resulting in a decrease in the number of years with fires, with a simul- taneous increase in mean age of forest over the study area. At the same time, the size of the fire-burned area increases because the conditional probability of igniting a cell adjacent to the one already burning in- creases. In other words it is more difficult to ignite an older forest but if it is ignited it burns over larger areas.

Finally, Fig. 14.8d shows the dependence of mean total biomass of forest on mean seasonal air temperature. One should take these results with caution: being largely concerned with the simulation of spatial forest dynamics over large areas, we neglected a detailed description of a single cell's growth, particularly its dependence on climatic conditions. There- fore changes in total biomass presented in Fig. 14.8d are linked with changes in mean age of forest. This means that our model does not take into account possible increases in biomass of a single cell caused by increase in mean seasonal air temperature. These increases could signifi- cantly change the shape of the curve presented in Fig. 14.8d.

Nevertheless we consider these results to be a valuable first step toward predicting changes in patterns of large forested areas caused by possible climatic changes. We hope that the next stage in the develop- ment of our model will include a more detailed description of a single cell's growth, making the predictions more accurate.

14.8. Conclusions

Of the findings detailed in this chapter we consider three aspects to be of particular importance.

First, the model successfully simulates the effect of pulsing of a fire- burned area. This effect is reflected in the bimodal shape of the distri- bution of area burned per year and corresponds to a fire regime in which, during some years only, a small area of forest is burned, with years of major fires occurring irregularly. This regime is the most important feature of the dynamics of large forested areas; its correct description requires the use of spatial models incorporating interaction of different forest cells during dynamics.

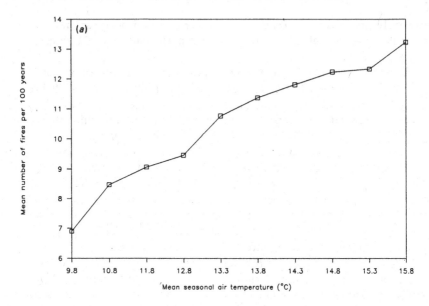

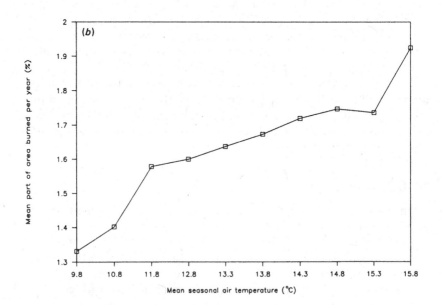

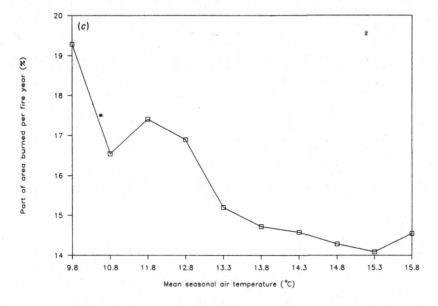

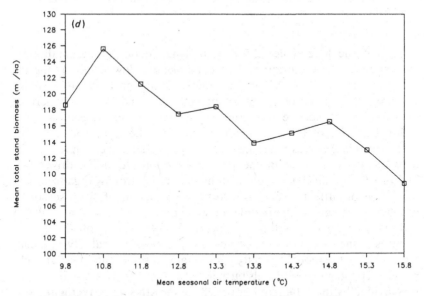

Fig. 14.8. (a) Dependence of mean number of fires per 100 years on mean air temperature. (b) Dependence of mean percentage area burned per year on mean air temperature. (c) Dependence of mean percentage area burned per fire year on mean air temperature. (d) Dependence of mean total stand biomass on mean seasonal air temperature.

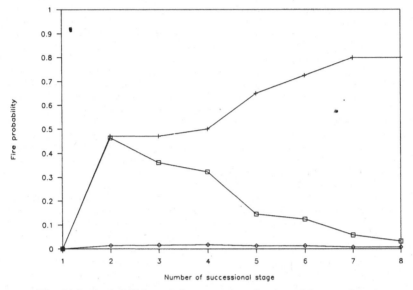

Fig. 14.9. Probabilities of fire maturity, fire transition, and real burning. Squares, fire maturing; pluses, fire spread; diamonds, real burning.

Second, the dependence of forest fire dynamics on climatic conditions was taken into consideration. This dependence allowed the linking of forest dynamics with such general climatic parameters as mean seasonal air temperature, seasonal sum of precipitation and maximum period between two successive rains during one season. Being general, the future behavior of these parameters and thus the future behavior of forests over large areas can more easily be predicted. An example of such a prediction was given in the previous section. A model driven by hydrometeorological parameters for use by fire-controlling organizations would be of little use because detailed prediction of their seasonal changes is unlikely to be possible for some time.

Finally, a further finding is concerned with the use of 'fire probabilities'. One of the model outputs was the mean probability of a cell burning over the model trajectory; to obtain these probabilities, at each model step the number of cells burned in the kth stage (for $k = 1, \ldots, 8$) was divided by the total number of cells in the kth stage. Averaging this ratio over the complete model trajectory gave us mean probabilities of fire burning; these probabilities are shown in Fig. 14.9. It can be seen that they differ significantly from the probabilities of fire maturity and fire spread used directly in the model. This difference reflects the difference between local and spatial models of forest fire dynamics. Indeed, probabilities of fire maturity and fire spread refer to the local scale and reflect the state of a single cell. When being used in spatial models with interaction of cells they lead to completely different probabilities of a cell's burning, which already reflects spatial aspects of forest dynamics. Two warnings follow from this difference. The first is connected with discussion on the shape of dependence of fire probabilities on forest age; examples of such discussion can be found in Heinselman (1981*b*) and Van Wagner (1983). In these discussions it is necessary to distinguish carefully between 'local' probabilities of a cell's burning (which are in fact probabilities of fire maturity and can be calculated from field experiments and meteorological data) and 'spatial' probabilities of a cell's burning, which can be obtained from observed wildfires. As our example shows, shapes of dependence of these probabilities on forest age are completely different. The second warning is connected with the use of results obtained from locally distributed long-term models dealing with wildfires. Were these results to be expanded to represent mean dynamics of a large area, then observed probabilities of a cell's burning should be used; the use of local probabilities of fire maturity would lead to correct simulation of stand dynamics but would distort patterns of large forested areas.

The authors wish to express their gratitude to Dr. P. A. Kolosov and Dr. M. D. Korzukhin, from the Natural Environment and Climate Monitoring Laboratory, USSR, for their invaluable help with this work.

15 A simulation analysis of environmental factors and ecological processes in North American boreal forests

Gordon B. Bonan

Introduction

The boreal forest biome is a broad circumpolar mixture of cool coniferous and deciduous tree species that covers approximately 17% of the world's land surface area. Throughout this region, the landscape is a mosaic of forest vegetation types. For example, in interior Alaska above-ground tree biomass ranges from 26 t ha^{-1} for black spruce (*Picea mariana* (Mill.) B.S.P.) forests growing on cold, wet, permafrost soils to 250 t ha^{-1} for white spruce (*Picea glauca* (Moench) Voss) forests on warm, permafrost-free sites (Van Cleve *et al.* 1983*b*). In northern Quebec, above-ground tree biomass averages 20 t ha^{-1} in nutrient-poor black spruce–lichen woodlands and 108 t ha^{-1} in more fertile black spruce–moss forests (Moore & Verspoor 1973). In the warmer jack pine (*Pinus banksiana* Lamb.) forests of New Brunswick, above-ground tree biomass ranges from 0.8 t ha^{-1} for recent burns to 91.1 t ha^{-1} for more mature stands (MacLean & Wein 1976).

Numerous researchers have examined specific aspects of boreal forests, but no one has formulated a unifying model of the boreal forest biome, a paradigm to link the pattern of forest vegetation with causal environmental factors. Our most detailed understanding of the ecology of boreal forests comes from interior Alaska, where interactions among soil temperature, permafrost, soil moisture, the forest floor, litter quality, nutrient availability, and fire largely control forest productivity and vegetation patterns (Lutz 1956*a*; Heilman 1966; Viereck 1973, 1975; Van Cleve, Barney & Schlentner 1981; Van Cleve & Viereck 1981; Van Cleve *et al.* 1983*a,b*; Van Cleve & Yarie 1986). Unproductive black spruce stands are found on cold, wet, nutrient-poor, north-facing and bottomland sites with permafrost. Low soil temperatures and poor litter quality restrict decomposition and nutrient availability, reducing tree growth while promoting the buildup of a thick forest floor. With its low thermal conductivity and high water-holding capacity, this thick forest floor is an important factor reinforcing cold, wet conditions. In contrast, white spruce, birch (*Betula papyrifera* Marsh.), aspen (*Populus tremuloides* Michx.), and balsam poplar (*Populus balsamifera* L.) form more productive stands on warm, well-drained, permafrost-free sites. In these forests, higher soil temperatures and better litter quality result in rapid organic matter decomposition and nutrient mineralization.

The control of forest productivity and vegetation patterns by interactions among soil temperature, permafrost, soil moisture, the forest floor, litter quality, nutrient availability, and fire is a recurring theme in the ecology of the circumpolar boreal forest (Bonan & Shugart 1989). Using data from upland boreal forests of interior Alaska, I have quantified these relationships in a series of simulation analyses (Bonan 1988*a*, 1989*a,b*, 1990*a,b*; Bonan & Korzukhin 1989; Bonan, Shugart & Urban 1990). In this chapter, I synthesize key results from these analyses to show that: (1) current hypotheses concerning the environmental factors and ecological processes controlling the structure and function of upland boreal forests in interior Alaska (see Van Cleve *et al.* (1986) for a summary) are an adequate understanding of the ecology of these forests; (2) these ecological relations may be extended to other biogeographic regions of the boreal forest and may be the basis for a mechanistic rather than floristic biogeographical classification of the circumpolar boreal forest; and (3) though the boreal forest is floristically simple, it is complex in terms of the environmental factors and ecological processes required to understand forest productivity and landscape vegetation patterns.

Background

Subsequent discussions will be based on results from previously published models; for clarity, the development and important features of each model are summarized below. Each model was designed to address certain key issues and though each was based on prior work, each differs in various respects from the others. Obviously, the most recent analyses are the most detailed quantification of the ecology of boreal forests.

Bonan (1988*a*, 1989*a*) quantified the important environmental relations to examine the factors regulating the soil moisture and soil thermal regimes in boreal forests. This model has been extensively described by Bonan (1988*a*, 1989*a*); implications are summarized in Chapter 4 (this volume). The model uses easily obtainable climatic, topographic, soils, forest canopy, and forest floor data to simulate

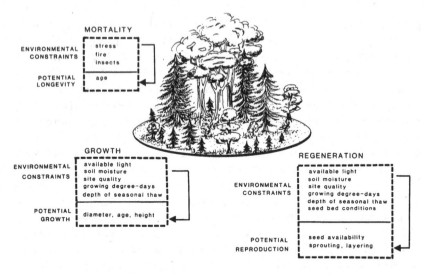

Fig. 15.1. Schematic diagram of boreal forest gap model (from Bonan 1989*b*).

monthly solar radiation, soil moisture, and depths of soil freezing and thawing. The interactions among these variables are complex. For example, depth to permafrost is a direct function of heat load. Mineral soil water content is also important because the thermal conductivity and the latent heat of fusion of the soil are functions of soil moisture. Shallow depth to permafrost, in turn, impedes soil drainage, creating saturated soil conditions. With its high water-absorbing capacity, a thick forest floor maintains high soil moisture conditions. This layer also greatly impedes heat flow because of its low bulk density and low thermal conductivity. Likewise, by reducing the flux of heat into the soil profile, the forest canopy is an important factor maintaining low soil temperatures and a shallow soil active layer. Fires are important not for direct heating effects, which are minimal, but rather for indirect effects caused by the removal of the forest canopy and the forest floor.

Bonan (1988*a*, 1989*b*) combined this environmental model with the 'gap' model approach (Botkin, Janak & Wallis 1972; Shugart & West 1977; Shugart 1984) to examine the factors controlling forest dynamics and vegetation patterns (Fig. 15.1). In this model, rather than being static input parameters, the forest canopy and forest floor were dynamic functions of site conditions. Individual tree growth on a 1/12 ha forest plot under optimal site conditions was a function of a species-specific growth rate, tree diameter, and tree height (Fig. 15.2) and was decreased multiplicatively to the extent that site conditions (growing season tem-

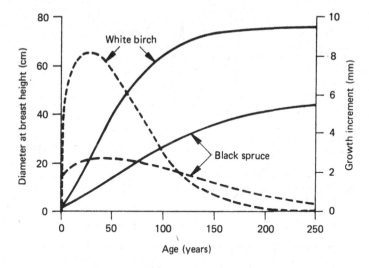

Fig. 15.2. Optimal simulated individual tree growth evaluated for birch and black spruce. Broken line, growth; solid line, diameter.

perature sums, available light, soil moisture, site quality, depth of soil thaw) were less than optimal. Regeneration was constrained by these same site conditions and by other effects such as a thick forest floor, seed availability following forest fires, and vegetative reproduction. Mortality was a function of potential longevity, stress caused by sub-optimal site conditions, and forest fires (Fig. 15.3). Forest floor organic matter accumulation was a function of annual input and decomposition, scaled for site conditions. Simulated fire intensity (i.e. the heat generated by the fire) was a function of fuel buildup. Fire severity (i.e. the depth of burn in the forest floor) was a function of forest floor moisture content and pre-burn thickness. This model did not explicitly simulate nutrient avail-ability. Low nutrient availability is associated with permafrost and the buildup of a thick forest floor (Van Cleve & Viereck 1981; Van Cleve, Barney & Schlentner 1981; Van Cleve et al. 1983a,b; Van Cleve & Yarie 1986); rather than explicitly considering the effect of nutrient availability on forest dynamics, this effect was assumed to be correlated with permafrost and the status of the forest floor.

Bonan & Korzukhin (1989) reformulated the moss growth algorithm to examine the ecological consequences of moss and tree interactions. The forest floor was treated as a two-compartment system consisting of moss and decomposing organic matter (i.e. humus). The annual growth of moss was the difference between carbon assimilation and respiration

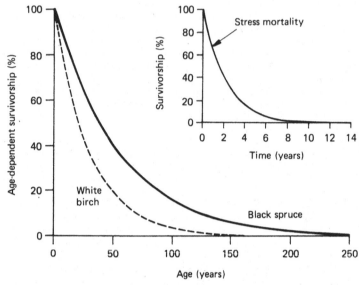

Fig. 15.3. Simulated age-dependent mortality evaluated for birch and black spruce. Inset: stress mortality (from Bonan 1990*b*).

and decay. Carbon assimilation was directly proportional to moss biomass and was decreased to the extent that available light within the moss layer reduced assimilation. Respiration and decay were directly proportional to moss biomass. The annual growth of the humus layer was the difference between organic matter input and decomposition. Humus decomposition was an exponential function of depth of soil thaw.

Bonan (1990*a,b*) modified the model to include nitrogen cycling. Nitrogen availability was simulated using the approach developed by Aber, Melillo & Federer (1982) and Pastor & Post (1986), but modified to include factors such as fire and the soil thermal regime. All annual litter cohorts entered a litter pool where they immobilized nitrogen during decomposition (Fig. 15.4). Annual litter and humus decomposition were a function of nitrogen concentration and depth of soil thaw (Fig. 15.5). When nitrogen reached a critical concentration, the litter cohort was transferred to the humus pool where mineralization occurred (Fig. 15.4). Nitrogen mineralization was proportional to humus mass loss.

Fires, when they occurred, mineralized nitrogen contained in the forest floor and above-ground biomass. The direct effects of fire on carbon and nitrogen cycling included consumption of biomass during burning, loss of nutrients through volatilization, and transfer of mineral elements contained in biomass to the ash layer (Fig. 15.6). Post-fire effects included loss of nutrients as runoff and increased organic matter decomposition and nitrogen mineralization under a warmer soil thermal regime.

As in the previous models, optimal individual tree growth was constrained for sub-optimal site conditions by a multiplicative function of site conditions such as growing degree-day temperature sums, soil moisture, and available light (Fig. 15.7). However, in contrast to the previous models, the effect of low soil temperatures on individual tree growth was partitioned into reduced plant metabolism (i.e. the depth of soil thaw growth multiplier) and reduced nutrient availability (i.e. the nitrogen availability growth multiplier).

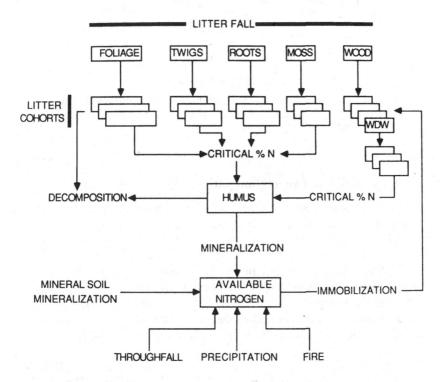

Fig. 15.4. Simulated nitrogen cycling in the absence of fire. Litter cohorts immobilize nitrogen during decomposition. When nitrogen concentrations reach a critical level, the litter is transferred to the humus pool, where nitrogen is mineralized during decomposition. Woody litter is first transferred to well-decayed wood before becoming humus. Additional nitrogen inputs include mineral soil mineralization, throughfall, precipitation, and nitrogen mineralized during fire (from Bonan 1990a).

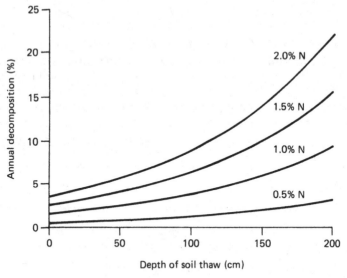

Fig. 15.5. Simulated annual litter and humus decomposition evaluated for different depths of thaw and nitrogen concentrations (from Bonan 1990*b*).

Forest dynamics and vegetation patterns in interior Alaska

Testing has shown that the model simulates post-fire successional dynamics and site characteristics that are consistent with observed data in the uplands of interior Alaska (Bonan 1989*a*). Moreover, the model preserves the observed distribution of forest types in the landscape (Fig. 15.8). Black spruce is found primarily on cold, wet sites, and white spruce, birch, and aspen occur on warmer, drier sites (Viereck *et al.* 1983). In the simulated forest landscape, cold, wet sites with shallow depth of thaw were dominated exclusively by black spruce. Drier sites with deeper active layers were dominated by a mixture of white spruce, birch, and aspen.

With the addition of nitrogen cycling to the model, Bonan (1990*a*) confirmed the hypothesis of Van Cleve & Viereck (1981), Van Cleve *et al.* (1981, 1983*a,b*), and Van Cleve & Yarie (1986) that litter quality and the soil thermal regime control nutrient availability and stand productivity in the upland boreal forests of interior Alaska. Low soil temperatures encountered in black spruce forests underlain with permafrost result in slow organic matter decomposition and nutrient mineralization. Higher soil temperatures on permafrost-free sites enhance productivity through more rapid organic matter decomposition and nutrient mineralization.

Substrate quality interacts with soil temperature to enhance or restrict nutrient availability. Cold, wet black spruce sites have forest floors with low nitrogen and high lignin concentrations, which further slow organic matter decomposition. Warm sites dominated by white spruce, birch, and aspen have forests floors with higher nitrogen and lower lignin concentrations.

Bonan (1990a) used the model to examine these relations. In addition to the interactive effects of substrate quality and the soil thermal regime

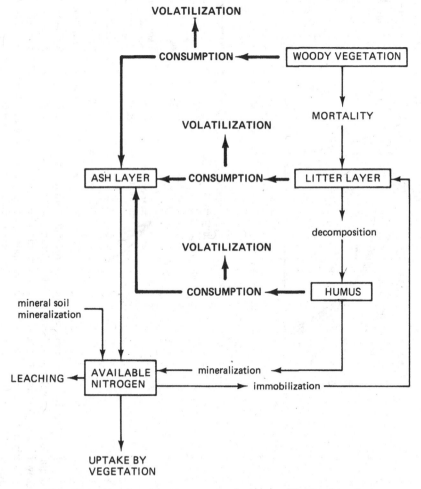

Fig. 15.6. Simulated carbon and nitrogen redistribution during (boldface lettering) and following (normal lettering) fire. Lower case lettering indicates post-fire processes influenced by the soil thermal regime (from Bonan 1990a).

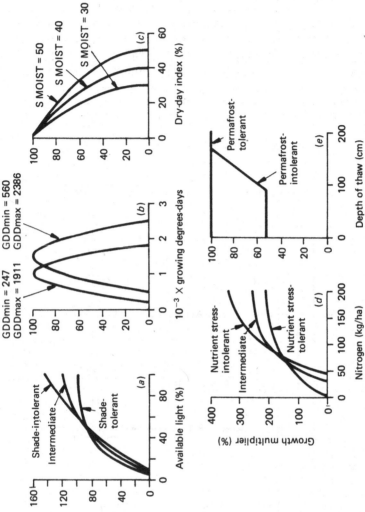

Fig. 15.7. Relative growth multipliers used to scale optimal individual tree growth for sub-optimal site conditions. (*a*) available light growth multiplier for shade-tolerant, intermediate, and shade-intolerant species; (*b*) growing degree-days growth multiplier as a function of minimum and maximum observed growing degree-days (GDDmin and GDDmax, respectively); (*c*) soil moisture growth multiplier as a function of drought tolerance (SMOIST), (*d*) available nitrogen growth multiplier for nutrient stress-tolerant, intermediate, and intolerant species; (*e*) depth of soil thaw growth multiplier for permafrost-tolerant and permafrost-intolerant species (from Bonan 1990*b*).

on organic matter decomposition (Fig. 15.5), observed data were used to develop decomposition equations based on substrate quality only and the soil thermal regime only. Each decomposition equation was used in the model to simulate black spruce forests growing on permafrost and black spruce, white spruce, and birch forests growing on permafrost-free soils.

No one decomposition equation resulted in simulated forests that were entirely consistent with observed data, but over the range of forests

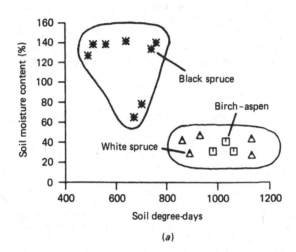

(a)

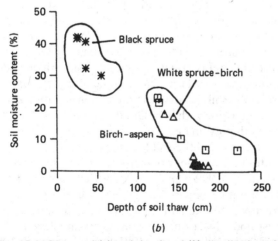

(b)

Fig. 15.8. Observed (a) and simulated (b) distribution of forest types in the uplands of interior Alaska in relation to soil moisture and the soil thermal regime (from Bonan 1989b).

Table 15.1. *Observed and simulated range of mature white spruce forests growing on permafrost-free soils near Fairbanks, Alaska*

Observed data are from forests 70–250 years old. Simulated data are for forests within this age span.

		Decomposition equations		
	Observed data	Substrate quality	Soil thermal regime	Interactive effects
Age (yr)	70–250	71–129	121–214	103–242
Tree biomass (t ha^{-1})				
above-ground	61.5–245.9	116.8–149.6	103.8–165.8	73.2–170.9
litter	0.1–3.2	1.2–1.5	0.5–1.4	0.6–1.4
Basal area (m^2 ha^{-1})	30–60	21–28	18–30	14–28
Density (ha^{-1})	550–1000	1068–2460	240–2136	168–2640
Forest floor				
mass (t ha^{-1})	47.1–105.3	99.5–102.4	16.2–31.1	32.0–61.8
residence time (yr)	19	44–50	8–10	16–23
depth (cm)	5–18	16–17	3–6	6–12
nitrogen (%)	0.66–1.04	0.97–1.07	0.94–1.23	0.83–0.96
Moss biomass (t ha^{-1})	4.5	0.0–3.4	1.0–4.4	2.3–4.6

tested, the equation that combined the effects of substrate quality and the soil thermal regime resulted in simulated forests that were most consistent with observed data. For example, the substrate quality decomposition equation resulted in simulated white spruce forest floor characteristics that were inconsistent with observed data (Table 15.1). In particular, the substrate quality equation simulated slow rates of decomposition that resulted in too large an accumulation of forest floor biomass. Conversely, the soil thermal decomposition equation simulated faster rates of decomposition that resulted in too little forest floor biomass. Simulations that estimated decomposition from the interactive effects of substrate quality and the soil thermal regime resulted in more appropriate estimates of forest floor biomass and turnover rates.

The results of Bonan's (1990a) analyses supported the hypothesis that productivity and nutrient cycling in the uplands of interior Alaska are a response to interactions among soil temperature and permafrost, soil moisture, the forest floor, litter quality, and fire (Van Cleve & Viereck 1981; Van Cleve, Barney & Schlentner 1981; Van Cleve *et al.* 1983a,b; Van Cleve & Yarie 1986). On the warm, permafrost-free south slope, simulated organic matter decomposition and nitrogen mineralization were rapid in birch forests with good litter quality. High soil temperatures and large amounts of available nitrogen sustained rapid tree growth with little accumulation of biomass in the forest floor. As white spruce began to dominate the simulated forest, mosses became abundant and the forest floor biomass accumulated because of the poor litter quality.

In the black spruce forests growing on permafrost soils, low soil temperatures resulted in little organic matter decomposition and nutrient mineralization, reducing tree growth while promoting the accumulation of a thick forest floor composed of moss and humus. With its low bulk density and thermal conductivity, the thick forest floor reduced soil temperatures, creating a cold soil with a shallow depth of thaw that prevented drainage. The poor quality of black spruce litter further reduced decomposition rates. Over time, the forest floor became the principal reserve of nitrogen and biomass.

Fires are thought to interrupt this sequence, removing the forest canopy and consuming forest floor biomass, thereby increasing post-fire soil temperatures and depths of soil thaw and mineralizing nutrients contained in the forest floor (Lutz 1956; Viereck 1973, 1983; Viereck & Schandelmeier 1980; Van Cleve & Viereck 1981; Dyrness, Viereck & Van Cleve 1986). To test this hypothesis, simulations of the black spruce forest growing on permafrost were repeated for 500 years with fire occurring every 100 years. The occurrence of fire significantly affected above-ground tree biomass, forest floor biomass, depth of soil thaw, and nitrogen availability (Fig. 15.9). Recurring forest fires partly consumed the thick forest floor. This, combined with removal of the forest canopy, resulted in increased post-fire depths of soil thaw. Mineralization during fire of nitrogen contained in the forest floor and woody plants and faster rates of decomposition caused by the warmer post-fire soil thermal regime resulted in large amounts of available nitrogen. The improved thermal and nutrient regimes created conditions favorable for stand regeneration, and the dynamics were repeated.

Biogeographical vegetation patterns

Two geographic tests of the model have indicated that ecological relations in the uplands of interior Alaska are valid for other biogeographical regions of the boreal forest. Bonan (1989b) demonstrated that the model adequately simulates: (1) stand structure of unproductive black spruce–lichen woodlands growing on nutrient-poor soils and productive black spruce–moss forests growing on fertile soils in subartic Quebec; (2) stand structure of 70-year-old aspen and jack pine stands in the Boundary Waters Canoe Area of Minnesota and the landscape distribution of vegetation types in this region; and (3) successional dynamics for birch,

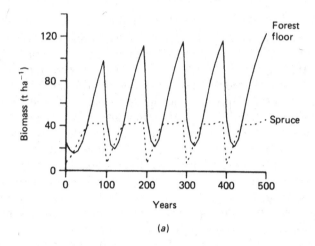

(a)

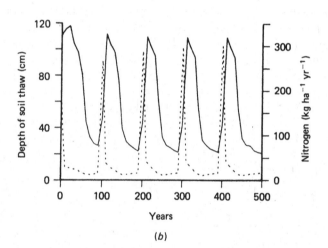

(b)

Fig. 15.9. Above-ground tree biomass, forest floor biomass, depth of
soil thaw, and available nitrogen (broken line in (b)) for an interior
Alaska black spruce forest growing on permafrost with fires every
100 years (from Bonan 1990a).

black spruce, and balsam fir (*Abies balsamea* (L.) Mill.) forests in
Newfoundland.

With the inclusion of nitrogen cycling in the model, Bonan (1990b)
conducted additional geographic tests of the model. In addition to black
spruce, white spruce, and birch stands in the uplands of interior Alaska,
test sites included jack pine stands in New Brunswick, black spruce and
balsam fir stands in Newfoundland, black spruce stands in central

Table 15.2. *Observed and simulated jack pine forests in New Brunswick*

Observed data are the range for stands 13–57 years old. Simulated data are the range for stands 4–53 years old.

	Observed	Simulated
Age (yr)	13–57	4–53
Tree biomass (t ha^{-1})		
above-ground	0.8–91.1	23.9–124.6
litter (foliage + twig)	0.1–2.0	1.8–3.7
Basal area (m^2 ha^{-1})	0.2–28.4	7.3–27.3
Density (ha^{-1})	2200–6560	744–6348
Forest floor		
biomass (t ha^{-1})	74.1–129.6	19.5–91.6
depth (cm)	3–7	4–21
nitrogen (%)	0.72–1.11	0.87–1.09

Source: Bonan (1990*b*).

Quebec, and black spruce-lichen woodlands in subarctic Quebec. Test comparisons included above-ground tree biomass, basal area, density, litterfall, moss and lichen biomass, and forest floor biomass, turnover, thickness, nitrogen concentration, and nitrogen mineralization. Example simulations are shown in Table 15.2 for jack pine and Table 15.3 for balsam fir.

The model correctly simulated 60 (76%) of the 79 variables tested. Above-ground tree biomass was correctly simulated in all six tests. Foliage biomass was correctly simulated in both available tests. Foliage and twig litter biomass were correctly simulated in 8 of 9 tests. The model correctly simulated 81% of the 31 forest floor variables tested. To the degree that forest floor organic matter turnover rates were available, they were well simulated by the model. The biomass and nitrogen concentration of moss and lichen proved to be the one major recurring problem with the model.

These geographic tests have identified the critical processes and parameters required to understand the ecology of boreal forests. Climate, solar radiation, soil moisture, soil temperature, permafrost, the forest floor, nutrient availability, litter quality, and forest fires interacted to contribute to the mosaic pattern of forest types and the wide range in stand productivity of the observed forests. Different environmental factors were important in different biogeographical regions. Vegetation

Table 15.3. *Observed and simulated balsam fir forests in Newfoundland*

Observed data are means or 95% confidence intervals for a 65 year-old stand. Simulated data are the range for stands 50–81 years old.

	Observed	Simulated
Age (*yr*)	65	50–81
Basal area ($m^2\,ha^{-1}$)	42–44	26–36
Foliage litter		
biomass ($t\,ha^{-1}$)	1.7–3.3	1.0–2.1
nitrogen ($kg\,ha^{-1}$)	13.3–25.7	7.3–14.8
Root litter		
biomass ($t\,ha^{-1}$)	2.2	1.3–2.7
nitrogen ($kg\,ha^{-1}$)	20.6	12.5–25.5
Forest floor		
biomass ($t\,ha^{-1}$)	46–84	61–91
nitrogen ($kg\,ha^{-1}$)	878	638–1036
depth (cm)	10	12–18
Moss layer		
biomass ($t\,ha^{-1}$)	1.9–2.5	3.8–4.0
nitrogen ($kg\,ha^{-1}$)	23.4–30.8	17.4–18.4

Source: Bonan (1990*b*).

patterns were most complex in the uplands of interior Alaska, where forest fires, the forest floor, soil moisture, and the presence or absence of permafrost interacted to create the mosaic pattern of forest vegetation found in the landscape.

Fire was an important determinant of the mosaic pattern of forest types in all regions. The role of fire in boreal forests has been reviewed by others (Heinselman 1973, 1981*b*; Rowe & Scotter 1973; Viereck 1973, 1983; Larsen 1980; Viereck & Schandelmeier 1980; Dyrness, Viereck & Van Cleve 1986). These reviews indicate that post-fire vegetation is a complex function of climate, pre-burn vegetation type and age, time of burn, fire severity, topography, and the presence or absence of permafrost. In the biogeographical testing, life-history characteristics that allowed mass regeneration following fires were particularly important. The influence of pre-burn stand composition on seed availability following fire was also an important feature of the model. Field observations confirm that, following a catastrophic fire, regeneration is frequently

Table 15.4. *Average simulated environmental variables*

Numbers in parentheses are the effect of each environmental factor on tree
growth expressed as a percentage of growth under optimal conditions

Forest type	Available light (%)	Growing degree-days (°C days)	Dry-day index (%)	Nitrogen availability (kg ha^{-1})	Depth of thaw (cm)
interior Alaska					
black spruce	47 (87)	612 (69)	0 (100)	15 (19)	22 (100)
white spruce	20 (51)	904 (95)	0 (100)	36 (42)	150 (89)
birch	14 (9)	904 (92)	0 (100)	39 (46)	167 (99)
subarctic Quebec					
black spruce	55 (91)	472 (47)	0 (100)	24 (31)	146 (100)
central Quebec					
black spruce	18 (46)	870 (94)	0 (100)	31 (38)	171 (100)
Newfoundland					
black spruce	15 (38)	1046 (100)	0 (100)	32 (39)	174 (100)
balsam fir	9 (17)	1046 (78)	0 (100)	67 (65)	163 (97)
New Brunswick					
jack pine	33 (38)	1262 (86)	0 (100)	40 (47)	156 (93)

Source: Bonan (1990*b*).

dominated by pre-burn species (Carleton & Maycock 1978; Larsen 1980;
Heinselman 1981*b*; Viereck 1983). Moreover, in interior Alaska and
subarctic Quebec, where a thick moss or lichen layer prevented stand
regeneration in the later stages of succession, the complete or partial
removal of the forest floor during fire was an important factor allowing
stand regeneration.

   With the addition of nitrogen availability to the model, available light,
growing degree-days, soil moisture, nitrogen availability, and depth of
soil thaw interacted to constrain tree growth for less than optimal
conditions (Fig. 15.7). When standardized relative to optimal conditions,
these growth factors indicated which environmental factor was most
limiting in each simulated forest. Available nitrogen was the most
consistently sub-optimal growth factor, ranging from 19% of optimal
conditions in the cold, wet black spruce forest growing on permafrost soil
to 65% of optimal conditions in the balsam fir forest (Table 15.4). The
direct effects of the soil thermal regime on tree growth were minimal, but
depth of thaw was important in determining nitrogen availability. The
growing degree-day growth factor was most limiting in the subarctic

lichen woodland. Soil moisture did not restrict tree growth on any of the sites. Stand structural constraints, as indicated by the available light on the forest floor, were most important in the birch forest and least important in the open black spruce–lichen woodland and black spruce forest growing on permafrost.

The ability of the model to simulate forest structure and landscape vegetation patterns in different bioclimatic regions of the North American boreal forest indicates that ecological relations from the uplands of interior Alaska can be extended to other regions of the North American boreal forest. Interior Alaska contains all the major North American boreal forest vegetation types except jack pine, balsam fir, and black spruce–lichen woodlands. However, if one defines the boreal forest by key environmental features such as soil moisture, soil temperature, nutrient availability, and the fire regime, interior Alaska spans the range of conditions found in the North American boreal forest.

Soil moisture ranges from xeric upland soils to lowland bogs; soil temperature ranges from cold, permafrost soils to warm, permafrost-free soils (Viereck *et al.* 1983). Nitrogen mineralization ranges from 8–9 kg ha^{-1} yr^{-1} in cold, wet black spruce forests (Van Cleve, Barney & Schlentner 1981) to 24–58 kg ha^{-1} yr^{-1} in birch forests on warmer, permafrost-free soils (Flanagan & Van Cleve 1983). The fire recurrence interval varies from 50 to 200 years depending on topography and forest type (Viereck & Schandelmeier 1980), but in extremely dry deciduous forests, the fire cycle may be as short as 26 years (Yarie 1981).

The boreal forest biome is the major soil organic matter reserve and is second only to broadleaved humid forests in terms of phytomass carbon storage (Lashof 1989). Not surprisingly, the boreal forest appears to have a significant role in the seasonal dynamics of atmospheric CO_2 (D'Arrigo, Jacoby & Fung 1987). In the Northern hemisphere, the amplitude of seasonal atmospheric CO_2 concentrations has increased with time, and this is thought to reflect increased biological activity of vegetation in northern latitudes (Bacastow, Keeling & Whorf 1985; Gammon, Sundquist & Fraser 1985; Houghton 1987). Moreover, based on current estimates of oceanic uptake of carbon, a mid- to high-latitude terrestrial carbon sink is required to simulate atmospheric CO_2 concentrations that are consistent with observed data (Tans, Fung & Takahashi 1990). Given these concerns and other interest in the ecological consequences of CO_2-induced climate change for boreal forests, the pronounced local gradients in soil moisture, soil temperature, nutrient availability, and fire and the apparent generality of ecological relations may make interior Alaska a useful area to examine the sensitivity of ecological relations to climate change.

Ecological complexity in boreal forests

These simulation analyses have indicated that though boreal forests are not floristically complex, they are complex in terms of the environmental factors and ecological processes required to understand their ecology (Fig. 15.10). The complex interactions among these factors ensure that a multifunctional approach is needed to understand the ecology of boreal forests. Moreover, the response of boreal forests to climatic change can only be understood in terms of the sensitivity of each of these factors and their interactions to climatic parameters. These conclusions are best illustrated by two model experiments that examined the sensitivity of black spruce-permafrost forests in interior Alaska to: (1) climatic parameters and (2) moss and tree interactions.

The potential ecological consequences of climatic warming in boreal regions has been the subject of much speculation (Emanuel, Shugart & Stevenson 1985*a,b*; Solomon 1986; Pastor & Post 1988). While these analyses have been useful in focusing attention on the ecological conse-

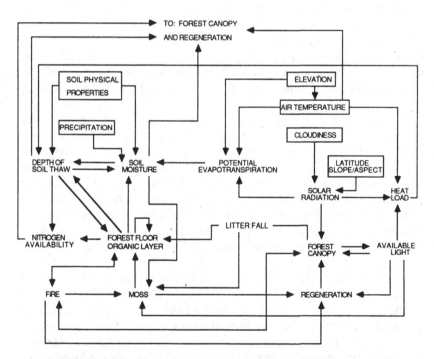

Fig. 15.10. Schematic diagram of environmental factors and ecological processes controlling forest dynamics and vegetation patterns in the upland boreal forests of North America. Boxes indicate required input parameters (from Bonan 1990*b*).

Table 15.5. *Simulated above-ground woody biomass ($t\,ha^{-1}$) for a 100-year-old north-slope black spruce stand at Fairbanks, Alaska, in response to climatic warming and precipitation increases and without recurring forest fires*

Precipitation	Air temperature			
	base	1 °C	3 °C	5 °C
base	42.3	52.2	60.9	50.9
120%	—	58.4	64.4	49.3
140%	—	75.0	69.8	60.0
160%	—	96.5	96.1	67.1

Source: Bonan, Shugart & Urban (1990).

quences of projected climatic changes, they have not explored the full complex of processes controlling the productivity and composition of boreal forests. Bonan, Shugart & Urban (1990) used Bonan's (1989b) model of boreal forest dynamics to examine the sensitivity of forests in the uplands of interior Alaska to climatic parameters. A factorial design was used to increase mean monthly air temperature by 1 °C, 3 °C, and 5 °C and monthly precipitation by 120%, 140%, and 160% from current values.

As discussed in Chapter 4, in the absence of recurring fires, soil temperatures in black spruce forests growing on permafrost did not increase with climatic warming unless there was an accompanying increase in precipitation (Table 4.4). Without increased precipitation, the forest floor became drier and its thermal conductivity, a linear function of moisture content, decreased, thereby impeding the flow of additional heat into the soil. Precipitation increases that allowed the forest floor to remain moist (i.e. with a higher thermal conductivity) resulted in significant soil thawing. These changes in depth of thaw were reflected in stand biomass (Table 15.5). However, when recurring fires were included in the simulations, there was substantial soil thawing (Fig. 15.11), and the overall long-term effect of climatic warming was to increase the biomass of the white spruce, birch, and aspen (Table 15.6). Again, increased potential evapotranspiration demands with a warmer climate resulted in drier forest floors, but depth of burn in the forest floor was a function of moisture content, and recurring fires consumed more of the forest floor. Depth of soil thaw is directly related to the forest floor thickness (Bonan 1989a); the thinner forest floor resulted in increased depths of soil thaw. Warmer soils increased the biomass of aspen, birch, and white spruce.

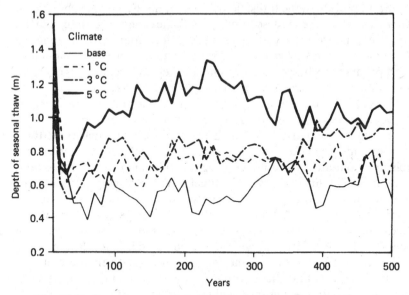

Fig. 15.11. Long-term simulated depth of soil thaw as a function of climatic warming for an interior Alaska north-slope forest with recurring forest fires (from Bonan, Shugart & Urban 1990).

Table 15.6. *Simulated equilibrium above-ground woody biomass for a north-slope forest at Fairbanks, Alaska, in response to air temperature increases and with recurring forest fires*

Air temperature	Biomass (t ha^{-1})			
	Black spruce	White spruce	Aspen	Birch
base	22.8	1.8	1.6	8.3
1 °C	21.8	4.6	2.5	16.4
3 °C	17.9	10.8	7.1	38.4
5 °C	15.4	6.4	14.6	37.5

Source: Bonan, Shugart & Urban (1990).

These results indicate two important responses of boreal forests to climatic change. First, the ecological response of boreal forests to increased air temperatures may not occur instantaneously, but rather might occur as a lagged response. In the simulated black spruce sites, climatic warming did not cause soils to warm until a fire had removed a significant portion of the forest floor. Second, the system must be studied *in toto*. The response of these forests to climatic change was the combined response to interactions among moisture, soil temperature, fire, and the effects of these on tree growth and regeneration. Correlative models of forest productivity based on extant climate–vegetation relations may give misleading and erroneous responses to climatic change.

Throughout the taiga, the ground cover of many coniferous forests is composed of a thick moss-organic layer (Bonan & Shugart 1989). With its low thermal conductivity and high water-absorbing capacity, a thick moss–organic layer lowers soil temperatures and maintains high soil moisture contents, thereby reducing organic matter decomposition, nutrient availability, and stand productivity. Cold, wet, nutrient-poor site conditions, in turn, promote the productivity of some mosses (Skre & Oechel 1979; Oechel & Van Cleve 1986). The moss ground cover also affects the ability of trees to regenerate. The germination and establishment of many boreal tree species is hindered by a thick moss layer (Bonan & Shugart 1989). However, some species can vegetatively reproduce by layering when low branches covered by organic matter develop adventitious roots and produce new individuals (Zasada 1986).

Trees, in turn, affect moss growth. The highest productivity of some feathermosses occurs near the edge of the tree canopy (Tamm 1953; Abolin' 1974; Tarkova & Ipatov 1975). This has been attributed to favorable light and nutrient availability (Tamm 1953), inhibitory effects of litter fall beneath the canopy (Tarkova & Ipatov 1975), and insufficient precipitation under the canopy (Abolin' 1974; Busby, Bliss & Hamilton 1978). Yet the productivity of feathermosses is also apparently reduced on open sites (Tamm 1953; Weetman 1968b), perhaps because of increased desiccation with a more open forest canopy (Hytteborn, Packham & Verwijst 1987). Dense deciduous leaf litter can also be an important factor limiting the growth and establishment of mosses (Tamm 1953; Oechel & Van Cleve 1986).

The numerous positive and negative feedbacks among mosses, trees, and site conditions imply a complex dynamical system in which forest structure, forest floor moss–organic matter, and site conditions may be highly sensitive to these feedbacks. Bonan & Korzukhin (1989) used their model to examine the ecological importance of these interactions in black spruce forests growing on permafrost in interior Alaska. Specific feedbacks included: (1a) colder, wetter soils as the forest floor accumulated; (1b) reduced decomposition and greater forest floor accumulation with colder soils; (2a) increased heat load onto the soil with a more open forest canopy; (2b) more productive tree growth with warmer soils; and (3a) reduced moss growth under a more open forest canopy; (3b) an inability for trees to sexually regenerate on thick moss-covered soils.

Their analyses indicated that these forests are a tightly coupled system in which forest dynamics is highly sensitive to the interactions among site conditions, mosses, and trees. The effect of mosses on the soil thermal regime was a particularly important feedback. Direction interactions between mosses and trees that affected the development of a thick forest floor were also important. In particular, shading of moss by trees, reduced tree regeneration on moss-covered soils, and reduced moss growth with open forest canopies were important determinants of forest succession. For example, when all moss and tree interactions were included in the model, simulations from various initial conditions con-verged on a single successional trajectory which resulted in a stable equilibrium point of approximately 85 t ha^{-1} forest floor and 30 t ha^{-1} spruce (Fig. 15.12). When the inability of black spruce to sexually reproduce on a thick moss layer was excluded from the model, all trajectories converged on a stable equilibrium point of approximately 20 t ha^{-1} forest floor and 105 t ha^{-1} spruce (Fig. 15.13). The ability of trees to sexually reproduce on a thick moss layer enabled the development and maintenance of a dense forest canopy that shaded and eventually prec-luded moss growth. The low forest-floor biomass significantly increased depth of soil thawing, increasing tree growth while further decreasing forest floor accumulation because of faster rates of decomposition. These

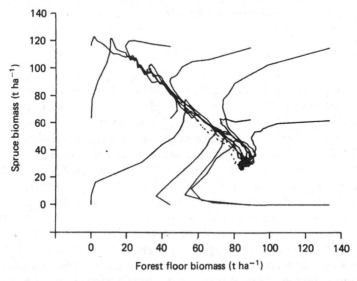

Fig. 15.12. Forest floor and black spruce biomass for a cold, wet, permafrost site in interior Alaska. Each line represents the relation-ship between forest floor and spruce biomass for 500 years from a variety of initial conditions (from Bonan & Korzukhin 1989).

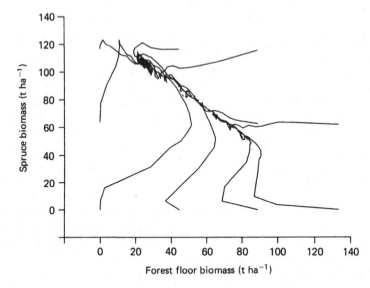

Fig. 15.13. Same as Fig. 15.12, but with sexual regeneration of spruce on moss-covered soil not precluded (from Bonan & Korzukhin 1989).

complex feedbacks ensure that an ecosystem approach is needed to understand the ecology of boreal forests.

Conclusion

Simulation models have provided a useful framework to examine linkages among boreal forest vegetation patterns and causal environmental and ecological factors. Analyses in the upland boreal forests of interior Alaska have confirmed our current understanding of the ecology of these forests. Biogeographical tests have shown that ecological relations from interior Alaska can be extended to other regions of the North American boreal forest. Moreover, these tests have begun to identify important parameters and processes required to understand vegetation patterns in different biogeographic regions and thus may provide the basis for a mechanistic rather than floristic classification of the boreal forest. The generality of ecological relations from interior Alaska and the wide range of local site conditions, which span the range of conditions in the North American boreal forest, suggest that interior Alaska may be a useful area to examine the sensitivity of these relations to climatic change. Finally, model experiments of black spruce growing on permafrost have highlighted the complexity of these forests. Though they are not floristically diverse, these forests are complex in terms of the environmental factors and ecological processes required to understand their ecology. Model experiments with other boreal forest types have shown similar features. For example, interactions among litter quality, nitrogen availability, and tree growth may result in cyclic birch–white spruce stand dynamics on permafrost-free soils (Pastor *et al.* 1987). This complexity ensures that a multidisciplinary, systems approach is needed to understand the ecology of boreal forests.

This manuscript was prepared while the author was a post-doctoral fellow in the Advanced Study Program at the National Center for Atmospheric Research. The National Center for Atmospheric Research is sponsored by the National Science Foundation.

16 The biological component of the simulation model for boreal forest dynamics

Rik Leemans

Introduction

The extent of the circumpolar boreal forest strongly corresponds to macroclimate. Within its climatic limits, the system functions as a complex interrelation between solar radiation, soil moisture, the forest floor organic layer, nutrient availability, forest fires, insect outbreaks and vegetation patterns (Bonan & Shugart 1989). Bonan (1989*a* and Chapter 15 of this volume) has specified a model for the environmental regimes, which sets the limits driving boreal forest dynamics. He has linked this model with a forest succession model, a gap model, which simulates the demographic processes of tree populations through time within the environmental constraints, and a model of moss dynamics. The combined model mimics the large-scale dynamics of boreal forest (Bonan 1989*a*; Bonan & Korzukhin 1989).

Bonan's model simulated different stands well with respect to species composition, biomass and density. The ability to simulate such trends in quantitative characteristics of tree species is a robust feature of gap models in general (Shugart 1984; Leemans & Prentice 1987). This robustness is largely a result of the coupling of growth responses of individual trees to environmental factors. If the annual growth of a tree declines as a result of environmental conditions, its chances of dying increase. Thus, the individual tree is removed from the plot, leaving room for better-adapted individuals. This aspect of the traditional gap models appears to provide a robust ability to reproduce composition, biomass and density. If more precisely defined forest structures are used to test such models, the models are often less successful.

Gap models of forest succession were initially developed for forest undergoing so-called 'gap-phase replacement' (Jones 1945; Watt 1947),

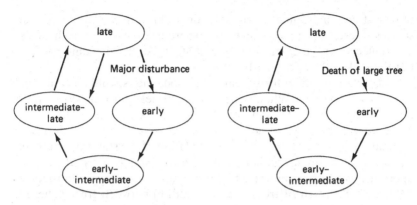

Fig. 16.1. Simplified flowchart of gap-phase dynamics (left) and directional dynamics (right).

in which single trees eventually grow large enough to dominate their immediate surroundings. The death of such trees creates gaps in the canopy which are large enough to initiate a local succession, starting with the establishment of saplings of early successional (often equivalent to shade-intolerant) species. The first simulation model based on this dynamic feature was JABOWA (Botkin, Janak & Wallis 1972). Forest succession models based on JABOWA clearly display gap-phase dynamics in all their simulations. A recent example can be seen from Overpeck, Rind & Goldberg (1990), where early-successional species are present throughout the whole simulation of many plots within a forest region in southern Quebec. These results contrast our present understanding of boreal forest dynamics in which a directional succession scheme with shade-tolerant (late-successional) species eventually replacing shade-intolerant (early-successional) species occurs in the absence of any disturbances acting over a large area (e.g. Sernander 1936; Bergeron & Dubuc 1989; Leemans 1989).

Boreal forest dynamics differ from the traditional gap-phase dynamics (Fig. 16.1). Similar deviations from cyclic gap-phase dynamics have been reported for more temperate forests (Trimble 1973; Mayer, Neumann & Summer 1980). Boreal deep-crowned conifers with a small lateral extent do not dominate their surroundings. After the death of a single tree, the resultant gap is too small to allow the establishment of early successional species (Leemans 1991a). The dynamics observed are directional. Early-successional species are only capable of re-entering the system after a large disturbance. This type of dynamics is observed within a large area of the circumpolar boreal forest (USSR: Polikarpov, Tchebakova & Nazimova 1986; North America: Bergeron & Dubuc 1989; Scandinavia: Steijlen & Zackrisson 1987; Leemans 1991a). Departures of this type are observed in environments suitable only for a few shade-intolerant species. Here a succession develops, where age-cohorts of such species are replaced by younger cohorts and strong cyclic dynamics develop (Zyabchenko 1982; Shugart 1987).

A simulation model for boreal forest dynamics should simulate such directional dynamics. When using the gap-model approach, this requires a re-evaluation of the traditional JABOWA-based gap models. In an early effort, the gap model FORSKA version 0 was developed by Leemans & Prentice (1987) for the simulation of the stand structure and dynamics of a small southern boreal forest stand in Sweden. Their approach was to parameterize the most important biological processes, determining growth of trees and competition between trees. These in turn set limits on the dynamical and structural properties of vegetation, such as species composition, population structure, canopy profiles and successional pathways. The underlying philosophy is that most character-istics of vegetation are predictable from the lower-level mechanisms by which individual plants affect and respond to their immediate environ-ment (Prentice 1986a; Shugart et al. 1988; Roberts 1989; Prentice & Leemans 1990).

In this chapter I will present the structure of a more elaborate version of the FORSKA model (Leemans & Prentice 1989). In the description of the concepts of the model I will elaborate on the differences between the JABOWA–FORET-type forest succession models (hereafter referred to as traditional gap models) and FORSKA, clarify the underlying assump-tions, and discuss why these are important in a more general boreal forest simulator. I will then discuss a way of linking this model with models defining the environmental constraints and/or other important parts of the ecosystem, and finally will present a comparison between FORSKA and a traditional gap model.

Model structure of FORSKA

Gap models generally simulate a forest stand as an array of independent plots. Each plot has its own tree population consisting of one to many tree individuals. Each individual is characterized by its species, diameter at breast height (DBH), bole height (height from forest floor to the lower extent of the tree's crown), leaf area and age. All resources (nutrients, light, moisture, etc.) are assumed to be distributed homogenously and horizontally over each plot, which means that each tree competes and interacts with all other individuals on the plot. The calculation of light absorption through the canopy and the different species' responses to reduced light levels make gap models spatially explicit in the vertical direction.

The extent of a plot is commonly defined to be 0.1 ha, which corresponds with the influence area of large single trees. Plot size is a critical parameter in defining the outcome of the simulation (Shugart & West 1979) and defines different successional pathways (Prentice & Leemans 1990). A complete stand description is built up from the single plots and includes many aspects of forest structure, such as size distributions, canopy profiles, and stand characteristics such as species composition and basal area, production and biomass. Application to management might emphasize basal area, stemwood production and size distributions, while ecological applications might stress species composition changes over time.

The dynamics on a stand level are simulated by annual time steps and involve effects of shading, crowding and resource depletion on growth, regeneration and mortality. Species differ in the way in which they can utilize the actual resources during different phases of the simulation. Regeneration and mortality are stochastic processes with probabilities defined by the prevailing conditions on a plot and the vitality of each individual. Growth is modeled deterministically as annual increment of the dimensions of each individual tree. Tree characteristics at any simulated time are used to project the next year's characteristics. Although many of the formulas used in the models are presented as differential equations, the models are solved as difference equations with a unit time step of, most commonly, one year.

Regeneration

Many different regeneration routines have been used in gap models. Regeneration is modeled as the ingrowth of new saplings on the plot. These saplings start with a pre-defined small DBH and a height of 1.3 m. Germination, seedling establishment and the growth of small saplings (<1.3 m) are thus not explicitly modeled, although sometimes Leslie matrix submodels, which generate and simulate seedling cohorts through time and contribute an appropriate sapling number, have been included.

Before new saplings of a species are 'planted', the conditions on the plot are screened to determine if they are appropriate for growth and establishment of that species. Filters that define eligible species include, for example, the presence of mineral soil, presence of leaf litter, climatic conditions and seed sources. If conditions are appropriate then a small number of saplings are drawn randomly from the eligible species list and added to the population on the plot.

In many gap models, light is the most important requirement for sapling establishment. If light levels at the forest floor are high, large numbers of saplings of shade-intolerant species are planted until light levels drop again. This mimics the explosive occurrence of early-successional species after the formation of a gap by the death of a mature tree. For boreal forests, however, high establishment rates occur only after the formation of a large gap on the scale of one to several plots and even then it could take several decades before the forest floor is covered again. It appears sufficient for a model of boreal forest dynamics to use a simple stochastic regeneration routine.

In FORSKA the prevailing light conditions at the forest floor are determined and all species are checked to see if they are able to grow, i.e. if their compensation point (see below) is below the computed light intensity. A random number of saplings for the eligible species are drawn from a Poisson distribution and planted on the plot. The Poisson distribution simulates the sampling variance in annual establishment on a plot and is loaded with a species-specific intensity, the species establishment rate per unit area, and time. The saplings are planted with a small but varying DBH and leaf area (following Eqn (16.5)). Their bole height and age are set to zero.

Light interception and tree growth

Photosynthesis is the ultimate physiological process which provides an individual tree with energy and carbon for its biomass production. This process is driven by the interception of photosynthetically active radiation (PAR) by a tree's leaf area. The intercepted light is then converted into biomass via photosynthesis and metabolic processes. The simulation of growth should thus describe and parameterize these processes in a consistent way.

The growth function of most traditional gap models is merely an empirical description of annual growth. The growth equation is based on the assumption that tree height (H, m), leaf area (L, $m^2 \times m^{-2}$) and tree volume (D^2H, m^3) are simple allometric functions of diameter at breast height (D, cm). Growth is defined as an annual volume increment, scaled by a growth rate parameter and specified in the tree's actual and maximum size characteristics (Botkin, Janak & Wallis 1972; Shugart 1984). The growth rate parameter is defined by assuming that there is no growth when a tree reaches its maximum age and that the optimal growth curve is defined so that a tree will reach two thirds of its maximum diameter at half of its maximum age. The maximum growth obtained is then decreased by light, climatic and moisture limitations.

This parameterization of growth has proven useful in many different applications. However, by defining growth in this way, the critical parameters in the model become maximum age and diameter of a species, which in turn define the growth rate parameter. This has been demonstrated by using a sensitivity analysis of the growth function (Dale *et al.* 1988). Although data on maximum age and diameter are frequently reported, these values are often anecdotal and may not reveal important information on physiological and structural properties of a tree species. In the boreal forest region, with its large variance in tree characteristics due to considerable differences in local environments, a precisely defined growth function is required. Models used for single stands can often, by slight adaptations and changes of model parameters, be moved to other stands. If the modeling effort, however, is aimed at construction of a correct model for a whole biome, the simulation of growth processes must then be based on ecophysiological principles. This is the principal motivation for rederiving the traditional growth function used in gap models.

FORSKA differs from most of its predecessors in that, instead of situating all leaf area at the top of the bole, leaf area is distributed uniformly from the top height to the bole height (Fig. 16.2). Thus individual crowns overlap. This is especially important in parameterizing growth processes of the deep-crowned conifers of the boreal forests. The bole height is zero at first (when a sapling is placed onto the plot) but rises irreversibly (self-pruning) when light levels become too low at that level in the canopy of the plot.

Light extinction down into the canopy follows the Lambert–Beer law (Monsi & Saeki 1953; Kasanaga & Monsi 1954; Kira, Shinozaki & Hozumi 1969):

$$I_z = I_0\, e^{-kF_z},\qquad(16.1)$$

where the available light at a particular depth in the canopy (I_z) is a function of the cumulative leaf area index (F_z) of the canopy above and the solar radiation (I_0) at the top of the canopy. Light interception is scaled by the extinction coefficient (k) with typical values of c. 0.4–

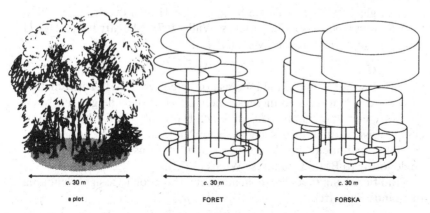

Fig. 16.2. Vertical canopy distribution for the different gap models.

$0.6\,\mu\text{mol m}^{-2}\,\text{s}^{-1}$ for most forests. Solar radiation is a function of latitude and cloudiness and can be computed by using the algorithm developed by Weiss & Norman (1985).

The response of net assimilation to prevailing light levels in the canopy is defined by the light response curve. This curve is an asymptotic function similar to the Michaelis–Menten equations often used to model enzyme-dependent processes:

$$P_z = \frac{kI_z - c}{kI_z + a - c}. \tag{16.2}$$

P_z gives the proportion of maximum possible annual assimilation achieved by leaves at depth z in the canopy; c is the compensation point, and a the half-saturation point (i.e. the light intensity at which net assimilation reaches half its maximum value). The parameters a and c define a species' shade tolerance.

The net assimilation integrated through the crown of a tree determines its growth. Growth is defined as an increase in the tree volume, D^2H:

$$\frac{\mathrm{d}}{\mathrm{d}t}(D^2H) = \left(1 - \frac{W_{\text{tot}}}{W_{\text{max}}}\right)\int_B^H S_L(\gamma P_z - \delta z)\,\mathrm{d}z, \tag{16.3}$$

where D is DBH, H is tree height, W is biomass, B is bole height, S_L is the vertical density of leaf area, γ is the species-specific growth-scaling factor, and δ is a species-specific factor for maintaining the tree's actual size. Equation (16.3) states that total carbon gain is proportional to leaf area and corrected for shading (Eqn (16.2)), while total carbon loss due to non-photosynthetic tissue is proportional to sapwood volume, which is in its turn proportional to the product of leaf area and crown depth. It is assumed that a fixed fraction of the tree's carbon budget is allocated to stemwood.

Most gap models relate tree height to DBH by a quadratic function (Ker & Smith 1955; Botkin, Janak & Wallis 1972; Shugart 1984). Here I use an asymptotic function (Meyer 1941):

$$H = 1.3 + (H_{\text{max}} - 1.3)(1 - \exp\left[-sD/(H_{\text{max}} - 1.3)\right]), \tag{16.4}$$

where s stands for the initial increase of height vs. DBH. This function allows individual trees to increase their DBH as their height reaches a species-specific maximum (H_{max}), which is probably limited by physical characteristics of a tree such as water uptake capacity. An annual diameter increment can be computed from Eqns (16.3) and (16.4), analogously to Botkin, Janak & Wallis (1972).

Traditional gap models use an allometric function to assign a leaf area to an individual tree:

$$L_i = CD^2, \tag{16.5}$$

where C is a species-specific constant (Botkin, Janak & Wallis 1972; Shugart 1984). This function applies only when all stemwood consists of functioning sapwood supporting that leaf area (Shinozaki *et al.* 1964*a*,*b*; Waring & Schlesinger 1985; Mäkelä 1986). This is only true for small saplings. FORSKA uses a more dynamic implementation of the so-called 'pipe model' (Shinozaki *et al.* 1964*a*,*b*) and assumes a fixed rate of sapwood conversion to heartwood.

$$\frac{\mathrm{d}L}{\mathrm{d}t} = C\frac{\mathrm{d}}{\mathrm{d}t}D^2 - tL, \tag{16.6}$$

where t determines the species-specific sapwood turnover and C is the constant from Eqn (16.5). From the calculated annual DBH increment and Eqn (16.6) it is possible to approximate the new leaf area increment and leaf area density. Leaf area thus increases at a slower rate than DBH. The total leaf area of a tree can be further reduced if assimilation by the lower leaves should become negative through excessive shading. These leaves are dropped irreversibly, and the bole height rises correspondingly.

Mortality and vegetative reproduction

In the traditional gap models there are two kinds of mortality. The first type is related to aging and is implemented in such a way that only a small proportion (1%; Shugart & West 1977) reaches maximum age. The second type of mortality is related to growth and is enhanced when the annual growth increment drops below a pre-defined DBH increment, i.e. slow or declined growth. When such an event occurs the probability of the death of the tree increases. Generally such trees only survive one or two subsequent years with reduced growth. The rationale for DBH increment as an index of vigor is that if a tree cannot maintain a certain minimum growth rate it will be more susceptible to factors that cause mortality, such as disease.

Such parameterization of mortality is less suitable for the boreal forest, where many tree species are able to maintain slow growth under suppressed conditions for long periods. FORSKA's mortality rate is based on a different rationale, the growth efficiency, i.e. stem volume per unit leaf area (Waring & Schlesinger 1985). This growth efficiency is reduced by shading (including self-shading), competition for nutrients, and the increasing cost of maintaining leaves at greater heights. Such an index may be a better indicator of the vigor of any individual. Mortality always occurs at a low rate (cf. the first type described above) and increases if vigor falls below a specified threshold.

Newly dead trees are checked to see if they are of the right species to be capable of vegetative reproduction (i.e. sprouting). If so, and if the

Table 16.1. *Model and site parameters used in FORSKA*

Symbol	Description	Dimension
I_0	Growing season-average incident light intensity	$\mu\text{mol m}^{-2}\text{ s}^{-1}$
k	Light extinction coefficient	
b	Stemwood biomass conversion factor	$\text{kg cm}^{-2}\text{ m}^{-1}$
W_{max}	Site-specific maximum biomass	Mg ha^{-1} .
Δz	Vertical integration step through the canopy	m
D_0	Initial DBH for saplings	cm
t	Length of the time step	yr

Table 16.2. *Species parameters used in FORSKA*

Symbol	Description	Dimension
E'	Sapling establishment rate	$\text{ha}^{-1}\text{ yr}^{-1}$
C	Leaf area to DBH^2 ratio	$\text{m}^2\text{ cm}^{-2}$
H_{max}	Maximum tree height	m
s	Initial slope of diameter vs. height	m cm^{-1}
c	Light compensation point	$\mu\text{mol m}^{-2}\text{ s}^{-1}$
α	Half saturation point	$\mu\text{mol m}^{-2}\text{ s}^{-1}$
γ	Growth scaling factor	$\text{cm}^2\text{ m}^{-1}\text{ yr}^{-1}$
δ	Sapwood maintenance cost factor	$\text{cm}^2\text{ m}^{-2}\text{ yr}^{-1}$
t	Sapwood turnover rate	yr^{-1}
U_0	Intrinsic mortality rate	yr^{-1}
U_1	Mortality rate due to suppression	yr^{-1}
θ	Threshold value for index of vigor	

individual is within the dimensions permitting sprouting, new individuals replace the dead tree with sizes as if they were newly established saplings.

Implementation of the model

The model described above is implemented in FORTRAN 77 in a modular way. This allows for an easy linkage with environmental sub-models (see below), the incorporation of new or updated processes, and adaptation to specific requirements. For example, the model for long-term forest fire dynamics (Antonovski, Ter-Mikaelian & Furyaev, Chapter 14 of this book) can be coupled to the biological model without problems. The model structure and input and output requirements are described by Leemans & Prentice (1989) and the input parameters necessary to run the model and define species characteristics are listed in Tables 16.1 and 16.2. The source code is available from the author on request.

The linkage of the forest dynamics model with the environment

The FORSKA model for boreal forest dynamics is based on a conceptual carbon balance of individual trees. Growth is a function of assimilation and respiration rates. The annual assimilation rate is determined explicitly by available light in each successive layer of the forest canopy and is thus corrected for shading within the forest canopy and within each individual tree. The annual growth of an individual tree is a function of net assimilation in its vertical leaf area profile scaled by the growth rate scaling factor (γ, Eqn (16.3)) and decreased by a specific carbon cost to maintain its size. The growth scaling factor is specific for each separate species and is also an explicit function of the total environment. In a separate sensitivity analysis of FORSKA, the growth scaling factor together with the light response parameters proved to be the most critical parameters in the model (Leemans 1991b). From this sensitivity analysis it follows that both the environment and current canopy structure largely determine the annual growth. This is in agreement with our current understanding of forest systems (Shugart 1984; Kimmins 1987). Environmental constraints all influence the value of the growth scaling factor and repress growth in a direct way. I shall call any factor that limits potential maximum growth and thus alters the actual value of the growth scaling factor a *growth multiplier*. Growth multipliers generally have a value in the interval 0–1 and are most commonly combined in a multiplicative way with the growth scaling factor (Botkin, Janak & Wallis 1972; Shugart 1984; Pastor & Post 1985). This means that the actual value estimated for the growth scaling constant probably results in a much larger growth than observed at any place in a natural system because growth under optimal conditions for all multipliers rarely occurs. Defining the value of the growth scaling factor on the basis of the observed maximum age and DBH of a species (e.g. Botkin, Janak & Wallis 1972) may lead to an under-estimation of its theoretical value and this should be corrected by assuming a departure from the multiplicative way of determining the actual value of the growth scaling constant. (Sometimes only the minimum value of all available growth multipliers is taken, which is then justified mistakenly as an implementation of Liebig's (1840) law.)

The most commonly used growth multiplier in all traditional gap models is the shading multiplier (Fig. 16.3a), defining growth responses to available light at any vertical level in the canopy. Each species has a specific response function, which determines its shade tolerance and thus defines if a species can grow at a particular canopy level. Growth decline of species (especially shade-intolerant ones) can be fairly steep owing to the effects of this multiplier because in this case all leaf area is situated at the top of a tree's bole, which exaggerates the asymmetric competition

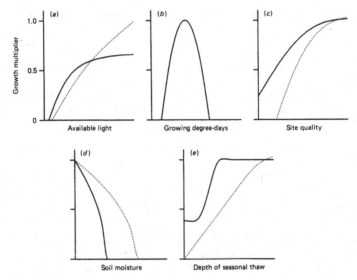

Fig. 16.3. Growth multipliers to be combined with the growth scaling factor of gap models: (*a*) shading; (*b*) climate as defined by growing degree-days; (*c*) site quality; (*d*) soil moisture; (*e*) permafrost.

for light. FORSKA does not need this multiplier because growth is determined by the explicit simulation of net assimilation driven by the decreasing light levels from top to bottom of the forest canopy.

A multiplier that defines growth response to changing climatic conditions over a landscape or over time is defined by a species' climatic limits (Fig. 16.3*b*). The definition of this multiplier most often used stems directly from JABOWA (Botkin, Janak & Wallis 1972) and is implemented as a simple quadratic response function on growing degree-days. The assumption that maximum growth occurs in the center of a species' climatic limits justifies such a simple function. For the boreal forest this assumption is violated. Although the northern limit of most boreal forest species coincides well with a growing degree-days isocline (Tuhkanen 1984), the southern limit is more often defined by moisture limits or does not clearly coincide with a particular growing degree-day isocline. Further, optimal growth most often does not occur in the center of a species' range, but declines sharply along its southern border and remains fairly constant over a large part of its distribution. To simulate the response of boreal forest species to climate, another multiplier for climate response should be defined. I. C. Prentice & M. T. Sykes (personal communication) have used temperature ranges over the total growing season and obtained promising results. They used different values for the lowest temperature allowing growth for evergreen and deciduous trees and so simulated competitive advantages of evergreen trees in the boreal climate.

In constructing a general model for boreal forest dynamics which simulates reliable species distributions for the whole circumpolar region, several other climate indices must be included. Species distributions in this region are limited not only by their northern and southern limits, but also by eastern and western limits. This demands incorporating species response functions of climatic parameters such as continentality and absolute minimum and maximum temperature, to define growth response. Until now such growth multipliers have not been defined or used.

Other growth multipliers define site quality (Fig. 16.3c). Such multipliers define the response to different levels of nutrient availability. This concept was first included in gap models to simulate fertilization effects (Aber, Botkin & Melillo 1979; Weinstein, Shugart & West 1982) and was later complemented with a more complete nutrient cycling routine by Pastor & Post (1985). A similar approach is adopted for the generalized boreal forest model.

Different models have been developed which drive the available soil water (Pastor & Post 1985; Cramer & Prentice 1988; Bonan 1989). They are coupled to monthly temperature and precipitation values and specific water-holding capacities of the forest soil. While linked to such physical models, gap models are capable of simulating the response to different averaged weather scenarios over longer timespans. The species-specific response function to available soil moisture is linked to gap models by a multiplier, based on a dry-day index (Fig. 16.3d) (Bonan 1988a).

The model by Bonan (1989a; Chapter 15) defining the constraints of a specific boreal forest stand, simulates the actual depth of permafrost and its dynamics over a year, as influenced by soil characteristics, solar radiation and prevailing weather patterns. Species tolerate permafrost in different ways. The important variable which links permafrost dynamics to growth is the seasonal depth of thaw, and characteristic response functions are derived for both tolerant and intolerant species (Fig. 16.3e) (Bonan 1988a).

The different multipliers described above are the most commonly used. Phipps (1979) describes a multiplier that defines growth responses to the depth of the soil water table. Such a multiplier could be useful in some wet or bog forest stands in the boreal forest, but is probably not necessary for the whole boreal forest region.

Growth can be limited by several other factors during different parts of the life cycle of a tree. In applying the concepts of growth multipliers, it is assumed that these do not vary during the complete life cycle of individual trees. This simplification is probably true for the general multipliers described above but there are seveal processes that act only on certain well-defined parts of a species' life cycle, such as seed germination and establishment. These factors should not be implemented in a gap model as a growth multiplier, but directly on the appropriate routine that defines and simulates that part of the life cycle. Bonan & Korzukhin (1989), for example, simulate forest floor dynamics and apply explicitly the response of a species' regeneration on the thickness of the moss layer.

Comparison with a traditional gap model

There have been several applications of gap models to simulate aspects of boreal forest ecosystems (e.g. Solomon 1986; Bonan & Korzukhin 1989; Overpeck, Rind & Goldberg 1990). Most often the dynamics of these forest systems are presented as summarized charts of biomass per species over time, averaged over tens to hundreds of simulated plots. Species composition is generally simulated correctly with the traditional gap models. This may result from the pre-defined growth response to climate, which is directly derived from a species' actual geographical distribution limits. Gap models are sensitive for parameters defining and influencing the growth function (Kercher & Axelrod 1984a; Dale et al. 1988; Leemans 1991b), and relating growth to distribution may contribute to the models' ability to simulate compositional patterns.

A more rigorous test of a model's capacities is the simulation of one or a few single patches and comparison of the dynamics through time with those of an actual stand of which the history is well known. Such a test is difficult to perform because of the lack of appropriate data, such as time series of forest development and complete and thorough forest structure descriptions. Most relevant studies are only published in a largely summarized way and the primary data are not available from the literature. Such a test has been done for the different versions of the FORSKA model (Leemans & Prentice 1987, 1989) with data collected by myself (Leemans & Prentice 1987; Leemans 1989, and unpublished), but I do not know of any published small-scale test for the traditional gap models. Here I present the initial result of the test with one of the latest scions of the gap-model family, ZELIG (Smith & Urban 1988; Urban 1989) using the same comparison data from the unmanaged forest stand of Fiby Urskog, Sweden, which were used in the FORSKA test. In this comparison, the FORSKA model performance is being compared with a JABOWA–FORET traditional gap model.

Fiby Urskog (59°54' N, 17°22' E) is typical of the southern edge of the boreal forest zone, with several boreonemoral species in the forest floor vegetation. Its dynamics over the last two centuries have been well studied by Hytteborn & Packham (1985, 1987), Leemans (1989) and

Sernander (1936). The present forest is thought to be the result of a succession after a large disturbance by a severe windstorm at the end of the seventeenth century. At present all canopy species (*Betula pendula* Roth., *B. pubescens* Ehrh., *Populus tremula* L., *Pinus sylvestris* L. and *Picea abies* (L.) Karst.) are declining and many minor gaps are created in which only saplings of *Picea* establish (Leemans 1991a).

The traditional gap model ZELIG is designed in such a way that individual users are easily able to adapt the model to their own objectives, making it a most suitable model for a critical comparison with FORSKA. Most other gap models needed a larger implementation effort to become comparable with only the biological processes of FORSKA. ZELIG includes all the features that have made former gap models successful, and it has added a more elaborate spatial arrangement and interaction between single plots (Smith & Urban 1988). I designed a stripped version of ZELIG, which was comparable with the structure of FORSKA, defining a simulation as a single array of independent patches, and with no growth reduction due to moisture, nutrients or climatic limits, but I left the characteristic growth function, allometric relations and establishment and mortality routines intact. This resulted in a model similar in the simulated processes, but different in their definitions. The estimation of most model and species parameters was taken from values given for the species involved by Kienast (1987) and Helmisaari & Nikolov (1989, and Chapter 2 of this volume). The parameters for the species' dimensions were taken from regressions on collected material from Fiby Urskog (Leemans & Prentice 1987). This parameter estimation (Table 16.3) should allow for an accurate comparison.

FORSKA and the traditional gap model were both applied to simulate a period of 190 years with the standard five 0.1 ha patches and six 0.08 ha patches respectively. The resulting forest after 190 years as measured in the biomass distribution among species is correctly simulated by both models, although the traditional gap model underestimates the total biomass on the stand. The height distributions (Fig. 16.4) display a much larger difference between the two models. The distribution for spruce (shade-tolerant) is comparable with the actual distribution of the real stand for both models, although the real stand simulates a more pronounced bimodal distribution than the models. This could be due to the lack of small saplings (smaller than 1.3 m) in the simulation of the real forest.

The largest discrepancy in the traditional gap model as compared with the real forest and FORSKA lies in the simulation of the height distributions of the shade-intolerant species (*Pinus*, *Betula* and *Populus*). The height distributions of these species are inaccurate, owing to the deviation of the actual forest dynamics in the boreal zone, where succession is

Table 16.3. *Parameter values for (a) FORSKA and (b) the traditional gap model in the comparative run*

(*a*) FORSKA

Site parameters			Model parameters	
I_0	400	μmol m^{-2} s^{-1}	Δz	0.5 m
k	0.4		D_0	1 cm
W_{max}	600	Mg ha^{-1}	t	2 yr
b	0.035	kg cm^{-2} m^{-1}		

Species parameters	Pinus	Picea	Betula	Populus
H_{max}	36.0	33.0	32.0	35.0
s	0.62	0.66	0.86	1.27
α	100	330	330	100
c	21.7	46.3	41.1	25.4
γ	9.1	23.0	37.8	9.6
t	0.004	0.004	0.004	0.004
C	0.330	0.260	0.224	0.304
δ	0.048	0.062	0.071	0.053
E'	5	8	12	3
θ	0.025	0.025	0.025	0.025
U_0	0.0132	0.0046	0.0132	0.0115
U_1	0.46	0.46	0.46	0.46

(*b*) Traditional gap model

Species parameters	Pinus	Picea	Betula	Populus
Age_{max}	700	450	150	200
D_{max}	55	45	35	65
H_{max}	35.1	42.9	24.1	28.3
Growth factor	175.0	150.0	190.0	180.0
Shade tolerance[a]	5	2	5	4
Establishment	5	8	13	3

Note: Shade tolerance class (1 = very tolerant, . . ., 5 = very intolerant)

directed from shade-intolerant species toward more shade-tolerant ones. The event causing the return of shade-intolerant species in the traditional gap model can be easily traced to the simulation between year 140 and 150 (Fig. 16.5). The death of some large trees results in a drop in both biomass and basal area, but density suddenly increases. This indicates the start of a successful regeneration of shade-intolerant species as observed later in the size distributions. In the actual forest stand such regeneration is only observed in large gaps, with sizes comparable to or larger than the simulated plot size (Leemans 1991*a*). Such a return of early successional species can be simulated with a pre-defined disturbance regime, but I did

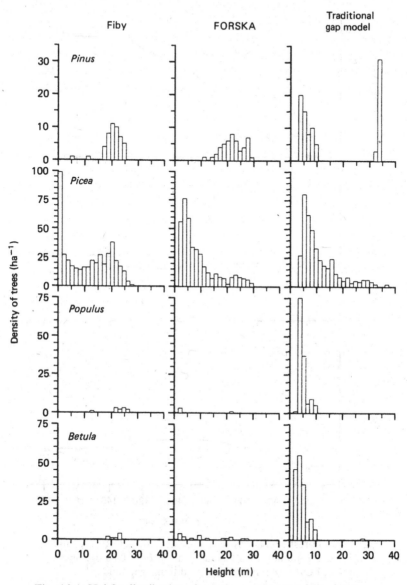

Fig. 16.4. Height distributions from the real stand in Fiby Urskog, Sweden (left), and those simulated by FORSKA (center) and a traditional gap model (right).

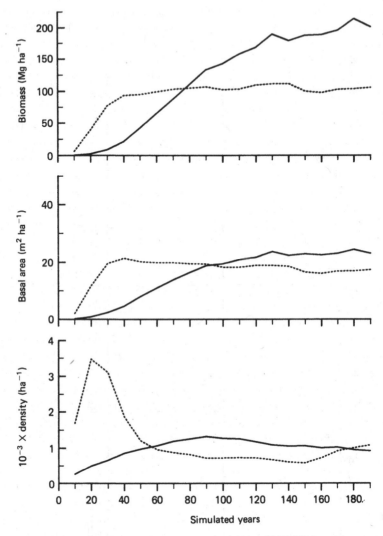

Fig. 16.5. Forest dynamics as simulated by FORSKA and a
traditional gap model for a 0.5 ha stand in Fiby Urskog, Sweden.
(Solid line, FORSKA; dotted line, traditional gap model.)

not use one in these simulations, so that the shade-intolerant species
should eventually disappear completely from the simulated plots.

These results indicate that the traditional gap models (sensu Shugart
1984) are less suitable for a simulation of the actual observed small-scale
dynamics of boreal forests. The main reason for this is their simplified
vertical canopy distribution in which all leaf area is positioned at the top

of a tree's bole. This leads to an exaggeration of the asymmetric competition for light, in which large trees have a considerable advantage over small trees. The vertical canopy distribution is probably a suitable approximation for the more temperate and tropical forests, where sun angles are steeper, although FORSKA is also capable of grasping the proper dynamics of these forests (Leemans 1989). The capability of FORSKA to simulate boreal forest dynamics properly is also due to the dynamical assignment of leaf area using the pipe model and the implementation of self-pruning (see above), which results in an effective and realistic definition of the vertical canopy and tree-crown structure. The different characteristics of individual trees leave little room for the development of single-sized, single-aged tree cohorts, which allows for an improved simulation of a stand and its individual successional history.

A large difference between the two establishment routines of the models can also be seen in Fig. 16.4. Although it is of minor concern for the final results of the models after 190 years, FORSKA grows the initial forest much more slowly, owing to pre-defined yearly establishment rates for each species. The traditional gap model adds new individuals to the plot until limitation of light does not allow more new individuals. This results in an explosion of new individuals as soon as favorable conditions occur. Such a parameterization of establishment is less suited to boreal forest dynamics, where regeneration processes are generally slow, and this routine would lead to a much too fast recovery of the canopy layer after gap formation.

The results described above affect boreal forest modeling. I have shown that the traditional gap models cannot be applied directly to simulate the specific dynamics of boreal forest systems. The importance of asymmetric competition for light within the canopy is exaggerated and establishment and mortality are too vigorous. This leads to the simulation of unrealistic size distributions of the species involved and thus inadequate representations of the simulated forests through time. I have developed a more accurate succession model for the boreal forest, which has overcome most disadvantages of the traditional gap models. When this improved model becomes coupled to environmental models (e.g. Bonan 1989 and Chapter 15) and models of disturbance regimes (e.g. Antonovski et al., Chapter 14), it will allow us to assess more precisely the short- and long-term impacts of, for example, climate change. Such a model will also be more generally applicable to different parts of the boreal forest.

17 Role of stand simulation in modeling forest response to environmental change and management interventions

Peter Duinker, Ola Sallnäs & Sten Nilsson

Introduction

Strong advances in understanding forest ecology and forest-ecosystem responses to disturbance and environmental change have come through the application of systems analysis and simulation in studies of small forested ecosystems. For example, the whole series of gap models of forest succession (e.g. JABOWA (Botkin, Janak & Wallis 1972), FORET (Shugart & West 1977), FORTNITE (Aber & Melillo 1982), LINKAGES (Pastor & Post 1985), FORENA (Solomon 1986) and FORSKA (Leemans 1989); see Shugart (1984) for an overview of several of these models) examines forest ecosystems at the spatial scale of one or several large trees, i.e. about 0.1 ha. These models represent areas of sufficient spatial extent to represent adequately important processes such as inter- and intra-specific competition, and soil–vegetation–atmosphere interactions. Results from such models might reasonably be scaled up spatially to the level of the stand, where stand is defined as an ecosystem with a relatively homogenous community of trees and relatively homogenous site conditions compared with neighboring ecosystems. However, generalization of results from stand-level models (as the gap and other microcosm models will now be called) to forest ecosystems containing many stands, without the use of specially formulated forest simulation models, can be dangerous, if not absolutely misleading.

Ecologists generally recognize that ecosystem boundaries are more or less arbitrary. Some such boundaries are easy to assign and are ecologically very meaningful, such as the perimeter of an island in a lake, or the perimeter of a farm woodlot surrounded by field crops. Other boundaries, such as the demarcation of individual stands in a boreal forest, are less objectively set, yet are still meaningful because of obvious differences

in stand composition and site conditions between neighboring communities of trees. Those who study or manage forests of extents reaching into
the thousands to hundreds of thousands of hectares are accustomed to
classifying the forest into specific ecosystem types and partitioning maps
of the forested landscape into units assigned to these types. For forest-
management and inventory purposes, forests are divided into stands, as
defined above, often ranging in size from about one hectare to several
hundred hectares.

The spatial scale of interest for many forest ecologists and certainly for
all forest-management agencies, including governments, forest-products
companies and other landowners, is that of the forest, often extending to
hundreds of thousands of hectares. Thus, systems analyses that attempt
to model *forest* response to environmental change or to management
interventions must explicitly recognize spatial variability in vegetation
communities and site conditions across broad areas. The most generally
useful type of these models is of hierarchical construction, where individual stands or groups of similar stands are modeled separately and the
results appropriately aggregated to generate whole-forest results. Such
aggregations are seldom simple sums, because there are many ecological
processes (e.g. fire, pest occurrences) and management regimes (e.g.
harvest scheduling) where the occurrence of events in some stands affects
the occurrence of similar and other events in other stands.

Considering stands to be relatively small, homogeneous, forested
ecosystems, and forests to be relatively large ecosystems comprised of
many, perhaps thousands, of stands, the basic principle behind this
chapter is that *forest-level* models are required to gain insight into *forest-
level* responses to environmental change and management interventions.
The question we pose and address in this chapter is the following: how can
stand simulators help in modeling forest responses? We begin by reviewing some basic concepts in forest management that are especially relevant
to boreal forests. The critical role of stand-level models in forest-level
analyses will be described, using two examples of common types of forest
simulators.

Forest management

There are as many definitions of forest management as there are writers
on the subject. We prefer a definition that explicitly recognizes the
relations between stands and the forest: forest management is the
planning and implementation of stand-level (or stand-group) actions
(e.g. harvest, renewal, tending, protection, access) to reach forest-level
goals. A related view is that forest management consists of control of

448 P. Duinker, O. Sallnäs and S. Nilsson

patterns of stand development over the long term across a forest (Basker-ville 1986). Silviculture, then, is the suite of technical, stand-level tools (actions) that are used to alter natural stand development to help contribute to forest-level goals. In other words, foresters practice silviculture to try to fix forest-level problems. Thus, when a forest will not naturally provide specified desired benefit flows over time, silviculture is planned and implemented to alter forest structure and development so that the desired benefit flows might be realized.

We have defined stand above adequately for our purposes, but the concept of forest requires additional treatment. 'Forest' could be used in an ecological sense, such as the boreal forest, in a familiar sense, where it means the woods or bush (to use typical Canadian synonyms), or in a management sense where it means any arbitrarily defined forested landscape that is managed more or less as a single unit. A description of the management scene in the boreal forest of the Province of Ontario, Canada, may help clarify the concept of forest. Ontario has a land area of 88.5 million ha, of which 56.6 million ha is considered productive forest land. Some 35 million ha is boreal forest, with almost 9 million ha classified as boreal barrens and some 12 million ha temperate mixedwood–hardwood. More than 90% of the productive land area is publicly owned, with management responsibility vested in the Government of Ontario. Besides areas set aside as parks and wilderness areas, the provincial forest is divided into about 115 so-called forest-management units ranging in size from under 100 000 ha to over 1 000 000 ha. Each unit is managed as a discrete entity with its own goals, problems, budget and sets of activities. Some units are managed by specific forest-products companies under contract with the provincial government, while others are managed directly by the Ontario Ministry of Natural Resources (OMNR). The latter are usually harvested under cutting licenses to mill owners and jobbers.

Thus, in the management sense, when we speak of a forest on public land in Ontario, we speak of a forest-management unit. In the Forest Resource Inventory (FRI) undertaken for each management unit by OMNR (OMNR 1978), stands are normally delineated into parcels of land ranging in size from under one hectare to several hundred hectares, with the average near 10 ha. Forests could therefore consist of thousands of stands. Each stand in the FRI is described by location (on a map), areal extent, stand composition, stand age, and merchantable volume of timber.

Until recently, forest managers used rather crude sets of equations for projecting forest inventories into the future to determine annual allow-able cuts and forest development in terms of age-class structure and growing stock (i.e. total standing volume of timber). There were no sophisticated computers readily available, so wood-supply studies used highly aggregated inventories to simplify matters. Nowadays, with computational power exceeding most needs, and with widespread wood-supply shortages, managers are becoming much more sophisticated with inventory and wood-supply projection. Detailed inventory projection models are routinely used in planning harvests and silviculture at the management-unit level. Even for the largest of forest-management units, there is ample opportunity with such models, especially through use of geographic information systems, to account for and keep track of the development of every stand in the forest.

Stand development patterns

A broad range of assumptions needs to be made when managers and researchers model whole forests. Among these, a key assumption, perhaps *the* key assumption, concerns the long-term development patterns of individual stands and stand types. This assumption is of vital importance because: (a) forest models are usually used to project forest development over periods of many decades, often as long into the future as a century; and (b) most forest models generate forest-level results by aggregating responses from individual stands or stand types. There are three main sources of information on what the future development patterns of stands in any forest might be:

1. Empirical models (e.g. yield tables or curves). Yield tables or curves are generally based on field measurements, and thus are most applicable to the forests in which the measurements were made. Sometimes stand development has been measured in permanent plots over several decades, but more often the yield tables or curves are built upon measurements in similar stands at different ages (so-called chrono-sequences). The application of yield tables or curves based on such measurements in studies of future forest development assumes that future stand-development patterns will be precisely identical to past development patterns. This can be a problematic assumption for studies in which the objective is to explore potential forest responses to new management strategies or a changing atmospheric environment.

2. Process-based stand-level simulation models. These models are normally developed by the research community for a variety of purposes, often expressly for defining plausible future stand-development patterns for use in forest studies. Such models usually have regional applicability. They assume that basic ecological and physiological processes will remain unchanged in the future, but that the stand-level outcomes of such processes may not. These models are particularly well suited to providing information for studies of forest response to management regimes and a

changing atmospheric environment, as long as they have been structured with such objectives in mind.

3. Intuition and informed judgment. In the absence of local measurements or appropriate stand simulators, forest managers have often had to resort to professional judgment in determining how stands might behave over future time. This approach to generating evidence for future stand-development patterns is probably the weakest of the three.

Forest managers and researchers would do well to apply all three sources of information in undertaking forest studies that explore management strategies and a changing environment. Thus, stand simulators would be constructed and used to generate the required yield tables or curves, which would then be checked against measured yield tables or curves and against the professional judgment of local foresters and researchers. In studies where the objective is explicitly to investigate future forest responses under alternative management strategies or under a changing environment, use of stand yield tables or curves developed without the aid of specially constructed stand simulators is a risky proposition. Results of such studies will hardly be defensible.

Two examples of simulation models for studies of long-term forest responses to management strategies and environmental change

Example 1. Management-unit modeling in eastern Canada

Background

Like Ontario, other eastern Canadian provinces such as Newfoundland, New Brunswick, Nova Scotia and Québec have divided their publicly owned forests into forest-management units. In the early 1980s, a potpourri of hundreds of cutting licenses in the New Brunswick forests was replaced by ten new management units and ten new industrial management contracts. This restructuring of timber allocation was prompted by a growing realization on the part of managers and researchers in the 1970s that serious wood-supply shortages were imminent. The realization, in turn, sprang initially from a large study that attempted to determine the need to continue to protect New Brunswick forests from the ravages of spruce budworm (*Choristoneura fumiferana*). The study included a sophisticated simulation model of the forests of the entire province (see case study in Holling (1978)). When researchers became aware that wood-supply shortages were projected by even this model, the purpose of which was not specifically wood-supply forecasting but rather forest-inventory projection for budworm research, they began to look more closely at the wood-supply situation through use of volume-based wood-supply simulators (Hall 1981). Then followed a series of developments in personal-computer-based wood-supply simulators for use both in operational forest-management planning and in professional educational programs. The latest simulator in this series is a very transparent, easy-to-use, widely shared model named FORMAN (Wang, Erdle & Roussell 1987).

Model structure

FORMAN is described as an inventory projection tool for creating and evaluating forest-management strategies (Fig. 17.1). Its main functions, like those of all detailed forest-inventory simulators, are bookkeeping (i.e. keeping track of individual stands or groups of stands) and updating (i.e. calculating new conditions for the forest based on previous conditions and specified rules for change). FORMAN handles initial forest conditions by accepting an aggregated-inventory data file. In this file, the user specifies the types of stands (called forest classes) that should be separately accounted for, based on any criteria the user can and wants to invoke, e.g. stand composition, site class, stand age, stand location.

The basic updating mechanism in FORMAN is a set of user-input stand-development curves. The curves are simply plots of expected stand behavior, in terms of a few common management-oriented variables, over the entire life of a stand from regeneration through immaturity and maturity to overmaturity and decadence. The current version of FORMAN (Wang, Erdle & Roussell 1987) requires at least one set of stand-development curves for 'primary volume', i.e. that commercial timber volume of primary importance to the user. Additional optional curves can be specified for secondary volume (e.g. hardwood yield in primarily coniferous stands), product percentage (particularly useful for specifying sawlog yields), harvest cost, and an open category for any variable or criterion desired.

For each forest class, there are two main sets of curves. The 'current' curves are followed when: (a) stands are not subject to management intervention, and environmental conditions either are considered irrelevant or remain unchanged in the future; and (b) stands are harvested and left to regenerate naturally with the assumption that future development patterns on these sites will follow the historical ones (again, with environmental conditions irrelevant or constant over time). 'New' curves can be specified for natural regeneration, planted stands, and thinned stands. The new curves for naturally regenerated stands permit assumptions about, for example, regeneration delays and stand conversions from one type to another. New curves for planting and thinning allow users to explore the wood-supply consequences of implementing various levels of these two types of silvicultural treatment.

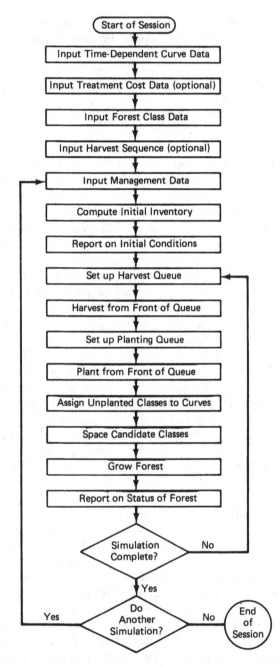

Fig. 17.1. Flowchart for the FORMAN wood-supply simulator (adapted from Wang, Erdle & Roussell 1987).

A very attractive feature of FORMAN for its use by field foresters is the input format for stand-development curves. The curves are not specified in equation form, but rather as a set of vertices (i.e. x and y coordinates where x is year after disturbance) which define major turning points in the shape of the curve. The model then interpolates straight lines between adjacent vertices. This feature makes foresters, most of whom (like many reseachers too) are quite daunted by complex equations, quite comfortable with the model. Specification of stand-development patterns in this way may seem crude to many researchers and biometricians who are accustomed to equations to describe such relations, but there is additional justification for the vertex approach. Stand-development patterns are generally poorly known from any source of evidence, and are very unconfidently predicted over long time frames. Moreover, the development patterns used in forest simulation models are 'average' patterns applying to classes of stands within which there is usually considerable yet poorly described variability. Therefore, the precise approach of equation-fitting, compared with the seemingly less precise approach of specifying turning points in development patterns over time, is not necessarily a more accurate description of reality in the forest. To enhance use of such models by field foresters, the simple approach, which is philosophically no less rigorous, is preferable.

FORMAN permits the user to specify alternative regimes for harvest, planting and spacing (and any other treatments that can be described in terms of area treated and changes in stand-development patterns). To show the range of possibilities in the current version (Wang, Erdle & Roussel 1987), the following harvest rules can be invoked:

1. minimize losses of unharvested primary volume;
2. maximize primary volume ha^{-1} harvested;
3. minimize harvest of secondary volume ha^{-1};
4. maximize harvest of secondary volume ha^{-1};
5. maximize product percentage (e.g. sawlogs);
6. minimize harvest cost per cubic meter; and
7. maximize the criterion on the fifth development curve.

While FORMAN has become a widespread forest-planning tool in eastern Canada, it has also been subject to a number of improvements to the basic model by various forest-sector analysts. For example, the FORMAN capacity for forest classes (assemblages of stands of equivalent type, age, structure, and development patterns) is insufficient to be able to track separately each stand in a large forest. This is a desirable feature for forest models in order to be able to link them in a useful way to geographic information systems (GISs). The connection of forest simulators to GISs is vital in investigating the implementability of management strategies for large forests where (a) access roads need to be created, and (b) individual stands are smaller than normal operating blocks.

Conclusion

Obviously, FORMAN projections of the forest inventory and harvest levels are vitally dependent on the characteristics of the stand-development curves specified by the user. The user must somehow ensure that the curves specified have some correspondence with reality. Unfortunately, data collected from real forests, no matter how well measured and voluminous, are all data about past development patterns of stands, yet the forest-inventory projection tool is to be used to explore future development patterns of forests. If field data are used to specify stand-development curves, then the analyst makes the assumption that future patterns will repeat past patterns. For investigation of alternative management strategies, this may be a reasonable assumption, *if* the analyst has been able to obtain field data on stand responses to all the kinds of management interventions being explored. Often, though, this is not the case.

A similar problem occurs when one wishes to investigate forest responses to a changing environment in the future. There simply are no field data available to chart stand-development patterns under a changing environment, especially long-term air pollution (e.g. acid rain) and climate change. The analyst who would investigate such questions has no choice but to conjecture what the stand-development patterns might be under scenarios of a changing environment, or to invoke stand simulators built expressly for the purpose of forecasting potential stand development under changing environments.

Let us imagine a proposed study to address the following question: how might management for a forest in northern Ontario appropriately change over the next decade to anticipate a possible dramatic climate change over the next half to full century? Such a study must make the following assumption: since the future will not repeat the past with respect to stand-development patterns, all empirically based yield curves are not applicable to the long-term future. The approach would be to prepare a set of inventory data from an appropriate management unit for a GIS-based version of a FORMAN-like model. A stand simulator of appropriate structure (e.g. the boreal-stand simulator discussed in this volume) would be used to develop a library of stand-development curves for all stand types in the forest under conditions of: (a) no management and a stable climate; (b) no management and a changing climate; (c) management of various intensities and a stable climate; and (d) management of various intensities and a changing climate. It is important to track stands under conditions of no management because in forests in northern Ontario, and indeed throughout the boreal forest, many stands see no intervention for their entire life because of factors such as unmerchantable types or inaccessibility.

Drawing upon appropriate selected sets of stand-development curves from the library, many forest-level simulations would be run to see how management and a changing climate would interact. The forest simulator of course provides the forecasts, and the GIS provides spatial information on the implementability of alternative management regimes and on the spatial patterns of forest-structure evolution. In addition, if the relationships can be established, the GIS can also be used to investigate broad-scale processes that are expected to be altered by a changing climate, such as insect infestations and wildfire.

In conclusion, forest-inventory simulators, of which FORMAN is a particularly simple yet powerful example, must be used for investigating potential patterns in future forests. If one believes that the future atmospheric environment will change significantly over the timespan of a typical rotation of a boreal forest stand, and that the atmospheric environment has a strong influence on stand-development patterns, then an appropriate process-based stand simulator must be used to develop credible forecasts for stand behavior for use in the forest simulator.

Example 2. Future forest resources of Europe

Background

Within the Environment Program at the International Institute for Applied Systems Analysis in Austria, the Biosphere Dynamics Project seeks to examine long-term, large-scale interactions between the world's economy and its environment. The Project conducts its work through a variety of basic research efforts and applied case studies. The immediate focus of the Forest Study of the Biosphere Dynamics Project (Nilsson, Attebring & Sallnäs 1987; Nilsson, Duinker & Sallnäs 1990) is on the future development of the forest resources in Europe. Possible effects of continued air pollution is one of the major concerns.

A key task in creating the analytical framework for the Forest Study was construction of a model for simulating standing and harvested wood volume over time, differentiated according to country, species group and age. The subject and area under study, the forests of Europe, are very heterogeneous and rich in variation. Few existing simulation systems are suitable for such a large-area analysis (Binkley 1987; Brooks 1987). In addition, elaborating the effects of a changing environment implies incorporation of knowledge not yet formalized.

Conditions for a timber assessment for Europe, and data accessibility

In selecting an appropriate model structure for a Europe-wide timber assessment in the Forest Study, actual forest-resource conditions and data availability were taken into account. There are large variations and differences among European countries along several dimensions of forestry, including:

1. the natural potential for forest production;
2. forest-management and silvicultural traditions;
3. economic importance of forest resources; and
4. ownership patterns and objectives of owners.

Throughout Europe, the environmental and social benefits of forests are increasing in importance compared with industrial benefits. Unfortunately, the non-timber benefits of forests are difficult to measure and model quantitatively. The threat from air pollutants, climate change and fire are increasing and pose great uncertainties in analyzing forest-resource futures. In addition, the public increasingly wants direct involvement in setting forest policies. Along with the expected large-scale conversions of agricultural land in Europe to forest cover in the coming decades, these factors suggest that forestry over the near future will be characterized by large political complications.

Notwithstandingt the above conditions, the most important factor in designing a consistent, dynamic timber-assessment model for all of Europe is the availability of forest-resources data. While some countries have databases rather well suited for timber-assessment studies (e.g. Finland, Poland), others are lacking in basic information (e.g. Spain, Luxembourg). In general terms, there are some fairly widespread data problems regarding European forests and their management (Nilsson, Attebring & Sallnäs 1987):

1. insufficient data on young forests and forests comprised of uneven-aged stands;
2. inconsistent definitions of forest land;
3. insufficient information about the dynamics of potential forest land;
4. sparse documentation of silviculture programs; and
5. outdated inventories.

The modeling approach

The analyses of the Forest Study focus on two different levels of the forest production system. One level involves projected development of some basic forest units under different perturbations, for example management activities and environmental change. On the second, more aggregated level, the focus is on the administration of the set of units. One major issue here is how to allocate activities over the units when the original information is highly aggregated. Thus, the basic demands on a timber-assessment simulation model for use in the Forest Study are that it must properly depict forest dynamics under a wide range of exogenously determined management regimes, and that it reacts in a reasonable way to changes in the environment.

The modeling approach adopted is based on forest area and its characteristics. The basic concept is that the forest area under study (again typically of the order of tens of thousands or hundreds of thousands of hectares) can be regarded as a set of biologically independent units. The forest area belonging to a specific forest type (see below) is described by age and standing volume. A matrix defined by 10 intervals for the volume dimension and 6–15 intervals for the age dimension is defined. The forest state is depicted by an area distribution over this matrix. Dynamics in terms of aging and volume increment are introduced as transitions of areas among the fixed states in the matrix. Basically, the model is a first-order Markov model (Fig. 17.2).

Harvest and regeneration activities are introduced via controlled transitions. Thinnings are expressed as the fraction of the area residing in a cell of the age–volume matrix that is thinned. This area is moved one step down in the volume dimension, thus simulating the harvest of the difference in volume between the cells, whereafter the area grows in a normal way. An area unit that is clearcut is moved to a bare-land class, the transitions out of which are controlled by a 'young forest coefficient'. This coefficient can then be regarded as expressing the intensity and quality of regeneration efforts. A more extensive discussion of the area matrix model and its characteristics is found in Sallnäs (1989).

In the area matrix model, an age–volume matrix is established for every 'forest type'. The concept of forest type is in this context defined by a stratum that can be defined by: country; geographical region; owner; forest structure (e.g. high forest, coppice); site class; and species. The level of aggregation into forest types is of course dependent on available data. For the different countries of Europe, the number of forest types used ranges from 2 to 130. A forest type is distinguished if proper data can be found to provide the state-descriptive parameters for the forest type. A minimum demand is that the total area of the forest type can be separated into age classes for which areas and standing volumes are available.

Management is controlled at two levels in the model. First, a basic management program is defined for each forest type. In this program the activities of thinning, final felling and regeneration are included. Thinnings are expressed as a percent of growth in each forecast period. This percentage of growth is then converted in the model to percentage of the

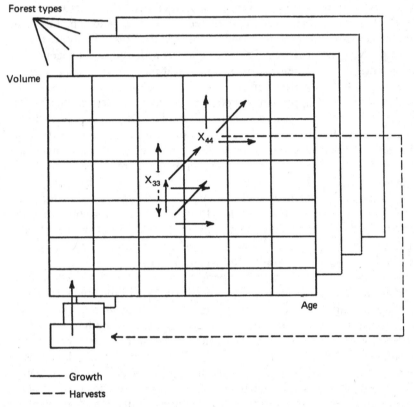

Fig. 17.2. Basic structure of an area matrix wood-supply simulator (from Sallnäs 1989).

area in a cell to be thinned. The thinning percentages are extracted mainly from yield tables and vary with age, species and site. Final felling programs are defined by the fraction of area in every cell that should be clearcut harvested. On the second level, the structure defined by the basic management programs can be shifted upwards or downwards in order to meet a demanded harvest level.

The transition matrices constitute, of course, the kernel of the model. Different matrices can be established for different situations. The units can then be allowed to flow between the different sets of matrices, relating to a changing situation. The model was designed originally to be used only for Swedish forests, and different transition matrices were used for undisturbed growth, thinnings and fertilization (Sallnäs 1989). For the Forest Study, a new set of matrices was established to correspond to a situation involving growth reductions attributed to air pollutants.

Parameter estimation for the model

Three sets of parameters must be estimated or given for the model:

1. parameters describing the forest state, such as area, standing volume, etc., including data about the external changes to the initial conditions, such as changes in the landbase allocated to forestry;
2. parameters governing the biological dynamics, such as growth, site quality, etc.; and
3. parameters describing activities and external factors influencing the dynamics, such as air pollution.

The description of the forest state can be deduced from alternative sources, while the estimation of the dynamic part of the model is a crucial point. The discussion above on scarcity of data relates indeed to data concerning the dynamics of the forest. Normally, Markov-type models are estimated from remeasured inventory plots (e.g. Hool 1966; Buongiorno & Michie 1980). However, if the model is large, the demands on the data set are often overwhelming. It is rare that data are available in sufficient amount and quality. Another problem is that in most cases, disturbances that have taken place, for example harvests, are not recorded. This lack of basic data is of course further pronounced when it comes to assessing the effects of air pollution. An alternative estimation procedure, indicated by Kaya & Buongiorno (1987), suggests use of ordinary growth functions associated with a measure of the variation. This option has been used by, among others, Haigth & Getz (1987). In the Forest Study, a rather rough estimation method had to be used, owing to the inadequate structure of the basic data available.

Volume distribution

The forest-type definitions imply that only figures for area and standing volume are at hand at the age-class level. The remaining task for the state description is then to generate the distribution over the volume dimension. This means that to be able to use the model, a procedure has to be applied to produce a distribution of area over volume per age class. When elaborating this procedure, two assumptions were made:

1. the standard deviation of the mean volume per hectare for different areas with similar types of forests is approximately the same; and
2. the variance in volume per hectare increases with age.

When the area distribution over volume classes is calculated, three variables are used: the mean volume per hectare, the coefficient of variation in volume per hectare, and the correlation between volume per hectare and age or transformations of age. The latter two are deduced from areas where good data are available.

Functions for volume increment

The percent volume increment for an age class in a forest type is expressed by functions of the following simple form:

$$I_v = a_0 + \frac{a_1}{T} + \frac{a_2}{T^2},$$

where I_v is 5-year volume increment in percent of the standing volume, T is the total stand age, and a_0, a_1 and a_2 are coefficients.

The functions are estimated from data on age and percent volume increment. Data are obtained either directly from inventory information or from yield tables or existing growth functions. Each function is associated with a data series of standing volume over age for which the function is valid. Since an area distribution over each volume class is created in the matrix, the mean volume in an age–volume cell will deviate from the mean volume series. Accordingly, the percentage volume increment will also deviate from the value given by the function, which means that some correction must be made. The correction is made according to:

$$I_{va} = I_{vf}[V_m/V_a]^\beta,$$

where I_{va} is 5-year percentage volume increment for actual standing volume, I_{vf} is 5-year percentage volume increment given by the function, V_a is actual standing volume (m^3 ha^{-1}) and V_m is standing volume (m^3 ha^{-1}) from the mean volume series. The parameter β describes the relation between relative standing volume and relative volume increment. Studies of this relation in yield tables and data available to the Forest Study reveal a range for the value of the parameter β from 0.25 to 0.45, depending on species, site quality and the type of data used when constructing the yield tables (J. Attebring, S. Nilsson & O. Sallnäs, unpublished data). By using the growth functions and the procedure for adjusting growth in relation to standing volume, growth figures for every individual cell in the state-descriptive matrices could be estimated. In a second step, these figures were converted to transition rates.

Another issue is the estimation of the change of growth rates that should be introduced as an assumed response to air pollution. Here, scarcity of data is compounded by a severe lack of basic knowledge, However, in the former German Democratic Republic (GDR), extensive work has been devoted to the task of modeling the stand-level response to pollutants (Bellman *et al.* 1988). In this work, the so-called PEMU model has been developed for analyzing the effects of air pollution on forest stands. The model is based on a rather long series of records from a grid-based set of field plots in the GDR. This simulator was used to estimate the growth reduction that can be expected for different forest units, differentiated into species groups, ages and site conditions. Since the basic transition matrices are based mainly on yield tables which cannot be expected to include any effects of forest decline, the growth-reduction figures derived from the PEMU model were applied to the basic matrices, thus establishing new matrices corresponding to conditions of forest decline.

An application with aggregated results

Clearly, the forest-inventory projection model employed in the Forest Study is unable to forecast credibly the potential forest-resource developments in Europe, under conditions of continued air pollution, without the assistance of a stand simulator such as PEMU in modifying the basic transition matrices. We have recently completed a set of simulation analyses for all of Europe except the Soviet Union, based on such modified transition matrices (Nilsson, Duinker & Sallnäs 1990). Below, we summarize the assumptions behind our scenarios and the key results.

For our analyses we implemented several scenarios, six of which are reported here. The base year for the simulations is 1985. All of the simulations have a time horizon of 100 years (up to year 2085).

Three basic scenarios assume no forest decline, and are named and described as follows.

Handbook Basic Scenario
In this scenario, the forests of each country are treated strictly in accordance with the silviculture programs that have been defined as ideal for each country. Handbook silviculture means the ideal silvicultural programs that analysts and researchers claim should be practised under ideal conditions in the forests of each country. Results from using this approach show the degree to which the actual forest structure in each country matches an 'ideal' structure, which in turn indicates the degree to which existing forest policies incorporating ideal silviculture have been implemented in the various countries.

ETTS-IV Basic Scenario
In this scenario, we have constrained total wood supply taken from the forests at the high estimates in the fourth European Timber Trend Study (UN 1986). We use this scenario to explore the forest-dynamics effects of implementing the harvest level of ETTS-IV (official country estimates) up to year 2020. The harvest level of 2020 is also used for the remainder of the simulation period up to year 2085.

Forest-Study Basic Scenario

The objective with this scenario is to strive for consistently high levels of both growing stocks and harvest levels over the total simulation period.

Three scenarios incorporating pollution-induced forest decline were designed to complement the basic scenarios. The same principles have been followed as for the no-decline scenarios. Following the pattern above, these scenarios are called, respectively, the Handbook Decline Scenario, the ETTS-IV Decline Scenario, and the Forest-Study Decline Scenario. The models have been adjusted according to several assumptions about forest decline, as indicated by the PEMU model (see Nilsson, Duinker & Sallnäs 1990).

It is important to underline that we are projecting biological potential wood supplies in the scenarios and not the market supply. In reality there are factors that can restrict actual harvests, such as roundwood prices, behavior of forest owners, and restrictions for non-wood benefits. These factors have not been taken into consideration in our scenarios.

Our results show that, as expected intuitively, the 'handbook' scenarios give the highest harvest potentials (Table 17.1). In most regions (see Table 17.1 for the countries of each region) there are no problems to reach the harvest levels suggested by ETTS-IV (Table 17.1). By comparing the 100-year average harvests from the Forest-Study Basic Scenario with actual harvests in 1987, it can be seen that there is a potential to increase long-term sustainable harvests throughout Europe by about 110 million m^3 yr^{-1}.

In all our scenarios, growing stocks increase strongly in most regions over time (Table 17.2). This means that the potential harvest levels presented in Table 17.1 are rather conservative and cautious.

The regions most affected by forest decline attributed to air pollutants are the Eastern and Central Regions, followed by EEC-9 (Table 17.3). The total loss of potential harvest caused by air pollutants expected to be emitted in Europe up to 2000–05 (we have very conservatively assumed no pollutant emissions from this period up to year 2085) is estimated to be about 16% of the total potential timber harvest in Europe. This represents a loss of about 85 million m^3 yr^{-1} averaged over 100 years. If we calculate expected decline effects as a proportion of actual 1987 harvests (Table 17.3), the decline effects are seen to be even higher.

Conclusion

If the Forest-Study model is to be used to explore potential impacts of climate change on the forest resources of Europe, then again it will need to be supported by stand-level simulation to create appropriate adjustments to the transition matrices. Basic transition matrices built from

Table 17.1. *Aggregated simulation results on potential harvest*

Scenario data are 100-year averages in million $m^3 yr^{-1}$ over bark for all species.

	Region[a]					
	Nordic	EEC-9	Central	Southern[b]	Eastern	Europe
Basic scenarios						
Handbook	153.4	159.3	26.6	—	130.7	—
ETTS-IV	150.8	129.5	25.8	—	108.4	—
Forest study	155.3	150.1	24.8	78.4	126.0	534.5
Decline scenarios						
Handbook	155.2	146.4	25.4	—	123.2	—
ETTS-IV	149.9	122.7	25.8	—	104.6	—
Forest study	144.2	126.2	18.9	71.0[c]	91.7	452.1
1987 Actual harvest[d]						
	120.7	109.3	22.3	72.1	99.5	423.9

Notes: [a] Countries in each region are: Nordic Region – Finland, Norway, Sweden; EEC-9 Region – Belgium, Denmark, France, Federal Republic of Germany, Ireland, Italy, Luxembourg, Netherlands, United Kingdom; Central Region – Austria, Switzerland; Southern Region – Greece, Portugal, Spain, Turkey, Yugoslavia; Eastern Region – Bulgaria, Czechoslovakia, German Democratic Republic, Hungary, Poland, Romania.
[b] There has been no possibility of calculating Handbook Basic and ETTS-IV Basic scenarios for the Southern Region and total Europe, owing to insufficient data.
[c] Decline effects for Spain are not included owing to insufficient data.
[d] Source: FAO statistics.

Source: Nilsson, Duinker & Sallnäs (1990).

classic yield tables would be decidedly inappropriate. The most defensible approach to devising suitable new matrices is stand-level, process-based simulation. Despite the fact that most of the forests of Europe are not boreal, the basic structure of the boreal-stand simulators reported in this volume undoubtedly makes them suitable for use in this regard.

Summary

It is unfortunate that there have not as yet been, to our knowledge, any simulation-based, forest-level studies of the interaction of future climate change with forest management. The best we could do by way of example was to describe two of many forest simulators that could be used in such studies. There is little more to say in this essay but to summarize the main principles and points:

Table 17.2. *Aggregated simulation results on the development of growing stock*

Data are year-0 and year-100 figures in m³/ha over bark for all species.

	Region[a]					
	Nordic	EEC-9	Central	Southern[b]	Eastern	Europe
Basic scenarios						
Handbook	93–154	152–184	298–342	—	169–202	—
ETTS-IV	93–147	152–222	298–315	—	169–239	—
Forest study	93–147	152–185	298–350	82–127	169–206	122–166
Decline scenarios						
Handbook	93–125	152–161	298–215	—	169–148	—
ETTS-IV	93–124	152–170	298–179	—	169–147	—
Forest study	93–139	152–190	298–369	82–124[b]	169–200	122–164

Notes: [a] There has been no possibility of calculating Handbook Basic and ETTS-IV Basic scenarios for the Southern Region and total Europe owing to insufficient data.
[b] Decline effects for Spain are not included owing to insufficient data.

Source: Nilsson, Duinker & Sallnäs (1990).

Table 17.3. *Potential harvest losses caused by air pollutants*

Data are percentages. Scenario data used to calculate the percentages are 100-year averages. 'Basic' is from the Forest Study Basic Scenario, 'Decline' is from the Forest Study Decline Scenario, and 1987 are actual harvests from FAO statistics.

Region	(Basic − Decline)/Basic × 100	(Basic − Decline)/1987 × 100
Nordic	7	9
EEC-9	16	22
Central	23	26
Southern[a]	13	14
Eastern	27	34
Europe	16	20

Note: [a] No decline effects have been calculated for Spain.

1. In studies of future forest resources over the next half to full century, we can be sure that growing conditions, especially atmospheric conditions, will be significantly different from past conditions. In this respect, the future will not and cannot repeat the past.

2. Studies of potential forest response to a changing environment, and changing management regimes to accommodate a changing environment, must rely on forest-inventory simulators to generate credible and useful results.

3. Forest-level simulators all rely on basic assumptions about behavior of the component stands making up the forest. With a changing environment, those assumptions cannot be based on traditional yield curves or tables, but must be developed using process-based stand-level simulation models built expressly for the purpose of forecasting stand responses to a changing environment.

18 Concluding comments

Herman H. Shugart, Rik Leemans & Gordon B. Bonan

In the Introduction, we expressed a hope that this book would represent a point of departure for subsequent studies of the world's boreal forest ecosystems. We have presented the physical (Chapter 15) and biological (Chapter 16) elements of a computer model designed to simulate the local changes in any forest in the boreal zone. For purposes of future identification, we will refer to the merged version of this model as the BOFORS model and we expect these versions of the model to be updated as further work and information are incorporated into the model structure. Detailed information on the model is available in Bonan (1989a) for the biophysical features of the model and in Leemans & Prentice (1989) for the biological and silvicultural features of the model. Listings of the computer program for the BOFORS model are available on request from the University of Virginia's Science and Technology Library (Charlottesville, Virginia, USA) on interlibrary loan. The model is available in two implementations: Bonan (1990c) is a version of the model with all of the physical subroutines included (see discussions in Chapter 15); Leemans (1990) is a version of the model that has been used to simulate several locations in the USSR and Fennoscandinavia (see Chapter 16). The FORSKA model (Leemans & Prentice 1989) is available (in English) upon request from: Meddelanden från Växtbiologiska Institutionen, Uppsala University, Uppsala, Sweden, and on interlibrary loan from the University of Virginia library mentioned above.

The BOFORS model is one of a set of models designed to emphasize the growth, birth, and death of individual trees on small elements of the landscape (*c.* 0.1 ha). The minimum time resolution of this model is 1 year (the model computational step) and the maximal time resolution is limited by the degree to which phenomena not included in the model become important at longer and longer time scales. Earlier models of the same *genre* have been used to simulate the changes in vegetation in response to climatic change over the past 20 000 years (see Chapters 11 and 12). At a minimum, the spatial resolution of the BOFORS model is the computational element scale. Most of the applications of this class of models have been at the scale that is represented by 50 to 100 computational elements, i.e. of the order of a few tens of ha or less. Complementing the BOFORS model, we also provide discussion on related models based on population dynamics (Chapter 13) and landscape-level processes (Chapter 14). Sources for these other models of boreal forests discussed herein are indicated in the chapters in which they are discussed.

Some future directions

As a closing comment, we would like to posit opinions on what we feel are promising directions for future studies. The world's boreal forests are under increasing pressures from a variety of sources. Because the forests in many cases have borders with or surround major industrial regions, the impacts of a wide range of atmospheric pollutants can potentially effect these forests. Further, the boreal forests in several regions are potential recipients of long-range pollutant transport manifested as precipitation with a decreased pH reading ('acid rain'). The assessment of anthropogenic stress on forests that have the complex dynamics that have been described in this book are daunting. There is a clear need for a continuation of process-oriented comparative studies in polluted and non-polluted regions of the boreal forests to better understand these effects. It is clear from the reviews of actual observations and experimental evidence from the boreal forest and from the models presented in this text that the landscape response of boreal forests to stress is complex and not easily obtained from static measurements. Further, the feedback complexities in the boreal forest ecosystem suggest a multiple research program of experimentation, modeling and observation may lead to a better understanding of the forest dynamics under stress or novel situations than more unidimensional research programs.

The next logical steps in improving the simulation capabilities of the boreal models are several fold but in many cases could be attacked simultaneously. In no particular order, these steps include:

1. *Improved characterization of soil surface features and surface dynamics.* The nature of the ground surface can have a fairly profound effect on the heat transfer to and from the boreal soils and the associated temperature effects are important with respect to any number of processes. The dynamics of moss, lichen and litter layers in the forests are extremely important and an increased knowledge-base is important.

2. *Increased understanding of feedbacks in biogeochemical cycles.* Several of the chapters have touched upon the importance of nutrients in understanding the dynamics of boreal forests. Chapter 8 and nutrient-cycling forest simulation models (Pastor & Post 1986, 1988) point to the possibility of multiple-stable equilibria in the southern transition between boreal and temperate deciduous forests. The understanding of the effects of nutrient availability on tree growth and/or form for a variety of species would be an important addition.

3. *Coupling of more explicit plant physiology into the stand simulation models.* Using plant physiology (particularly leaf level responses typically resolved at time scales of minutes) to predict the annual growth of trees has been pursued for several decades. The linking of canopy models that incorporate a considerable degree of fundamental plant physiology to stand simulation models appears to be an attainable objective at this time. The interaction between the two models is in both directions. Canopy physiology models attempt to represent fundamental biophysical processes (or reasonable proxies thereof) to predict the carbon fixation of a layered, horizontally homogeneous canopy. The models can provide an estimate of total carbon fixed by a canopy over a period of time but (because they represent the forest as an homogeneous, aggregated system) they have no internal features to alter canopy structure over time. Stand simulation models would be improved by the addition of total productivity and because they grow individual trees based on indices that include competition and other factors can deallocate a given degree of productivity into tree growth and canopy dynamics. The assessment of the direct effects of increased ambient CO_2 in the atmosphere will probably continue to make the interfacing of plant physiological models and stand dynamics models an important topic.

4. *An increased ability to observe large-scale pattern.* The feedback dynamics that have emerged in many of the preceding chapters in many cases imply patterns on the boreal landscape. Remote sensing (Chapter 10) offers an important family of methodologies for detecting such pattern, particularly for ecosystems like the boreal forests, in which large tracts are relatively inaccessible.

5. *Basic studies to connect the surface features of the boreal forests to other geospheric systems, particularly the atmosphere.* The annual variation in the CO_2 concentration by latitude is greatest in the higher northern latitudes and appears to be well correlated with the uptake of photosynthetically active radiation in boreal regions. The role of the world's boreal forests as major terrestrial repositories of carbon, the importance of the carbon cycle, and our need to better understand global carbon dynamics all point to a need for a better understanding of the interaction between the forest surface in boreal zones and the atmos-

phere. The hopeful scenario is for these studies (some of which are being drafted currently) to also provide valuable data for step 3 (above).

6. *Testing and application of the models over large areas and over long time scales.* The model testing step hopefully will provide some appreciation of the potential applications of the models and of the next information needed. It is our hope that by making the models available to our colleagues this testing over large areas will take us to a next step in ordering research needs. Tests over long time scales with these models will likely be retrospective simulations developed in conjunction with paleoecological studies (see Chapter 11).

7. *Incorporation of animal–plant interactions with vegetation dynamics models.* Chapter 6 provides some initial background as to the importance of insects in particular on boreal forest dynamics and Chapter 8 identifies important interactions among forest composition, element cycles and large herbivore grazing. These two chapters both point to major feedback systems involving animals. The animal-involved interactions with boreal forest ecosystems appear to be able to change the spatial grain, the dynamics and the stability of boreal forests.

8. *Continued effort in coupling of wildfire dynamics and forest dynamics in boreal models.* The importance of wildfire in shaping the forests of the boreal zone has been discussed in several of the chapters in this book (Chapter 5 in particular), and the modeling of the vegetation–wildfire interaction has been discussed in Chapter 14. The ecology of the boreal forest is so shaped by fire and the fire has such a regular presence on the boreal landscape that continued work to improve our knowledge base and the representation of fire in models is imperative. From a modeling standpoint, consideration of fire has several important aspects: fire initiation and spatial propagation, the coupling of tree-level phenomena with landscape-level wildfire behavior, and the quantification of boreal fires as sources of CO, CO_2 and other organic compounds to the atmosphere.

Epilogue

In ending a book, it is appropriate and traditional to apologize for errors intended or otherwise, to thank one's reviewers, and to make it clear that one's mistakes are, indeed, one's own. We comply with that tradition in the fullest degree.

Beyond this closing sentiment, we would like to convey an additional impression that has been with us through the development of this project. We feel that it is important for ecologists to rise to the intellectual challenge thrown before us by the increased interest in understanding global systems dynamics. The challenge impels us to apply at larger scales, even at the global scale, the implications of what we think we understand about ecological systems. The approach presented in this book has been to bring together an international confederation of scientific colleagues and to interweave a combination of field studies, observations, and mathematical models.

At this time we feel that we can, with some degree of confidence, project dynamic change in response to specified environmental conditions for an arbitrary location anywhere in the boreal forest zone. To the degree one can consider the boreal forests of the world as being approximated by a collection of such locations, then an initial capability of simulating the boreal zone has also been developed. If we may have made errors of omission or commission in our attempt to develop such a capability, we encourage our peers to build upon what is presented here or to start afresh. In either case, it is important to remember that the exciting challenge is the understanding of the global ecosystem.

References

Aber, J. D., Botkin, D. B. & Melillo, J. M. (1979). Predicting the effects of different harvesting regimes on productivity and yield in northern hardwoods. *Canadian Journal of Forest Research*, **9**, 10–14.

Aber, J. D. & Melillo, J. M. (1982). *FORTNITE: A computer model of organic matter and nitrogen dynamics in forest ecosystems.* Madison: University of Wisconsin, Research Bulletin R3130.

Aber, J. D., Melillo, J. M. & Federer, C. A. (1982). Predicting the effects of rotation length, harvest intensity, and fertilization on fiber yield from northern hardwood forests in New England. *Forest Science*, **28**, 31–45.

Abolin, R. I. (1914). An attempt at an epigenological classification of mires. *Peatland Science*, **3**, 1–55. (In Russian.)

Abolin', A. A. (1974). Change of the structure of the moss cover in relation to the distribution of precipitation under the forest canopy. *Soviety Journal of Ecology*, **5**, 243–47.

Ackerman, R. F. (1957). *The effect of various seedbed treatments on the germination and survival of white spruce and lodgepole pine seedlings.* Canadian Department of Northern Affairs and Natural Resources, Forestry Branch, Forest Research Division, Technical Note No.63.

Ageenko, A. S. ed. (1969). *Forests of the Far East.* Moscow: Lesnaya promyshlenost. (In Russian.)

Ager, T. A. (1983). Holocene vegetational history of Alaska. In *Late-Quaternary Environments of the United States*, Vol II, *Holocene*, ed. H. E. Wright, Jr., pp. 128–41. Minneapolis: University of Minnesota Press.

Ager, T. A. & Brubaker, L. B. (1985). Quaternary palynology and vegetational history of Alaska. In *Pollen Records of Late-Quaternary North American Sediments*, ed. V. M. Brynat, Jr. & R. G. Holloway, pp. 353–384. Dallas: American Association of Stratigraphic Palynologists Foundation.

Ågren, G. I. & Axellson, B. (1980). PY – a tree growth model. *Ecological Bulletin* (Stockholm), **32**, 525–36.

Ahlgren, C. E. (1959). Some effects of fire on forest reproduction in northeastern Minnesota. *Journal of Forestry*, **57**, 194–200.

Ahlgren, C. E. & Ahlgren, I. (1984). *Lob Trees in the Wilderness*. Minneapolis: University of Minnesota Press.

Ahti, T. (1959). Studies on the caribou lichen stands of Newfoundland. *Annales Botanici Societatis Zoologicae Fennicae 'Vanamo'*, **30**, 1–44.

Ahti, J., Hämet-Ahti, L. & Jalas, J. (1968). Vegetation zones and their sections in northwestern Europe. *Annales Botanici Fennici*, **5**, 169–211.

Aikman, D. P. & Watkinson, A. R. (1980). A model for growth and self-thinning in even-aged monocultures of plants. *Annals of Botany*, **45**(4), 419–27.

Alaback, P. B. (1982). Dynamics of understorey biomass in Sitka spruce–western hemlock forests of southeast Alaska. *Ecology*, **63**(6), 1932–48.

Aleksandrova, V. D. (1980). *The Arctic and Antarctic: Their Division into Geobotanical Areas*. Translated by D. Löve. Cambridge University Press.

Allen, T. F. H. & Hoekstra, T. W. (1984). Nested and non-nested hierarchies: a significant distinction for ecological systems. In *Proceedings of the Society for General Systems Research. I. Systems Methodologies and Isomorphies*, ed. A. W. Smith, pp. 175–80. Lewiston, N.Y.: Intersystems Publications, Coutts Library Service.

Allen, T. F. H. & Starr, T. B. (1982). *Hierarchy: Perspectives for Ecological Complexity*. Chicago: University of Chicago Press.

Alverson, W. S., Waller, D. M. & Solheim, S. L. (1988). Forests too deer: edge effects in northern Wisconsin. *Conservation Biology*, **2**, 348–58.

Aminoff, F.(1909). Våra skogsträd 1. Granen. (Our forest trees 1. The spruce.) *Skogsvådsföreningens folkskrifter* No. 17, pp. 1–33.

Aminoff, F. (1912). Våra skogsträd 4. Tallen. (Our forest trees 4. The pine.) *Skogsvådsföreningens folkskrifter* No.29.

Anderson, M. C. (1971). Radiation and crop structure. In *Plant Photosynthetic Production*, ed. Z. Sestak, J. Catsley & P. G. Jarvis, pp. 412–66. The Hague: Dr. W. Junk.

Anderson, R. C. & Loucks, O. L. (1979). White-tail deer (*Odocoileus virginianus*) influence on structure and composition of *Tsuga canadensis* forests. *Journal of Applied Ecology*, **16**, 855–61.

Anderson, T. W. (1985). Late-Quaternary pollen records from Eastern Ontario, Québec, and Atlantic Canada. In *Pollen Records of Late-quaternary North American Sediments*, ed. V. M. Brynat, Jr. & R. G. Holloway, pp. 261–326. Dallas: American Association of Stratigraphic Palynologists Foundation.

Andersson, F. (1990). Global geographic distribution of atmospheric pollutant loads. In *The Challenge of Modeling Global Biospheric Change*, ed. A. M. Solomon. Laxenburg, Austria: International Institute for Applied Systems Analysis (in press).

Andreev, V. M. & Aleksandrova, V. D. (1981). Geobotanical division of the Soviet arctic. In *Tundra Ecosystems: a Comparative Analysis*, ed. L. C. Bliss, O. W. Heal & J. J. Moore, pp. 25–34. Cambridge University Press.

Anonymous. (1964). *Fiziko-geograficheskiy atlas mira*. Plates 240–241. Moscow: Academy of Sciences of USSR/State Geological Committee of USSR.

Antonovski, M. Ya., Korzukhin, M. D.& Matskiavichus, V. K. (1989). *Periodic behavior of an age-distributed population of trees*. Working Paper WP-89-39. International Institute for Applied Systems Analysis, Laxenburg, Austria.

Antonovski, M. Ya., Korzukhin, M. D.& Ter-Mikaelian, M. T. (1987). *Model of the optimal development of a plant taking into account defense and competition.* Working Paper WP-87-106. International Institute for Applied Systems Analysis, Laxenburg, Austria.

Antonovski, M. Ya.& Ter-Mikaelian, M. T. (1987). *On spatial modeling of long-term forest fire dynamics.* Working Paper WP-87-105. International Institute for Applied Systems Analysis, Laxenburg, Austria.

Archibold, O. W. (1979). Buried viable propagules as a factor in postfire regeneration in northern Saskatchewan. *Canadian Journal of Botany*, **57**, 54–8.

Archipov, C. A. & Votah, M. P. (1980). Palynological characteristics and absolute age of a peat bog in the outfall of the River Tom. In *Paleopalynology of Siberia*, pp. 118–22. Moscow: Science. (In Russian.)

Archipov, S. A., Vdovin, V. V., Mizerov, V. B. & Nickolaev, V. A. (1970). History of evolution of relief in Siberia and Far East USSR. In *West Siberian Plain*. Moscow: Science. (In Russian.)

Archipov, S.S. (1932). *Bogging and Forest Types in Kotlas Forest Management.* Moscow: Goslesbumizdat. (In Russian.)

Arris, L. L. & Eagleson, P. A. (1989). Evidence of a physiological basis for the boreal–deciduous forest ecotone in North America. *Vegetatio*, **82**, 55–8.

Arsenault, P. (1979). *Effets allélopathiques causés par des lichens fruticuleux terricoles sur* Picea mariana. Quebéc: M.Sc. Thesis, Université Laval.

Assmann, E. (1970). *The Principles of Forest Yield Study.* Oxford: Pergamon Press.

Aswar, G., Myneni, R. B. & Kanemasu, E. T. (1989). Estimation of plant-canopy attributes from spectral reflectance measurements. In *Theory and Applications of Optical Remote Sensing*, ed. G. Aswar, pp. 252–95. New York: John Wiley and Sons.

Atkinson, K. (1981). Vegetation zonation in the Canadian subarctic. *Area*, **15**, 13–17.

Attema, E. P. W. & Ulaby, F. T. (1978). Vegetation modeled as a cloud. *Radio Science*, **13**, 357–64.

Auclair, A. N. D. (1983). The role of fire in lichen-dominated Tundra and Forest-Tundra. In *The Role of Fire in Northern Circumpolar Ecosystems*, ed. R. W. Wein & D. A. MacLean, pp. 235–56. Toronto: John Wiley & Sons.

Auclair, A. N. D. (1985). Postfire regeneration of plant and soil organic pools in a *Picea mariana-Cladonia stellaris* ecosystem. *Canadian Journal of Forest Research*, **15**, 279–91.

Auclair, A. N. & Goff, F. G. (1971). Diversity relations of upland forests in the western Great Lakes area. *American Naturalist*, **105**, 499–528.

Bärner, J. (1961). *Die Nutzhölzer der Welt.* Bd. I-IV. (In German.).

Bacastow, R. B., Keeling, C. D. & Whorf, T. P. (1985). Seasonal amplitude increase in atmospheric CO_2 concentration at Mauna Loa, Hawaii, 1959–1982. *Journal of Geophysical Research*, **90**, 10529–40.

Baes, C. F., Jr., Bjorkstrom, A. & Mulholland, P. J. (1985). Uptake of carbon dioxide by the oceans. In *Atmospheric carbon dioxide and the global carbon cycle*, ed. J. R. Trabalka, pp. 81–111. DOE/ER-0239, U.S. Department of Energy, Washington, DC.

Bailey, E. L. & Dell, T. R. (1973). Quantifying diameter distributions with the Weibull function. *Forest Science*, **19**, 97–104.

Bajenov, V. V. (1982). Modelling forest fire spread and its localization. *In Modelling of Processes in Nature-economics Systems*, pp. 72–9. Novosibirsk: Nauka. (In Russia.)

Baker, D. J. & Wilson, W. S. (1987). Spaceborne observations in support of earth science. *Oceans*, **29**(4), 76–85.

Baker, W. L. (1989). Landscape ecology and the nature reserve design in the Boundary Waters Canoe Area, Minnesota. *Ecology*, **70**(1), 23–35.

Ball, T. F. (1986). Historical evidence and climatic implications of a shift in the boreal forest–tundra transition in central Canada. *Climatic Change*, **8**,121–34.

Barney, R. J. & Stocks, B. J. (1983). Fire frequencies during the suppression period. In *The Role of Fire in Northern Circumpolar Ecosystems*, ed. R. W. Wein & D. A. MacLean, pp. 45–62. SCOPE 18, Toronto: J. Wiley & Sons.

Barry, T. A. (1969). Origins and distribution of peat-types in the bogs of Ireland. *Irish Forestry*, **26**(2), 40–52.

Baryshnikov, M. K. (1929). Sedge-hypnum bogs of West Vasuganye (Narim region). *Report of Institute of Meadow and Mire Culture*, after W. R. Williams, **2**, 1–38. (In Russian.).

Baskerville, G. L. (1975). Spruce budworm: super silviculturalist. *Forestry Chronicle*, **51**, 138–140.

Baskerville, G. L. (1986). Understanding forest management. *Forestry Chronicle*, **62**, 339–47.

Bauch, J. (1975). *Dendrologie der Nadelbäume und übrigen Gymnospermen.* Berlin and New York: de Gruyter. (In German.)

Baumgartner, A. (1979). Climatic variability and forestry. In *World Meteorological Organization. Proceedings of the World Climate Conference*, pp. 581–607. Geneva: WMO-No. 537.

Bazilevich, N. I., Drozdov, A. D. & Rodin, L. E. (1971). World forest productivity, its basic regularities and relationship with climatic factors. In *Productivity of Forest Ecosystems: Proceedings of the Brussels Symposium*, ed. P. Duvigneau, pp. 345–53. Paris: UNESCO & IBP.

Bazzaz, F. A. & Pickett, S. T. A. (1980). Physiological ecology of tropical succession: a comparative review. *Annual Review of Ecology and Systematics*, **11**, 287–310.

Beh, I. A. (1971). About dynamics of the cedar pine forests on the south of the cedar pine distribution range in Priobe. In *Utilization and Regeneration of Cedar Pine Forests*, ed. Anonymous, pp. 206–24. Novosibirsk: Nauka. (In Russian.)

Beh, I. A. (1974). *Cedar Pine Forests of the South Priobia.* Novosibirsk: Nauka. (In Russian.)

Beh, I. A. & Tarpan, I. V. (1979). *The Siberian Miracle Tree.* Novosibirsk: Nauka. (In Russian.)

Beisser, L. (1891). *Handbuch der Nadelholzkunde.* (In German.)

Bella, I. E. (1986). Logging practices and subsequent development of aspen stands in east central Saskatchewan. *Forestry Chronicle*, **62**, 81–3.

Bellman, K., Lasch, P., Hofman, G., Andes, S. & Schulz, H. (1988). The PEMU forest-interaction model FORST K. A pine stand decline and wood supply model. In *Mathematical Research: Systems and Simulation* 1988:*II*, Volume 47, ed. A. Sydow, S. G. Tzafestas & R. Vichnevestksy, pp. 45–50. Berlin: Akademia Verlag Berlin.

Bennett, K. D. (1983). Postglacial population expansion of forest trees in Norfolk, U.K. *Nature*, **303**, 164–7.

Bennett, K. D. (1988). Post-glacial vegetation history: Ecological considerations. In *Vegetation History*, ed. B. Huntley & T. Webb III, pp. 699–724. Kluwer Academic Publishers.

Benninghoff, W. S.(1952). Interaction of vegetation and soil frost phenomena. *Arctic*, **5**, 34–44.

Berg, B., Ekbohm, G. & McClaugherty, C. A. (1985). Lignin and holocellulose relations during long-term decomposition of some forest litters. *Canadian Journal of Botany*, **62**, 2540–50.

Berger, A., Imbrie, J., Hays, J., Kukla, G. & Saltzman, B. ed. (1984). *Milankovitch and Climate*. NATO Advanced Studies Institute Series, vol. 126. Dordrecht/Boston: D. Reidel Publishing Company.

Bergeron, Y. & Dubuc, M. (1989). Succession in the southern part of the Canadian boreal forest. *Vegetatio*, **79**, 51–63.

Bergerud, A. T. (1971). Abundance of forage on the winter range of Newfoundland caribou. *Canadian Field-Naturalist*, **85**, 39–52.

Bergman, F. (1981). Seed availability, cone collection, and natural regeneration. In *Forest regeneration at high latitudes. Experience from Northern Sweden*, ed. M. Murray, pp. 21–32. USDA Forest Service General Technical Report PNW-132.

Bergman, H. F. & Stallard, H. (1916). The development of climax formations in Minnesota. *Minnesota Botanical Studies*, **4**, 333– 78.

Berryman, A. A. (1982). Mountain pine beetle outbreaks in Rocky Mountain lodgepole pine forests. *Journal of Forestry*, **80**, 410–13.

Berryman, A. A. (1987). The theory and classification of outbreaks. In *Insect Outbreaks*, ed. P. Barbosa & J. C. Schultz, pp. 3–30. San Diego: Academic Press, Inc.

Berryman, A. A. & Millstein, J. A. (1989). Are ecological systems chaotic – and if not, why not? *Trends in Ecology and Evolution*, **4**, 26–8.

Berryman, A. A., Stenseth, N. C. & Isaev, A. S. (1987). Natural regulation of herbivorous forest insect populations, *Oecologia*, **71**, 174–84.

Berryman, A. A., Stenseth, N. C. & Wollkind, D. J. (1984). Metastability of forest ecosystems infested by bark beetles. *Researches in Population Ecology*, **26**, 13–29.

Bertalanffy, L. (1957). Quantitative laws in metabolism and growth. *Quarterly Biological Review*, **32**, 217–31.

Bichele, Z. N., Moldau, H. A. & Ross, J. K. (1980). *Mathematical Modeling of Plant Transpiration and Photosynthesis under Soil Moisture Stress*. Leningrad: Hydromet Publishers. (In Russian.)

Binkley, C. S. (1987). Economic models of timber supply. In *The Global Forest Sector: An Analytical Perspective*, ed. M. Kallio, D. P. Dykstra & C. S. Binkley, pp. 109–36. Chichester, UK: John Wiley & Sons.

Bjorkbom, J. C. (1971). *Production and germination of paper birch seed and its dispersal into a forest opening*. USDA Forest Service Research Paper NE-209. Northeastern Forest Experiment Station, Upper Darby, PA.

Bjorkbom, J. C., Marquis, D. A. & Cunningham, F. E. (1965). *The variability of paper birch seed production, dispersal and germination*. USDA Forest Service

Research Paper NE-41. Northeastern Forest Experiment Station, Upper Darby, PA.

Black, R. A. & Bliss, L. C. (1978). Recovery sequence of *Picea mariana-Vaccinium uliginosum* forest after burning near Inuvik, Northwest Territories, Canada. *Canadian Journal of Botany*, **56**, 2020–30.

Black, R. A. & Bliss, L. C. (1980). Reproductive ecology of *Picea mariana* (Mill.) BSP., at the tree line near Inuvik, Northwest Territories, Canada. *Ecological Monographs*, **50**, 331–54.

Blagovidov, A. K. (1984). *Changes in* Pinus *stand state under defoliation after phytophage impact*. PhD thesis, Moscow Forest Technology Institute. (In Russian.)

Blais, J. R. (1968). Regional variation in susceptibility of eastern North American forests to budworm attack based on history of outbreaks. *Forestry Chronicle*, **44**, 17–23.

Blomqvist, A. G. (1887). Iakttagelser angående sibiriska lärkträdet, pictagranen och cembratallen i deras hemland samt om deras forstliga förhållanden derstädes (Observations on the Siberian larch, spruce and pine in their natural habitat and their use in forestry). *Finska Forstfören. Meddelanden*, **5**, 149–81. (In Swedish.)

Bonan, G. B. (1988*a*). *A simulation model of environmental processes and vegetation patterns in boreal forests: test case Fairbanks, Alaska*. Working Paper WP-88-63. International Institute for Applied Systems Analysis.

Bonan, G. B. (1988*b*). *Environmental controls and stand dynamics in boreal forest ecosystems*. PhD dissertation, University of Virginia, Charlottesville, VA.

Bonan, G. B. (1988*c*). *Environmental processes and vegetation patterns in boreal forests*. PhD thesis, University of Virginia, Charlottesville, VA.

Bonan, G. B. (1989*a*). A computer model of the solar radiation, soil moisture, and soil thermal regimes in boreal forests. *Ecological Modelling*, **45**, 275–306.

Bonan, G. B. (1989*b*). Environmental factors and ecological processes controlling vegetation patterns in boreal forests. *Landscape Ecology*, **3**, 111–30.

Bonan, G. B. (1990*a*). Carbon and nitrogen cycling in North American boreal forests. I. Litter quality and soil thermal effects in interior Alaska. *Biogeochemistry*, (in press).

Bonan, G. B. (1990*b*). Carbon and nitrogen cycling in North American boreal forests. II. Biogeographic patterns. *Canadian Journal of Forest Research* (in press).

Bonan, G. B. (1990*c*). A listing of the biophysical configuration of the BOFORS (Version 1b) model. Science and Technology Library, The University of Virginia, Charlottesville, VA.

Bonan, G. B. & Korzukhin, M. D. (1989). Simulation of moss and tree dynamics in the boreal forest of interior Alaska. *Vegetatio*, **84**, 31–44.

Bonan, G. B. & Shugart, H. H. (1989). Environmental factors and ecological processes in boreal forests. *Annual Review of Ecology and Systematics*, **20**, 1–28.

Bonan, G. B., Shugart, H. H. & Urban, D. L. (1990). The sensitivity of some high-latitude boreal forests to climatic parameters. *Climatic Change*, **16**, 9–29.

Bone, R. A., Lee, D. W. & Norman, J. M. (1985). Epidermal cells functioning as lenses in leaves of tropical rain-forest shade plants. *Applied Optics*, **24**, 1408–12.

Bonnemann, A. & Röhrig, E. (1971). *Der Wald als Vegetationstyp und seine Bedeutung für den Menschen*. (The Forest as Vegetational Type and its Importance for Humans). Hamburg: Paul Parey Verlag. (In German.)

Bonnemann, A. & Röhrig, E. (1972). *Waldbau*. Teil 1 und 2, Hamburg and Berlin: P. Parey. (In German.)

Bormann, F. H. & Likens, G. E. (1979a). *Pattern and Process in a Forested Ecosystem*. New York: Springer-Verlag.

Bormann, F. H. & Likens, G. E. (1979b). Catastrophic disturbance and the steady state in northern hardwood forests. *American Scientist*, **67**, 660–9.

Bormann, F. H. & Likens, G. E. (1981). *Patterns and Process in Forested Ecosystem*. New York: Springer-Verlag.

Botkin, D. B., Janak, J. F. & Wallis, J. R. (1972). Some ecological consequences of a computer model of forest growth. *Journal of Ecology*, **60**, 849–73.

Box, E. O. (1981). *Macroclimate and Plant Forms: An Introduction to Predictive Modeling in Phytogeography*. The Hague: Dr. W. Junk.

Brakke, T. W., Kanemasu, E. T., Steiner, J. L., Ulaby, F. U. & Wilson, E. (1981). Microwave radar response to canopy moisture, leaf-area index and dry weight of wheat, corn, and sorghum. *Remote Sensing of Environment*, **11**, 207–20.

Brakke, T. W., & Smith, J. A. (1987). A ray tracing model for leaf bidirectional scattering studies. *Proceedings of the IGARSS 1987 Symposium*, Ann Arbor, Michigan, 18–21 May, pp. 643–8.

Brand, D. G. & Janas, P. S. (1988). Growth and acclimation of planted white pine and white spruce seedlings in response to environmental conditions. *Canadian Journal of Forest Research*, **18**, 320–9.

Braun, E. L. (1950). *Deciduous Forests of Eastern North America*. Philadelphia: Blakiston.

Bray, J. R. (1971). Vegetational distribution, tree growth and crop succession in relation to recent climatic change. *Advances in Ecological Research*, **7**, 177–223.

Brink, C. H. (1964). *Spruce seed as a food of the squirrels*, Tamiasciurus hudsonicus *and* Glaucomys sabrinus *in interior Alaska*. M.S. thesis, University of Alaska, Fairbanks, AK.

Britton, N. F. (1982). A model for suppression and dominance in trees. *Journal of Theoretical Biology*, **97**, 691–8.

Broecker, W. S., Takahashi, T., Simpson, H. H. & Peng, T. H. (1979). Fate of fossil fuel carbon dioxide and the global carbon budget. *Science*, **206**, 409–17.

Brooks, D. J. (1987). Modeling forest dynamics. In *The Global Forest Sector: An Analytical Perspective*, ed. M. Kallio, D. P. Dykstra & C. S. Binkley, pp. 91–108. Chichester, UK: John Wiley & Sons.

Brown, K. (1983). *Growth and reproductive ecology of* Larix laricina *in interior Alaska*. M.S. thesis, Oregon State University, Corvallis, OR.

Brown, K., Zobel, D. B. & Zasada, J. C. (1988). Seed dispersal, seedling emergence, and early survival of *Larix laricina* (Du Roi) K. Koch in the Tanana Valley, Alaska. *Canadian Journal of Forest Research*, **18**, 306–14.

Brown, R. J. E. (1963). Influence of vegetation on permafrost. In *Permafrost: Proceedings of the First International Conference*, pp. 20–5. Washington, D.C.: National Academy of Science.

Brown, R. J. E. (1969). Factors influencing discontinuous permafrost in Canada. In *The Periglacial Environment*, ed. T. L. Pewe, pp. 11–53. Montreal: McGill-Queens' University Press.

Brown, R. J. E. (1970). Permafrost as an ecological factor in the subarctic. In *Ecology of the Subarctic Regions*, pp. 129–40. Paris: UNESCO.

Brown, R. J. E. (1983). Effects of fire on the permafrost ground thermal regime. In *The Role of Fire in Northern Circumpolar Ecosystems*, ed. R. W. Wein & D. A. MacLean, pp. 97–110. New York: Wiley.

Brown, R. J. E. & Pewe, T. L. (1973). Distribution of permafrost in North America and its relationship to the environment: a review, 1963–1973. In *Permafrost: Proceedings of the Second International Conference* (*North American Contribution*), pp. 71–100. Washington, D.C.: National Academy of Science.

Brown, R. T. & Curtis, J. T. (1952). The upland conifer–hardwood forests of northern Wisconsin. *Ecological Monographs*, **22**, 217–34.

Brown, R. T. & Mikola, P. (1974). The influence of fruticose soil lichens upon mycorrhizae and seedling growth of forest trees. *Acta Forestalia Fennica*, **141**, 1–22.

Brown-Macpherson, J. (1982). Postglacial vegetational history of the eastern Avalon Peninsula, Newfoundland, and Holocene climatic change along the eastern Canadian seaboard. *Géographie Physique et Quaternaire*, **36**, 175–96.

Brubaker, L. B., Garfinkel, H. L. & Edwards, M. E. (1983). A Late-Wisconsin and Holocene vegetation history from the central Brooks Range: implications for Alaskan paleoecology. *Quaternary Research*, **20**, 194–214.

Bryant, J. P. (1987). Feltleaf willow–snowshoe hare interactions: plant carbon/nutrient balance and floodplain succession. *Ecology*, **68**, 1319–27.

Bryant, J. P. & Chapin, F. S., III (1986). Browsing-woody plant interactions during boreal forest plant succession. In *Forest Ecosystems in the Alaskan Taiga*, ed. K. Van Cleve, F. S. Chapin, III, P. W. Flanagan, L. A. Viereck & C. T. Dyrness, pp. 213–25. New York: Springer-Verlag.

Bryant, J. P., Chapin, F. S., III & Klein, D. R. (1983). Carbon/nutrient balance of boreal plants in relation to vertebrate herbivory. *Oikos*, **40**, 357–68.

Bryant, J. P. & Kuropat, P. J. (1980). Selection of winter forage by subarctic browsing vertebrates: the role of plant chemistry. *Annual Review of Ecology and Systematics*, **11**, 261–85.

Bryant, J. P., Tuomi, J. & Niemela, P. (1988). Environmental constraints of constitutive and long-term inducible defenses in woody plants. In *Chemical Mediation of Coevolution*, ed. K. C. Spencer, pp. 367–89. San Diego: Academic Press.

Bryant, J. P., Wieland, G. D., Clausen, T. P. & Kuropat, P. J. (1985). Interactions of snowshoe hares and feltleaf willow (*Salix alaxensis*) in Alaska. *Ecology*, **66**, 1564–73.

Bryson, R. A. (1966). Air masses, stream lines, and the boreal forest. *Geographical Bulletin*, **8**, 228–69.

Bryson, R. A., Irving, W. N. & Larsen, J. A. (1965). Radiocarbon and soil evidence of former forest in the southern Canadian tundra. *Science*, **147**, 46–8.

Bryson, R. A. & Wendland, W. M. (1967). Tentative climatic patterns for some late-glacial and post-glacial episodes in central North America. In *Life, Land and Water*, ed. W. J. Mayer-Oakes, pp. 271–98. Winnepeg, MN: University of Manitoba Press.

Buchman, R. G., Pederson, S. P. & Walters, N. R. (1983). A tree survival model with application to species of the Great Lakes region. *Canadian Journal of Forest Research*, **13**, 601–8.

Budyko, M. I. (1977). *Global Ecology*. Moscow: Thought Publishers. (In Russian.)

Budyko, M. I. & Izrael, Yu.A. ed. (1987). *Anthropogenic Climatic Changes*. Leningrad: Gidrometeoisdat. (In Russian.)

Buell, M. F. & Gordon, W. E. (1945). Hardwood–conifer forest contact zone in Itasca Park, Minnesota. *American Midland Naturalist*, **34**, 433–9.

Buongiorno, J. & Bruce, M. (1980). A matrix model of uneven-aged forest management. *Forest Science*, **26**, 609–25.

Burenckov, V. A., Kotscheev, A. L. & Malchevskya, N. N. (1934). Results of studying the bogging processes of full cuttings in Lisinsk forest state management. *Transactions of Institute of Leningrad Forestry Academy*, Goslestechizdat, **4**(42), 10–8. (In Russian.)

Busby, J. R., Bliss, L. C. & Hamilton, C. D. (1978). Microclimate control of growth rates and habitats of the boreal mosses. *Tomenthypnum nitens* and *Hylocomium splendens*. *Ecological Monographs*, **48**, 95–110.

Busing, R. T. & Clebsch, E. E. C. (1987). Application of a spruce–fir forest canopy gap model. *Forest Ecology and Management*, **20**, 151–69.

Buzykin, A. I. (1970). About geographical and edapho-cenotical factors of forest productivity. *Questions of Forestry*, **1**, 80–91. Krasnoyarsk: Forest Institute Publishers. (In Russian.)

Caesar, J. C. & Macdonald, A. D. (1984). Shoot development in *Betula papyrifera* V. Effect of male inflorescence formulation and flowering on long shoot development. *Canadian Journal of Botany*, **62**, 1708–13.

Cahalan, R. F. & Joseph, J. H. (1988). Fractal statistics of cloud fields. *Monthly Weather Review*, **117**, 257–68.

Cajander, A. K. (1913). Studien über die Moore Finnlands. *Acta Forestalia Fennica*, **2**(3), 1–208.

Campbell, G. S. (1977). *An Introduction to Environmental Biophysics*. New York: Springer-Verlag.

Campbell, R. W. (1975). The analysis of numerical change in gypsy moth populations. *Forest Science Monographs*, **15**, 1–33.

Campbell, R. W. (1979). Gypsy moth: forest influence. *Agricultural Information Bulletin*, **423**, 1–44.

Campbell, R. W., Beckwith, R. C. & Torgerson, T. T. (1983). Numerical behavior of some western spruce budworm (Lepidoptera: Tortricidae) populations in Washington and Idaho. *Environmental Entomology*, **12**, 1360–6.

Canham, C. D. & Loucks, O. L. (1984). Catastrophic windthrow in the presettlement forests of Wisconsin. *Ecology*, **65**, 803–9.

Cannell, M. G. R. (1982). *World Forest Biomass and Primary Production Data*. London: Academic Press.

Carleton, T. J. (1982*a*). The composition, diversity, and heterogeneity of some jack pine (*Pinus banksiana*) stands in northeastern Ontario. *Canadian Journal of Botany*, **60**, 2629– 36.

Carleton, T. J. (1982*b*). The pattern of invasion and establishment of *Picea mariana* (Mill.) BSP. into the subcanopy layers of *Pinus banksiana* Lamb. dominated stands. *Canadian Journal of Forest Research*, **12**, 973–84.

Carleton, T. R. & Maycock, P. F. (1978). Dynamics of boreal forest south of James Bay. *Canadian Journal of Botany*, **56**, 1157–73.

Carroll, S. B. & Bliss, L. C. (1982). Jack pine–lichen woodland on sandy soils in northern Saskatchewan and northeastern Alberta. *Canadian Journal of Botany*, **60**, 2270–82.

Carter, R. N. & Prince, R. D. (1981). Epidemic models used to explain biogeographic distribution limits. *Nature*, **293**, 644– 5.

Cattelino, P. J., Noble, I. R., Slatyer, R. O. & Kessell, S. R. (1979). Predicting the multiple pathways of plant succession. *Environmental Management*, **3**, 41–50.

Cayford, J. H., Hildahl, V., Hairn, L. D. & Wheaton, M. P. H. (1959). Injury to trees from winter drying and frost in Manitoba and Saskatchewan in 1958. *Forestry Chronicle*, **35**(4), 282–90.

Cayford, J. H. & McRae, D. J. (1983). The ecological role of fire in jack pine forests. In *The Role of Fire in Northern Circumpolar Ecosystems*, ed. R. W. Wein & D. A. MacLean, pp. 183–99. Toronto: John Wiley & Sons.

Cennel, M. G. R., Rothery, P. & Ford, E. D. (1984). Competition within stands of *Picea sitchensis*, and *Pinus contorta*. *Annals of Botany* (London), **53**, 349–62.

Chapin, F. S. (1986). Controls over growth and nutrient use by taiga forest trees. In *Forest Ecosystems in the Alaskan Taiga*, ed. K. Van Cleve, F. S. Chapin, P. W. Flanagan, L. A. Viereck & C. T. Dyrness, pp. 96–111. New York: Springer-Verlag.

Chapin, F. S., III, Vitousek, P. M. & Van Cleve, K. (1986). The nature of nutrient limitation in plant communities. *American Naturalist*, **127**, 48–58.

Cherkashin, A. K. (1980). A boreal forest dynamic model. In *Optimal Management of Natural-economic Systems*, ed. G. A. Gurman, pp. 132–41. Moscow: Science Publishers. (In Russian.)

Chertovsky, V. G. (1979). *Spruce Forests of the European USSR*. Moscow: Lesnaya promyshlenost. (In Russian.)

Chindyaev, A. S. (1987). Freezing and thawing of drained peat soils in forests of the middle Urals. *Soviet Forest Science*, **1987**(1), 68–72.

Chjan, D. & Chjao, S. (1985). Investigation of density laws during forest autoregulation. *Scientia Silvae Sinicae*, **21**, 369–74. (In Chinese.)

Clark, J. S. (1988*a*). Stratigraphic charcoal analysis on petrographic thin sections: application to fire history in northwestern Minnesota. *Quaternary Research*, **30**, 81–91.

Clark, J. S. (1988*b*). Effect of climate change on fire regimes in northwestern Minnesota. *Nature*, **334**, 233–5.

Clark, W. C. (1986). Sustainable development of the biosphere: themes for a research program. In *Sustainable Development of the Biosphere*, ed. W. C. Clark & R. E. Munn, pp. 5–48. Cambridge University Press.

Clark, W. C., Jones, D. D. & Holling, C. S. (1979). Lessons for ecological policy design: a case study of ecosystem management. *Ecological Modelling*, 7, 1–53.

Clark, W. C., & Munn, R. E. ed. (1986). *Sustainable Development of the Biosphere*. Cambridge University Press.

Clautice, S. F., Zasada, J. C. & Neiland, B. J. (1979). Autecology of first year postfire tree regeneration. In *Ecological effects of the Wickersham Dome fire near Fairbanks, Alaska*, ed. L. A. Viereck & C. T. Dyrness, pp. 50–3. USDA Forest Service General Technical Report PNW-90. Pacific Northwest Forest and Range Experiment Station, Portland, OR.

Clayden, S. & Bouchard, A. (1983). Structure and dynamics of conifer–lichen stands on rock outcrops south of Lake Abitibi, Québec. *Canadian Journal of Botany*, 61, 850–71.

CLIMAP Members (1981). *Albedo maps of land for the modern and last glacial maximum*. Maps 9A and 9B, Map and Chart Series, Geological Society of America, Boulder, CO.

Clymo, R. S. (1978). A model of peat bog growth. In *Production Ecology of British Moors and Montane Grasslands*, Ecological Studies, Vol. 27, pp. 187–223. Berlin/New York: Springer-Verlag.

Cogbill, C. V. (1985). Dynamics of the boreal forests of the Laurentian Highlands, Canada. *Canadian Journal of Forest Research*, 15, 252–61.

Coldwell, B. B. & DeLong, W. A. (1950). Studies of the composition of deciduous forest leaves before and after partial decomposition. *Scientific Agriculture*, 30, 456–66.

Coley, P. D., Bryant, J. P. & Chapin, F. S., III (1985). Resource availability and plant herbivore defense. *Science*, 230, 895–9.

Colwell, J. E. (1974). Vegetation canopy reflectance. *Remote Sensing of Environment*, 3, 175–83.

Comtois, P. (1982). Histoire holocène du climat et de la vegetation à Lanoraie (Québec) *Canadian Journal of Earth Sciences*, 19, 1938–52.

Cooper, W. S. (1913). The climax forest of Isle Royale and its development. *Botanical Gazette*, 55, 1–44, 115–40, 189–235.

Costanza, R., Sklar, F. H. & White, M. L. (1990). Modeling coastal landscape dynamics. *BioScience*, 40, 91–107.

Covault, C. (1989). Mission to planet earth. *Aviation Week and Space Technology*, March 13, pp. 34–50.

Cowan, I. R. (1982). Regulation of water use in relation to carbon gain in higher plants. In *Physiological Plant Ecology, Encyclopedia of Plant Physiology*, Vol. 12B, ed. O. L. Lange, P. S. Noble, C. B. Osmond, and H. Ziegler, pp. 549–587. Berlin: Springer.

Cowan, I. R. (1986). Economics of carbon fixation in higher plants. In *On the Economy of Plant Form and Function*, ed. T. J. Givnish, pp. 133–70. Cambridge University Press.

Cowles, S. (1982). Preliminary results of investigating the effect of lichen ground cover on the growth of black spruce. *Naturaliste Canadien*, 109, 573–81.

Cram, W. H. (1951). Spruce seed viability: dormancy of seed from four species of spruce. *Forestry Chronicle*, **27**(4), 349–57.

Cramer, W. & Prentice, I. C. (1988). Simulation of regional soil moisture deficits on a European scale. *Norsk Geografish Tidskrift*, **42**, 149–51.

Curtis, J. T. (1959). *The Vegetation of Wisconsin*. Madison, Wisconsin: University of Wisconsin Press.

Cwynar, L. C. (1978). Recent history of fire and vegetation from laminated sediment of Greenleaf Lake, Algonquin Park, Ontario. *Canadian Journal of Botany*, **56**, 10–21.

Czerepanov, S. K. (1981). *Plantae Vasculares of the USSR*. Leningrad: Nauka. (In Russian and Latin.)

Dahlberg, B. L. & Guettinger, R. C. (1956). *The White-tailed Deer in Wisconsin*. Madison, Wisconsin: Wisconsin Conservation Department.

Dale, V. H., Doyle, T. W. & Shugart, H. H. (1985). A comparison of tree growth-models. *Ecological Modelling*, **29**, 145–69.

Dale, V. H. & Hemstrom, M. A. (1984). *CLIMACS: A computer model of forest stand development for western Oregon and Washington*. USDA Forest Service Research Paper PNW-327.

Dale, V. H., Jager, H. I., Gardner, R. H. & Rosen, A. E. (1988). Using sensitivity and uncertainty analyses to improve predictions of broad-scale forest developments. *Ecological Modelling*, **42**, 165–78.

Damman, A. W. H. (1964). *Some Forest Types of Central Newfoundland and their Relation to Environmental Factors*. Ottawa: Forest Science Monograph no. 28.

Damman, A. W. H. (1971). Effect of vegetation changes on the fertility of a Newfoundland forest site. *Ecological Monographs*, **41**, 253–70.

Damman, A. W. H. (1983). An ecological subdivision of the island of Newfoundland. In *Biogeography and Ecology of the Island of Newfoundland*, ed. G. R. South, pp. 163–206. The Hague: Dr. W. Junk Publishers.

Danilov, D. N. 1952). *Seeding Years Frequencies and Spatial Distribution of Seed Crops for Coniferous Tree Species*. Moscow: Goslesbumizdat. (In Russian.)

Darlington, H. T. (1930). *Vegetation of the Porcupine Mountains, Northern Michigan*. Michigan Academy of Sciences, Arts, and Letters Paper 13.

D'Arrigo, R., Jacoby, G. C. & Fung, I. Y. (1987). Boreal forests and atmosphere–biosphere exchange of carbon dioxide. *Nature*, **329**, 321–23.

Daubenmire, R. & Prusso, D. C. (1963). Studies of the decomposition of tree litter. *Ecology*, **44**, 589–92.

Daughtry, C. S. T. & Ranson, K. J. (1986). *Measuring and modeling biophysical and optical properties of diverse vegetative canopies*. LARS Report No. 043086, Purdie University, West Lafayette, IN 47907.

Davis, M. B. (1981). Quaternary history and the stability of forest communities. In *Forest Succession: Concepts and Applicants*, ed. D. C. West, H. H. Shugart & D. B. Botkin, pp. 132–53. New York: Springer-Verlag.

Davis, M. B. (1983). Quaternary history of deciduous forests of eastern North America and Europe. *Annals of the Missouri Botanical Gardens*, **70**, 550–63.

Davis, M. B. (1984). Holocene vegetational history of the eastern United States. In *Late-Quaternary Environments of the United States*, ed. H. E. Wright, Jr., pp. 166–81. London: Longman Group Limited.

Davis, M. B. (1986). Climatic instability, time lags and community disequilibrium. In *Community Ecology*, ed. J. Diamond & T. Case, pp. 269–84. New York: Harper and Row.

Davis, M. B. & Botkin, D. B. (1985). Sensitivity of cool-temperate forests and their fossil pollen record to rapid temperature change. *Quaternary Research*, **23**, 327–40.

Davis, M. B., Woods, K. D., Webb, S. L. & Futyma, D. (1986). Dispersal versus climate: Expansion of *Fagus* and *Tsuga* into the Upper Great Lakes region. *Vegetatio*, **67**, 93–103.

Davis, M. B. & Zabinski, C. (1990). Changes in geographical range resulting from greenhouse warming effects on biodiversity in forests. In *Consequences of Greenhouse Warming to Biodiversity*, ed. R. L. Peters. Washington, DC: World Wildlife Institute (in press).

Davis, R. B., Bradstreet, T. E., Stuckenrath, R., Jr. & Borns, H. W., Jr. (1975). Vegetation and associated environments during the past 14,000 years near Moulton Road, Maine, *Quaternary Research*, **5**, 435–65.

DeAngelis, D. L., Gardner, R. H. & Shugart, H. H. (1981). Productivity of forest ecosystems studied during the IBP: the woodland dataset. In *Dynamic Properties of Forest Ecosystems*, ed. D. E. Reichle, pp. 567–672. Cambridge University Press.

Delaney, B. B. & Cahill, M. J. (1978). A pattern of forest types of ribbed moraines in eastern Newfoundland. *Canadian Journal of Forest Research*, **8**, 116–20.

Delcourt, H. R., Delcourt, P. A. & Webb, T. III. (1983). Dynamic plant ecology: The spectrum of vegetation change in space and time. *Quaternary Science Review*, **1**, 153–75.

Delcourt, P. A. & Delcourt, H. R. (1987). *Long-Term Forest Dynamics of the Temperate Zone: a Case Study of Late-Quaternary Forests in Eastern North America*. New York: Springer-Verlag.

DeLiocourt, F. (1898). De la ménagement des sapinières. *Bulletin de la Société Forestieres de Franches-Comte et Belfort* **4**(6), 396–409.

Delkov, N. (1984). *Textbook of Dendrology*. Sofia: Zemizdad. (In Bulgarian.)

Dementev, P. I. (1959). *Forester's Notes*. Ministry of Agriculture of the USSR. (In Russian.)

Denslow, J. S. (1980). Gap partitioning among tropical rainforest trees. *Biotropica*, **12**, 47–55 (suppl.).

Densmore, D. (1980). *Vegetation and forest dynamics of the upper Dietrich River Valley, Alaska*. M.S. thesis, North Carolina State University, Durham, NC.

Densmore, R. (1979). *Aspects of seed ecology of woody plants of the Alaska taiga and tundra*. PhD thesis, Department of Botany, Duke University, Durham, NC.

Densmore, R. & Zasada, J. (1983). Seed dispersal and dormancy patterns in northern willows: ecological and evolutionary significance. *Canadian Journal of Botany*, **61**, 3207–16.

Desponts, M. (1990). *Ecologie du pin gris* (Pinus banksiana *Lamb*) á *limite nord au Québec nordique*. Québec: PhD thesis, Université Laval, Québec.

Develice, R. L. (1988). Test of a forest dynamics simulator in New Zealand. *New Zealand Journal of Botany*, **26**(3), 387–92.

de Vries, D. A. (1975). Heat transfer in soils. In *Heat and Mass Transfer in the Biosphere. I. Transfer Processes in Plant Environment*, ed. D. A. de Vries & N. H. Afgan, pp. 5–28. Washington, D.C.: Scripta.

Diggle, P. J. (1976). A spatial stochastic model of interplant competition. *Journal of Applied Probability*, **13**, 662–71.

Dimo, V. M. (1969). Physical properties and elements of the heat regime in permafrost meadow-forest soils. In *Permafrost Soils and their Regime*, pp. 119–91. New Delhi: Indian National Scientific Documentation Center.

Dingman, S. L. & Koutz, F. R. (1974). Relations among vegetation, permafrost, and potential insolation in Central Alaska. *Arctic and Alpine Research*, **6**, 37–42.

Dix, R. L. & Swan, J. M. A. (1971). The role of disturbance and succession in upland frost at Candle Lake, Saskatchewan. *Canadian Journal of Botany*, **49**, 657–76.

Dobbs, R. C. (1972). *Regeneration of white and Engelmann spruce: a literature review with special reference to the British Columbia interior.* Canadian Forestry Service, Information Report BC-X-69. Pacific Forest Research Centre, Victoria, BC.

Dobbs, R. C. (1976). White spruce seed dispersal in central British Columbia. *Forestry Chronicle*, **52**(5), 225–8.

Dobson, M. C. & Ulaby, F. T. (1986). Active microwave soil moisture research. *IEEE Transactions on Geoscience and Remote Sensing*, **GE-24**, 23–36.

Dokturovski, V. S. (1932). *Peat bogs: A course of lectures for peatland science* Moscow/Leningrad: State Science and Technology Publication. (In Russian.)

Dolgushin, I. Yu. (1972). Present day mire formation processes in middle taiga Priobye. *Transactions of Science Academy of USSR: Ser. Geogr.*, **2**, 26–38. (In Russian.)

Dolgushin, I. Yu. (1973). Patterns of effect of rains in paludification and bogged forests in West Siberia. *Transactions of Science Academy of USSR: Ser. Geogr.*, **4**, 70–9. (In Russian.)

Doyle, T. W. (1981). The role of disturbance in the gap dynamics of a montane rain forest: An application of a tropical forest succession model. In *Forest Succession: Concepts and Application*, ed. D. C. West, H. H. Shugart & D. B. Botkin, pp. 56–73. New York: Springer-Verlag.

Drakenberg, B. (1981). *Kompendium i allmän dendrologi samt barrträds och barrvirkesegenskaper.* Umeå: SLU Inst. för skoglig ståndortslära. (In Swedish.)

Dranitsin, D. A. (1914). On some zonal forms of the relief in North Siberia. *Soil Science*, **4**, 21–68. (In Russian.)

Ducruc, J. P., Zarnovican, R., Gerardin, V. & Jurdant, M. (1976). Les Régions écologiques du Territoire de la Baie-James: caractéristiques dominantes de leur couvert végétal. *Cahiers de Géographie de Québec*, **20**, 365–91.

Duncan, D. P. & Hodson, A. C. (1958). Influence of the forest tent caterpillar upon the aspen forests of Minnesota. *Forest Science*, **4**, 71–93.

Dunn, C. P., Guntenspergen, G. R. & Dorney, J. R. (1983). Catastrophic wind disturbance in an old-growth hemlock–hardwood forest, Wisconsin. *Canadian Journal of Botany*, **61**, 211–17.

Durden, S. L., Van Zyl, J. J. & Zebker, H. A. (1989). Modeling and observation of the radar polarization of forested areas. *IEE Transactions on Geosciences and Remote Sensing*, **27**, 290–301.

Dyke, A. S. & Prest, V. K. (1987). *Paleogeography of Northern North America*, 18 000–5 000 *years ago*. Ottawa: Geological Survey of Canada. Map 1703A.

Dylis, N. V. (1961). *The larch of Eastern Siberia and the Far East*. Moscow: Academy of Sciences of the USSR. (In Russian.)

Dylis, N. V. (1981). *The Larch*. Moscow: Lesnaya promyshlenost. (In Russian.)

Dyrenkov, S. A. & Gorovaya, E. N. (1980). Probabilistic modeling of multi–aged stand dynamics. In *Economy-Mathematical Modeling of Forest-Industrial Actions*, pp. 113–20. Leningrad: Forest Industry Publishers. (In Russian.)

Dyrness, C. T. (1982). *Control of depth to permafrost and soil temperature by the forest floor in black spruce/feathermoss communities*. U.S. Forest Service Research Note PNW-396.

Dyrness, C. T. & Norum, R. A. (1983). The effects of experimental fires on black spruce forest floors in interior Alaska. *Canadian Journal of Forest Research*, **13**, 879–93.

Dyrness, C. T., Viereck, L. A., Foote, M. J. & Zasada, J. C. (1988). *The effect on vegetation and soil temperature of logging flood-plain white spruce*. USDA Forest Service Research Paper PNW-RP-392. Pacific Northwest Research Station, Portland, OR.

Dyrness, C. T., Viereck, L. A. & Van Cleve, K. (1986). Fire in taiga communities of interior Alaska. In *Forest Ecosystems in the Alaskan Taiga*, ed. K. Van Cleve, F. S. Chapin, III, P. W. Flanagan, L. A. Viereck & C. T. Dyrness, pp. 74–86. New York: Springer-Verlag.

Eagleson, P. S. (1986). The emergence of global-scale hydrology. *Water Resources Research*, **22**, 65–145.

Edwards, D. G. W. (1980). Maturity and quality of tree seeds; a state of the art review. *Seed Science and Technology*, **8**, 625–57.

Eggler, W. A. (1938). The maple–basswood forest type in Washburn County, Wisconsin. *Ecology*, **19**, 243–63.

Eis, S. (1967). Establishment and early development of white spruce in the interior of British Columbia. *Forestry Chronicle*, **43** (2), 174–7.

Eis, S. & Inkster, J. (1972). White spruce cone production and prediction of cone crops. *Canadian Journal of Forest Research*, **2**, 460–6.

Eiselt, M. G. & Schröder, R. (1977). *Laubgehölze*. Vienna: Verlag J. Newmann-Neudamm. (In German.)

Ek, A. R. & Monserud, R. A. (1974). *FOREST: a computer model for simulating the growth and reproduction of mixed forest stands*. Research Report A 2635, pp. 1–13 and three appendices. College of Agricultural and Life Sciences, University of Wisconsin, Madison.

Ek, A. R. & Monserud, R. A. (1979). Performance and comparison of stand growth models based on individual tree and diameter-class growth. *Canadian Journal of Forest Research*, **9**, 232–44.

El Bayoumi, M. A., Shugart, H. H. & Wein, R. W. (1984). Modeling succession of the eastern Canadian mixed-wood forest. *Ecological Modelling*, **21**, 175–98.

Ellenberg, H. (1952). Physiologishes und Ökologisches Verhalten derselben Pflanzenarten. *Berichte der Deutsche Botanische Gesellschaft*, **65**, 350–61. (In German.)

Ellenberg, H. (1978). *Vegetation Mitteleuropas mit den Alpen – in Ökologischer Sicht.* (The Vegetation of Middle Europe including the Alps – from an ecological point of view.) Stuttgart: Eugen Ulmer Verlag. (In German.)

Elliot, D. L. (1979*a*). The current regenerative capacity of the northern Canadian trees, Keewatin, N.W.T., Canada: some preliminary observations. *Arctic and Alpine Research*, **11**, 243–51.

Elliot, D. L. (1979*b*). *The stability of the northern Canadian tree limit: current regenerative capacity.* PhD thesis, University of Colorado, Boulder, CO.

Elton, C. (1927). *Animal Ecology.* New York: Macmillan.

Emanuel, W. R., Fung, I.Y-S, Killough, G. G., Moore, B. and Peng, P. H. (1985). Modeling the global carbon cycle and changes in the atmosphere CO_2 levels. In *Atmospheric CO_2 and the Global Carbon Cycle*, ed. J. R. Trabalka, pp. 141–173. Carbon Dioxide Research Division, US-DOE, Washington, DC.

Emanuel, W. R., Killough, G. G., Post, W. M. & Shugart, H. H. (1984). Modeling terrestrial ecosystems in the global carbon cycle with shifts in carbon storage capacity by land-use change. *Ecology*, **65**, 970–83.

Emanuel, W. R., Shugart, H. H. & Stevenson, M. P. (1985*a*). Climate change and the broad-scale distribution of terrestrial ecosystem complexes. *Climatic Change*, **7**, 29–43.

Emanuel, W. R., Shugart, H. H. & Stevenson, M. P. (1985*b*). Climate change and the broad-scale distribution of terrestrial ecosystem complexes: response to comment. *Climatic Change*, **7**, 457–60.

Engelmark, O. (1984). Forest fires in the Muddus National Park (northern Sweden) during the past 600 years. *Canadian Journal of Botany*, **62**, 893–98.

Engstrom, D. R. & Hansen, B. C. S. (1985). Postglacial vegetational change and soil development in southeastern Labrador as inferred from pollen and chemical stratigraphy. *Canadian Journal of Botany*, **63**, 543–61.

Enright, N. & Ogden, J. (1979). Application of transition matrix models in forest dynamics: *Araucaria* in Papua New Guinea and *Nothofagus* in New Zealand. *Australian Journal of Ecology*, **4**, 3–23.

Eom, H. J. & Fung, A. K. (1986). Scattering from a random layer embedded with dielectric needles. *Remote Sensing of Environment*, **19**, 139–49.

Eronen, M. (1979). The retreat of pine forest in Finnish Lapland since the Holocene climatic optimum: a general discussion with radiocarbon evidence from subfossil pines. *Fennia*, **157**, 93–114.

Eronen, M. & Hyvärinen, H. (1982). Subfossil pine dates and pollen diagrams from northern Fennoscandia. *Geologiska föreningens i Stockholm förhandlingar*, **103**, 437–45.

Eurola, S. (1968). Über die Ökologie der nordfinnischen Moorvegetation im Herbst, Winter und Frühling. *Annales botanici Fennici*, **5**(2), 83–97.

Eurola, S., Antti, H., Huttunen, M. & Paasovaara, P. (1982). Kaksi rinnesuota riisitunturin kansallispuistesta. *Suo.*, **33**(3), 75–9.

Eyre, S. R. (1968). *Vegetation and Soils: a World Picture.* 2nd ed. London: E. Arnold.

Falaleev, E. M. (1982). *The Fir.* Moscow: Lesnaya promyshlenost. (In Russian.)

Farquhar, G. D. and Sharkey, T. D. (1982). Stomatal conductance and photosynthesis. *Annual Review of Plant Physiology*, **33**, 317–45.

Farquhar, G. D. and von Caemmerer, S. (1982*a*). Modeling of climate recorded in the Holocene deposits of Québec. *Nature*, **309**, 543–6.

Farquhar, G. D. & von Caemmerer, S. (1982*b*). Modeling of photosynthetic response to environmental conditions. In *Physiological Plant Ecology, Encyclopedia of Plant Physiology (NS)*, *Vol. 12B*, ed. O. L. Lange, P. S. Noble, C. B. Osmond & H. Ziegler, pp. 549–587. Berlin: Springer-Verlag.

Filion, L. (1984). A relationship between dunes, fire and climate as recorded in the Holocene deposits of Québec. *Nature*, **309**, 543–6.

Fires, J., ed. (1974). *Growth Models for Tree and Stand Simulation.* Research Notes 30. Stockholm: Royal College of Forestry.

Firsov, L. V., Volckova, V. A., Levina, T. P., Nockolaeva, I. V., Orlova, L. A., Panytchev, B. A. & Volkov, I. A. (1982). Stratigraphy, geochronology, and standard spore–pollen diagram of Holocene peatland of 'Boloto Gladkoye' in Novosibirsk (Pravye Tchemyi). In *Problems of Stratigraphy, and Paleography of Pleistocene in Siberia*, pp. 96–107. Novosibirsk: Science, Siberian Branch.

Fisher, R. F. (1979). Possible allelopathic effects of reindeer-moss (*Cladonia*) on jackpine and white spruce. *Forest Science*, **25**, 256–60.

Flanagan, P. W. & Van Cleve, K. (1977). Microbial biomass, respiration, and nutrient cycling in a black spruce taiga ecosystem. In *Soil Organisms as Components of Ecosystems*, ed. U. Lohm & T. Persson. *Ecological Bulletins* (Stockholm), **25**, 261–73.

Flanagan, P. W. & Van Cleve, K. (1983). Nutrient cycling in relation to decomposition and organic matter quality in taiga ecosystems. *Canadian Journal of Forest Research*, **13**, 795–817.

Flerov, A. F. (1899). Botanic–geographical essays. II. Mire formation and overgrowing of lakes in north-west part of Vladimir gubernia. *Pedology* (Zemlevedenie), Book I–II, 1–16. (In Russian.)

Flinn, M. A. & Wein, R. W. (1977). Depth of underground plant organs and theoretical survival during fire. *Canadian Journal of Botany*, **55**, 2550–4.

Foote, M. J. (1976). *Classification, Description and Dynamics of Plant Communities Following Fire in the Taiga of Interior Alaska.* Fairbanks: US Forest Service, Bureau of Land Management, Fire Effects Study.

Ford, J. & Bedford, B. L. (1987). The hydrology of Alaskan wetlands, U.S.A.: a review. *Arctic and Alpine Research*, **19**, 209–29.

Ford, R. H., Sharik, T. L. & Feret, P. P. (1983). Seed dispersal of the endangered round-leafed birch (*Betula uber*). *Forest Ecology and Management*, **6**, 115–28.

Foster, D. R. (1983). The history and pattern of fire in the boreal forest of southeastern Labrador. *Canadian Journal of Botany*, **61**, 2459–71.

Foster, D. R. (1984). Phytosociological description of the forest vegetation of southeastern Labrador. *Canadian Journal of Botany*, **62**, 899–906.

Foster, D. R. (1985). Vegetation development following fire in *Picea mariana* (black spruce)–*Pleurozium* forests of south-eastern Labrador, Canada. *Journal of Ecology*, **73**, 517–34.

Foster, J. R. & Reiners, W. A. (1983). Vegetation patterns in a virgin subalpine forest at Crawford Notch, White Mountains, New Hampshire. *Bulletin of the Torrey Botanical Club*, **110**, 141–53.

Fowells, H. A. (1965). *Silvics of Forest Trees of the United States*. Agricultural Handbook No. 271, USDA Forest Service, Government Printing Office, Washington, D.C.

Fox, J. F. & Van Cleve, K. (1983). Relationships between cellulose decomposition, Jenny's *k*, forest floor nitrogen, and soil temperature in Alaskan taiga forests. *Canadian Journal of Forest Research*, **13**, 789–94.

Fralish, J. (1975). Ecological aspects of aspen succession in northern Wisconsin. *Transactions of the Wisconsin Academy of Sciences, Arts, and Letters*, **63**, 54–65.

Francline, J. F. & Waring, R. M. (1979). Distinctive features of the north-western coniferous forest: development, structure and function. In *Forests: Fresh Perspectives from Ecosystem Analysis*, ed. R. M. Waring, pp. 59–85. Corvallis, Oregon: Oregon State University Press.

Fraser, E. M. (1956). *The Lichen Woodlands of the Knob Lake Area of Québec–Labrador*. Montréal: McGill Subarctic Research Paper 1.

Fraser, J. W. (1976). Viability of black spruce seed in or on a boreal forest seedbed. *Forestry Chronicle*, **52**(5), 229–31.

Fredskild, B. (1967). Palaeobotanical investigations at Sermermiut, Jakobshavn, West Greenland. *Meddelelser om Gronland*, **178**, 1–54.

Frelich, L. E. (1986). *Natural disturbance frequencies in the hemlock–hardwood forests of the Upper Great Lakes Region*. PhD thesis, University of Wisconsin, Madison.

Frelich, L. E. & Lorimer, C. G. (1985). Current and predicted long-term effects of deer browsing in hemlock forests in Michigan, USA. *Biological Conservation*, **34**, 99–120.

Fries, C. (1985). The establishment of seed-sown birch (*Betula verrucosa* Ehrh. and *B. pubescens*) on clear-cuttings in Sweden. In *Broadleaves in boreal silviculture – an obstacle or an asset?*, ed. B. Hagglund & G. Peterson, pp 111–25. Swedish University of Agricultural Sciences, Department of Silviculture, Umeå, Sweden, Rapporter Nr 14.

Frissell, S. S. (1973). The importance of fire as a natural ecological factor in Itasca State Park, Minnesota. *Quaternary Research*, **3**, 397–407.

Fulton, M. (1990). A fast forest succession model: design and initial tests. *Forest Ecology and Management*, (in press).

Fung, A. K. & Ulaby, F. T. (1984). Matter–energy interactions in the microwave region. In *Manual of Remote Sensing*, 2nd ed., Vol. II, pp. 115–64. American Society of Photogrammetry and Remote Sensing, Falls Church, VA.

Furyaev, V. V. & Kireev, D. M. (1979). *A Landscape Approach in the Study of Postfire Forest Dynamics*. Novosibirsk: Nauka. (In Russian.)

Gagnon, R. & Payette, S. (1981). Fluctuations Holocènes de la limite des forêts de mélèzes, Rivière-aux-Feuilles, Nouveau-Québec: une analyse macrofossile en milieu tourbeux. *Géographie Physique et Quaternaire*, **35**, 57–72.

Gajewski, K., Winkler, M. G. & Swain, A. M. (1985). Vegetation and fire history from three lakes with varved sediments in northwestern Wisconsin (USA). *Review of Palaeobotany and Palynology*, **44**, 277–92.

Galkina, E. A. (1946). Mire landscapes and principles of classification. In *Collection of scientific works made by V. L. Komarov Botanical Institute of Science Academy of USSR for the three year period of the 2nd World War (1941-1943)*, pp. 139–56. Leningrad: Publication of Science Academy of USSR. (In Russian.)

Galkina, E. A. (1959). Mire landscapes of Karelia and principles of their classification. *Transactions of Karelian branch of Science Academy of USSR, Petrozavodsk*, **15**, 3–48. (In Russian.)

Galloway, G. & Worrall, J. (1979). Cladoptsis: a reproductive strategy in black cottonwood. *Canadian Journal of Forest Research*, **9**, 122–5.

Gammon, R. H., Sundquist, E. T. & Fraser, P. J. (1985). History of carbon dioxide in the atmosphere. In *Atmospheric Carbon Dioxide and the Global Carbon Cycle*, ed. J. R. Trabalka, pp. 27–62. Report DOE/ER-0239. US Department of Energy, Carbon Dioxide Research Division, Washington, D.C.

Ganns, R. C. (1977). *Germination and survival of artificially seeded white spruce on prepared seedbeds on an interior Alaskan floodplain site.* M.S. thesis, University of Alaska, Fairbanks, AK.

Gardner, A. C. (1986). *Natural regeneration of white spruce following harvesting of alluvial floodplain sites in the Liard River Drainage.* M.S. thesis, Department of Forest Science, University of Alberta, Edmonton.

Garfinkel, H. L. & Brubaker, L. B. (1980). Modern climate–tree-growth relationships in sub-arctic Alaska. *Nature*, **286**, 872–4.

Gates, D. M. & Benedict, C. M. (1963). Convection phenomena from plants in still air. *American Journal of Botany*, **50**, 563–73.

Gates, F. C. (1912). The vegetation of the region of the vicinity of Douglas Lake, Cheboygan County, Michigan. *Michigan Academy of Sciences, Arts, and Letters Annual Report*, **14**, 46–106.

Gerardin, V. (1980). *Les régions écologiques et la vegetation des sols minéraux.* L'inventaire du Capital-Nature du Territoire de la Baie-James. Ottawa: Environnement Canada.

Gerardin, V. (1981). *Inventaire du Capital-Nature de la Côte-Nord, Les régions écologiques provisoires (2e approximation)*, Rapport no.8. Québec: Service des Inventaires écologiques, Environnement-Québec.

Gill, D. (1972). The evolution of a discrete beaver habitat in the Mackenzie River Delta, Northwest Territories. *Canadian Field-Naturalist*, **86**, 233–9.

Gleason, H. A. (1924). The structure of the maple–beech association in northern Michigan. *Michigan Academy of Sciences, Arts, and Letters Paper* 4, 285–96.

Gleason, H. A. (1926). The individualistic concept of the plant association. *Bulletin of the Torrey Botanical Club*, **57**, 7–26.

Gleason, H. A. (1927). Further views on the succession concept. *Ecology*, **8**, 299–326.

Glebov, F. Z. (1988). *Interrelations between Forest and Mire in the Taiga Zone.* Novosibirsk: Science, Siberian Branch. (In Russian.)

Glebov, F. Z. & Alexandrova, S. R. (1973). Phytocenotical characterization, hydrothermal regime and soil microflora of some types of bogged forests at Tomsk field station in relation to the microrelief. In *The Complex Evaluation of Bogs and Boggy Forests with Reference to their Amelioration*, pp. 44–94. Novosibirsk: Science, Siberian Branch. (In Russian.)

Glebov, F. Z., Gorozhankina, S. M., Kireev, D. M. & Karpenko, L. V. (1978). Experiment for studying the structure and genesis of forest-mire complexes. In *Specificities of Forest-Mire Ecosystems in West Siberia*, pp. 14–9. Krasnoyarsk: Forest and Wood Institute, Siberian Branch of Science Academy of USSR. (In Russian.)

Glebov, F. Z. & Kobyakov, M. V. (1984). Influence of bogging and exposure to fire of forest stands on their productivity and development. In *Productivity of Forest Ecosystems,* ed. A. I. Buzykin, pp. 133–42. Krasnoyarsk: Science Publishers. (In Russian.)

Glebov, F. Z. & Korzukhin, M. D. (1985). Biogeocenotic model of interchanges of forest and mire. *Journal of General Biology,* **46**(3), 409–21. (In Russian.)

Glebov, F. Z. & Korzukhin, M. D. (1988). Biogeocenotic model of forest paludification. *Journal of General Biology,* **49**(5), 592–600. (In Russian.)

Glebov, F. Z. & Korzukhin, M. D. (1990). Ecological models in bog dynamics. In *Experiment and Mathematical Modeling in Investigation of Forests and Bogs,* pp. 74–92. Moscow: Science Publishers.

Glebov, F. Z., Toleiko, L. S., Starikov, E. V. & Zhidovblenko, V. A. (1980). History of forest-mire interrelations on the basis of the paleobotanical analysis of West Siberian peat bogs. In *Problems of Forest Biogeocenology,* pp. 115–40. Moscow: Science. (In Russian.)

Glover, G. R. & Hool, J. N. (1979). A basal area ratio predictor of loblolly pine plantation mortality. *Forest Science,* **25**, 275–82.

Gode, I. D. (1986). *Bäume und Sträucher.* (In German.)

Godman, R. M. & Krefting, L. W. (1960). Factors important to yellow birch establishment in Upper Michigan. *Ecology,* **41**, 18–28.

Godman, R. M. & Mattson, G. A. (1976). *Seed crops and regeneration problems of 19 species in northeastern Wisconsin.* U.S. Department of Agriculture Forest Service Research Paper NC-118.

Goel, N. S., Strebel, D. E. & Thompson, R. L. (1984). Inversion of vegetation canopy reflectance models for estimating agronomic variables. II. Use of angle transforms and error analysis as illustrated by the Suits model. *Remote Sensing of Environment,* **14**, 77–111.

Goff, F. G. & West, D. C. (1975). Canopy–understorey interaction effects on forest population structure. *Forest Science,* **21**, 98–108.

Goldstein, G. H., Brubaker, L. B. & Hinckley, T. M. (1985). Water relations of white spruce (*Picea glauca* (Moench) Voss) at tree line in north central Alaska. *Canadian Journal of Forest Research,* **15**, 1080–7.

Goodwin, C. W., Brown, J. & Outcalt, S. I. (1984). Potential responses of permafrost to climatic warming. In *The Potential Effects of Carbon Dioxide-Induced Climatic Changes in Alaska,* ed. J. H. McBeath, G. P. Juday, G. Weller & M. Murray, pp. 92–105. Miscellaneous Publication 83-1, School of Agricultural and Land Resource Management, University of Alaska, Fairbanks.

Gordon, A. G. (1983). Nutrient cycling dynamics in differing spruce and mixed wood ecosystems in Ontario and the effects of nutrient removals through harvesting. In *Resources and Dynamics of the Boreal Zone,* ed. R. W. Wein, R. R. Riewe & I. R. Methven, pp. 97–118. Ottawa: Association of Canadian Universities for Northern Studies.

Gordon, A. M. & Van Cleve, K. (1983). Seasonal patterns of nitrogen mineralization following harvesting in the white spruce forests of interior Alaska. In *Resources and Dynamics of the Boreal Zone,* ed. R. W. Wein, R. R. Riewe & I. R. Methven, pp. 119–30. Ottawa: Association of Canadian Universities for Northern Studies.

Gordyagin, A.Ya. (1901). Data for the identification of soils and vegetation in West Siberia. *Transactions of Nature Investigators of Kazan University*, **34**(3), 1–222. (In Russian.)

Gore, A. J. P. (1983). Introduction. In *Ecosystems of the World, Mires: Swamp, Bog, Fen and Moor*, ed. A. J. P. Gore, pp. 1–34. Vol. 4a, General Studies. Amsterdam: Elsevier.

Gorodkov, B. N. (1916a). An experiment to separate the West Siberian Plain into botanic–geographical provinces. *Annual Issue of Tobolsk Museum*, **27**, 1–56. (In Russian.)

Gorodkov, B. N. (1916b). Observations on growth of the Siberian cedar pine in Ural Mountains. *Papers of the Botanical Museum of Academy of Sciences of the USSR*, **16**, 153–72. (In Russian.)

Gorozhankina, S. M. & Konstantinov, V. D. (1978). *Geography of West Siberian Taiga*. Novosibirsk: Science, Siberian Branch. (In Russian.)

Gortchakovsky, P. L. (1956). Distribution range of the Siberian cedar pine in Ural mountains. In *For the 75th Jubilee of Acad. V. N. Sukatchev*, ed. Anonymous, pp. 131–41. Academy of Sciences of the USSR. (In Russian.)

Gosz, J. G., Likens, G. E. & Bormann, F. H. (1973). Nutrient release from decomposing leaf and branch litter in the Hubbard Brook forest, New Hampshire. *Ecological Monographs*, **43**, 173–91.

Goward, S. N. (1985). Shortwave infrared detection of vegetation. *Advances in Space Research*, **5**(5), 41–50.

Goward, S. N. (1989). Satellite bioclimatology. *Journal of Climate*, **2**, 710–20.

Goward, S. N., Kerber, A., Dye, D. & Kalb, V. (1987). Comparison of North and South American biomes from AVHRR observations. *Geocarto International*, **2**, 27–39.

Goward, S. N., Tucker, C. J. & Dye, D. G., (1985). North American vegetation patterns observed with the NOAA-7 advanced very high resolution radiometer. *Vegetatio*, **64**, 3–14.

Graebner, P. (1904). *Handbuch der Heidekultur*. Leipzig.

Graham, S. A. (1941). The climax forests of the upper peninsula of Michigan. *Ecology*, **22**, 355–62.

Graham, S. A. (1954). Changes in northern Michigan forests from browsing by deer. *Transactions of the North American Wildlife Conference*, **19**, 526–33.

Graham, S. A., Harrison, R. P. & Westell, C. E. (1963). *Aspen, Phoenix Trees of the Lake States*. Ann Arbor, Michigan: University of Michigan Press.

Grandtner, M. M. (1966). *La vegetation forestière du Québec méridional*. Québec: Les Presses de l'Université Laval.

Granström, A. (1982). Seed banks in five boreal forest stands originating between 1810 and 1963. *Canadian Journal of Botany*, **60**, 1815–21.

Granström, A. (1987). Seed viability of fourteen species during five years of storage in forest soil. *Journal of Ecology*, **75**, 321–31.

Granström, A. & Fries, C. (1985). Depletion of viable seeds of *Betula pubescens* and *B. verrucosa* sown onto some north Swedish forest soils. *Canadian Journal of Forest Research*, **15**, 1176–80.

Grant, M. L. (1934). Climax forest community in Itasca County, Minnesota. *Ecology*, **15**, 243–57.

Green, D. G. (1981). Time series and postglacial forest ecology. *Quaternary Research*, **15**, 265–77.

Green, D. G. (1982). Fire and stability in the post–glacial forests of southwest Nova Scotia. *Journal of Biogeography*, **9**, 29–40.

Green, D. G. (1987). Pollen evidence for the postglacial origins of Nova Scotia's forests. *Canadian Journal of Botany*, **65**, 1163–79.

Gregory, R. A. (1966). The effect of leaf litter upon establishment of white spruce beneath paper birch. *Forestry Chronicle*, **42**(3), 251–5.

Grigal, D. F. & Ohmann, L. F. (1975). Classification, description, and dynamics of upland plant communities within a Minnesota wilderness area. *Ecological Monographs*, **45**, 389–407.

Griggs, R. F. (1937). Timberlines as indicators of climatic trends. *Science*, **85**, 251–5.

Grime, J. P. (1979). *Plant Strategies and Vegetation Processes*. New York: Wiley.

Grimm, E. C. (1983). Chronology and dynamics of vegetation change in the prairie–woodland region of southern Minnesota. *New Phytologist*, **93**, pp. 311–35.

Grisebach, A. (1838). Uber den Einfluss des Klimas auf die Begrenzung der Floren. *Linnaea*, **12**, 159–200.

Gross, H. L. (1972). Crown deterioration and reduced growth associated with excessive seed production by birch. *Canadian Journal of Botany*, **50**, 2431–7.

Grubb, P. J. (1977). The maintenance of species-richness in plant communities: The importance of the regeneration niche. *Biological Reviews*, **52**, 107–45.

Guetter, P. J. & Kutzbach, J. E. (1990). A modified Koeppen classification applied to model simulations of glacial and interglacial climates. *Climatic Change*, **16**, 193–215.

Gurtin, M. E. & MacCamy, R. C. (1979). Some simple models for nonlinear age-dependent population dynamics. *Mathematical Biosciences*, **43**, 199–211.

Gurtzev, A. I. & Korzukhin, M. D. (1988). Crown and root competition in a linear pine plantation. *Problems of Ecological Monitoring and Ecosystem Modeling*, **12**, 206–23. Leningrad: Gydromet Publishers. (In Russian.)

Haavisto, V. F. (1975). Peatland black spruce seed production and dispersal in northeastern Ontario. In *Black spruce symposium*, pp. 250–64. Canadian Forestry Service, Sault Saint Marie, Ontario, Symposium Proceedings 0-P-4.

Haavisto, V. F. (1978). Lowland black spruce seedfall: viable seedfall peaks in mid-April. *Forestry Chronicle*, **54**(4), 213–5.

Hadley, J. L. & Smith, W. K. (1983). Influence of wind exposure on needle desiccation and mortality for timberline conifers in Wyoming, USA. *Arctic and Alpine Research*, **15**, 127–35.

Hadley, J. L. & Smith, W. K. (1986). Wind effects on needles of timberline conifers: seasonal influence on mortality. *Ecology*, **67**, 12–9.

Hadley, J. L. & Smith, W. K. (1987). Influence of krummholz mat microclimate on needle physiology and survival. *Oecologia*, **73**, 82–90.

Haglund, E. (1912). Die Brandtheorie: Ein neuer Beitrag zur Frage nach der Entstehung der Hochmoore. *Mitt. Balt. Moorver*, **2**, 105.

Haigth, R. G. & Getz, W. M. (1987). A comparison of stage-structured and single-tree models for projecting forest stands. *Natural Resource Modeling*, **2**, 279–98.

Haines, D. A. & Sando, R. W. (1969). *Climatic conditions preceding historically great fires in the North Central Region*. U.S. Department of Agriculture Forest Service Research Paper NC-34.

Hall, F. G., Sellers, P. J., MacPherson, I., Kelly, R. D., Verma, S., Markham, B., Blad, B., Wang, J. & Strebel, D. E. (1989). FIFE: Analysis and results – A review. *Advances in Space Research*, **9**, 275–93.

Hall, F. G., Strebel, D. E., Goetz, S. J., Woods, K. D. & Botkin, D. B. (1987). Landscape pattern and successional dynamics in the boreal forest. *Proceedings of the IGARSS 1987 Symposium*, Ann Arbor, Michigan, 18–21 May, pp. 473–82.

Hall, F. G., Strebel, D. E., Heummrich, K. F., Goetz, S. J., Nickeson, J. E. & Woods, K. D. (1990). *Radiometric and biophysical observations of a boreal ecosystem: The COVER experiment*. NASA Technical Memorandum, Goddard Space Flight Center, Code 923, Greenbelt, MD 20771 (in press).

Hall, F. G., Strebel, D. E. & Sellers, P. J. (1988). Linking knowledge among spatial and temporal scales: Vegetation, atmosphere, climate and remote sensing. *Landscape Ecology*, **2**, 3–22.

Hall, T. H. (1981). Forest management decision making: art or science. *Forestry Chronicle*, **57**, 233–8.

Hämet-Ahti, L. (1963). Zonation of the mountain birch forest in northern Fennoscandia. *Annales Botanici Societatis Vanamo*, **34**, 1–127.

Hämet-Ahti, L. (1981). The boreal zone and its biotic subdivisions. *Fennia*, **159**, 69–75.

Hämet-Ahti, L. & Ahti, T. (1969). The homologies of the Fennoscandian mountain and coastal birch forests in Eurasia and North America. *Vegetatio*, **19**, 208–19.

Hamilton, D. A., Jr. (1980). Modeling mortality: a component of growth and yield modeling. *Proceedings of the Workshop Forecasting Forest Stand Dynamics*, Thunder Bay, Ontario, Canada, 1979, pp. 82–9.

Hamilton, G. D., Drysdale, P. D. & Euler, D. L. (1980). Moose winter browsing patterns on clear-cuttings in northern Ontario. *Canadian Journal of Zoology*, **58**, 1412–6.

Hansen, J. E., Lacis, A., Rind, D., Russel, G., Stone, P., Fung, I., Ruedy, R. & Lerner, J. (1984). Climate sensitivity: Analysis of feedback mechanisms. In *Climate processes and sensitivity*, ed. J. E. Hansen & T. Takahashi, pp. 130–63. Maurice Ewing Series No. 5, American Geophysical Union, Washington, D.C.

Harcombe, P. A. & Marks, P. L. (1978). Tree diameter distributions and replacement processes in southeast Texas forests. *Forest Science*, **24**, 153–66.

Hardy, Y. J., Lafond, A. & Hamel, L. (1983). The epidemiology of the current spruce budworm outbreak in Québec. *Forest Science*, **29**, 715–25.

Hare, F. K. (1950). Climate and zonal division of the boreal forest formations in eastern Canada. *Geographical Review*, **40**, 615–35.

Hare, F. K. (1959). *A Photo-Reconnaissance Survey of Labrador-Ungava*. Ottawa: Canada Department of Mines and Technical Surveys, Geographical Branch, Memoir **6**, 1–83.

Hare, F. K. & Hay, J. E. (1974). The climate of Canada and Alaska. In *Climates of North America. World Survey of Climatology Volume* 11, ed. R. A. Bryson & F. K. Hare, pp. 49–192. New York: Elsevier.

Hare, F. K. & Ritchie, J. C. (1972). The boreal bioclimates. *Geographical Review*, **62**, 333–65.

Hare, F. K. & Thomas, M. K. (1979). *Climate Canada*. Toronto: University of Toronto Press. 2nd edn.

Hari, P., Raunemma, T. & Hautojarvi, A. (1986). The effects on forest growth of air pollution from energy production. *Atmospheric Environment*, **20**, 129–37.

Haritonovitsh, F. M. (1968). *Biology and Ecology of Tree Species*. Moscow: Lismaya promyshlenost. (In Russian.)

Harlow, W. M., Harrar, E. S. & White, F. M. (1979). *Textbook of Dendrology*, 6th edn. New York: McGraw-Hill Book Company.

Harper, J. L. (1977). *Population Biology of Plants*. London: Academic Press.

Harrison, E. A. & Shugart, H. H. (1990). Evaluating performance on an Appalachian oak forest dynamics model. *Vegetatio*, **86**, 1–13.

Hartshorn, G. S. (1975). A matrix model of tree population dynamics. In *Tropical Ecological Systems*, pp. 41–51. New York: Springer-Verlag.

Haukioja, E. (1980). On the role of plant defenses in the fluctuation of herbivore populations. *Oikos*, **35**, 202–13.

Haukioja, E. & Hakala, T. (1975). Herbivore cycles and periodic outbreaks. Formulation of a general hypothesis. *Report of Kevo Subarctic Research Station*, **12**, 1–9.

Haukioja, E., Kapiainen, K., Niemela, P. & Tuomi, J. (1983). Plant availability hypothesis and other explanations of herbivore cycles: complementary or exclusive alternatives? *Oikos*, **40**, 419–32.

Hayes, A. J. (1965). Studies in the decomposition of coniferous leaf litter. I. Physical and chemical changes. *Journal of Soil Science*, **16**, 121–40.

Hedlin, A. F. (1974). *Cone and seed insects of British Columbia*. Environment Canada, Canadian Forest Service Publication No. BC-X-90. Pacific Forest Research Center, Victoria, BC.

Heikenheimo, O. (1932). Matäpuiden siementämiskyvystä I (in Finnish). Deutsches referat: Uber die Besamungsfähigkeit der Waldbaume. *Communications Instituti Forestalis Fenniae*, **17**(3), 1–61.

Heikurainen, L. (1967). Effect of cutting on the ground-water level on drained peatlands. In *Forest Hydrology*, pp. 345–54. Oxford: Pergamon Press.

Heilman, P. A. (1966). Change in distribution and availability of nitrogen with forest succession on north slopes of interior Alaska. *Ecology*, **47**, 825–34.

Heinselman, M. L. (1973). Fire in the virgin forests of the Boundary Waters Canoe Area, Minnesota. *Quaternary Research*, **3**, 329–82.

Heinselman, M. L. (1981a). Fire intensity and frequency as factors in the distribution and structure of northern ecosystems. In *Fire regimes and ecosystem properties*, ed. H. A. Mooney, T. M. Bonnicksen, N. L. Christensen, J. E. Lotan & W. A. Reiners. U.S. Department of Agriculture Forest Service General Technical Report WO-26.

Heinselman, M. L. (1981b). Fire and succession in the conifer forests of northern North America. In *Forest Succession: Concepts and Application*, ed. D. C. West, H. H. Shugart & D. B. Botkin, pp. 374–405. New York: Springer-Verlag.

Hellum, A. K. (1972). *Germination and early growth of white spruce on rotten woods and peat moss in the laboratory and nursery.* Canadian Forestry Service Report NOR-x-39.

Helmisaari, H. & Nikolov, N. (1989). *Survey of ecological characteristics of boreal forest tree species in Fennoscandia and the USSR.* WP-89-65. International Institute for Applied Systems Analysis, Laxenburg, Austria.

Helvey, J. D. & Patric, J. H. (1965). Canopy and litter interception of rainfall by hardwoods of eastern United States. *Water Resources Research,* 1, 193–206.

Hempel, G. & Wilhelm, K. (1897). *Die Bäume und Sträucher des Waldes.* (Forest Trees and Shrubs). Vienna: Verlag Ed. Hölzel. (In German.)

Henry, J. D. & Swan, J. M. A. (1974). Reconstructing forest history from live and dead plant material – an approach to the study of forest succession in southwest New Hampshire. *Ecology,* 55, 772–83.

Henry, R. M. & Barnes, B. V. (1977). Comparative reproductive ability of bigtooth and trembling aspen and their hybrid. *Canadian Journal of Botany,* 55, 3093–8.

Henttonen, H., Kanninen, M., Nygren, M. & Ojansuu, R. (1986). The maturation of *Pinus sylvestris* seeds in relation to temperature climate in northern Finland. *Scandinavian Journal of Forest Research,* 1, 243–9.

Hesselman, H. (1907). Om tvenne nybildade tjarnar: Afdalens Kronopark. *Geologiska föreningens i Stockholm förhandlingar,* 29, 23–37.

Hesselman, H. (1912). Våra skogsträd 2. Aspen. *Skogsvådsföreningens folkskrifter.* No. 21.

Hesselman, H. (1931). Om aspen. *Skogsägaren,* 7, 185–8.

Hett, J. M. & Loucks, O. L. (1976). Age structure models of balsam fir and eastern hemlock. *Journal of Ecology,* 64, 1029–44.

Hirano, S. & Kira, T. (1965). Intraspecific competition among higher plants. XII. Influence of autotoxic root exudation on the growth of higher plants grown at different densities. *Journal of Biology,* 16, 27–44. Osaka City University, Osaka.

Hodanova, D., (1985). Leaf optical properties. In *Photosynthesis During Leaf Development,* ed. Z. Sestak, pp. 107–27. Boston: W. Junk Publishers.

Hoffer, R. M., Lozano–Garcia, D. F. & Gillespie, D. D. (1986). *Mapping forest cover with SIR-B data.* Technical Paper, ASPRS-ACSM Fall Convention, Anchorage, AK, Sept 28-Oct 3, pp. 407–15.

Hohrin, A. V. (1970). About the climate response of Siberian cedar pine plantations in the Central Urals. *Papers of the Institute for Plant and Animal Ecology,* 67, 312–22. (In Russian.)

Holdridge, L. R. (1947). Determination of world plant formations from simple climatic data. *Science,* 105, 367–8.

Holdridge, L. R. (1964). *Life Zone Ecology.* San Jose, Costa Rica: Tropical Science Center.

Holdridge, L. R. (1967). *Life Zone Ecology* (revised edn). San Jose, Costa Rica: Tropical Science Center.

Holling, C. S. (1973). Resilience and stability of ecological systems. *Annual Review of Ecology and Systematics,* 4, 1–23.

Holling, C. S. ed. (1978). *Adaptive Environment Assessment.* Chichester, UK: John Wiley & Sons.

Holling, C. S. (1980). Forest insects, forest fires and resilience. In *Fire regimes and ecosystem properties*, ed. H. Mooney, J. M. Bonnicksen, N. L. Christensen, J. E. Lotan & W. A. Reiners, pp. 445–64. USDA Forest Service General Technical Report.

Holling, C. S. (1986). Resilience of ecosystems; local surprise and global change. In *Sustainable Development of the Biosphere*, ed. W. C. Clark & R. E. Munn, pp. 292–317. Cambridge: Cambridge University Press.

Holling, C. S. (1988). Temperate forest insect outbreaks, tropical deforestation and migratory birds. *Memoirs of the Entomological Society of Canada*, **146**, 21–32.

Holling, C. S., Jones, D. D. & Clark, W. C. (1979). Ecological policy design: a case study of forest and pest management. In *Pest Management*, ed. G. A. Norton & C. S. Holling, pp. 13–90. Oxford: Pergamon Press.

Holloway, R. G. & Bryant, V. M. Jr. (1985). Late-Quaternary pollen records and vegetational history of the Great Lakes Region: United States and Canada. In *Pollen Records of Late-Quaternary North American Sediments*, ed. V. M. Bryant, Jr. & R. G. Holloway, pp. 205–44. Dallas: American Association of Stratigraphic Palynologists Foundation.

Hom, J. L. & Oechel, W. C. (1983). The photosynthetic capacity, nutrient content, and nutrient use efficiency of different needle age-classes of black spruce (*Picea mariana*) found in interior Alaska. *Canadian Journal of Forest Research*, **13**, 834–9.

Hool, J. N. (1966). A dynamic programming–Markov chin approach to forest production control. *Forest Science Monograph*, **12**, 1–26.

Horn, H. S. (1975a). Forest succession. *Scientific American*, **232**, 90–8.

Horn, H. S. (1975b). Markovian properties of forest succession. In *Ecology and Evolution in Communities*, ed. M. L. Cody & J. M. Diamond, pp. 196–211. Cambridge, Massachusetts: Harvard University Press.

Horn, H. S. (1976). Succession. In *Theoretical Ecology*, ed. R. M. May, pp. 187–204. Oxford: Blackwell.

Horner, J. D., Gosz, J. R. & Cates, R. G. (1988). The role of carbon-based secondary plant metabolites in decomposition in terrestrial ecosystems. *American Naturalist*, **132**, 869–83.

Hough, A. F. (1932). Some diameter distributions in forests stands of northwestern Pennsylvania. *Journal of Forestry*, **30**, 933–43.

Houghton, R. A. (1987). Biotic changes consistent with the increased seasonal amplitude of atmospheric CO_2 concentration. *Journal of Geophysical Research*, **92**, 4223–30.

Hozumi, K. (1980). Ecological and mathematical considerations of self-thinning in even-aged pure stands. II. Growth analysis of self-thinning. *Botanical Magazine*, **93**, 149–66.

Huete, A. R. (1989). Soil influences in remotely sensed vegetation-canopy spectra. In *Theory and Applications of Optical Remote Sensing*, ed. G. Aswar, pp. 107–41. New York: John Wiley and Sons.

Huikari, O. (1957). Untersuchungen über den Anteil der primären Versumpfung an der Entstehung der finnischen Moore. *Communicationes Instituti Forestalis Fenniae*, **46**(6), 1–79.

Humboldt, A. von (1807). *Ideen zu einer Geographie der Pflanzen*. Tubingen: Cotta.

Huntley, B. & Webb, T. III. ed. (1988). *Vegetation History*. Dordrecht/Boston: Kluwer Academic Publishers.

Hustich, I. (1939). Notes on the coniferous forest and tree limit on the east coast of Newfoundland–Labrador, including a comparison between the coniferous forest limit in Labrador and in northern Europe. *Acta Geographica*, **7**, 5–77.

Hustich, I. (1949). On the forest geography of the Labrador peninsula. A preliminary synthesis. *Acta Geographica*, **10**, 3–63.

Hustich, I. (1950). Notes on the forests on the east coast of Hudson Bay and James Bay. *Acta Geographica*, **11**, 3–83.

Hustich, I. (1951*a*). Forest–botanical notes from Knob Lake area in the interior of Labrador Peninsula. *National Museum of Canada Bulletin*, **123**, 166–217.

Hustich, I. (1951*b*). The lichen woodlands in Labrador and their importance as winter pastures for domesticated reindeer. *Acta Geographica*, **12**, 3–48.

Hustich, I. (1958). On the recent expansion of Scots pine in northern Europe. *Fennia*, **82**, 1–25.

Hustich, I. (1966). On the forest–tundra and the northern tree-lines. *Annales Universitatis Turkuensis Series A II*, **36**, 7–47.

Hustich, I. (1979). Ecological concepts and biogeographical zonation in the north: the need for a generally accepted terminology. *Holarctic Ecology*, **2**, 208–17.

Huston, M., DeAngelis, D. L. & Post, W. M. (1988). New computer models unify ecological theory. *BioScience*, **38**, 682–91.

Huston, M. A. & Smith, T. (1987). Plant succession: life history and competition. *American Naturalist*, **130**, 168–98.

Hytteborn, H. & Packham, J. R. (1985). Left to nature: forest structure and regeneration in Fiby Urskog, Central Sweden. *Arboricultural Journal*, **9**, 1–11.

Hytteborn, H. & Packham, J. R. (1987). Decay rate of *Picea abies* and the storm gap theory: a re-examination of Sernander plot III, Fiby Urskog, Central Sweden. *Arboricultural Journal*, **11**, 299–311.

Hytteborn, H., Packham, S. R. & Verwijst, T. (1987). Tree population dynamics, stand structure and species composition in the montaine virgin forest of Vallibacken, northern Sweden. *Vegetatio*, **72**, 3–19.

Hyvärinen, H. & Ritchie, J. C. (1975). Pollen stratigraphy of Mackenzie pingo sediments, N.W.T., Canada. *Arctic and Alpine Research*, **7**, 261–72.

Ignatenko, I. V., Knorre, A. V., Lovelius, N. V. & Norin, B. N. (1973). Phytomass stock in typical plant communities in the Ary-Mas forest. *Soviet Journal of Ecology*, **4**, 213–17.

Ilvessalo, L. (1917). Tutkimuksia mantymetsien uudistumisvuosista Etela-ja Keski-Suomessa. Referat: Studien uber die Verjungungsjahre der Kieferwalder in Sud- und Mittelfinnland. *Acta Forestalia Fennica*, **6**, 1–83.

International Council of Scientific Unions (1986). *The International Geosphere–Biosphere Program: A Study of Global Change*. Report No. 1, Final report of the Ad Hoc Planning Group. ICSU 21st General Assembly, Bern, Switzerland.

Irons, J. R., Ranson, K. J., Williams, D. L., Irish, R. R. & Huegel, F. B. (1990). An off-nadir pointing imaging spectroradiometer for terrestrial ecosystem studies. *IEEE Transactions on Geoscience and Remote Sensing*, **29**, 66–74.

Isaev, A. S. ed. (1985). *Cedar Pine Forests of Siberia*. Novosibirsk: Nauka.

Isaev, A. S., Khlebopros, R. G., Nedorezov, L. V., Kondakov, Y. P. & Kiselev, V. V. (1984). *Population Dynamics of Forest Insects*. Novosibirsk: Nauka Publishing House, Siberian Division. (In Russian.)

Isaev, A. & Krivosheina, N. (1976). The principles and methods of the integrated protection of Siberian forests from destructive insects. In *Man and the Boreal Forest*, ed. C. O. Tamm. *Ecological Bulletins* (Stockholm), 21, 121–24.

Ivanov, K. E. (1975). *Water Exchange in Mire Landscapes*. Leningrad: Hydrometeoizdat. (In Russian.)

Ivanov, K. E. (1976). Effect of large-scale drainage amelioration of bogs on thermal regime and ground freezing. In *Bogs of West Siberia: their Configuration and Hydrological Regime*, pp. 321–2. Lengingrad: Hydrometeoizdat. (In Russian.)

Ivanov, K. E. & Shumkova, E. L. (1967). Hydrological evaluation and estimation of forest devastation and expansion in areas of natural paludification under flooding from river systems. *Transactions of Institute of Hydrology*, Hydrometeoizdat, 145, 3–26. (In Russian.)

Izotov, V. F. (1968). Pattern of soil freezing and thawing in the waterlogged forests of the northern taiga subzone. *Soviet Soil Science*, 1968(6), 807–13.

Jablanczy, A. & Baskerville, G. L. (1969). *Morphology and development of white spruce and balsam fir seedlings in feather moss*. Canadian Forestry Service Report M-x-19.

Jacobson, G. L., Webb, T. III & Grimm, E. C. (1987). Patterns and rates of change during the deglaciation of eastern North America. In *North America and adjacent oceans during the last deglaciation*, ed. W. F. Ruddiman & H. E. Wright, pp. 277–88. Vol. K-3 of *The Geology of North America*, Geological Society of America, Boulder, CO.

Jaeger, J. (1988). *Developing policies for responding to climatic change: a summary of discussions and recommendations of workshops at Vollach and Bellagio in 1987*. WMO/TD No. 225, World Meteorological Organization Secretariat, Geneva, Switzerland.

Jaeger, J. (1990). Climate: Approaches to projecting temperature and moisture changes. In *Toward ecological sustainability in Europe*, ed. A. M. Solomon & L. Kauppi, pp. 7–29. IIASA Research Report, International Institute for Applied Systems Analysis, Laxenburg, Austria, (in press).

Jarvis, J. M., Steneker, G. A., Waldron, R. M. & Lees, J. C. (1966). *Review of silvicultural research – white spruce and trembling aspen cover types, mixedwood forest section, boreal forest region Alberta–Saskatchewan–Manitoba*. Canadian Department of Forestry and Rural Development, Forestry Branch Department Publication 1156.

Jarvis, P. G., Edwards, W. R. N. & Talbot, M. (1981). Models of plant and crop water use. In *Mathematics and Plant Physiology*, pp. 151–94. New York: Academic Press.

Jarvis, P. G. & McNaughton, K. G. (1986). Stomatal control of transpiration: scaling up from leaf to region. *Advances in Ecological Research*, 15, 1–49.

Jeannson, E., Bergman, F., Elving, B., Falck, J. & Lundquist, L. (1989). *Natural regeneration of pine and spruce – proposal for a research program*. Swedish University of Agricultural Sciences, Department of Silviculture, Report 25.

Jenny, H. (1980). *The Soil Resource*. New York: Springer-Verlag.

Johnson, E. A. (1975). Buried seed populations in the subarctic forest east of Great Slave Lake, Northwest Territories. *Canadian Journal of Botany*, **53**, 2933–41.

Johnson, E. A. (1979). Fire recurrence in the subarctic and its application for vegetation composition. *Canadian Journal of Botany*, **57**, 1374–9.

Johnson, E. A. (1981). Vegetation organization and dynamics of lichen woodland communities in the Northwest Territories, Canada. *Ecology*, **62**, 200–15.

Johnson, E. A. & Rowe, J. S. (1975). Fire in the subarctic wintering ground of the Beverley Caribou Herd. *American Midland Naturalist*, **94**, 1–14.

Johnson, E. A. & Rowe, J. S. (1977). *Fire and vegetation change in the western subarctic*. Canadian Department of Indian Affairs and Northern Development, ALUR Report 1975–1976, Ottawa.

Johnson, E. A. & Van Wagner, C. E. (1985). The theory and use of two fire history models. *Canadian Journal of Forest Research*, **15**(1), 214–20.

Johnston, C. A. & Naiman, R. J. (1990a). Browse selection by beaver: effects on riparian forest composition. *Canadian Journal of Forest Research*, (in press).

Johnston, C. A. & Naiman, R. J. (1990b). Aquatic patch creation in relation to beaver population trends. *Ecology*, (in press).

Johnston, W. F. (1972). *Seeding black spruce on brushy lowland successful if vegetation density kept low*. USDA Forest Service Research Note NC-139. North Central Forest Experiment Station, St. Paul, MN.

Jones, D. D. (1979). The budworm site model. In *Pest Management*, ed. G. A. Norton & C. S. Holling, pp. 91–159. Oxford: Pergamon Press.

Jones, E. W. (1945). The structure and reproduction of the virgin forest of the north temperate zone. *New Phytologist*, **44**, 130–48.

Jones, H. E. & Gore, A. J. P. (1978). A simulation of production and decay in Blanket Bog. In *Production Ecology of British Moors and Montane Grasslands*, Ecological Studies, Vol. 27, pp. 160–86. Berlin/New York: Springer-Verlag.

Jones, P. D., Wigley, T. M. L. & Kelly, P. M. (1982). Variations in surface air temperatures: part 1. Northern Hemisphere, 1881–1980. *Monthly Weather Review*, **110**, 59–70.

Jones, P. D., Wigley, T. M. L. & Wright, P. B. (1986). Global temperature variation between 1861 and 1984. *Nature*, **322**, 430–4.

Justice, C. O. ed. (1986). Monitoring the grasslands of semi-arid Africa using NOAA AVHRR data. Special Issue, *International Journal of Remote Sensing*, **7**, 1383–622.

Justice, C. O., Townshend, J. R. G., Holben, B. N. & Tucker, C. J. (1985). Analysis of the phenology of global vegetation using meteorological satellite data. *International Journal of Remote Sensing*, **6**, 1271–318.

Kabanov, N. E. (1977). *Coniferous Tree and Shrub Species of the Far East*. Moscow: Nauka. (In Russian.)

Kaks, A. P. (1914). Mires of vicinities of Dulova Lake. *Results of studies in mire region of Pskov district*, **7**, 1–76. (In Russian.)

Kalinin, V. I. (1965). *The Larch of the European North*. Moscow: Lesnaya promyshlenost. (In Russian.)

Kallio, P. & Lehtonen, J. (1973). Birch forest damage caused by *Oporina automna* (Bkh.) in 1965–66 in Utsjoki, N. Finland. *Report from the Kevo Subarctic Research Station*, **10**, 55–69.

Kampfmann, G. (1980). Eiche, Glas und Kartofell. *Natur und Mus.*, **110**(8), 225–41.

Kapur, J. N. (1982). Some mathematical models for optimal management of forests. *Indian Journal of Pure and Applied Mathematics*, **13**, 273–86.

Karavaeva, N. A. (1982). *Paludification and Soil Evolution*. Moscow: Science. (In Russian.)

Kareiva, P. and Andersen, M. (1988). Spatial aspects of species interactions: The wedding of models and experiments. In *Community Ecology*, Lecture Notes in Biomathematics 77, ed. A. Hastings, pp. 35–50. Berlin: Springer-Verlag.

Karlén, W. (1976). Lacustrine sediments and tree limit variations as indicators of Holocene climatic fluctuation in Lapland, northern Sweden. *Geografiska Annaler*, **58a**, 1–34.

Karlén, W. (1983). Holocene fluctuations of the Scandinavian alpine tree-limit. *Nordicana*, **47**, 55–9.

Kasanaga, H. & Monsi, M. (1954). On the light-transmission of leaves and its meaning for the production of dry matter in plant communities. *Japanese Journal of Botany*, **14**, 302–24.

Katayeva, K. V. & Korzukhin, M. D. (1987). *Dynamics of Dark-coniferous Siberian Pine Forests*. Moscow: Laboratory of Environmental Monitoring Publishers. (In Russian.)

Kauppi, P. & Posch, M. (1985). Sensitivity of boreal forests to possible climatic warming. *Climatic Change*, **7**, 45–54.

Kauppi, A., Rinne, P. & Ferm, A. (1987). Initiation, structure and sprouting of dormant basal buds in *Betula pubescens*. *Flora*, **179**, 55–83.

Kauppi, A., Rinne, P. & Ferm, A. (1988). Sprouting ability and significance for coppicing of dormant buds on *Betula pubescens* Ehrh. stumps. *Scandinavian Journal of Forest Research*, **3**, 343–54.

Kaya, I. & Buongiorno, J. (1987). Economic harvesting of uneven-aged northern hardwood stands under risk: a Markovian decision model. *Forest Science*, **33**, 889–907.

Kazimirov, N. I. (1971). *Karelian Picea Forests*. Leningrad: Science Publishers. (In Russian.)

Kazimirov, N. I. (1983). *The Spruce*. Moscow: Lesnaya promyshlenost. (In Russian.)

Keane, R. E., Arno, S. F. & Brown, J.K. (1990). Simulation of cumulative fire effects in Ponderosa Pine/Douglas-Fir forests. *Ecology*, **71**, 189–203.

Kell, L. L. (1938). Effect of the moisture-retaining capacity of soils on forest succession in Itasca Park, Minnesota. *American Midland Naturalist*, **20**, 682–94.

Kellomäki, S. (1987). *Metsäekologia*. Silva Carelica 7. University of Joensuu. (In Finnish.)

Kelly, D. G. (1978). *Population density, territoriality and foraging ecology of red squirrels* (Tamiasciurus hudsonicus) *in black and white spruce forests of interior Alaska*. M.S. thesis, University of Alaska, Fairbanks, AK.

Ker, J. W. & Smith, J. H. G. (1955). Advantages of the parabolic expression of height–diameter relationships. *Forestry Chronicle*, **31**, 235–46.

Kercher, J. R. & Axelrod, M. C. (1984*a*). Analysis of SILVA: a model for forecasting the effects of SO_2 pollution and fire on western coniferous forests. *Ecological Modelling*, **23**, 165–84.

Kercher, J. R. & Axelrod, M. C. (1984b). A process model of fire ecology and succession in a mixed-conifer forest. *Ecology*, **65**(6), 1725–42.

Kershaw, K. A. (1977). Studies on lichen-dominated systems. XX. An examination of some aspects of the northern boreal lichen woodlands in Canada. *Canadian Journal of Botany*, **55**, 393–410.

Kershaw, K. A. (1978). The role of lichens in boreal tundra transition areas. *Bryologist*, **81**, 294–306.

Kershaw, K. A. & Rouse, W. R. (1976). *The Impact of Fire on Forest and Tundra Ecosystems*. Ottawa: Department of Indian Affairs and North. ALUR 75-76-63.

Kessell, S. R. (1976). Gradient modeling: A new approach to fire modeling and wilderness resource management. *Environmental Management*, **1**, 39–48.

Kessell, S. R. (1979a). *Gradient Modeling: Resource and Fire Management*. New York: Springer-Verlag.

Kessell, S. R. (1979b). Phytosociological inference and resource management. *Environmental Management*, **3**, 29–40.

Kessell, S. R. (1982). Creation of generalized models of secondary succession of plants. In *Biosphere Reserves, Proceedings of Second American–Soviet Symposium*, pp. 183–213. Leningrad: Gidrometeoisdat. (In Russian.)

Kessell, S. R. & Potter, M. W. (1980). A quantitative succession model for nine Montana forest communities. *Environmental Management*, **4**, 227–40.

Khotinskiy, N. A. (1984). Holocene vegetation history. In *Late Quaternary Environment of the Soviet Union*. ed. V. V. Velichko, H. E. Wright & C. W. Barnosky (eds. of the English-language edition), pp. 170–200. Minneapolis: University of Minnesota Press.

Kienast, F. (1987). *FORECE – a forest succession model for southern Central Europe*. Oak Ridge, Tennessee: Oak Ridge National Laboratory, ORNL/TM–2989.

Kimes, D. S. & Kirchner, J. A. (1982). Radiative transfer model for heterogeneous 3-D scenes. *Applied Optics*, **21**, 4119–29.

Kimmins, J. P. (1987). *Forest Ecology*. New York: Macmillan Publishing Company.

Kingsbury, C. M. & Moore, T. R. (1987). The freeze-thaw cycle of a subarctic fen, northern Quebec, Canada. *Arctic and Alpine Research*, **19**, 289–95.

Kira, T., Shinokzaki, K. & Hozumi, K. (1969). Structure of forest canopies as related to their primary productivity. *Plant Cell Physiology*, **10**, 129–42.

Kirsanov, V. A. (1974). *Establishment and growth of cedar pine forests in North Urals and Transuralia*. PhD thesis, Sverdlovsks. (In Russian.)

Kittredge, J. A. (1938). The interrelationships of habitat, growth rate, and associated vegetation in the aspen community of Minnesota and Wisconsin. *Ecological Monographs*, **8**, 152–245.

Kleman, J. (1986). The spectral reflectance of stands of Norway spruce and Scotch pine, measured from a helicopter. *Remote Sensing of Environment*, **20**, 253–65.

Knuchel, H. (1953). *Planning and Control in the Managed Forest*. Edinburgh: Oliver and Boyd.

Kohyama, T. (1989). Simulation of the structural development of warm-temperature rain forest stands. *Annals of Botany* (London), **63**, 625–34.

Kolesnikov, B. P. (1956). *Pinus sibirica* forests in the Far East. *Proceedings of the Far East Branch of the Academy of Sciences, Botanical Series, Vladivostok*, **2**(4), 261 pp. (In Russian.)

Kolesnikov, B. P. & Smolonogov, E. P. (1960). On some appropriateness in regeneration of cedar forests in Zauralskoye Priobye region. In *Problems of Cedar*, pp. 21–31. Novosibirsk. (In Russian.)

Kolosov, P. A. & Liseev, A. A. (1987). Assessment method for probability distribution curve by short-period climate observations. *Problems of Ecological Monitoring and Ecosystem Modeling*, **10**, 225–37. Leningrad: Gidrometeoisdat. (In Russian.)

Koop, H. & Hilgen, P. (1987). Forest dynamics and regeneration mosaic shifts in unexploited beech (*Fagus sylvatica*) stands at Fontainebleau (France). *Forest Ecology and Management*, **20**, 135–50.

Koppen, W. (1931). *Grundriss der Klimakunde*. Berlin: W. De Gruyter.

Korotaev, A. A. (1987). Effect of soil temperature and moisture content on root growth in coniferous cultures. *Soviet Forest Science*, **1987**(2), 52–61.

Korsunov, V. M. & Vedrova, E. F. (1980). On degree of soil podzolic content of autonomous taiga landscapes. In *Genesis and Geography of Forest Soils*, pp. 104–23. Moscow: Science. (In Russian.)

Korsunov, V. M. & Vedrova, E. F. (1982). *Diagnostics of Soil Formation in Zonal Forest Soils*. Novosibirsk: Science, Siberian Branch. (In Russian.)

Korzukhin, M. D. (1980). Age dynamics of strongly edifying tree populations. *Problems of Ecological Monitoring and Ecosystem Modeling*, **3**, 162–78. Leningrad: Gydromet Publishers. (In Russian.)

Korzukhin, M. D. (1985). Some ontogenesis strategies of plant under competition. *Problems of Ecological Monitoring and Ecosystem Modeling*, **3**, 162–78. Leningrad: Gydromet Publishers. (In Russian.)

Korzukhin, M. D. (1986). An ecophysiological model of forest dynamics. *Problems of Ecological Monitoring and Ecosystem Modeling*, **9**, 259–76. Leningrad: Gydromet Publishers. (In Russian.)

Korzukhin, M. D., Antonovski, M. Ya. & Matskiavichus, V. K. (1989). Periodicity in multi-aged tree populations. *Problems of Ecological Monitoring and Ecosystem Modeling*, **12**, 122–38. Leningrad: Gydromet Publishers. (In Russian.)

Korzukhin, M. D. & Sedych, V. N. (1983). On background monitoring of West Siberian forests. *Problems of Ecological Monitoring and Ecosystem Modeling*, **6**, 122–30. Leningrad: Gidrometeoisdat. (In Russian.)

Korzukhin, M. D., Sedych, V. N. & Ter-Mikaelian, M. T. (1987). Formulation of a predicting model of age-regeneration forest dynamics. *Proceedings of the Siberian Branch of the Academy of Sciences, Biological Series*, **3**(20), 58–67. (In Russian.)

Korzukhin, M. D., Sedych, V. N. & Ter-Mikaelian, M. T. (1988). Application of forecast model of age-regeneration dynamics to *Pinus sibirica* forests in central Ob region. *Ibid.*, 34–44. (In Russian.)

Korzukhin, M. D. & Ter-Mikaelian, M. T. (1982). Competition for light and dynamics of model trees independently distributed on surface. *Problems of Ecological Monitoring and Ecosystem Modeling*, **5**, 242–8. Leningrad: Gydromet Publishers. (In Russian.)

Korzukhin, M. D. & Ter-Mikaelian, M. T. (1987). Optimal development of a plant taking into account defense and competition. *Problems of Ecological Monitoring and Ecosystem Modeling*, **10**, 244–55. Leningrad: Gydromet Publishers. (In Russian.)

Koshkarova, V. L. (1986). *Seed Floras of Peat Lands of Siberia*. Novosibirsk: Science, Siberian Branch. (In Russian.)

Koski, V. (1973). On self-pollination, genetic load, and subsequent inbreeding in conifers. *Communicationes Instituti Forestalis Fenniae*, **78**, 1–42.

Koski, V. & Tallquist, R. (1978). Results of long-time measurements of the quality of flowering and seed crop of trees. *Folia Forestali*, **364**, 1–60.

Kotscheev, A. L. (1955). *Paludification of Fells and Measures Against It*. Moscow: Publication of Science Academy of USSR. (In Russian.)

Kovalev, P. V., Klenov, B. M. & Arslanov, H. A. (1972). Problems of radiocarbon dating of organic matter of soddy podzolic soils with second humus horizon in Tomsk Priobye. *Proceedings of Siberian Branch of Science Academy of USSR: Ser. Biol.*, **3**(15), 6–9. (In Russian.)

Kozin, E. K. (1982). About the periodicity in development of Sihote-Ahlin natural forests. *Forestry* (Russian), N3, 24–31. (In Russian.)

Kozlowski, T. T. (1971a). *Growth and Development of Trees, Vol. I. Seed Germination, Ontogeny and Shoot Growth*. New York: Academic Press.

Kozlowski, T. T. (1971b). *Growth and Development of Trees, Vol.II, Cambial Growth, Root Growth and Reproductive Growth*. New York: Academic Press.

Kramer, P. J. & Kozlowski, T. T. (1979). *Physiology of Woody Plants*. New York: Academic Press.

Krasny, M. E., Vogt, K. A. & Zasada, J. C. (1984). Root and shoot biomass and mycorrhizal development of white spruce seedlings naturally regenerating in interior Alaska floodplain communities. *Canadian Journal of Forest Research*, **14**(4), 554–8.

Krasny, M. E., Vogt, K. A. & Zasada, J. C. (1988). Establishment of four Salicaceae species on riverbars in interior Alaska. *Holarctic Ecology*, **11**, 210–9.

Krasny, M. E., Zasada, J. C. & Vogt, K. A. (1988). Adventitious rooting of fur Salicaceae species along the Tanana River, Alaska. *Canadian Journal of Botany*, **66**, 2579–98.

Krause, H. H. (1981). *Factorial experiments with nitrogen, phosphorus, and potassium in spruce and fir stands of New Brunswick*: 10-year results. Canadian Forestry Service, Maritime Forest Research Center Information Report M-X-123.

Krebs, J. S. & Barry, R. G. (1970). The arctic front and the tunda–taiga boundary in Eurasia. *Geographical Review*, **60**, 548–54.

Krilov, G. V. (1961). *Forests of West Siberia*. Moscow: Publication of Science Academy of USSR. (In Russian.)

Kriuchkov, V. V.(1968a). Tussock tundras. *Botanicheskii Zhurnal*, **53**, 1716–30. (In Russian, quoted by Wein (1975).)

Kriuchkov, V. V. (1968b). Soil of the far north should be conserved. *Priroda*, **12**, 72–4. (In Russian, quoted by Wein (1975).)

Krugman, S. L., Stein, W. I. & Schmitt, D. M. (1974). Seed biology. In *Seeds of*

Woody Plants in the United States, ed. C. S. Schopmeyer, pp. 5–40. USDA Forest Service Agricultural Handbook 450.

Krummel, J. R., Gardener, R. H., Sugihara, G., O'Neill, R. V. & Coleman, P. R. (1987). Landscape patterns in a disturbed environment. *Oikos*, **48**, 321–4.

Krussman, G. (1971). *Handbuch der Nadelhölze*. Berlin: P. Parey. (In German.)

Krylov, G. V. (1957). The nature of western Siberian forests and recommendations for their proper utilisation and regeneration. In *Investigations in Siberian Forestry*, ed. Anonymous, pp. 136–46. Novosibirsk: Nauka. (In Russian.)

Krylov, G. V., Talantsev, N. K. & Kozakova, N. F. (1983). *The Cedar Pine*. Moscow: Lesnaya promyshlenost. (In Russian.)

Kryuchkov, V. V. (1973). The effect of permafrost on the northern tree line. In *Permafrost: Proceedings of the Second International Conference (USSR Contribution)*, pp. 136–8. Washington, D.C.: National Academy of Science.

Kuchler, A. W. (1978). Natural vegetation map. In *Goude's World Atlas*, 15th ed., ed. E. B. Espenshade, Jr. & J. L. Morrison, Chicago: Rand McNally & Co.

Kulagin, Y.Z. (1972). Problems of the adaptation of trees to northern conditions. *Soviet Journal of Ecology*, **3**, 485–7.

Kull, K. & Kull, O. (1984). Ecophysiological model of spruce growth. *Proceedings of the Estonian Academy of Sciences, Biological Series*, **33**, 268–77. (In Russian.)

Kullman, L. (1983). Past and present treeline of different species in the Handölan Valley, Central Sweden. *Nordicana*, **47**, 25–45.

Kullman, L. (1985). Late Holocene reproductional patterns of *Pinus sylvestris* and *Picea abies* at the forest limit in central Sweden. *Canadian Journal of Botany*, **64**, 1682–90.

Kullman, L. (1987a). Sequences of Holocene forest history in the Scandes, inferred from megafossil *Pinus sylvestris*. *Boreas*, **16**, 21–6.

Kullman, L. (1987b). Little Ice Age decline of cold marginal *Pinus sylvestris* forest in the Swedish Scandes. *New Phytologist*, **106**, 567–84.

Kullman, L. (1987c). Long-term dynamics of high altitude populations of *Pinus sylvestris* in the Swedish Scandes. *Journal of Biogeography*, **14**, 1–8.

Kutzbach, J. E. & Guetter, P. J. (1986). The influence of changing orbital parameters and surface boundary conditions on climatic simulations for the past 18 000 years. *Journal of the Atmospheric Sciences*, **43**, 1726–59.

Kuzmina, M. S. (1949). Evolution of vegetation in mires of Barabinsk lowland in relation to their drainage and natural drying. In *Transactions of Jubilee session honoring 100 years from the birth of V. V. Dokuchaev*, pp. 588–96. Moscow/Leningrad: Publication of Science Academy of USSR. (In Russian.)

Kuznetsov, N. I. (1915a). Description of vegetation over Narim region in Tomsk gubernia. *Transactions of Soil-botanical Expedition*, 2: Botanical investigations in 1911, St. Petersburg, 162 pp. (In Russian.)

Kuznetsov, N. I. (1915b). Results of investigations of soils and vegetation in middle part of Tomsk gubernia. *Transactions of Soil-botanical Expedition*, 2: Botanical investigations in 1912, St. Petersburg, 48 pp.

Laessle, A. M. (1965). Spacing and competition in natural stands of sand pine. *Ecology*, **46**, 65–72.

Lag, J. (1959). Influence of forest stand and ground cover vegetation on soil formation. *Agrochimica*, **4**(1), 72–7.

Lagerberg, T. & Sjörs, H. (1972). *Trädkännedom*. I & II. Ibid. (In Swedish.)

Laine, K. & Henttonen, H. (1982). The role of plant production in microtine cycles in northern Fennoscandia. *Oikos*, **40**, 407–18.

Lakari, O. J. (1921). Tutkimuksia mantymetsien uudistumisvuosista Etela-ja Keski-Suomessa. Referat: Studien uber die Verjungungsjahre der Fichtenwin Sud- und Mittelfinnland. *Communicationes Instituti Forestalis Fenniae*, **4**, 1–58.

Lamb, H. F. (1980). Late Quaternary vegetational history of southeastern Labrador. *Arctic and Alpine Research*, **12**, 117–35.

Lamb, H. F. (1985). Palynological evidence for postglacial change in the position of tree limit in Labrador. *Ecological Monographs*, **55**, 241–58.

Lamb, H. H. (1977). *Climate, Present, Past and Future*, Vol. 2: *Climatic History and the Future*. London: Methuen.

Larcher, W. (1975). *Physiological Plant Ecology*. New York: Springer-Verlag.

La Roi, G. H. (1967). Ecological studies in the boreal spruce–fir forests of the North American taiga. *Ecological Monographs*, **37**, 229–53.

La Roi, G. H. & Stringer, M. H. L. (1976). Ecological studies in the boreal spruce–fir forests of the North American taiga. II. Analysis of the bryophyte flora. *Canadian Journal of Botany*, **54**, 619–43.

Larsen, J. A. (1965). The vegetation of Ennadai Lake area, N.W.T.: Studies in subarctic and arctic bioclimatology. *Ecological Monographs*, **35**, 37–59.

Larsen, J. A. (1974). Ecology of the northern continental forest border. In *Arctic and Alpine Environments*, ed. J. D. Ives & R. G. Barry, pp. 341–69. London: Methuen.

Larsen, J. A. (1980). *The Boreal Ecosystem*. New York: Academic Press.

Larsen, J. A. (1989). *The Northern Forest Border in Canada and Alaska. Biotic Communities and Ecological Relationships*. Ecological Studies, 70. New York: Springer-Verlag.

Larson, B. C. & Oliver, C. D. (1979). Forest dynamics and fuelwood supply of the Stehekin Valley, Washington. *Proceedings of the Second Conference of Scientific Researchers in the National Parks Service, San Francisco, CA*, 1979, pp. 276–92. U.S. Department of Interior, Parks Service, Volume 10.

Lashof, D. A. (1989). The dynamic greenhouse: feedback processes that may influence future concentrations of atmospheric trace gases and climatic change. *Climatic Change*, **14**, 213–42.

Lavrenko, E. M. & Sochava, V. B. ed. (1954). *Geobotanicheskoy karte SSSR*. Moscow. (Reproduced by Tikhomirov (1960).)

Lawrence, W. T. & Oechel, W. C. (1983a). Effects of soil temperature on the carbon exchange of taiga seedlings. I. Root respiration. *Canadian Journal of Forest Research*, **13**, 840–9.

Lawrence, W. T. & Oechel, W. C. (1983b). Effects of soil temperature on the carbon exchange of taiga seedlings. II. Photosynthesis, respiration, and conductance. *Canadian Journal of Forest Research*, **13**, 850–9.

Leak, W. B. (1964). An expression of diameter distribution for unbalanced, uneven-aged stands and forests. *Forest Science*, **10**, 39–50.

Leak, W. B. (1973). *Species and structure of a virgin hardwood stand in New Hampshire*. USDA Forest Service Research Note NE-181.

Leak, W. B. (1975). Age distribution in virgin red spruce and northern hardwoods. *Ecology*, **56**, 1451–4.

Lee, S. C. (1924). Factors controlling forest succession at Lake Itasca, Minnesota. *Botanical Gazette*, **78**, 129–74.

Leemans, R. (1986). Structure of the primaeval coniferous forest of Filby. In: *Forest Dynamics Research in Western and Central Europe*, ed. J. Fanta, pp. 221–30. Wageningen: PUDOC.

Leemans, R. (1989). Description and simulation of stand structure and dynamics in some Swedish forests. *Comprehensive Summaries of Uppsala Dissertations from the Faculty of Science*, **221**, 1–44.

Leemans, R. (1990). A listing of the biological configuration of the BOFORS (Version 1a) model. Science and Technology Library, The University of Virginia, Charlottesville, Virginia.

Leemans, R. (1991*a*). Canopy gaps and establishment patterns of spruce (*Picea abies* (L.) Karst.) in two old-growth coniferous forests in a central Sweden. *Vegetatio*, (in press).

Leemans, R. (1991b). Sensitivity analysis of a forest succession model. *Ecological Modelling*, **53**, 1–16.

Leemans, R. & Cramer, W. (1990). *The IIASA database for mean monthly values of temperature, precipitation, and cloudiness of a global terrestrial grid*. IIASA Working Paper, International Institute for Applied Systems Analysis, Laxenburg, Austria, Working Paper WP-90-41.

Leemans, R. & Prentice, I. C. (1987). Description and simulation of a tree-layer composition and size distributions in a primaeval *Picea–Pinus* forest. *Vegetatio*, **69**, 147–56.

Leemans, R. & Prentice, I. C. (1989). FORSKA: A General Forest Succession Model. *Meddelanden*, **2**, 1–45. Växtbiologiska institutionen, Uppsala, Sweden.

Lees, J. (1964). *A test of harvest cutting methods in Alberta's spruce-aspen region*. Canadian Department of Forestry, Forest Research Branch, Publication 1042.

Lefkovitch, L. P. (1965). The study of population growth in organisms grouped by ages. *Biometrics*, **21**, 1–18.

Legere, A. & Payette, S. (1981). Ecology of a black spruce (*Picea mariana*) clonal population in the hemiarctic zone, northern Quebec: population dynamics and spatial development. *Arctic and Alpine Research*, **13**(3), 261–76.

Lehto, J. (1956). Tutkimusksia ma annyn luontaisesta uudistumisesta Etela a-Suomen kangasmailla. Summary: Studies on the natural reproduction of Scots pine on the upland soils of southern Finland. *Acta Forestalia Fennica*, **66**, 1–106.

Lehtonen, J. (1977). Recovery and development of birch forests damaged by *Epirrita automna* in Utsjoski area, North Finland. *Report from the Kevo Subarctic Research Station*, **20**, 35–9.

Leibundgut, H. (1982). *Europaische Urwalder der Bergstufe*. Bern: P. Haupt.

Leibundgut, H. (1984). *Unsere Waldebäume*. Switzerland: Frauenfeld, Huber. (In German.)

Lenhart, J. D. & Clutter, J. L. (1971). *Cubic-foot yield tables for old-field loblolly pine plantations in the Georgia Piedmont.* Georgia Forest Research Council Report 22, Series 3.

Leslie, P. H. (1945). On the use of matrices in certain population mathematics. *Biometrika*, **33**, 183–212.

Levin, S. A. & Paine, R. T. (1974). Disturbance, patch formation and community structure. *Proceedings of the National Academy of Sciences*, **71**, 2744–7.

Levitskii, L. N. (1910). On the problem of mire evolution of the Amur district. *Soil Science*, **1**, 82–90. (In Russian.)

Li, X. & Strahler, A. H. (1985). Geometric-optical modeling of a conifer forest canopy. *IEEE Transactions on Geoscience and Remote Sensing*, **GE-23**, 705–21.

Li, X. & Strahler, A. H. (1986). Geometric–optical bidirectional reflectance modeling of a conifer forest canopy. *IEEE Transactions on Geoscience and Remote Sensing*, **GE-24**, 906–18.

Liebig, J. (1840). *Chemistry and its Application to Agriculture and Physiology.* London: Taylor and Walter.

Lindeman, R. L. (1942). The trophic-dynamic aspect of ecology. *Ecology*, **23**, 399–418.

Lindholm, T. & Tiainen, I. (1982). Dispersal and establishment of an introduced conifer, the Siberian fir *Abies sibirica*, in a nemoral forest in Southern Finland. *Annales Botanica Fennica*, **19**, 235–45.

Linell, K. A. (1973). Long-term effects of vegetative cover on permafrost stability in an area of discontinuous permafrost. In *Permafrost: Proceedings of the Second International Conference (North American Contribution)*, pp. 688–93. Washington, DC.: National Academy of Science.

Liss, O. L. & Berezina, N. A. (1981). *Mires of West Siberian Plain.* Moscow: Moscow State University. (In Russian.)

Little, E. L. (1971). *Atlas of United States trees.* US Department of Agriculture Forest Service, Miscellaneous Publication No. 1146.

Lonsdale, W. M. & Watkinson, A. R. (1983). Plant geometry and self-thinning. *Journal of Ecology*, **71**, 285–97.

Lorimer, C. G. (1977). The presettlement forest and the natural disturbance cycle of northeastern Maine. *Ecology*, **58**, 141–8.

Lorimer, C. G. (1983). Tests of age-independent competition indices for individual trees in natural hardwood stands. *Forest Ecology and Management*, **6**, 343–60.

Lorimer, C. G. & Gough, W. R. (1982). *Number of days per month of moderate and extreme drought in Northeastern Wisconsin, 1864–1979.* University of Wisconsin Forestry Research Note 248, Madison, Wisconsin.

Loucks, O. L. (1970). Evolution of diversity, efficiency, and community stability. *American Zoologist*, **10**, 17–25.

Louisier, J. D. & Parkinson, D. (1978). Litter decomposition in a cool temperate forest. *Canadian Journal of Botany*, **54**, 419–36.

Lovejoy, S. (1982). Area perimeter relation for rain and cloud areas. *Science*, **216**, 185–7.

Lovelock, J. (1988). *The Ages of GAIA. A Biography of Our Living Earth.* New York: W. W. Norton, & Co.

Lucarotti, C. (1976). *Post-fire change in mycoflora species and mesofauna populations in lichen-woodland soils, Schefferville, Québec.* Montréal: M.Sc. Thesis. McGill University.

Ludwig, D., Jones, D. D. & Holling, C. S. (1978). Qualitative analysis of insect outbreak systems: the spruce budworm and forest. *Journal of Animal Ecology*, **44**, 315–32.

Lunardini, V. J. (1981). *Heat Transfer in Cold Climates.* New York: Van Nostrand Reinhold.

Lutz, H. F. (1956a). *Ecological effects of forest fires in the interior of Alaska.* USDA Technical Bulletin Number 1133.

Lutz, H. F. (1956b). *Effects of red squirrels on crown form of black spruce in interior Alaska.* USDA Forest Service, Alaska Forest Research Center Technical Note No. 42. Alaska Forest Research Center, Juneau, AK.

Lvov, Yu.A. (1976). Characteristics and mechanisms for paludification of territory in the Tomsk region. In *Theory and Practice of Forest Peat Land Science and Hydroforest Amelioration*, pp. 36–44. Krasnoyarsk: Forest and Wood Institute, Siberian Branch of Science Academy of USSR. (In Russian.)

MacDonald, G. M. (1987). Postglacial vegetation history of the MacKenzie River Basin. *Quaternary Research*, **28**, 245–62.

MacDonald, G. M. & Cwynar, L. S. (1985). A fossil pollen based reconstruction of the late Quaternary history of lodgepole pine (*Pinus contorta* spp. *latifolia*) in the western interior of Canada. *Canadian Journal of Forest Research*, **15**, 1039–44.

Machaniček, J. (1973). Economic collection of cones of forest conifers on the basis of preceding estimation of cone crop. In *Seed Problems*. IUFRO Symposium on Seed Processing, Bergen, Norway, 1973, Paper No. 12. Royal College of Forestry. Stockholm, Sweden.

MacKay, W. A. (1967). *The Great Canadian Skin Game.* Toronto: Macmillan of Canada.

MacLean, D. A. & Wein, R. W. (1976). Biomass of jack pine and mixed hardwood stands in northeastern New Brunswick. *Canadian Journal of Forest Research*, **6**, 441–47.

MacLean, D. A., Woodley, S. J., Weber, M. G. & Wein, R. W. (1983). Fire and nutrient cycling. In *The Role of Fire in Northern Circumpolar Ecosystems*, ed. R. W. Wein & D. A. MacLean, pp. 111–32. New York: Wiley.

Mahendrappa, M. K. & Salonius, P. O. (1982). Nutrient dynamics and growth response in a fertilized black spruce stand. *Soil Science Society of America Journal*, **46**, 127–333.

Maikawa, E. & Kershaw, K. A. (1976). Studies on lichen-dominated systems. XIX. The postfire recovery sequence of black spruce-lichen woodland in the Abitau Lake Region. *Canadian Journal of Botany*, **54**, 2679–87.

Mäkelä, A. (1986). Implications of the pipe model theory on dry matter partitioning and height growth in trees. *Journal of Theoretical Biology*, **123**, 103–20.

Mäkelä, A. (1988). *Models of pine stand development: an ecophysiological systems analysis.* Research Notes 62. University of Helsinki, Department of Silviculture.

Mäkelä, A. & Hari, P. (1986). Stand growth model based on carbon uptake and allocation in individual trees. *Ecological Modeling*, **33**, 205–29.

Malik, L. K. (1978). *Hydrological Problems of Transformation of Nature in West Siberia*. Moscow: Science. (In Russian.)

Mälkönen, E. (1971). Fertilizer treatment and seed crop of *Picea abies*. *Communicationes Instituti Forestalis Fenniae*, **73**(4), 1–16.

Malmström, C. (1932). Om faren for skogsmarkens försumping in Norland. *Medd. Statens skogsförsöksanst* (Stockholm), **26**, 1–126.

Malyanov, A. P. (1939). Soil paludification in clear fells. *Soil Science*, **5**, 67–77. (In Russian.)

Manabe, S. & Stouffer, R. J. (1980). Sensitivity of a global climate model to an increase of CO_2 concentration in the atmosphere. *Journal of Geophysical Research*, **85**, 5529–54.

Manabe, S. & Wetherald, R. T. (1987). Large-scale changes of soil wetness induced by an increase in atmospheric carbon dioxide. *Journal of Atmospheric Science*, **44**, 1211–35.

Mandelbrot, B. B. (1977). *Fractals, Form, Chance and Dimension*. San Francisco: W. H. Freeman and Co.

Mandelbrot, B. B. (1983). *The Fractal Geometry of Nature*. San Francisco: W. H. Freeman and Co.

Manko, Ju.I. (1987). *The Ajanensian Spruce*. Leningrad: Nauka. (In Russian.)

Manko, Ju.I. & Voroshilov, V. P. (1978). *Spruce forests in Kamchatka*. Moscow: Nauka. (In Russian.)

Marchand, P. J. & Chabot, B. F. (1978). Water relations of tree-line plant species on Mount Washington, New Hampshire. *Arctic and Alpine Research*, **10**, 105–16.

Marie-Victorin, Fr. (1938). Phytogeographical problems of eastern Canada. *American Midland Naturalist*, **19**, 489–558.

Marks, P. L. (1974). The role of pin cherry (*Prunus pensylvanica* L.) in the maintenance of stability in northern hardwood ecosystems. *Ecological Monographs*, **44**, 73–88.

Marschner, F. J. (1974). *The original vegetation of Minnesota*. Map, US Department of Agriculture Forest Service.

Marsden, M. A. (1983). Modelling the effect of wild fire frequency on forest structure and succesion in the Northern Rocky Mountains. *Journal of Environmental Management*, **16**(1), 45–62.

Martell, D. L. (1980). The optimal rotation of a flammable forest stand. *Canadian Journal of Forest Research*, **10**, 30–4.

Martin, C. (1978). *Keepers of the Game: Indian-animal Relations and the Fur Trade*: Berkeley: University of California Press.

Mathieu, C., Payette, S. & Morin, H. (1987). Chronologie ^{14}C et développement des combes à neige du Lac a l'Eau Claire, Québec nordique. *Géographie physique et Quaternaire*, **41**, 97–108.

Matlack, G. (1989). Secondary dispersal of seeds across snow in *Betula lenta*, a gap colonizing species. *Journal of Ecology*, **77**, 853–69.

Maxwell, B. D. (1990). *The population dynamics and growth of salmonberry* (Rubus spectabilis) *and thimbleberry* (R. parviflorus). PhD thesis, Oregon State University, Corvalis, OR.

May, R. M. (1976). Simple mathematical models with very complicated dynamics. *Nature*, **261**, 459–67.

510   *References*

Maycock, P. F. & Curtis, J. T. (1960). The phytosociology of boreal conifer–hardwood forests of the Great Lakes Region. *Ecological Monographs*, **30**, 1–35.

Maycock, P. F. & Matthews, B. (1966). An arctic forest in the tundra of northern Ungava, Québec. *Arctic*, **19**, 114–44.

Mayer, H. (1909). *Waldbau auf naturgesetzlicher Grundlage*. Berlin. (In German.)

Mayer, H., Neumann, M. & Summer, H. G. (1980). Bestandsaufbau und Verjüngungs Dynamik unter dem Einfluß naturlicher Wilddichten im Kroatischen Urwald Reservat Cordova Uwala, Plitvicer Seen. *Schweizerischen Zeitschrift Forstwesen*, **131**, 45–70.

McAndrews, J. H. & Samson, G. (1977). Analyse pollinique et implications archéologiques et géomorphologiques, lac de la Hutte Sauvage (Mushuau Nipi), Nouveau-Québec. *Géographie physique et Quaternaire*, **31**, 177–83.

McBeath, J. H. (1981). Rust disease on white spruce in Alaska. *Agroborealis*, **13**, 41–3.

McClaugherty, C. A., Pastor, J., Aber, J. D. & Melillo, J. M. (1985). Forest litter decomposition in relation to soil nitrogen dynamics and litter quality. *Ecology*, **66**, 266–75.

McCune, B. & Allen, T. F. H. (1984a). Will similar forests develop in similar sites? *Canadian Journal of Botany*, **63**, 367–75.

McCune, B. & Allen, T. F. H. (1984b). Forest dynamics in the Bitterroot Canyons, Montana. *Canadian Journal of Botany*, **63**, 377–83.

McDonough, W. T. (1979). *Quaking aspen-seed germination and early seedling growth*. USDA Forest Service Research Paper INT-234. Intermountain Forest and Range Experiment Station, Ogden, UT.

McFadden, G. & Oliver, C. D. (1988). Three-dimensional forest growth model relating tree size, tree number, and stand age: relation to previous growth models and to self-thinning. *Forest Science*, **34**, 662–76.

McGee, C. E. & Della-Bianca, L. (1967). *Diameter distributions in natural yellow-poplar stands*. USDA Forest Service Research Paper SE-25. Southeast Forest Experiment Station, Asheville, NC.

McLeod, J. M. (1979). Discontinuous stability in a sawfly life system and its relevance to pest management strategies. In *Selected Papers in Forest Entomology from the XV International Congress of Entomology*, ed. W. E. Walters, pp. 68–81. US Forest Service.

McMurtrie, R. & Wolf, L. (1983). A model of competition between trees and grass for radiation, water and nutrients. *Annals of Botany*, **52**, 449–58.

McNamee, P. J., McLeod, J. M. & Holling, C. S. (1981). The structure and behavior of insect/forest systems. *Researches in Population Ecology*, **23**, 280–98.

Meades, W. J. (1983). Heathlands. In *Biogeography and Ecology of the Island of Newfoundland*, ed. G. R. South, pp. 267–318. The Hague: Dr. W. Junk Publishers.

Melechov, I. S. & Goldobina, P. V. (1947). Change in the soil cover in relation to concentrated fekls. *Proceedings of Research Works of Archangelsk Forest Institute*, **9**, 119–40. (In Russian.)

Melillo, J. M., Aber, J. D. & Muratore, J. F. (1982). Nitrogen and lignin control of hardwood leaf litter decomposition dynamics. *Ecology*, **63**, 621–6.

Melin, E. (1930). Biological decomposition of some types of litter from North American forests. *Ecology*, **11**, 72–101.

Menabe, S. & Wetherald, R. T. (1986). Reduction in summer soil wetness induced by an increase in atmospheric carbon dioxide. *Science*, **232**, 626–8.

Meyer, H. A. (1941). A mathematical expression for height curves. *Journal of Forestry*, **38**, 415–20.

Meyer, H. A. (1952). Structure, growth and drain in balanced unevenaged forests. *Journal of Forestry*, **50**, 85–92.

Mezentsev, V. S. & Karnatsevich, I. V. (1969). *Moistening of West Siberian Plain*. Leningrad: Gidrometeoizdat. (In Russian.)

Micheev, V. S. & Dibtsov, B. N. (1975). Landscapes of Soswa Priobye. In *Soswa Priobye* (*Essays of Nature and Economy*, pp. 353–404. Irkutsk: Institute of Geography of Siberian and Far East Branches of Science Academy of USSR. (In Russian.)

Miller, H. G., Cooper, J. M., Miller, J. D. & Pauline, O. J. L. (1979). Nutrient cycles in pine and their adaptation to poor soils. *Canadian Journal of Forest Research*, **9**, 19–26.

Millet, J. & Payette, S. (1987). Influence des feux sur la déforestation des îles centrales du Lac á l'Eau Claire, Québec nordique. *Géographie physique et Quaternaire*, **41**, 79–86.

Mitchell, H. L. & Chandler, R. F. (1939). *The nitrogen nutrition and growth of certain deciduous trees of northeastern United States*. Black Rock Forest Bulletin No. 11.

Mitchell, J. F. B. (1983). The seasonal response of a general circulation model to changes in CO_2 and sea temperatures. *Quarterly Journal of the Royal Meteorological Society*, **109**, 113–52.

Mitchell, K. J. (1969). *Simulation of the growth of even-aged stands of White Spruce*. Yale University Bulletin 75. Yale University, School of Forestry.

Mitchell, K. J. (1975). Dynamics and simulated yield of Douglas-fir. *Forest Science Monographs*, **17**, 1–39.

Mitscherlich, G. (1970). *Wald, Wachstum, Umwelt: I. Form und Wachstum von Baum und Bestand*: Frankfurt: Sauerlander's.

Mladenoff, D. J. (1985). *Dynamics of soil seed banks, vegetation, and nitrogen availability in treefall gaps*. PhD thesis, University of Wisconsin, Madison, Wisconsin.

Mladenoff, D. J. (1987). Dynamics of nitrogen mineralization and nitrification in hemlock and hardwood treefall gaps. *Ecology*, **68**, 1171–80.

Mladenoff, D. J. (1990). The relationship of the soil seed bank and vegetation layers to treefall gaps in a northern hardwood–hemlock forest. *Canadian Journal of Botany*, **68**, 2714–21.

Mladenoff, D. J. & Howell, E. A. (1980). Vegetation change on the Gogebic Iron Range (Iron Country, Wisconsin) from the 1860s to the present. *Transactions of the Wisconsin Academy of Sciences, Arts, and Letters*, **68**, 74–89.

Mohler, C. L., Marks, P. L. & Sprugel, D. G. (1978). Stand structure and allometry of trees during self-thinning of pure stands. *Journal of Ecology*, **66**, 599–614.

Moloney, K. A. (1986). Wave and non-wave regeneration process in a subalpine *Abies balsamea* forest. *Canadian Journal of Botany*, **64**, 341–9.

Monserud, R. A. (1976). Simulation of forest tree mortality. *Forest Science*, **22**, 438–44.

Monsi, M. & Saeki, T. (1953). Über den Lichtfaktor in den Pflanzengelsellschaften und seine Bedeutung für die Stoffproduktion. *Japanese Journal of Botany*, **14**, 22–52.

Mooney, H. A. & Gulmon, S. L. (1982). Constraints on leaf structure and function in reference to herbivory. *BioScience*, **32**, 198–206.

Moore, B., III, Melillo, J. M., Peterson, B. J., Gildea, M. P. & Vorosmarty, C. J. (1986). Modeling of global biogeochemical cycles. *Proceedings III International Congress of Ecology*, Syracuse, N. Y., August 10–16, p. 243.

Moore, J. M. & Wein, R. W. (1977). Viable seed populations by soil depth and potential site recolonization after disturbance. *Canadian Journal of Botany*, **55**, 2408–12.

Moore, P. D. (1982). Beneath the blanket bogs of Britain. *Natural History*, **91**(11), 49–54.

Moore, P. D. & Bellamy, D. J. (1974). *Peatlands*. London: Elek.

Moore, T. R. (1980). The nutrient status of subarctic woodland soils. *Arctic and Alpine Research*, **12**, 147–60.

Moore, T. R. (1981). Controls on the decomposition of organic matter in subarctic spruce–lichen woodland soils. *Soil Science*, **131**, 107–13.

Moore, T. R. (1984). Litter decomposition in a subarctic spruce–lichen woodland, eastern Canada. *Ecology*, **65**, 299–308.

Moore, T. R. & Verspoor, E. (1973). Aboveground biomass of black spruce stands in subarctic Quebec. *Canadian Journal of Forest Research*, **3**, 596–98.

Moran, J. M. (1972). *An analysis of periglacial climatic indicators of late-glacial time in North America*. PhD dissertation, University of Wisconsin, Madison, WI.

Morin, A. & Payette, S. (1984). Expansion récente du mélèze à la limite des forêts (Québec nordique). *Canadian Journal of Botany*, **62**, 1404–8.

Morneau, C. & Payette, S. (1989). Post-fire lichen-spruce woodland recovery at the limit of the boreal forest in northern Québec. *Canadian Journal of Botany*, **67**, 2770–82.

Morozov, G. F. (1947). *Principles of Forest Ecology*. Moscow: Gosizdat. (In Russian.)

Morris, D. M. & Farmer, R. F., Jr. (1985). Species interactions of seedling populations of *Populus tremuloides* and *Populus balsamifera*. Canadian Journal of Forest Research, **15**, 595.

Morris, R. F. (1963). The dynamics of epidemic spruce budworm populations. *Memoirs of the Entomological Society of Canada*, **31**, 1–332.

Morrison, M. E. S. (1955). The water balance of the raised bog. *Irish Naturalist Journal*, **11**, 303–8.

Moser, J. W. (1972). Dynamics of an uneven-aged forest stand. *Forest Science*, **18**, 184–91.

Moskvin, Yu.P. (1974). Investigations of the thawing of the active soil layer in the permafrost zone. *Soviet Hydrology*, **1974**(5), 323–8.

Moss, E. H. (1953). Marsh and bog vegetation in Northwestern Alberta. *Canadian Journal of Botany*, **31**(4), 448–70.

Múnro, D. D. (1974). Forest growth models: A prognosis. In *Growth Models for Tree and Stand Simulation*, ed. J. Fries, pp. 7–21. Research Note 30, Royal College of Forestry, Stockholm.

Mutch, R. W. (1970). Wildland fires and ecosystems – a hypothesis. *Ecology*, **51**, 1046–51.

Myers, V. I. (1983). Remote sensing applications in agriculture. In *Manual of Remote Sensing*, 2nd ed., Vol. II, pp. 2111–228. American Society of Photogrammetry and Remote Sensing, Falls Church, VA.

Nadelhoffer, K. J., Aber, J. D. & Melillo, J. M. (1983). Leaf-litter production and soil organic matter dynamics along a nitrogen-availability gradient in southern Wisconsin (U.S.A.). *Canadian Journal of Forest Research*, **13**, 12–21.

Naiman, R. J., Johnston, C. A. & Kelley, J. C. (1988). Alteration of North American streams by beaver. *BioScience*, **38**, 753–62.

Naiman, R. J., Melillo, J. M. & Hobbie, J. E. (1986). Ecosystem alteration of boreal forest streams by beaver (*Castor canadensis*). *Ecology*, **67**, 1254–69.

NASA. (1984). *Earth Observing System. Science and Mission Requirements Working Group Report*, Vol. I. NASA TM 86129. Goddard Space Flight Center, Greenbelt, MD 20771.

NASA. (1986). *Earth System Science Overview*. Earth System Sciences Committee, NASA Advisory Council. NASA, Washington, D.C. 20546.

National Geographic Society (NGS). (1989). Endangered earth map supplement. *National Geographic*, December 1988, 174:910A.

Natural Regimes of the Middle Taiga of West Siberia. (1977). Novosibirsk: Science, Siberian Branch.

Nazarov, A. D., Rasskazov, N. M., Udododv, P. A. & Shwartsev, S. L. (1977). Hydrological conditions for mire formation. In *Scientific Prerequisites for Exploration of Mires in West Siberia*, pp. 93–103. Moscow: Science. (In Russian.)

Neiland, B. J., Zasada, J. C., Densmore, R., Masters, M. A. & Moore, N. (1981). *Investigations of techniques for large-scale reintroduction of willows in Arctic Alaska*. Final Report for Project Taps/41, Task Order 27. School of Agriculture and Land Resources Management, University of Alaska, Fairbanks, AK.

Nekrasova, T. P. (1960). Biological aspects of the cedar pine crop management in the Tomsk region. In *The Cedar Pine*, pp. 151–9. Novosibirsk: Nauka (In Russian.)

Nekrasova, T. P. (1962). About the seeding of the Siberian cedar pine. *Papers of the Tomsk Museum of Local Lore*, **6**(1), 25–9. (In Russian.)

Nemtchinov, A. A. (1957). The mire formation process and its exhibition in soddy-podzolic zone. *Issue of works of the Central Museum of Soil Science of Science Academy of USSR*, **2**, 57–101. Moscow: Publication of Academy of Sciences of USSR. (In Russian.)

Nepomilueva, N. I. (1974). *Siberian Cedar Pine* (Pinus sibirica *Du Tour*.) *in the North-East European Part of the USSR*. Leningrad: Nauka. (In Russian.)

Neustadt, M. I. (1965). Some results of deposit investigations of Holocene. In *Paleography and Chronology of Upper Pleistocene and Holocene after the Data of the Radio-carbon Method*, pp. 112–32. Moscow: Science. (In Russian.)

Neustadt, M. I. (1971). Paludification of the West Siberian Plain is a world natural phenomenon. *Transactions of Science Academy of USSR: Ser. Geogr.*, **1**, 21–34. (In Russian.)

Neustadt, M. I. & Malik, L. K. (1980). The past, present and future of West Siberian bogs. *Nature*, **11**, 24–35. (In Russian.)

Newman, P. C. (1985). *Company of Adventurers*. New York: Penguin Books.

Newnham, R. M. (1964). *The development of a stand model for Douglas Fir*. PhD thesis, University of British Columbia, Vancouver.

Newnham, R. M. & Smith, J. M. P. (1964). Development and testing of stand models for Douglas Fir and Lodgepole Pine. *Forestry Chronicle*, **40**, 494–502.

Nichols, G. E. (1935). The hemlock–white pine–northern hardwood region of eastern North America. *Ecology*, **16**, 403–20.

Nichols, H. (1969). Chronology of peat growth in Canada. *Paleogeography, Paleoclimatology and Paleoecology*, **6**(1), 61–5.

Nichols, H. (1974). Arctic North American palaeoecology: the recent history of vegetation and climate deduced from pollen analysis. In *Arctic and Alpine Environment*, ed. J. D. Ives & R. G. Barry, pp. 638–67. London: Methuen.

Nichols, H. (1975). *Palynological and paleoclimatic study of the late Quaternary displacement of the boreal forest–tundra ecotone in Keewatin and Mackenzie, N.W.T., Canada.* Occasional paper No. 15, Institute of Arctic and Alpine Research. Boulder: University of Colorado.

Nickiforov, B. B., Melnickova, N. I. & Kolymtsev, V. A. (1981). Paludification of forest ecosystems in Karelian Zaonezhye. In *Anthropogenous Disturbances and Natural Variations in Ecosystems*, pp. 76–92. Moscow: Institute for Evolutionary Morphology and Ecology of Animals, Science Academy of USSR. (In Russian.)

Nickolskya, M. V. (1982). Paleobotanical and paleoclimatic reconstruction of Holocene of Taimir. In *Anthropogen of Taimir*, pp. 148–57. Moscow: Science. (In Russian.)

Nikonov, M. N. (1955). Regionalization of peat bogs in relation to their use in state economy. *Transactions of Forest Institute of Science Academy of USSR*, **31**, 49–63. (In Russian.)

Nilsson, S., Attebring, J. & Sallnä, O. (1987). Forestry Study at IIASA – experience on timber assessment studies for Europe. *Proceedings of ad hoc FAO/ECE/Finnida Meeting of Experts on Forest Resource Assessment*, pp. 369–87. Helsinki: Finnish Forest Research Institute, Bulletin 284.

Nilsson, S., Duinker, P. & Sallnä, O. (1990). *Forest Decline in Europe – Forest Potentials and Policy Implications*. Contract report to Forestry Canada, Hull, Quebec, Canada.

Noble, I. R. & Slatyer, R. O. (1978). The effect of disturbances on plant succession. *Proceedings of the Ecological Society of Australia*, **10**, 135–45.

Noble, I. R. & Slatyer, R. O. (1980). The use of vital attributes to predict successional changes in plant communities subject to recurrent disturbances. *Vegetatio*, **43**, 5–21.

Nordenskjöld, O. & Mecking, L. (1928). *The geography of polar regions*. American Geographical Society, Special Publication No. 8.

Norman, J. M. & Jarvis, P. (1974). Photosynthesis in Sitka spruce (*Picea sitchensis*(Bong.)Carr.). III. Measurements of canopy structure and interception of radiation. *Journal of Applied Ecology*, **11**, 375–98.

Norman, J. M. & Welles, J. M. (1983). Radiative transfer in an array of canopies. *Agronomy Journal*, **75**, 481–8.

Odum, E. P. (1969). The strategy of ecosystem development. *Science*, **164**, 262–70.

Oechel, W. C. & Lawrence, T. W. (1985). Taiga. In *Physiological Ecology of the North American Plant Communities*, ed. B. F. Chabot & H. A. Mooney, pp. 66–94. New York: Chapman & Hall.

Oechel, W. C. & Van Cleve, K. (1986). The role of bryophytes in nutrient cycling in the taiga. In *Forest Ecosystems in the Alaskan Taiga*, ed. K. Van Cleve, F. S. Chapin, P. W. Flanagan, L. A. Viereck & C. T. Dyrness, pp. 121–37. New York: Springer-Verlag.

Ohmann, L. F. & Grigal, D. F. (1979). *Early vegetation and nutrient dynamics following the 1971 little Sioux forest fire in northeastern Minnesota*. Forest Science Monographs 21.

Ohmann, L. F. & Grigal, D. F. (1981). Contrasting vegetation responses following two forest fires in northeastern Minnesota. *American Midland Naturalist*, **106**, 54–64.

Ohmann, L. F. & Grigal, D. F. (1985). Biomass distribution of unmanaged upland forests in Minnesota. *Forest Ecology and Management*, **13**, 205–22.

Oja, T. (1985). Simple adaptive model of plant growth. I. Description of model. *Transactions of the Estonian Academy of Sciences, Biological Series*, **4**, 54–61. (In Russian.)

Oker-Blöm, P. (1986). Photosynthetic radiation regime and canopy structure in modelled forest stands. *Acta Forestalia Fennica*, **197**, 44 pp.

Oksanen, J. & Ahti, T. (1982). Lichen-rich pine forest vegetation in Finland. *Annales botanici Fennici*, **19**, 275–301.

Oliver, C. D. (1978). *The development of northern Red Oak in mixed species in Central New England*. Yale University Bulletin 91. Yale University, School of Forestry.

Oliver, C. D. (1981). Forest development in North America following major disturbances. *Forest Ecology and Management*, **3**, 153–68.

Oliver, C. D. & Stephens, E. P. (1977). Reconstruction of a mixed-species forest in central New England. *Ecology*, **58**, 562–72.

Olson, J. S., Watts, J. A. & Allison, L. J. (1983). *Carbon in live vegetation of major world ecosystems*. Oak Ridge, Tennessee: Oak Ridge National Laboratory Technical Report ORNL-5862.

OMNR. (1978). *Forest Inventory Procedure for Ontario*. Toronto: Ontario Ministry of Natural Resources, Edition III (Edition IV in preparation).

O'Neill, R. V., DeAngelis, D. L., Waide, J. B. & Allen, T. F. H. (1986). *A Hierarchical Concept of the Ecosystem*. New Jersey: Princeton University Press.

O'Neill, R. V., Krummel, J. R., Gardner, R. H., Jackson, B., DeAngelis, D. L., Milne, B. T., Turner, M. G., Zygmunt, B., Christensen, S. W., Dale, V. H. & Graham, R. L. (1990). Indices of landscape pattern. *Landscape Ecology*, (in Press).

Oosting, H. J. (1956). *The Study of Plant Communities*. San Franciso: W. H. Freeman & Co.

O'Regan, W. G., Kourtz, P. & Nozaki, S. (1976). Bias in the contagion analog to fire spread. *Forest Science*, **22**(1), 61–8.

Orlov, V. I. (1963). The zonation of forests. In *West Siberia*, pp. 331–76. Moscow: Publication of Science Academy of USSR. (In Russian.)

Orlov, V. I. (1968). *A Course of the Development of the Forest-mire Zone in West Siberia*. Leningrad: Nedra. (In Russian.)

Orlov, V. I. (1975). *Analysis of Dynamics of Natural Conditions and Resources*. Moscow: Science. (In Russian.)

Ormsby, J. P. & Soffen, G. A. ed. (1989). Special issue on the Earth Observing System (EOS). *IEEE Transactions on Geoscience and Remote Sensing*, 2, 106–242.

Osborne, J. G. & Schumacher, F. X. (1935). The construction of normal yield and stand tables for even-aged timber stands. *Journal of Agricultural Research*, 51, 547–64.

Oshima, Y., Kimura, M., Ivaki, H. & Kuriora, S. (1958). Ecological and physiological studies on vegetation of Mt. Shimigare. Preliminary survey of the vegetation of Mt. Shigimare. *Botanical Magazine*, 71, 289–300. Tokyo.

Ototskii, P. V. (1906). Ground waters and forests, mainly over plains of middle latitudes. *Transactions of experimental forest management*, St. Petersburg, 4, 1–300. (In Russian.)

Overpeck, J. T. & Bartlein, P. J. (1989). Assessing the response of vegetation to future climate change: Response surfaces and paleoecological model validation. In *The Potential Effects of Global Climate Change on the United States*, ed. J. Smith & D. Tirpak, pp. 1–32. US Environmental Protection Agency.

Overpeck, J. T., Rind, D. & Goldberg, R. (1990). Climate-induced changes in forest disturbance and vegetation. *Nature*, 343, 51–3.

Owens, J. N. & Molder, M. (1977). Bud development in *Picea glauca*. II. Cone differentiation and early development. *Canadian Journal of Botany*, 55, 2746–60.

Owens, J. N. & Molder, M. (1979). Sexual reproduction of white spruce (*Picea glauca*). *Canadian Journal of Botany*, 57, 152–69.

Owens, J. M., Molder, M. & Langer, H. (1977). Bud development in *Picea glauca*. I. Annual growth cycle of vegetative buds and shoot elongation as they relate to date and temperature sums. *Canadian Journal of Botany*, 55, 2728–45.

Päivänen, J. (1982). Hakkuun ja lannoituksen vaikutus vanhan metsäojitusalueen vesitalouteen. *Folia forest*, 516, 19.

Paine, R. T. (1974). Intertidal community structure: experimental studies on the relationship between a dominant competitor and its principal predator. *Oecologia*, 15, 93–120.

Paine, R. T. & Levin, S. A. (1981). Intertidal landscapes: disturbance and the dynamics of pattern. *Ecological Monographs*, 51, 145–78.

Pastor, J., Aber, J. D., McClaugherty, C. A. & Melillo, J. M. (1984). Aboveground production and N and P cycling along a nitrogen mineralization gradient on Blackhawk Island, Wisconsin. *Ecology*, 65, 25–8.

Pastor, J. & Bockheim, J. G. (1984). Distribution and cycling of nutrients in an aspen-mixed hardwood–spodosol ecosystem in northern Wisconsin. *Ecology*, 65, 339–53.

Pastor, J. & Broschart, M. (1990). The spatial pattern of a northern conifer–hardwood landscape. *Landscape Ecology*, (in press).

Pastor, J., Gardner, R. H., Dale, V. H. & Post, W. M. (1987). Successional changes in nitrogen availability as a potential factor contributing to spruce declines in boreal North America. *Canadian Journal of Forest Research*, **17**, 1394–400.

Pastor, J., Naiman, R. J., Dewey, B. & McInnes, P. F. (1988). Moose, microbes and the boreal forest. *BioScience*, **38**, 770–7.

Pastor, J. & Post, W. M. (1985). *Development of a linked forest productivity-soil process model*. Oak Ridge, Tennessee: Oak Ridge National Laboratory, ORNL/TM-9519.

Pastor, J. & Post, W. M. (1986). Influence of climate, soil moisture and succession on forest carbon and nitrogen cycles. *Biogeochemistry*, **2**, 3–28.

Pastor, J. and Post, W. M. (1988). Response of northern forests to CO_2-induced climate change. *Nature*, **334**, 55–8.

Patterson, W. A., Edwards, K. J. & Maguire, D. J. (1987). Microscopic charcoal as a fossil indicator of fire. *Quaternary Research*, **6**, 2–23.

Payette, S. (1974). Classification écologique des formes de croissance de *Picea glauca* (Moench.) Voss. et de *Picea mariana* (Mill.) BSP. en milieux subarctiques et subalpins. *Naturaliste canadien*, **101**, 893–903.

Payette, S. (1983). The forest-tundra and present tree-lines of the northern Québec–Labrador peninsula. In *Tree-line Ecology, Proceedings of the northern Québec Tree-Line Conference*, ed. P. Morisset & S. Payette, pp. 3–23. Québec: Nordicana **47**.

Payette, S., Deshaye, J. & Gilbert, H. (1982). Tree seed populations at the treeline in Riviere aux Feuilles area, northern Quebec, Canada. *Arctic and Alpine Research*, **14**(3), 215–21.

Payette, S. & Filion, L. (1985). White spruce expansion at the treeline and recent climatic change. *Canadian Journal of Forest Research*, **15**, 241–51.

Payette, S., Filion, L., Delwaide, A. & Bégin, C. (1989a). Reconstruction of tree-line vegetation response to long-term climate change. *Nature*, **341**, 429–32.

Payette, S., Filion, L., Gauthier, L. & Boutin, Y. (1985). Secular climate change in old-growth tree-line vegetation of northern Québec. *Nature*, **315**, 135–8.

Payette, S. & Gagnon, R. (1979). Tree-line dynamics in Ungava peninsula, northern Québec. *Holarctic Ecology*, **2**, 239–48.

Payette, S. & Gagnon, R. (1985). Late Holocene deforestation and tree regeneration in the forest-tundra of Québec. *Nature*, **313**, 570–2.

Payette, S., Morneau, C., Sirois, L. & Desponts, M. (1989b). Recent fire history of the Northern Quebec biomes. *Ecology*, **70**, 656–73.

Pearce, C. M., McLennan, D. & Cordes, L. D. (1988). The evolution and maintenance of white spruce woodlands on the Mackenzie Delta, N.W.T., Canada. *Holarctic Ecology*, **11**, 248–58.

Pearlstine, L., McKellar, H. & Kitchens, W. (1985). Modeling the impacts of river diversion on bottomland forest communities in the Santee River Floodplain, South Carolina. *Ecological Modelling*, **29**, 283–302.

Perala, D. A. (1979). *Regeneration and productivity of aspen grown on repeated short rotations*, USDA Forest Service Research Paper NE-176. North Central Forest Experiment Station, St. Paul, MN.

Perala, D. A. (1990). *Populus tremuloides*. In *Silvics of forest trees in the United States*. USDA Forest Service Agricultural Handbook, (in press).

Perala, D. A. & Alm, A. (1989). Regenerating paper birch in the Lake States with the shelterwood method. *Northern Journal of Applied Forestry*, **6**, 151–3.

Perala, D. A. & Laidly, P. R. (1989). *Growth of nitrogen-fertilized and thinned quaking aspen* (Populus tremuloides *Michx.*). USDA Forest Service Research Paper NC-286. North Central Forest Experiment Station, St. Paul, MN.

Peterman, R. M. (1978). The ecological role of the mountain pine beetle in lodgepole pine forests. In *Theory and Practice of Mountain Pine Beetle Management in Lodgepole Pine Forests*, ed. A. A. Berryman, G. D. Amman, R. W. Starek & D. L. Kibee, pp. 16–26. Moscow, Idaho: Forest Wildlife and Range Experiment Station, University of Idaho.

Peterman, R. M., Clark, W. C. & Holling, C. S. (1979). The dynamics of resilience. In *Population Dynamics*, ed. R. M. Anderson, B. D. Turner & L. R. Taylor, pp. 321–41. Oxford: Blackwell.

Peterson, D. L., Aber, J. D., Matson, P. A., Card, D. H., Swanberg, N., Wessman, C. & Spanner, M. (1988). Remote sensing of forest canopy and leaf biochemical contents. *Remote Sensing of Environment*, **24**, 85–106.

Peterson, R. L. (1955). *North American Moose*. Toronto: University of Toronto Press.

Phipps, R. L. (1979). Simulation of wetland forest vegetation dynamics. *Ecological Modelling*, **7**, 257–88.

Pianka, E. R. (1978). *Evolutionary Ecology*. New York: Harper and Row Publishers.

Pickett, S. T. A. & White, P. S. ed (1985). *The Ecology of Natural Disturbance and Patch Dynamics*. New York: Academic Press.

Pienaar, L. V. & Turnbull, K. J. (1973). The Chapman-Richards generalization of Von Bertalanffy's growth model for basal area growth and yield in even-aged stands. *Forest Science*, **19**, 2–22.

Pitovranov, S. & Jaeger, J. (1990). Climate projection: Grosswet-terlagen approaches. In *Toward ecological sustainability in Europe*, ed. A. M. Solomon & L. Kauppi, pp. 31–40. IIASA Research Report, International Institute for Applied Systems Analysis, Laxenburg, Austria, (in press).

Platonov, G. M. (1963). Mires of northern part of Ob and Tom watershed. In *Paludified Forests and Bogs of Siberia*, pp. 65–95. Moscow: Publication of Science Academy of USSR. (In Russian.)

Plotnikov, V. V. (1979). *Evolution of Plant Communities Structure*. Moscow: Science Publishers. (In Russian.)

Pobedinsky, A. B. (1979). *The Scotch Pine*. Moscow: Lesnaya promyshlenost. (In Russian.)

Pohtila, E. (1980). Climatic fluctuations and forestry in Lapland. *Holarctic Ecology*, **3**, 91–8.

Polikarpov, N. P. & Babintseva, P. M. (1963). Regeneration of the dark coniferous forests in west Sayan mountains. *Papers of the Institute for Forests and Wood*, **54**, 17–34. (In Russian.)

Polikarpov, N. P. & Nazimova, D. I. (1963). The dark coniferous forests of the northern part of the western Sayaan mountains. *Papers of the Institute for Forests and Wood*, **54**, 103–47. (In Russian.)

Polikarpov, N. P. & Tchebakova, N. M. (1982). Ecological estimation of biological productivity of forest forming tree species. In *Growth of Coniferous Sapling Stands*, ed. Anonymous, pp. 25–54. Novosibirsk: Nauka. (In Russian.)

Polikarpov, N. P., Tchebakova, N. H. & Nazimova, D. I. (1986). *Climate and Montane Forests in Southern Siberia*. Novosibirsk: Siberian Division, Academy of Sciences of the USSR. (In Russian.)

Poluectov, R. A., ed. (1974). *Dynamic Theory of Biological Populations*. Moscow: Science Publishers. (In Russian.)

Potter, M. W., Kessell, S. R. & Cattelino, P. J. (1979). FORPLAN: A FORest Planning LANguage and simulator. *Environmental Management*, 3, 59–72.

Povarnitsin, V. A. (1955). *Cedar Pine Forests of the USSR*. Krasnoyarsk: Nauka. (In Russian.)

Pozdnyakov, L. K. (1983). *Forests on Permafrost*. Novosibirsk: Nauka. (In Russian.)

Pozdnyakov, L. K. (1985). *Permafrost Forest Science*. Novosibirsk: Nauka. (In Russian.)

Pozdnyakov, L. K. (1986). *Permafrost Forestry*. Novosibirsk: Nauka. (In Russian.)

Pravdin, L. F. (1964). *The Scotch Pine. Genetic Variability, Systematics and Selection*. Moscow: Nauka. (In Russian.)

Pravdin, L. F. (1975). *The Norway and Siberian Spruces in the USSR*. Moscow: Nauka. (In Russian.)

Prentice, I. C. (1986a). Some concepts and objectives of forest dynamics research. In *Forest Dynamics Research in Western and Central Europe*, ed. J. Fanta, pp. 32–41. Wageningen: PUDOC.

Prentice, I. C. (1986b). Vegetation response to past climatic variation. *Vegetatio*, 67, 131–41.

Prentice, I. C. et al. (1989). *Developing a Global Vegetation Dynamics Model: Results of an IIASA Summer Workshop*. RR-89-7. International Institute for Applied Systems Analysis, Laxenburg, Austria.

Prentice, I. C. & Helmisaari, H. (1990). Silvics of north European trees: compilation, comparisons and implications for forest succession modelling. *Forest Ecology and Management*, (in press).

Prentice, I. C. & Leemans, R. (1990). Pattern and process and the dynamics of forest structure. *Journal of Ecology*, (in press).

Prentice, I. C., Monserud, R. A., Smith, T. M. & Emanuel, W. R. (1990). Modeling large-scale vegetation dynamics. In *The Challenge of Modeling Global Biospheric Change*, ed. A. M. Solomon. Laxenburg, Austria: International Institute for Applied Systems Analysis, (in press).

Prince, S. D. & Tucker, C. J. (1986). Satellite remote sensing of rangelands in Botswana. II. NOAA AVHRR and herbaceous vegetation. *International Journal of Remote Sensing*, 7, 1555–70.

Pukkala, T. (1987). Simulation model for natural regeneration of *Pinus sylvestris*, *Picea abies*, *Betula pendula*, and *Betula pubescens*. *Silva Fennica*, 21, 37–53.

Putman, W. E. & Zasada, J. C. (1986). Direct seeding techniques to regenerate white spruce in interior Alaska. *Canadian Journal of Forest Research*, 16, 660–4.

Pyatetskii, G. E. & Morozova, R. M. (1962). Changes in physical and chemical properties of forest soils in south Karelia in relation to forest fellings. In *Forest Soils of Karelia and their Change under Effect of Forest Economic Measures*, **34**, 71–92. (In Russian.)

Pyatkov, P. P. (1935). Paludification in forests of North of USSR. *Economy of North of USSR*, **5**, 56–66. (In Russian.)

P'yavchenko, N. I. (1953). Paludification of the forests in the region of the River Shecksna. *Transactions of Forest Institute of Science Academy of USSR*, **13**, 51–76. (In Russian.)

P'yavchenko, N. I. (1954). Conditions for paludification of forests of taiga zone. *Transactions of Forest Institute of Science Academy of USSR*, **23**, 277–87. (In Russian.)

P'yavchenko, N. I. (1956). Classification of paludified forests. In *For Academician V. N. Suckachev in honor of the 75th anniversary of his birth*, pp. 463–80. Moscow/Leningrad: Publication of Science Academy of USSR. (In Russian.)

P'yavchenko, N. I. (1963). *Forest Peatland Science*. Moscow: Publication of Science Academy of USSR. (In Russian.)

P'yavchenko, N. I. (1965). Types of bogged forests at Tomsk field station. In *Specificities of Mire Formation in some Forest and Foothill Regions of Siberia and Far East*, pp. 97–113. Moscow: Science. (In Russian.)

P'yavchenko, N. I. (1979). On the interrelation between forest and mire in taiga zone. *Transactions of Darwin State reserve, Vologda*, North-west book publication, **15**, 6–14. (In Russian.)

P'yavchenko, N. I. (1980). On the interrelation between forest and mire. *Forestry*, **3**, 24–33. (In Russian.)

P'yavchenko, N. I. (1985). *Peat Bogs, their Natural and Economic Significance*. Moscow: Science. (In Russian.)

Rachko, P. (1979). Simulation model of growth dynamics of a tree as an element of forest ecosystem. *Problems of Cybernetics*, **52**, 73–110. (In Russian.)

Raffa, K. F. & Berryman, A. A. (1983). The role of host plant resistance in the colonization, behavior and ecology of bark beetles (Coleoptera: Scolytidae). *Ecological Monographs*, **53**, 27–49.

Ranson, K. J. & Daughtry, C. S. T. (1986). Sun angle, view angle background effects on spectral response of simulated balsam fir canopies. *Photogrammetric Engineering and Remote Sensing*, **52**, 649–58.

Ranson, K. J. & Smith, J. A. (1990a). Airborne SAR experiment for forest ecosystem research – Maine 1989 Experiment. *Proceedings of the IGARSS 1990 Symposium*, College Park, Maryland, 20–24 May, pp. 861–4.

Ranson, K. J. & Smith, J. A. (1990b). Analysis of northern forest spatial patterns with fractal dimension. *IEEE Transactions on Geoscience and Remote Sensing*, (in press).

Raunkiaer, C. (1934). *The Life forms of Plants and Statistical Plant Geography*. Oxford: Clarendon.

Raup, H. M. (1941). Botanical problems in boreal North America. *Botanical Review*, **7**, 147–248.

Redko, G. I. (1978). *Density of Silvicultures*. Leningrad: Forest Academy Publishers. (In Russian.)

Reed, K. L. (1980). An ecological approach to modeling growth of forest trees. *Forest Science*, **26**, 35–50.

Reineke, L. H. (1933). Perfecting a stand-density index for even-aged forests. *Journal of Agricultural Research*, **46**, 627–38.

Reiners, W. A. & Lang, G. E. (1979). Vegetational patterns and processes in the balsam fir zone, White Mountains, New Hampshire. *Ecology*, **60**, 403–17.

Remröd, J. (1980). Experiences and practices related to forest regeneration in northern Sweden. In *Forest regeneration at high latitudes. Proceedings of an International Workshop*, Fairbanks, Alaska, November, 1979, ed. M. Murray & R. Van Veldhuizen, pp. 35–41. USDA Forest Service General Technical Report PNW-107.

Results of Experimental Works of the Forest Experimental Station of the Agricultural Academy by Timireaser, 1862–1962. Agricultural Academy Publishers. (In Russian.)

Reynolds, J. F., Bachelet, D., Leadley, P. and Moorhead, D. (1986). Response of vegetation to carbon dioxide. Assessing the effects of elevated carbon dioxide on plants: toward the development of a generic plant growth model. Progress Report 023 to U.S. Dept. of Energy.

Richard, P. J. H. (1977). *Histoire post-wisconsinienne de la vegetation du Québec méridional par l'analyse pollinique*. Québec: Service de la recherche, Ministère des Terres st Forêts du Québec (2 volumes).

Richard, P. J. H. (1979). Contribution à l'histoire postglaciaire de la végétation au Nord-Est de la Jamésie, Nouveau-Québec, Université du Québec. *Géographie physique et Quaternaire*, **33**, 93–112.

Richard, P. J. H. (1981*a*). *Paléophytogéographie postglaciaire en Ungava par l'analyse pollinique*. Montréal: Paleo-Québec, Université du Québec.

Richard, P. J. H. (1981*b*). Palaeoclimatic significance of the late-Pleistocene and Holocene pollen record in south-central Québec. In *Quaternary Paleoclimate*, ed. W. C. Mahaney, pp. 335–60. Norwich: Geoabstracts.

Richard, P. J. H., Larouche, A. & Bouchard, M. A. (1982). Age de la déglaciation finale et histoire postglaciare de la vegetation dans la partie centrale du Nouveau-Québec. *Géographie physique et Quaternaire*, **36**, 63–90.

Richards, F. J. (1959). A flexible growth function for empirical use. *Journal of Experimental Botany*, **10**, 290–300.

Rieger, S. (1983). *The Genesis and Classification of Cold Soils*. New York: Academic Press.

Rieger, S., Dement, J. A. & Sanders, D. (1963). *Soil survey of Fairbanks Area, Alaska*. USDA Series 1959, No. 25. Washington, DC.

Rinne, P., Kauppi, A. & Ferm, A. (1987). Induction of adventitious buds and sprouts on birch seedlings (*Betula pendula* Roth and *B. pubescens* Ehrh.). *Canadian Journal of Forest Research*, **17**, 545–55.

Riom, J. & LeToan, T. (1981). Relations entre des types de forets de pins maritimes et la retrodifusion radar en bande. *Proceedings of ISP International Colloquium on Spectral Signatures of Objects in Remote Sensing*. Avignon, France.

Risser, P. G. (1986). *Spatial and temporal variability of biospheric and geospheric processes: research needed to determine interactions with global environmental change*. Paris: International Council of Scientific Unions Press.

Ritchie, J. C. (1959). The vegetation of northern Manitoba. V. Establishing the major zonation. *Arctic*, **13**, 211–29.

Ritchie, J. C. (1962). *A geobotanical survey of northern Manitoba.* Arctic Institute of North America Technical Paper No. 9.

Ritchie, J. C. (1984). *Past and Present Vegetation of the Far Northwest of Canada.* Toronto: University of Toronto Press.

Ritchie, J. C. (1987*a*). *Postglacial Vegetation of Canada.* Cambridge University Press.

Ritchie, J. C. (1987*b*). Comparaison entre la végétation du Mackenzie et du nord Québecois á l'Holocene. *Géographie physique et Quaternaire*, **41**, 153–60.

Ritchie, J. C., Cwynar, L. C. & Spear, R. W. (1983). Evidence from north-west Canada for an early Holocene Milankovitch thermal maximum. *Nature*, **305**, 126–8.

Ritchie, J. C. & Hare, F. K. (1971). Late Quaternary vegetation and climate near the arctic tree line of northwestern North America. *Quaternary Research*, **1**, 331–42.

Roberts, D. W. (1987). A dynamical systems perspective on vegetation theory. *Vegetatio*, **69**, 27–33.

Roberts, D. W. (1989). Fuzzy systems vegetation theory. *Vegetatio*, **83**, 71–80.

Robinson, G. W., Hughes, D. O. & Roberts, E. (1949). Podzolic soils of Wales. *Journal of Soil Science*, **1**(1), 50–62.

Romme, W. H. (1980). Fire History Terminology: Report of the Ad Hoc Committee. In *Proceedings of the Fire History Workshop*, ed. M. A. Stokes & J. H. Dieterich, pp. 135–7. General Technical Report RM-81, Rocky Mountain Forest and Range Experiment Station. Fort-Collins: Forest Service, US Department of Agriculture.

Romme, W. H. (1982). Fire and landscape diversity in subalpine forests of Yellowstone National Park. *Ecological Monographs*, **52**, 199–221.

Rosenberg, N. J., Blad, B. L. & Verma, S. B. (1983). *Microclimate.* New York: Wiley.

Rosenzweig, C. & Dickinson, R. (1986). *Climate-vegetation interactions.* NASA Conference Publication 2440. Goddard Space Flight Center, Greenbelt, MD.

Rosenzweig, M. L. & Abramsky, Z. (1980). Microtine cycles: the role of habitat heterogeneity. *Oikos*, **34**, 141–6.

Ross, J. K. (1975). *Radiation Regime and Vegetation Cover Architecture.* Leningrad: Gydromet Publishers. (In Russian.)

Ross, M. S., Sharik, T. L. & Smith, D. W. M. (1982). Age-structure relationships of tree species in an Appalachian oak forest in southwest Virginia. *Bulletin of the Torrey Botanical Club*, **109**, 287–98.

Rousseau, J. (1952). Les zones biologiques de la péninsule Québec–Labrador et l'hémiarctique. *Canadian Journal of Botany*, **30**, 436–74.

Rousseau, J. (1968). The vegetation of the Québec–Labrador peninsula between 55° and 60° N. *Naturaliste canadien*, **95**, 469–563.

Rowe, J. S. (1961). Critique of some vegetational concepts as applied to forests of northwestern Alberta. *Canadian Journal of Botany*, **39**, 1007–17.

Rowe, J. S. (1972). *Forest regions of Canada.* Canadian Department of Environment, Canadian Forestry Service Publication No. 1300, Ottawa.

Rowe, J. S. (1983). Concepts of fire effects on plant individuals and species. In *The Role of Fire in Northern Circumpolar Ecosystems*, ed. R. W. Wein & D. A. MacLean, pp. 135–54. New York: Wiley.

Rowe, J. S. (1984). Lichen woodlands in northern Canada. In *Northern Ecology and Resources Management*, ed. R. Olson, R. Hastings & F. Geddes, pp. 225–37. Edmonton: The University of Alberta Press.

Rowe, J. S. & Scotter, G. W. (1973). Fire in the boreal forest. *Quaternary Research*, **3**, 444–64.

Rowe, J. S., Spittlehouse, D., Johnson, E. A. & Jasieniuk, M. (1975). *Fire Studies in the Upper Mackenzie Valley and Adjacent Precambrian Uplands*. Ottawa: Department of Indian Affairs and North. ALUR 74-75-61.

Rozenberg, V. A., Manko, Yu. I. & Vasiliev, N. G. (1972). Main specifics of deployment and dynamics of forests in Far East and Amur Region. *Transactions of the Animal and Plant Ecology Institute of the Academy of Sciences*, **84**, 116–23. (In Russian.)

Rumney, G. R. (1968). *Climatology and the World's Climates*. New York: Macmillan.

Runkle, J. R. (1982). Patterns of disturbance in some old-growth mesic forests of eastern North America. *Ecology*, **63**, 1533–46.

Running, S. W. & Coughlan, J. C. (1988). A general model of forest ecosystem processes for regional applications. I. Hydrologic balance, canopy gas exchange and primary production processes. *Ecological Modelling*, **42**, 125–54.

Running, S. W. & Nemani, R. R. (1988). Relating seasonal patterns of the AVHRR vegetation index to simulated photosynthesis and transpiration of forests in different climates. *Remote Sensing of Environment*, **24**, 347–67.

Ruuhijärvi, R. (1983). The Finnish mire types and their regional distribution. In *Mires: Swamp, Bog, Fen, and Moore – Regional Studies*, pp. 47–67. Amsterdam: Elsevier.

Ryden, B. E. & Kostov, L. (1980). Thawing and freezing in tundra soils. In *Ecology of a Subarctic Mire*, ed. M. Sonesson. *Ecological Bulletins* (Stockholm), **30**, 251–81.

Ryynänen, M. (1982). Individual variation in seed maturation in marginal populations of Scots pine. *Silvae Fennica*, **16**(2), 185–7.

Sader, S. A. (1987). Forest biomass, canopy structure and species composition relationships with multipolarization L-band synthetic aperture radar data. *Photogrammetric Engineering and Remote Sensing*, **53**, 193–202.

Saeta, B. A. (1971). Multiple management of cedar pine forests in the Altay mountains. In *Utilisation and Regeneration of Cedar Pine Forests*, ed. Anonymous, pp. 121–33. Novosibirsk: Nauka. (In Russian.)

Safford, L. O. (1973). *Fertilization increases diameter growth of birch–beech–maple trees in new Hampshire*. US Department of Agriculture Forest Service Research Note NE-182.

Safford, L. O., Bjorkbom, J. C. & Zasada, J. C. (1990). *Betula papyrifera*. In *Silvics of forest trees in the United States*. USDA Forest Service Agricultural Handbook, (in press).

Sallnäs, O. (1989). *The forest matrix model concept – a contribution to forest sector modeling?* Garpenberg: Department of Operational Efficiency, Swedish University of Agricultural Sciences, Research Notes No. 150.

Sambuk, F. P. (1932). Forests of the River Petchera. *Transactions of the Botanical Museum of Science Academy of USSR*, **24**, 63–250. (In Russian.)

Sammi, J. C. (1969). Graphics stand tables. *Journal of Forestry*, **67**, 498–500.

Sarvas, R. (1952). On the flowering of birch and the quality of seed crop. *Communicationes Instituti Forestalis Fenniae*, **40**(7).

Sarvas, R. (1962). Investigations on the flowering and seed crop of *Pinus sylvestris*. *Metsantutkimuslaitoksen Julkaisuja*, **53**(4).

Sarvas, R. (1964). *Havupuut*. Helsinki: Werner Söderström Osakeyhtiö. (In Finnish.)

Sarvas, R. (1968). Investigations on the flowering and seed crop of *Picea abies*. *Metsantutkimuslaitoksen Julkaisuja*, **67**(5).

Sarvas, R. (1972). Investigations on the annual cycle of development of forest trees. I. Active period. *Communicationes Instituti Forestalis Fenniae*, **76**(3).

Sarvas, R. (1974). Investigations on the annual cycle of development of forest trees. II. Autumn dormancy and winter dormancy. *Communicationes Instituti Forestalis Fenniae*, **84**(1).

Schaffer, W. M. (1984). Stretching and folding in lynx fur returns: evidence for a strange attractor in nature. *American Naturalist*, **124**, 798–820.

Schaffer, W. M. (1985). Order and chaos in ecological systems. *Ecology*, **66**, 93–106.

Schaffer, W. M. & Kot, M. (1986). Chaos in ecological systems: the coals that Newcastle forgot. *Trends in Ecology and Evolution*, **1**, 58–63.

Schier, G. A. (1972). Apical dominance in multishoot cultures from aspen roots. *Forest Science*, **18**, 147–9.

Schier, G. A. (1973a). Seasonal variation in sucker production from excised roots of *Populus tremuloides* and the role of endogenous auxin. *Canadian Journal of Forest Research*, **3**, 459–61.

Schier, G. A. (1973b). Origin and development of aspen root suckers. *Canadian Journal of Forest Research*, **3**, 45–53.

Schier, G. A. (1975). *Deterioration of aspen clones in the middle Rocky Mountains*. USDA Forest Service Research Paper INT-170. Intermountain Forest and Range Experiment Station, Ogden, UT.

Schier, G. A. & Campbell, R. B. (1976). Differences among *Populus* species in ability to form adventitious roots and shoots. *Canadian Journal of Forest Research*, **6**, 253–61.

Schier, G. A. & Campbell, R. B. (1980). *Variation among healthy and deteriorating aspen clones*. USDA Forest Service Research Paper INT-264. Intermountain Forest and Range Experiment Station, Ogden, UT.

Schlentner, R. L. & Van Cleve, K. (1985). Relationships between CO_2 evolution from soil, substrate temperature, and substrate moisture in four mature forest types in interior Alaska. *Canadian Journal of Forest Research*, **15**, 97–106.

Schlesinger, M. E. & Mitchell, J. F. B. (1987). Climate model simulations of the equilibrium climatic response to increased carbon dioxide. *Reviews of Geophysics*, **25**, 760–98.

Schmelz, D. V. & Lindsey, A. A. (1965). Size-class structure of old-growth forests in Indiana. *Forest Science*, **11**, 731–43.

Schmidt-Vogt, H. (1977). *Die Fichte*. Band I. Hamburg und Berlin: Verlag Paul Parey. (In German.)

Schopmeyer, C. S. ed. (1974). *Seeds of woody plants in the United States*. USDA Forest Service Agricultural Handbook 450.

Schotte, G. (1917). Lärken och dess betydelse för svensk skogshushållning (*Larix* and its importance for Swedish forestry). *Meddelanden Från Statens Skogsförsöksanstalt Häfte*, **13–14**, 529–788. Centraltryckeriet Stockholm. (In Swedish.)

Schreiner, E. J. (1974). *Populus* L. – Poplar. In *Seeds of woody plants in the United States*, ed. C. S. Schopmeyer, pp. 645–55. USDA Forest Service Agricultural Handbook 450.

Scott, P. A., Hansell, R. I. C. & Fayle, D. C. (1987). Establishment of white spruce populations and responses to climatic change at the tree-line, Churchill, Manitoba, Canada. *Arctic and Alpine Research*, **19**, 45–51.

Scotter, G. W. (1964). *Effects of forest fires on the winter range of barren-ground caribou in northern Saskatchewan*. Ottawa: Canadian Wildlife Service, Wildlife Management Bulletin, Series 1, no. 18.

Searing, G. F. (1975). *Aggressive behavior and population regulation of red squirrels* (Tamiasciurus hudsonicus) *in interior Alaska*. M.Sc. thesis, University of Alaska, Fairbanks, AK.

Sellers, P. (1986). Canopy reflectance, photosynthesis and transpiration. *International Journal of Remote Sensing*, **6**, 1335–72.

Sellers, P. (1987). Canopy reflectance, photosynthesis and transpiration II. The role of biophysics in the linearity of their interdependence. *Remote Sensing of Environment*, **21**, 143–83.

Semetchkin, I. V. (1970). Dynamics of stand age structure and relevant methods of its investigation. *Questions of Forest Science* (*Krasnoyarsk*), **1**, 422–45. (In Russian.)

Semevsky, F. N. & Semenov, S. M. (1982). *Mathematical Modeling of Ecological Processes*. Leningrad: Gydromet Publishers. (In Russian.)

Sepälä, M. & Rasta, J. (1980). Vegetation map of northernmost Finland with special reference to subarctic forest limits and natural hazards. *Fennia*, **158**, 41–61.

Sergievskaya, L. P. (1971). More attention to the cedar pine. In *Utilisation and Regeneration of Cedar Pine Forests*, ed. Anonymous, pp. 23–6. Novosibirsk: Nauka. (In Russian.)

Sernander, R. (1936). Granskär och Fiby urskog, en studie över stormluckornas och marbuskarnas betydelse i den svenska granskogens regeneration (The primitive forests of Granskär and Fiby). *Acta Phytogeographica Suecica*, **8**, 1–232.

Shabad, T., ed. (1965). Translation of Fiziko-geograficheskiy atlas mira. *Soviet Geography*, **5–6**, 1–403.

Sharpe, P. J. H., Walker, J., Penridge, L. K. and Wu, H. (1985). A physiologically based continuous-time Markov approach to plant growth modeling in semi-arid woodlands. *Ecological Modeling*, **29**, 189–213.

Sharpe, P. J. H., Walker, J., Penridge, L. K., Wu, H. and Rykiel, E. J. (1986). Spatial considerations in physiological models of tree growth. *Tree Physiology*, **2**, 403–21.

Shennikov, A. P. (1933). Geobotanical ranges of the Northern region and their significance in development of productivity potential. In *Results of the 2nd conference for study of the productivity potentials of the Northern region: Flora and soils*, pp. 10–96. Archangelsk: Northern region publishing. (In Russian.)

Shinozaki, K., Yoda, K., Hozumi, K. & Kira, T. (1964a). A quantitative analysis of plant form – the pipe model theory. I. Basic analysis. *Japanese Journal of Ecology*, **14**, 97–105.

Shinozaki, K., Yoda, K., Hozumi, K. & Kira, T. (1964b). A quantitative analysis of plant form – the pipe model theory. II. Further evidence of the theory and its application in forest ecology. *Japanese Journal of Ecology*, **14**, 133–9.

Short, S. K. & Nichols, H. (1977). Holocene pollen diagrams from subarctic Labrador–Ungava: vegetational history and climatic change. *Arctic and Alpine Research*, **9**, 265–90.

Shugart, H. H. (1984). *A Theory of Forest Dynamics: the Ecological Implications of Forest Succession Models*. New York: Springer-Verlag.

Shugart, H. H. (1987). Dynamic ecosystem consequences of tree birth and death patterns. *BioScience*, **37**, 596–602.

Shugart, H. H., Antonovski, M. Ya., Jarvis, P. G. and Sandford, A. P. (1986). CO_2, climatic change and forest ecosystems: Assessing the response of global forests to the direct effects of increasing CO_2 and climatic change. In *The Greenhouse Effect, Climatic Change and Ecosystems* (*SCOPE 29*), eds. B. Bolin, B. R. Döös, J. Jager and R. A. Warrick. New York: John Wiley.

Shugart, H. H., Bonan, G. B. & Rastetter, E. B. (1987). Niche theory and community organization. *Canadian Journal of Botany*, **66**, 2634–9.

Shugart, H. H., Crow, T. R. & Hett, J. M. (1973). Forest succession models: a rationale and methodology for modeling forest succession over large regions. *Forest Science*, **19**, 203–12.

Shugart, H. H., Michaels, P. J., Smith, T. M., Weinstein, D. A. & Rastetter, E. B. (1988). Simulation models of forest succession. In *Scales and Global Change*, ed. T. Rosswall, R. G. Woodmansee & P. G. Risser, pp. 125–51. Chichester: John Wiley & Sons.

Shugart, H. H. & Noble, I. R. (1981). A computer model of succession and fire response of the high altitude Eucalyptus forest of the Brindabella Range, Australian Capital Territory. *Australian Journal of Ecology*, **6**, 149–64.

Shugart, H. H. & West, D. C. (1977). Development of an Appalachian deciduous forest succession model and its application to assessment of the impact of the chestnut blight. *Journal of Environmental Management*, **51**, 161–79.

Shugart, H. H. & West, D. C. (1979). Size and pattern of simulated forest stands. *Forest Science*, **25**, 120–2.

Shugart, H. H. & West, D. C. (1980). Forest succession models. *BioScience*, **30**, 308–13.

Shugart, H. H., West, D. C. & Emanuel, W. R. (1981). Patterns and dynamics of forests: An application of forest succession models. In *Forest Succession: Concepts and Application*, ed. D. C. West, H. H. Shugart & D. B. Botkin, pp. 74–94. New York: Springer-Verlag.

Shul'gin, A. M. (1965). *The Temperature Regime of Soils*. Jerusalem: Israel Program for Scientific Translations.

Shumilova, L. V. (1969). Mire regions of West Siberia in ranges of Tyumen district. *Reports of Institute of Geography of Siberia and Far East* (*Science Academy of USSR*), **23**, 14–20. (In Russian.)

Shvydenko, A. Z., Strotchynsky, A. A., Savitch, Yu.N. & Kashpor, S. M. ed. (1987). *Yield and Growth Tables for the Ukrainian and Moldavian SSR*. Kiev: Urojai. (In Russian.)

Shwareva, Yu.O. (1963). Climate. In *West Siberia*, pp. 70–95. Moscow: Publication of Science Academy of USSR. (In Russian.)

Siccama, T. G. (1974). Vegetation, soil, and climate on the Green Mountains of Vermont. *Ecological Monographs*, **44**, 325–49.

Sieber, A. J. (1985). Forest signatures in microwave scatterometer data. *IEEE Geoscience and Remote Sensing Society Newsletter*, **9**(4), 4–10.

Silvertown, J. W. (1982). *Introduction to Plant Population Ecology*. New York: Longman.

Simak, M. (1969). Frostschaden on larchen in Schweden. *Zeitschriften des Schweizerischen Forstvereins*, **46**, 115–25.

Simak, M. (1980). Germination and storage of *Salix caprea* L. and *Populus tremula* L. seeds. In *Seed problems. Proceedings of the International Symposium on Forest Tree Seed Storage*, ed. B. P. Wang & J. A. Pitel, pp. 142–60. Canadian Forestry Service, Ottawa.

Sims, R. A., Towill, W. D., Baldwin, K. A. & Wickware, G. M. (1990). *Field guide to the forest ecosystem classification for northwestern Ontario*. Forestry Canada and the Ontario Ministry of Natural Resources, Ottawa, Canada.

Singh, G., Kershaw, A. P. & Clark, R. (1981). Quaternary vegetation and fire history in Australia. In *Fire and the Australian Biota*, ed. A. M. Gill, R. H. Groves & I. R. Noble, pp. 23–54. Canberra: The Australian Academy of Sciences.

Sinko, J. W. & Streifer, W. (1967). A new model for age–size structure of a population. *Ecology*, **48**, 910–18.

Siren, G. (1955). The development of spruce forest on raw humus site in northern Finland and its ecology. *Acta Forestalia Fennica*, **62**, 1–363.

Sirois, L. (1988). *La déforestation subarctique. Une analyse écologique et démographique*. PhD thesis, Université Laval, Québec.

Sirois, L. & Payette, S. (1989). Post-fire black spruce establishment in boreal and subarctic Québec. *Canadian Journal of Forest Research*, **19**, 1571–80.

Sirois, L. & Payette, S. (1991). Reduced post-fire tree regeneration along a Forest-Tundra transect in northern Québec. *Ecology*, (in press).

Sjörs, H. (1963). Amphiatlantic zonation, nemoral to arctic. In *North Atlantic Biota and their History*, ed. A. Löve & D. Löve, pp. 109–25. New York: Macmillan Company.

Skorupskii, B. V. & Shelyag-Sosonko, Y. R. (1982). Relationship between mean zonal climatic characteristics and vegetation. *Soviet Journal of Ecology*, **13**, 107–14.

Skre, O. & Oechel, W. C. (1979). Moss production in a black spruce *Picea mariana* forest with permafrost near Fairbanks, Alaska, as compared with two permafrost-free stands. *Holarctic Ecology*, **2**, 249–54.

Slaughter, C. W. & Viereck, L. A. (1986). Climatic characteristics of the taiga in interior Alaska. In *Forest Ecosystems in the Alaskan Taiga*, ed. K. Van Cleve, F. S. Chapin, P. W. Flanagan, L. A. Viereck & C. T. Dyrness, pp. 9–21. New York: Springer-Verlag.

Smilga, Ya. (1986). *Aspen*. Riga: Zinatne. (In Russian.)

Smith, J. A. (1984). Matter-energy interactions in the optical region. In *Manual of Remote Sensing*, 2nd ed., Vol II, pp. 62–113. American Society of Photogrammetry and Remote Sensing, Falls Church, VA.

Smith, J. A. & Oliver, R. E. (1972). Plant canopy models for simulating composite scene spectroradiance in the 0.4 to 2.05 micrometer region. *Proceedings of the Eighth International Symposium on Remote Sensing of Environment*, University of Michigan, Ann Arbor, Oct 2–6, **2**, 1333–53.

Smith, M. C. (1967). *Red squirrel* (Tamiasciurus hudsonicus) *ecology during spruce cone failure in Alaska*. M.S. thesis, University of Alaska, Fairbanks, AK.

Smith, S. H. & Bell, J. F. (1983). Using competitive stress index to estimate diameter growth for thinned Douglas-fir stands. *Forest Science*, **29**, 491–9.

Smith, T. M. & Urban, D. L. (1988). Scale and the resolution of forest structural pattern. *Vegetatio*, **74**, 143–50.

Smith, W. K. & Carter, G. A. (1988). Shoot structural effects on needle temperatures and photosynthesis in conifers. *American Journal of Botany*, **75**, 496–500.

Smolonogov, Ye.P. (1970). Age dynamics and economy-selective cuttings of deciduous–dark-coniferous forests in the water-protection zone of the Ufa River in Sverdlovsk Region. *Transactions of the Animal and Plant Ecology Institute of the Academy of Sciences, Sverdlovsk*, **77**, 117–34.

Snyder, J. D. & Janke, R. A. (1976). Impact of moose browsing on boreal-type forests of Isle Royale National Park. *American Midland Naturalist*, **95**, 79–92.

Sokolov, S. Ya., Svyaseva, O. A. & Kubly, V. A. (1977). *Distribution Ranges of the USSR's Tree and Shrub Species*. Vol. 1. Leningrad: Nauka. (In Russian.)

Solomon, A. M. (1982). Plant community response to decreased seasonability during full-glacial time. *7th Biennial Meeting (Seattle, WA), American Quaternary Association Abstracts*, pp. 18–19.

Solomon, A. M. (1984). Forest responses to complex interacting full-glacial environmental conditions. *8th Biennial Meeting (Boulder, CO), American Quaternary Association Abstracts*, pp. 120.

Solomon, A. M. (1986). Transient response of forest to CO_2-induced climate change: simulation modeling experiments in eastern North America. *Oecologia*, **68**, 567–79.

Solomon, A. M. (1988). Ecosystem theory required to identify future forest responses to changing CO_2 and climate. In *Ecodynamics: Contributions to Theoretical Ecology*, ed. W. Wolff, C. J. Soeder & F. R. Drepper, pp. 258–74. Berlin: Springer-Verlag.

Solomon, A. M. & Leemans, R. (1990). Climatic change and landscape ecological response: Issues and analysis. In *Proceedings of European Conference on Landscape Ecological Impact of Climatic Change*, Lunteren, The Netherlands, December 1989, ed. R. S. de Groot and M. M. Boer, (in press).

Solomon, A. M. & Shugart, H. H., Jr. (1984). Integrating forest-stand simulations with paleoecological records to examine long-term forest dynamics. In *State and Change of Forest Ecosystems-Indicators in Current Research*, ed. G. I. Agren, pp. 333–56. Uppsala, Sweden: Swedish University of Agricultural Science.

Solomon, A. M. & Tharp, M. L. (1985). Simulation experiments with late-Quaternary carbon storage in mid-latitude forest communities. In *The carbon cycle and atmospheric carbon dioxide: natural variations Archaen to present*, ed. E. T. Sundquist & W. S. Broecker, pp. 235–50. Geophysical Monographs, Series Volume 32, American Geophysical Union, Washington, D. C.

Solomon, A. M., Tharp, M. L., West, D. C., Taylor, G. M., Webb, J. M. & Trimble, J. C. (1984). *Response of unmanaged forests to CO₂-induced climatic change: available information, initial tests, and data requirements.* Report TR-009, U.S. Department of Energy, Washington, D.C.

Solomon, A. M. & Webb, T. III. (1985). Computer-aided reconstruction of late-Quaternary landscape dynamics. *Annual Review of Ecology and Systematics*, **16**, 63–84.

Solomon, A. M. & West, D. C. (1987). Simulating forest responses to expected climate change in eastern North America: Applications to decision-making in the forest industry. In *The Greenhouse Effect, Climate Change, and U.S. Forests*, ed. W. E. Shands & J. S. Hoffman, pp. 189–217. Washington, DC: Conservation Foundation.

Sonesson, M. & Hoogesteger, J. (1983). Recent tree-line dynamics (*Betula pubescens* Ehrh. ssp. *tortuosa* (Ledeb.) Nyman) in northern Sweden. *Nordicana*, **47**, 47–54.

Sorenson, C. J., Knox, J. C., Larsen, J. A. & Bryson, R. A. (1971). Paleosoils and the forest border in Keewatin, N.W.T. *Quaternary Research*, **1**, 468–73.

Soswa Priobye (*Essays of Nature and Economy*). (1975). Irkutsk: Institute of Geography of Siberian and Far East Branches of Science Academy of USSR. (In Russian.)

Southwood, T. R. E. (1977). Habitat, the templet for ecological strategies? *Journal of Animal Ecology*, **46**, 337–65.

Sparro, R. P. (1924). Effect of paludification and drying on vegetation. *Transactions of Scientific Amelioration Institute*, **1**, 1–11. (In Russian.)

Spear, R. W. (1983). Paleoecological approaches to a study of tree-line fluctuation in the Mackenzie Delta region, Northwest territories: preliminary results. *Nordicana*, **47**, 61–72.

Sprugel, D. G. (1976). Dynamic structure of wave-regenerated *Abies balsamea* forests in the northern-eastern United States. *Journal of Ecology*, **64**, 889–991.

Sprugel, D. G. (1984). Density, biomass, productivity, and nutrient cycling changes during stand development in wave-generated balsam fir forests. *Ecological Monographs*, **54**, 165–86.

Sprugel, D. G. (1989). The relationship of evergreenness, crown architecture, and leaf size. *American Naturalist*, **133**, 465–79.

Sprugel, D. G. & Bormann, F. H. (1981). Natural disturbance and the steady state in high altitude balsam fir forests. *Science*, **211**, 390–3.

Spurr, S. H. (1952). *Forest Inventory*. New York: Ronald Press.

Spurr, S. H. (1954). The forests of Itasca in the nineteenth century as related to fire. *Ecology*, **35**, 21–5.

Stanek, W. (1968). Development of black spruce layers in Quebec and Ontario. *Forestry Chronicle*, **44**, 25–8.

Stearns, F. W. (1949). Ninety years changes in a northern hardwood forest in Wisconsin. *Ecology*, **30**, 350–8.

Steele, J. H. (1985). A comparison of terrestrial and marine systems. *Nature*, **313**, 355–8.

Steele, J. H. (1989). A message from the oceans. *Oceanus*, **32**, 4–9.

Steijlen, I. & Zackrisson, O. (1987). Long-term regeneration dynamics and successional trends in a northern Swedish coniferous forest stand. *Canadian Journal of Botany*, **65**, 839–48.

Stewart, H. & Swan, D. (1970). *Relationships between nutrient supply, growth, and nutrient concentrations in the foliage of black spruce and jack pine.* Pulp and Paper Research Institute of Canada, Pointe Claire, Quebec, Woodlands Papers No. 19.

Stocks, B. J. & Street, R. B. (1983). Forest fire weather and wildfire occurrence in the boreal forest of northwestern Ontario. In *Resources and Dynamics of the Boreal Zone*, ed. R. W. Wein, R. R. Riewe & I. R. Methven, pp. 249–65. Ottawa: Association of Canadian Universities for Northern Studies.

Stoeckeler, J. H., Strothmann, R. O. & Krefting, L. W. (1957). Effect of deer browsing on reproduction in the northern hardwood-hemlock type in northeastern Wisconsin. *Journal of Wildlife Management*, **21**, 75–80.

Stokes, M. A. & Dieterich, J. H. ed. (1980). *Proceedings of the Fire History Workshop*. General Technical Report RM-81, Rocky Mountain Forest and Range Experiment Station. Fort Collins: Forest Service, US Department of Agriculture.

Stone, E. L. & Cornwell, S. M. (1968). Basal bud burls in *Betula populifolia*. *Forest Science*, **14**(1), 64–5.

Suffling, R. (1983). Stability and diversity in boreal and mixed temperate forests: a demographic approach. *Journal of Environmental Management*, **17**, 359–71.

Suits, G. H. (1972). The calculation of the directional reflectance of vegetative canopies. *Remote Sensing of Environment*, **2**, 117–25.

Suits, G. H. (1983). The extension of a uniform canopy reflectance model to include row effects. *Remote Sensing of Environment*, **13**, 113–29.

Sukachev, V. N. (1914a). Mires, their formation, evolution and characteristics. In *Collection of lectures of additional courses for foresters*, pp. 249–405. St. Petersburg. (In Russian.)

Sukachev, V. N. (1914b). About a border horizon of peat bogs in relation to the question of climatic fluctuations during post-glacial time. *Soil Science*, **16**, 47–74. (In Russian.)

Sukachev, V. N. (1928). Principles of classification of the spruce communities of European Russia. *Journal of Ecology*, **16**, 1–18.

Sukachev, V. N. (1934). *Dendrology with Basics of Forest Botany*. Moscow: Goslesbumizdat. (In Russian.)

Sukachev, V. N. & Poplavskaya, G. I. (1927). Flora of the Crimea State Reserve. In *Crimea State Reserve: Its Nature, History, Significance*, ed. V. N. Sukachev, pp. 145–71. Moscow: Central Science Publishers. (In Russian.)

Sutherland, J. R. (1981). Effects of inland spruce cone rust, *Chrysomyxa pirolata* Wint., on seed yield, weight, and germination. *Canadian Forestry Service Research Note*, **1**(2), 8–9.

Swain, A. M. (1973). A history of fire and vegetation in northeastern Minnesota as recorded in lake sediment. *Quaternary Research*, **3**, 383–96.

Swain, A. M. (1978). Environmental changes during the past 2000 years in north-central Wisconsin: analysis of pollen, charcoal and seeds from varved sediments. *Quaternary Research*, **10**, 55–68.

Swain, A. M. (1980). Landscape patterns and forest history in the Boundary Waters Canoe Area, Minnesota: a pollen study from Hug Lake. *Ecology*, **61**, 747–54.

Swartzman, G. L. & Kaluzny, S. P. (1987). *Ecological Simulation Primer*. New York: Macmillan Publishing Company.

Swift, E. (1948). *Wisconsin's deer damage to forest reproduction survey – final report*. Madison: Wisconsin Conservation Department Publication No. 347.

Swift, M. J., Heal, O. W. & Anderson, J. M. (1979). *Decomposition in Terrestrial Ecosystems*. Studies in Ecology, Vol. 5. Berkeley: University of California Press.

Sylven, N. (1916). *De Svenska Skogsträden I. Barrträden* (Trees in Swedish Forests 1. Coniferous trees). Stockholm: C. E. Fritzes Bokförlag AB. (In Swedish.)

Tahvanainen, J., Hell, E., Julkunen-Tiittoo, R. & Lavola, A. (1985). Phenolic compounds of willow bark as deterrents against feeding by mountain hare. *Oecologia*, **65**, 319–23.

Tait, D. E. (1988). The dynamics of stand development: a general stand model applied to Douglas-fir. *Canadian Journal of Forest Research*, **18**, 696–702.

Takhtajan, A. (1986). *Floristic Regions of the World*. Berkeley: University of California Press.

Talantsev, N. K. (1981). *The Cedar Pine*. Moscow: Lesnaya promyshlenost. (In Russian.)

Talantsev, N. K., Pryajnikov, A. N. & Mishukov, N. P. (1978). *Cedar Pine Forests*. Moscow: Lesnaya promyshlenost. (In Russian.)

Tamm, C. O. (1953). Growth, yield and nutrition in carpets of a forest moss (*Hylocomium splendens*). *Meddelanden Fran Statens Skogsforskningsinstitut*, **43**, 1–140.

Tamm, O. (1950). *Northern Coniferous Forest Soils*. Oxford: Scrivener Press.

Tande, G. F. (1979). Fire history and vegetation pattern of coniferous forests in Jasper National Park, Alberta. *Canadian Journal of Botany*, **57**, 1912–31.

Tanfilyev, G. I. (1888). About mires of Petersburg gubernia. *Transactions of economical society*, **15**, 50–80. (In Russian.)

Tans, P. P., Fung, I. Y & Takahashi, T. (1990). Observational constraints on the global atmospheric CO_2 budget. *Science*, **247**, 1431–8.

Tansley, A. G. (1935). The use and abuse of vegetational concepts and terms. *Ecology*, **16**, 284–307.

Tappeiner, J. C. (1982). Aspen root systems and suckering in red pine stands. *American Midland Naturalist*, **107**(2), 408–10.

Tappeiner, J. C. & Alm, A. A. (1975). Undergrowth vegetation effects on the nutrient content of litterfall and soils in red pine and birch stands in northern Minnesota. *Ecology*, **56**, 1193–200.

Taran, I. V. (1964). Drainage measures in forests of Novosibirsk district. *Transactions on forest economy of Siberia*, **8**, 19–25. (In Russian.)

Tarkova, T. N. & Ipatov, V. S. (1975). Effect of illumination and litter on the development of some moss species. *Soviet Journal of Ecology*, **6**, 43–8.

Tcelniker, Ju.L. (1978). *Physiological Foundations of Tree Shade Tolerance*. Moscow: Science Publishers. (In Russian.)

Terborgh, J. (1989). *Where Have All the Birds Gone?* Princeton: Princeton University Press.

Ter-Mikaelian, M. T. & Furyaev, V. V. (1988). Model of spatial-time forest dynamics under fire impact. *Problems of Ecological Monitoring and Ecosystem Modeling*, **11**, 260–75. Leningrad: Gidrometeoisdat. (In Russian.)

Terskov, I. A. & Terskova, M. I. (1980). *Growth of Even-aged Stands*. Novosibirsk: Science Publishers. (In Russian.)

Thalenhorst, W. (1958). *Grunduge der Populations dynamic des grossen Fitchenborkenkaffers* Ips typographus *L*. Goettingen: Schriftener. Forstl. Fak., University of Goettingen. (In German.)

The Mountain Forests of China. (1979). Institute of surveying and planning at the Ministry of Forestry. Forestry Printing House of China.

Thornthwaite, C. W. (1948). An approach toward a rational classification of climate. *Geophysical Review*, **38**, 55–94.

Tikhomirov, B. A. (1960). Plant geographical investigations of the tundra vegetation in the Soviet Union. *Canadian Journal of Botany*, **38**, 815–32.

Tikhomirov, B. A. (1961). The changes in biogeographical boundaries in the north of USSR as related with climatic fluctuations and activity of man. *Botanisk Tidsskrift*, **5**, 284–92.

Tikhomirov, B. A. (1963). Principal stages of vegetation development in northern USSR as related to climatic fluctuations and the activity of man. *Canadian Geographer*, **7**, 55–71.

Tikhomirov, B. A. (1971). Forest limits as the most important biogeographical boundary in the north. In *Ecology of the Subarctic Regions: Proceedings of Helsinki Symposium*, pp. 35–40. Paris: UNESCO.

Tilman, D. (1988). *Plant Strategies and the Dynamics and Structure of Plant Communities*. Princeton: Princeton University Press.

Timmer, V. R. & Weetman, G. F. (1969). *Humus temperatures and snow cover conditions under upland black spruce in northern Quebec*. Pulp and Paper Research Institute of Canada, Pointe Claire, Quebec, Woodlands Papers No. 11.

Tiren, L. (1935). Om grannens kottsa attning, dess periodicitet och samband med temperatur och nederbröd. Summary: On the fruit setting of spruce, its periodicity and relation to temperature and precipitation. *Meddelanden fran Statens Skogsforsoksanstalt*, **28**, 413–524.

Tkatchenko, M. E. (1955). *General Forestry*. Moscow: Lesnaya promyshlenost. (In Russian.)

Tolonen, K. (1983). The post-glacial fire record. In *The Role of Fire in Northern Circumpolar Ecosystems*, ed. R. W. Wein & D. A. MacLean, pp. 21–44. Toronto: John Wiley & Sons.

Tonu, O. (1983). Metsa suktsessiooni ja tasandilise struktuuri imiteerimiset. *Yearbook of the Estonian Naturalist Society*, **69**, 110–7.

Tranquillini, W. (1979). *Physiological Ecology of the Alpine Timberline-tree Existence at High Elevations with Special Reference to the European Alps*. Berlin: Springer-Verlag.

Trees and Shrubs of the USSR. (1956). Vol. 3. Academy of Sciences of the USSR, Moscow and Leningrad. (In Russian.)

Trimble, G. R. (1973). *The regeneration of Central Appalachian hardwoods with emphasis on the effects of site quality and harvesting practice*. Washington, DC: USDA Forest Service, NE–282.

Troll, C. (1948). Der asymmetrische Aufbau der Vegetationszonen und Vegetationsstufen auf der Nord-und Sudhalbkugel. *Ber. Geobat. Forsch. Inst. Rbel.* 1947, 46–83.

Trottier, G. C. (1978). Beaked hazel – a key browse species for moose in the boreal forest region of western Canada? *Alces*, **17**, 257–81.

Tryon, P. R. & Chapin, F. S. (1983). Temperature control over root growth and root biomass in taiga forest tress. *Canadian Journal of Forest Research*, **13**, 827–33.

Tseplyaev, V. P. (1961). *The Forests of the USSR*. Israel Program for Scientific Translations Ltd, 1965. Jerusalem.

Tsinzerling, Yu.D. (1929). Description of mire vegetation along middle part of the river Petchora. *Proceedings of the General Botanical Garden of the USSR*, **28**(1–2), 95–129. (In Russian.)

Tubbs, C. H. (1965). *Influence of temperature and early spring conditions on sugar maple and yellow birch germination in upper Michigan*. USDA–FS Lake States Forest Experiment Station Research Note LS–72.

Tucker, C. J. (1979). Red and photographic infrared linear combinations for monitoring vegetation. *Remote Sensing of Environment*, **8**, 127–50.

Tucker, C. J., Fung, I. Y., Keeling, C. D. & Gammon, R. H. (1986). Relationship between atmospheric CO_2 variations and a satellite derived vegetation index. *Nature*, **319**(6050), 195–9.

Tucker, C. J. & Sellers, P. (1986). Satellite remote sensing of primary productivity. *International Journal of Remote Sensing*, **7**, 1395–416.

Tucker, C. J., Townshend, J. R. G. & Goff, T. E. (1985). Continental land cover classification using NOAA-7 AVHRR data. *Science*, **227**, 369–75.

Tukhanen, S. (1980). *Climatic parameters and indices in plant geography*. Acta phytogeographica Suecica No. 67. Uppsala.

Tukhanen, S. (1984). A circumboreal system of climatic-phytogeographical regions. *Acta Botanica Fennica*, **127**, 1–50.

Tuomi, J., Niemela, P., Haukioja, E., Siren, S. & Neuvonen, S. (1984). Nutrient stress: an explanation for plant-herbivore response to defoliation. *Oecologia*, **61**, 208–10.

Tyrtikov, A. P. (1973). Permafrost and vegetation. In *Permafrost: Proceedings of the Second International Conference (USSR Contribution)*, pp. 100–4. Washington, DC: National Academy of Science.

Tyrtikov, A. P. (1978). Climatic consequences of the southward shifting of the polar forest boundary. *Byullten' Moskovskogo Obshchestva Ispytatelei Prirody, Otdel Biologicheskii*, **83**, 60–4. (In Russian with English abstract.)

Tyuremnov, S. P. (1949). *Peat Deposits and their Surveying*. 2nd edition. Moscow/Leningrad: State energy publication. (In Russian.)

Uanjun, Z. (1958). *Arboretum of China*. Forestry Printing House of China.

Uggla, E. (1959). *Ecological effects of fire on North Swedish forests*. Institute of Plant Ecology, University of Uppsala, Uppsala, Sweden.

Ulaby, F. T., Allen, C. T., Eger, G. & Kanemasu, E. (1984). Relating the microwave backscatter coefficient to leaf area index. *Remote Sensing of Environment*, **14**, 113–33.

Ulaby, F. T., Batlivala, P. B. & Dobson, M. C. (1978). Microwave backscatter dependence on surface roughness, soil moisture and soil texture: Part I-Bare soil. *IEEE Transactions on Geoscience Electronics*, **GE-16**, 286–95.

Ulaby, F. T., Moore, R. K., & Fung, A. K. (1982). *Microwave Remote Sensing. Volume II: Radar remote sensing and surface scattering and emission theory*. Reading, MA: Addison-Wesley.

UN (1986). European Timber Trends and Prospect to the Year 2000 and Beyond.

Geneva: United Nationals Economic Commission for Europe; and Rome: Food and Agriculture Organization of the United Nations.

Urban, D. L. (1989). *A versatile model to simulate forest pattern: a user's guide to ZELIG*. University of Virginia, Charlottesville, Virginia.

Urban, D. L., O'Neill, R. V. & Shugart, H. H. (1987). Landscape ecology. *Bioscience*, **37**, 119–27.

US Department of Agriculture. (1982). *An analysis of the timber situation in the United States*, 1952–2030. Forest Service Resource Report No. 23, US Department of Agriculture, Washington, DC.

US Department of Commerce (1968). *Climatic atlas of the United States*, Superintendent of Documents, Washington, DC.

Uspenskii, S. M. (1963). Warming up of the arctic and the fauna of high latitude. *Priroda*, **52**, 48–53. (In Russian, quoted by Bray (1971).)

Van Cleve, K., Barney, R. & Schlentner, R. (1981). Evidence of temperature control of production and nutrient cycling in two interior Alaska black spruce ecosystems. *Canadian Journal of Forest Research*, **11**, 258–73.

Van Cleve, K., Chapin, F. S., Flanagan, P. W., Viereck, L. A. & Dyrness, C. T. (1986). *Forest Ecosystems in the Alaskan Taiga*. New York: Springer-Verlag.

Van Cleve, K., Dyrness, C. T., Viereck, L. A., Fox, J., Chapin, F. S. & Oechel, W. (1983a). Taiga ecosystems in interior Alaska. *Bioscience*, **33**, 39–44.

Van Cleve, K. & Oliver, L. K. (1982). Growth response of postfire quaking aspen (*Populus tremuloides* Michx.) to N, P, and K fertilization. *Canadian Journal of Forest Research*, **12**, 160–5.

Van Cleve, K., Oliver, L., Schlentner, R., Viereck, L. A. & Dyrness, C. T. (1983b). Productivity and nutrient cycling in taiga forest ecosystems. *Canadian Journal of Forest Research*, **13**, 747–66.

Van Cleve, K. & Viereck, L. A. (1981). Forest succession in relation to nutrient cycling in the boreal forest of Alaska. In *Forest Succession: Concepts and Application*, ed. D. C. West, H. H. Shugart & D. B. Botkin, pp. 185–211. New York: Springer-Verlag.

Van Cleve, K. & Viereck, L. A. (1983). A comparison of successional sequences following fire on permafrost-dominated and permafrost-free sites in interior Alaska. In *Permafrost: Proceedings of the Fourth International Conference*, pp. 1286–91. Washington, DC: National Academy Press.

Van Cleve, K. & Yarie, J. (1986). Interaction of temperature, moisture, and soil chemistry in controlling nutrient cycling and ecosystem development in the taiga of Alaska. In *Forest Ecosystems in the Alaskan Taiga*, ed. K. Van Cleve, F. S. Chapin, P. W. Flanagan, L. A. Viereck & C. T. Dyrness, pp. 160–89. New York: Springer-Verlag.

Van Cleve, K. & Zasada, J. (1976). Response of 70-year-old white spruce to thinning and fertilization in interior Alaska. *Canadian Journal of Forest Research*, **6**, 145–52.

Van Daalen, J. C. & Shugart, H. H. (1989). OUTENIQUA – A computer model to simulate succession in the mixed evergreen forests of the southern Cape, South Africa. *Landscape Ecology*, **2**(4), 255–67.

Van den Dreissche, R. (1982). Seedling spacing in the nursery in relation to growth, yield, and performance of stock. *Forestry Chronicle*, **60**, 345–55.

van der Pijl, L. (1972). *Principles of Dispersal in Higher Plants*. 2nd ed. Berlin: Springer-Verlag.

Vane, G. ed. (1988). *Proceedings of the airborne visible/infrared imaging spectrometer (AVIRIS) performance evaluation workshop*. NASA, JPL Publication 88-38.

van Tongeren, O. & Prentice, I. C. (1986). A spatial simulation model for vegetation dynamics. *Vegetatio*, **65**, 163–73.

Van Wagner, C. E. (1978). Age-class distribution and the forest fire cycle. *Canadian Journal of Forest Research*, **8**, 220–7.

Van Wagner, C. E. (1983). Fire behavior in northern conifer forests and shrublands. In *The Role of Fire in Northern Circumpolar Ecosystems*, ed. R. W. Wein & D. A. MacLean, pp. 65–80. SCOPE 18. New York: John Wiley and Sons.

Vasiliev, N. G. & Kolesnikov, B. P. (1962). Black-fir–deciduous forests in southern Far East. *Proceedings of the Far East Branch of the Academy of Sciences, Biology Series*, **10**, 147 pp. (In Russian.)

Veblen, T. T. (1986). Age and size structure of subalpine forests in the Colorado Front Range. *Bulletin of the Torrey Botanical Club*, **113**(3), 225–40.

Viereck, L. A. (1973). Wildfire in the taiga of Alaska. *Quaternary Research*, **3**, 465–95.

Viereck, L. A. (1975). Forest ecology of the Alaska taiga. In *Proceedings of the Circumpolar Conference on Northern Ecology*, pp. 1–22. Ottawa: National Research Council of Canada.

Viereck, L. A. (1979). Characteristics of the tree-line communities in Alaska. *Holarctic Ecology*, **2**, 228–38.

Viereck, L. A. (1982). Effects of fire and firelines on active layer thickness and soil temperatures in interior Alaska. In *Proceedings of the Fourth Canadian Permafrost Conference*, pp. 123–35. Ottawa: National Research Council of Canada.

Viereck, L. A. (1983). The effects of fire in black spruce ecosystems of Alaska and northern Canada. In *The Role of Fire in Northern Circumpolar Ecosystems*, ed. R. W. Wein & D. A. MacLean, pp. 201–20. New York: Wiley.

Viereck, L. A. & Dyrness, C. T. ed. (1979). *Ecological effects of the Wickersham Dome fire near Fairbanks, Alaska*. USDA Forest Service General Technical Report PNW-90. Pacific Northwest Forest and Range Experiment Station, Portland, OR.

Viereck, L. A., Dyrness, C. T., Van Cleve, K. & Foote, M. J. (1983). Vegetation, soils, and forest productivity in selected forest types in interior Alaska. *Canadian Journal of Forest Research*, **13**, 703–20.

Viereck, L. A. & Foote, M. J. (1985). Shrub, tree, and herbaceous biomass after the 1983 Rosie Creek fire. In *Early results of the Rosie Creek fire research project – 1984*, ed. G. P. Juday & C. T. Dyrness, pp. 30–3. Agricultural and Forestry Experimental Station, University of Alaska, Fairbanks, AK. Miscellaneous Publication 85-2.

Viereck, L. A., Foote, J., Dyrness, C. T., Van Cleve, K., Kane, D. & Seifert, R. (1979). *Preliminary results of experiments in the black spruce type of interior Alaska*. USDA Forest Service Research Note PNW-332.

Viereck, L. A. & Schandelmeier, L. H. (1980). *Effects of fire in Alaska and adjacent Canada – a literature review.* Fairbanks: United States Department of the Interior, Bureau of Land Management, Technical Report 6.

Viereck, L. A., Van Cleve, K. & Dyrness, C. T. (1986). Forest ecosystem distribution in the taiga environment. In *Forest Ecosystems in the Alaskan Taiga*, ed. K. Van Cleve, F. S. Chapin, P. W. Flanagan, L. A. Viereck & C. T. Dyrness, pp. 22–43. New York: Springer-Verlag.

Vincent, J. S. (1989). Quaternary geology of the southeastern Canadian Shield. In *Quaternary Geology of Canada and Greenland*, ed. R. J. Fulton, J. A. Heginbottam & S. Funder, Chapter 3. Ottawa: Geological Survey of Canada.

Viro, P. (1967). Forest manuring on mineral soils. *Medd. Nor. Skogforsoksves,* **23**, 113–36.

Vitousek, P. M. & Matson, P. A. (1984). Mechanisms of nitrogen retention in forest ecosystems: a field experiment. *Science,* **225**, 51–2.

Von Foerster, G. (1959). Some remarks on changing populations. In *The Kinetics of Cellular Proliferation*, pp. 382–407. New York: Grune and Stratton.

Vorobiyov, O. Yu. & Dorrer, G. A. (1974). Probabilistic model of forest fire spread. In *Problems of Forest Pyrology*, pp. 118–34. Krasnoyarsk. (In Russian.)

Vorobiyov, O. Yu. & Valendik, E. N. (1978). *Probabilistic Modeling of Forest Fire Spread.* Novosibirsk: Nauka. (In Russian.)

Voropanov, P. V. (1950). *Spruce Forests of the North.* Moscow: Goslesbumizdat. (In Russian.)

Vowinckel, T., Oechel, W. C. & Boll, W. C. (1974). The effect of climate on the photosynthesis of *Picea mariana* at the subarctic tree line. 1. Field measurements. *Canadian Journal of Botany,* **53**, 604–20.

Vowinckel, T., Oechel, W. & Boll, W. (1975). The effect of climate on the photosynthesis of *Picea mariana* at the subarctic tree line. *Canadian Journal of Botany,* **53**, 604–20.

Waggoner, P. E. & Stephens, G. R. (1971). Transition probabilities for a forest. *Nature,* **225**, 93–114.

Wagina, T. A. (1982). Dynamics of evolution of plant cover within Barabinsk low land. In *Natural Cycles of Baraba and its Economic Importance*, pp. 65–78. Novosibirsk: Science, Siberian Branch. (In Russian.)

Waldron, R. M. (1961). Seeding white spruce at the base of aspen. *Forestry Chronicle,* **37**, 224–7.

Waldron, R. M. (1965). Cone production and seedfall in a mature white spruce stand. *Forestry Chronicle,* **41**, 314–29.

Waldron, R. M. (1966). *Factors affecting natural white spruce regeneration on prepared seedbeds at the Rising Mountain Forest Experimental Area.* Canada Department of Forestry and Rural Development, Forestry Branch, Departmental Publication No. 1169.

Waldrop, T. A., Buckner, E. R., Shugart, H. H. & McGee, C. E. (1986). FORCAT: A single tree model of stand development on the Cumberland Plateau. *Forest Science,* **32**, 297–317.

Walker, B. H. (1981). Is succession a viable concept in African savanna ecosystems? In *Forest Succession: Concepts and Application*, ed. D. C. West, H. H. Shugart & D. B. Botkin. New York: Springer-Verlag.

Walker, J. & Sharpe, P. J. H. (1989). Ecological field theory: the concept and field theory. *Vegetatio*, **83** (in press).

Walker, L. R. (1985). *The processes controlling primary succession on an Alaskan floodplain*. PhD thesis, University of Alaska, Fairbanks, AK.

Walker, L. R., Zasada, J. C. & Chapin, F. S. III. (1986). The role of life history processes in primary succession on an Alaska floodplain. *Ecology*, **67**, 1243–53.

Walter, H. (1960). *Einführung in die Phytologie*. Grundlagen der Pflanzenverbreitung. Stuttgart: Gustav Fisher. (In German.)

Walter, H. (1971). *Ecology of Tropical and Subtropical Vegetation*. Edinburgh: Oliver and Boyd.

Walter, H. (1974). *Die Vegetation Osteuropas, Nord- und Zentralasien*. Stuttgart: Gustav Fisher. (In German.)

Walter, H. (1979). *Vegetation of the Earth*. New York: Springer-Verlag.

Wang, E., Erdle, T. & Roussell, T. (1987). *FORMAN wood supply model user manual*. Fredericton, New Brunswick: N. B. Executive Forest Research Advisory Committee Inc., and N. B. Department of Natural Resources and Energy.

Waring, R. H. (1983). Growth efficiency. *Advances in Ecological Research*, **13**, 327–54.

Waring, R. H. & Schlesinger, W. H. (1985). *Forest Ecosystems. Concepts and Management*. Orlando: Academic Press, Inc.

Warming, E. (1909). *Oecology of Plants: An Introduction to the Study of Plant Communities*. Oxford: Oxford University Press.

Watt, A. S. (1925). On the ecology of British beechwoods with special reference to their regeneration. *Journal of Ecology*, **13**, 27–73.

Watt, A. S. (1947). Pattern and process in the plant community. *Journal of Ecology*, **35**, 1–22.

Weatherhead, P. J. (1986). How unusual are unusual events? *American Naturalist*, **128**, 150–4.

Weaver, J. E. & Clements, F. E. (1929). *Plant Ecology*. New York: McGraw-Hill.

Webb, T. III. (1985). *A global paleoclimatic data base for 6,000 yr B.P.* Report TR-018, US Department of Energy, Washington, DC.

Webb, T. III. (1986). Is the vegetation in equilibrium with climate? How to interpret late-Quaternary pollen data. *Vegetatio*, **67**, 75–91.

Webb, T. III. (1987). The appearance and disappearance of major vegetational assemblages: Long-term vegetational dynamics in eastern North America. *Vegetatio*, **69**, 177–87.

Webb, T. III. (1988). Eastern North America. In *Vegetation History*, ed. B. Huntley & T. Webb, III, pp. 385–414. Kluwer Academic Publishers.

Webb, T., III, Cushing, E. J. & Wright, H. E., Jr. (1984). Holocene changes in the vegetation of the Midwest. In *Late-Quaternary Environments of the United States*, ed. H. E. Wright, Jr., pp. 142–66. London: Longman Group Limited.

Weber, M. G. (1990). Response of immature aspen ecosystems to cutting and burning in relation to vernal leaf flush. *Forest Ecology and Management*, **31**, 15–23.

Weetman, G. F. (1962). *Nitrogen relations in a black spruce* (Picea mariana Mill.) *stand subject to various fertilizer and soil treatments.* Woodland Research Index Number 129, Pulp and Paper Research Institute of Canada, Pointe Claire.

Weetman, G. F. (1968*a*). *The nitrogen fertilization of three black spruce stands.* Pulp and Paper Research Institute of Canada, Pointe Claire, Quebec, Woodlands Papers No. 6.

Weetman, G. F. (1968*b*). The relationship between feathermoss growth and the nutrition of black spruce. In *Proceedings of the third international peat conference*, ed. C. Lafleur & J. Butler, pp. 366–70. Ottawa: National Research Council of Canada.

Weetman, G. F. & Nykvist, N. B. (1963). Some more humus, regeneration and nutrient problems and practices in north Sweden. *Forestry Chronicle*, **39**, 188–98.

Wein, R. W. (1975). *Vegetation recovery in arctic tundra and forest-tundra after fire.* Canadian Department Indian & Northern Affairs, ALUR Report 74-75-62, Ottawa.

Wein, R. W. & El-Bayoumi, M. A. (1983). Limitations to predictability of plant succession in northern ecosystems. In *Resources and Dynamics of the Boreal Zone*, ed. R. W. Wein, R. R. Riewe & I. R. Methven, pp. 214–25. Ottawa: Association of Canadian Universities for Northern Studies.

Wein, R. W. & MacLean, D. A. (1983). *The Role of Fire in Northern Circumpolar Ecosystems.* New York: J. Wiley and Sons.

Wein, R. W. & Moore, J. M. (1977). Fire history and rotations in the New Brunswick Acadian forest. *Canadian Journal of Forest Research*, **7**, 285–94.

Weinstein, D. A., Shugart, H. H. & West, D. C. (1982). *The long-term nutrient retention properties of forest ecosystems: a simulation investigation.* Oak Ridge, Tennessee: Oak Ridge National Laboratory, ORNL/TM-8472.

Weiss, A. & Norman, J. M. (1985). Partitioning solar radiation into direct and diffuse, visible and near-infrared components. *Agricultural and Forest Meteorology*, **34**, 205–13.

Wellington, W. G. (1952). Air mass climatology of Ontario north of Lake Huron and Lake Superior before outbreaks of the spruce budworm and the forest tent caterpillar. *Canadian Journal of Zoology*, **30**, 114–27.

Wendrov, S. L., Gerasimov, I. P., Kunitsin, L. F. & Neustadt, M.I. (1966). Moisture rotation over plains of West Siberia: its role in the formation of nature and ways of modification. *Transactions of Science Academy of USSR: Ser. Geogr.*, **5**, 3–18. (In Russian.)

Wendrov, S. L., Gluh, I. S. & Malik, L. K. (1967). On the questions of water supply and water regime of West Siberian plain. *Transactions of Science Academy of USSR: Ser. Geogr.*, **1**, 41–53. (In Russian.)

Werner, R. A. (1964). White spruce seed loss caused by insects in interior Alaska. *Canadian Entomologist*, **96**(11), 1462–4.

Werner, R. A. & Holsten, E. H. (1983). Mortality of white spruce during a spruce beetle outbreak on the Kenai Peninsula in Alaska. *Canadian Journal of Forest Research*, **13**, 96–101.

West, D. C., Shugart, H. H. & Botkin, D. B. (1981). *Forest Succession: Concepts and Application.* New York: Springer-Verlag.

West, P. W. (1987). A model for biomass growth for individual trees in forest monoculture. *Annals of Botany* (London), **60**, 571–7.

Westman, W. E. & Paris, J. F. (1987). Detecting forest structure and biomass with C-band multipolarization radar: Physical model and field tests. *Remote Sensing of Environment*, **22**, 249–69.

Westman, W. E. & Price, C. V., (1988). Spectral changes in conifers subjected to air pollution and water stress: Experimental studies. *IEEE Transactions on Geoscience and Remote Sensing*, **26**, 11–20.

White, C. M. & West, G. C. (1977). The annual lipid cycle and feeding behavior of Alaskan redpolls. *Oecologia* (Berl.) **27**, 227–38.

White, J. & Harper, J. L. (1970). Correlated changes in plant size and number in plant populations. *Journal of Ecology*, **58**, 467–85.

White, P. S. (1979). Pattern, process, and natural disturbance in vegetation. *Botanical Review*, **45**, 229–99.

Whitmore, T. C. (1975). *Tropical Rain Forests of the Far East*. Oxford University Press.

Whittaker, R.H. (1953). A consideration of climax theory: the climax as a population and a pattern. *Ecological Monographs*, **23**, 41–78.

Whittaker, R. H. and Levin, S. I. (1977). The role of mosaic phenomena in natural communities. *Theoretical Population Biology*, **12**, 117–39.

Wigley, T. M. L., Jones, P. D. & Kelley, P. M. (1986). Empirical climate studies. In *The Greenhouse Effect, Climatic Change, and Ecosystems*, ed. B. Bolin, B. R. Doos, J. Jaeger & R. A. Warrick, pp. 271–322. SCOPE 29. New York: J. Wiley and Sons.

Wilcove, D. S. (1988). Changes in the avifauna of the Great Smoky Mountains: 1947–1983. *Wilson Bulletin*, **100**, 256–71.

Wiliams, V. P. (1939). *Soil Science*. Moscow: Agricultural publication. (In Russian.)

Williams, B. L. (1972). Nitrogen mineralization and organic matter decomposition in Scots pine humus. *Forestry*, **45**, 177–88.

Williams, D. L. (1989). The radiative transfer characteristics of spruce (*Picea* spp.): Implications relative to the canopy microclimate. PhD dissertation, Department of Geography, University of Maryland at College Park, MD 20742.

Williams, D.L. (1990). A comparison of spectral reflectance properties at the needle, branch, and canopy level for a variety of conifer species. *Remote Sensing of Environment*, (in press).

Williams, J., Barry, R. G. & Washington, W. W. (1974). Simulation of the atmospheric circulation using the NCAR global circulation model with ice age boundary conditions. *Journal of Applied Meteorology*, **13**, 305–17.

Willson, M. F. (1983). *Plant Reproductive Ecology*. New York: John Wiley and Sons.

Wilton, W. C. (1963). Black spruce seedfall immediately following a fire. *Forestry Chronicle*, **39**, 477–8.

Winston, D. A. & Haddon, B. D. (1981). Effects of early cone collection and artificial ripening on white spruce and red pine germination. *Canadian Journal of Forest Research*, **11**, 817–26.

Wisotskii, G. N. (1925). Initial water measuring investigations at Zharnovskii plot of Belorussian Forest experimental station. *Transactions of Belorussian State Institute of Agricultural and Forest Economy*, **6**, 76–101. (In Russian.)

Wolff, J. O. (1980). The role of habitat patchiness in the population dynamics of snowshoe hare. *Ecological Monographs*, **50**, 111–30.

Wolff, J. O., West, S. D. & Viereck, L. A. (1977). Zylem pressure potential in black spruce in interior Alaska. *Canadian Journal of Forest Research*, **7**, 422–8.

Wolff, J. O. & Zasada, J. C. (1979). Moose habitat and forest succession on the Tanana River flood plain and Yukon–Tanana upland. *Proceedings of the North American Moose Conference and Workshop*, **15**, 213–44.

Woodward, F. I. (1987). *Climate and Plant Distribution*. Cambridge University Press.

Woodwell, G. M., Whittaker, R. H., Reiners, W. A., Likens, G. E., Delwiche, C. C. & Botkin, D. B. (1978). The biota and the world carbon budget. *Science*, **199**, 141–6.

Worrell, R. (1983). Damage by the spruce bark beetle in south Norway, 1970–80: a survey and factors affecting its occurrence. *Norsk. Inst. Skogforsk*, **38**, 1–34.

Wright, H. E. Jr. (1987). Synthesis: the land south of the ice sheets. In *North America and Adjacent Oceans During the Last Glaciation*, ed. W. F. Ruddiman & H. E. Wright, Jr., pp. 479–88. DNAG Volume K-3, Geological Society of America, Boulder, Colorado.

Wright, H. E. Jr. & Heinselman, M. L. ed.(1973). The ecological role of fire in natural conifer forests of western and northern North America. *Quaternary Research*, **3**, 317–513.

Wu, Hsin-I, Sharpe, P. J. H., Walker, J. and Penridge, L. K. (1985). Ecological field theory: a spatial analysis of resource interference among plants. *Ecological Modeling*, **29**, 215–43.

Yang Hanxi & Wu Yegang (1987). Tree composition, age structure and regeneration strategy of the mixed broadleaved/*Pinus koraiensis* (Korean pine) forest in Changbai Mountain Reserve. In *The Temperate Forest Ecosystem*, ed. Yang Hanxi, Wang Zhan, J. N. R. Jeffers & P. A. Ward, pp. 12–20. Merlewood, England: Institute of Terrestrial Ecology.

Yarie, J. (1981). Forest fire cycles and life tables: a case study from interior Alaska. *Canadian Journal of Forest Research*, **11**, 554–62.

Yarie, J. (1983). *Forest community classification of the Porcupine River Drainage, interior Alaska, and its application to forest management*. US Forest Service General Technical Report PNW-154.

Yli-Vakkuri, P. (1963). Kokeellisia tutkimuksia taimien syntymisestä ja ensi kehityksestä kuusikoissa ja männiköissä. Summary: Experimental studies on the emergence and initial development of tree seedlings in spruce and pine stands. *Acta Forestalia Fennica*, **75**(1), 1–122.

Yoda, K., Kira, T., Ogawa, H. & Hozumi, K. (1963). Intraspecific competition among higher plants. XI. Self-thinning in over-crowded pure stands under cultivated and natural conditions. *Journal of Biology*, **14**, 107–29. Osaka City University, Osaka.

Young, S. B. (1971). The vascular flora of St-Lawrence island with special reference to floristic zonation in the Arctic regions. *Contribution from Gray Herbarium, Harvard University*, **210**, 11–115.

Youngblood, A. P. & Zasada, J. C. (1991). White spruce regeneration options on river floodplains in interior Alaska. *Canadian Journal of Forest Research*, (in press).

Yurev, M. M. (1926). Shuvalovskii peat deposit 'Black Mountain'. *Transactions of Scientific Amelioration Institute*, **11–12**, 144–270. (In Russian.)

Zackrisson, O. (1977). Influence of forest fires on the north Swedish boreal forest. *Oikos*, **29**, 22–32.

Zak, D. R., Host, G. E. & Pregitzer, K. S. (1989). Regional variability in nitrogen mineralization, nitrification, and overstorey biomass in northern Lower Michigan. *Canadian Journal of Forest Research*, **19**, 1521–6.

Zak, D. R., Pregitzer, K. S. & Host, G. E. (1986). Landscape variation in nitrogen mineralization and nitrification. *Canadian Journal of Forest Research*, **16**, 1258–63.

Zasada, J. C. (1971). *Frost damage to white spruce cones in interior Alaska*. USDA Forest Service Research Note PNW-149. Pacific Northwest Forest and Range Experiment Station, Portland, OR.

Zasada, J. C. (1973). Effect of cone storage method and collection date on Alaskan white spruce (*Picea glauca*) seed quality. In *Seed Problems*. IUFRO Symposium on Seed Processing, Bergen, Norway, 1973, Paper No. 19. Royal College of Forestry, Stockholm, Sweden.

Zasada, J. C. (1980). Some considerations in the natural regeneration of white spruce in interior Alaska. In *Forest Regeneration at High Latitudes*, ed. M. Murray & R. Van Veldhuizen, pp. 25–9. USDA Forest Service General Technical Report PNW-107. Pacific Northwest Forest and Range Experiment Station, Portland, OR.

Zasada, J. (1985). Production, dispersal, and germination, and first year seedling survival of white spruce and birch in the Rosie Creek burn. In *Early results of the Rosie Creek research project – 1984*, ed. G. P. Juday & C. T. Dyrness. Agriculture and Forestry Experiment Station, University of Alaska, Fairbanks, AK. Miscellaneous Publication 85–2.

Zasada, J. C. (1986). Natural regeneration of trees and tall shrubs on forest sites in interior Alaska. In *Forest Ecosystems in the Alaskan Taiga*, ed. K. Van Cleve, F. S. Chapin, P. W. Flanagan, L. A. Viereck & C. T. Dyrness, pp. 44–73. New York: Springer-Verlag.

Zasada, J. C. (1988). Embryo growth in Alaskan white spruce seeds. *Canadian Journal of Forest Research*, **18**, 64–7.

Zasada, J. C. & Densmore, R. (1977). Changes in Salicaceae seed viability during storage. *Seed Science and Technology*, **5**, 509–17.

Zasada, J. C. & Densmore, R. (1979). A trap to measure *Populus* and *Salix* seed fall. *Canadian Field-Naturalist*, **93**(1), 77–9.

Zasada, J. C., Foote, M. H., Deneke, F. J. & Parkerson, R. H. (1978). *Case history of an excellent white spruce cone and seed crop in interior Alaska: Cone and seed production, germination, and seedling survival*. USDA Forest Service General Technical Report PNW-65. Pacific Northwest Forest and Range Experiment Station, Portland, OR.

Zasada, J. C. & Gregory, R. A. (1972). *Paper birch seed production in the Tanana Valley, Alaska*. USDA Forest Service Research Note PNW-177. Pacific Northwest Forest and Range Experiment Station, Portland, OR.

Zasada, J. C. & Grigal, D. F. (1978). The effects of silvicultural system and seed bed preparation on natural regeneration of white spruce and associated species in interior Alaska. *Proceedings of the Fifth North American Forest Biology Workshop*, Gainesville, FL, pp. 213–20.

Zasada, J. C., Holloway, P. & Densmore, R. (1977). Considerations for the use of hardwood stem cuttings in surface management programs. *Proceedings of the symposium on surface protection through the prevention of damage: focus on the Arctic slope*. USDA Bureau of Land Management, Alaska State Office, Anchorage, AK.

Zasada, J. C. & Lovig, D. N. (1983). Observations on primary dispersal of white spruce, *Picea glauca*, seed. *Canadian Field-Naturalist*, **97**(1), 104–6.

Zasada, J. C., Norum, R. A., Van Veldhuizen, R. M. & Teutsch, C. E. (1983). Artificial regeneration of trees and tall shrubs in experimentally burned upland black spruce/feather moss stands in Alaska. *Canadian Journal of Forest Research*, **3**, 903–13.

Zasada, J. C. & Phipps, H. M. (1990). *Populus balsamifera*. In *The silvics of forest trees in the United States*. USDA Forest Service Agricultural Handbook, (in press).

Zasada, J. C. & Schier, G. A. (1973). Aspen root suckering in Alaska: effect of clone, collection date and temperature. *Northwest Science*, **47**(2), 100–4.

Zasada, J. C. & Viereck, L. A. (1970). *White spruce cone and seed production in interior Alaska*, 1957–68. USDA Forest Service Research Note PNW-129. Pacific Northwest Forest and Range Experiment Station, Portland, OR.

Zasada, J. C., Viereck, L. A. & Foote, M. J. (1979). Black spruce and seed fall and seedling establishment. In *Ecological effects of the Wickersham Dome fire near Fairbanks, Alaska*, ed. L. A. Viereck & C. T. Dyrness, pp. 42–50. US Forest Service, General Technical Report PNW-90. Pacific Northwest Forest and Range Experiment Station, Portland, OR.

Zasada, J. C., Viereck, L. A., Foote, M. H., Parkerson, R. H., Wolff, J. O. & Lankford, L. A., Jr. (1981). Natural regeneration of balsam poplar following harvesting in the Susitna Valley, Alaska. *Forestry Chronicle*, **57**(2), 57–64.

Zasada, J. C. & Wurtz, T. (1990). Natural regeneration of white spruce on upland sites in interior Alaska. *Forest Science*, (in press).

Zebker, H. A., Van Zyl, J. J. & Held, D. N. (1987). Imaging radar polarimetry from wave synthesis. *Journal of Geophysical Research*, **92**, 683–701.

Zeide, B. B. (1987). Analysis of the 3/2 power law of self-thinning. *Forest Science*, **33**, 517–37.

Zemlis, P. I. & Shvirta, D. I. (1987). Possibilities of mathematical population modeling of forest stands. *Proceedings of the Lithuanian Academy of Sciences, Biology Series*. 4(100), 119–32. (In Russian.)

Zemtsov, A. A. (1976). *Geomorphology of West Siberian Plain (Northern and Central Parts)*. Tomsk: Publication of Tomsk University. (In Russian.)

Zhao Dachang (1987). Preliminary studies on volcanic eruptions and historical vegetation succession in the eastern mountain area of north-east China. In *The Temperate Forest Ecosystem*, ed. Yang Hanxi, Wang Zhan, J. N. R. Jeffers & P. A. Ward, pp. 27–8. Merlewood, England: Institute of Terrestrial Ecology.

Zhukov, V. M. (1977*a*). Climate and mire formation process. In *Scientific Prerequisites for Exploitation of Mires in West Siberia*, pp. 13–25. Moscow: Science. (In Russian.)

Zhukov, V. M. (1977*b*). Landscape reformation in mire zones by way of hydroforest amelioration. In *Scientific Prerequisites for Exploitation of Mires in West Siberia*, pp. 161–75. Moscow: Science. (In Russian.)

Zhukov, V. M. & Potapova, L. S. (1977). Climate as a factor of present day mire formation. In *Scientific Prerequisites for Exploitation of Mires in West Siberia*, pp. 73–93. Moscow: Science. (In Russian.)

Ziller, W. G. (1974). *The tree rusts of western Canada*. Canadian Forestry Service Publication No. 1329. Pacific Forest Research Center, Victoria, BC.

Zinke, P. J. (1962). The pattern of influence of individual trees on soil properties. *Ecology*, **43**, 130–3.

Zoltai, S. C. (1975*a*). Tree ring record of soil movements on permafrost. *Arctic and Alpine Research*, **7**, 331–40.

Zoltai, S. C. (1975*b*). Structure of subarctic forests on hummocky permafrost terrain in northwestern Canada. *Canadian Journal of Forest Research*, **5**, 1–9.

Zubarev, V. M. (1965). Structure of Siberian pine forests and possibility of their cutting. *Forest Industry*, **10**, 15–8. (In Russian.)

Zubov, S. A. (1960). About the cedar pine problem in the Central Urals. In *Investigations in Siberian Forestry*, ed. Anonymous, pp. 162–71, Siberian Division of Academy of Sciences of the USSR, Novosibirsk. (In Russian.)

Zvetkov, V. F. & Semenov, B. A. (1985). *Pine Forests of the European Forest-tundra Transition Zone*. Moscow: Agroprom. (In Russian.)

Zyabchenko, S. S. (1982). Age dynamics of scotch pine forests in the European North. *Lesovedenie*, **2**, 3–10.

Index

Printed in the United States
By Bookmasters